Fundamental Concepts of ENVIRONMENTAL CHEMISTRY

Fundamental Concepts of ENVIRONMENTAL CHEMISTRY

G.S. Sodhi

Alpha Science International Ltd.

G.S. Sodhi (M. Phil., Ph.D.)
Department of Chemistry, SGTB Khalsa College
University of Delhi, Delhi-110 007, India

Alpha Science International Ltd.
P.O. Box 4067, Pangbourne RG8 8UT, UK

ISBN 1-84265-012-2

Printed in India.

Preface

The term *Environmental Chemistry* has come to be regarded as synonymous with environmental pollution. Many conventional texts stick to this narrow definition and highlight only the problem of pollution resulting from the varied uses of chemicals. Yet it must be emphasised that environmental chemistry deals with a wider aspect of knowledge. The problem of pollution is just a small segment of this discipline.

A broader interpretation of the chemistry of the environment predates the pollution problem; it predates even the origin of universe. It initiates with the emergence of small atoms and molecules—the simplest chemical entities—and their organisation through various geochemical and geophysical processes into planets, including our earth. It then explains the evolution of earth's atmosphere and hydrosphere.

The hydrosphere provided a medium for the biochemical transformation of some of the simple chemical entities into molecules of life. With the emergence of life, each living species extracted a niche for itself. The habitat was gradually modified and thus emerged the biosphere.

Right from the time the man made the earth his home, he has tried to follow the hedonistic doctrine, that is, he has treated avoidance of pain and achievement of pleasure as the highest aim of mankind. In pursuit of this aim he developed a wide range of energy resources. The easy access to energy paved the way for modern methods of industrial and agricultural production.

The application of energy resources no doubt provided comfort and happiness to man, but it also started degrading his habitat. It was then that the problem of air and water pollution cropped up. It soon became apparent that the contamination of natural environment threatened the very existence of mankind. The hedonistic principle, therefore, demanded measures to check further ecological degradation so as to conserve the environment. Since then the philosophy of environmental conservation has come to stay with mankind.

It is in this broad perspective that this book is written.

G.S. Sodhi

Acknowledgements

This text has been prepared under the Book Writing Scheme of the University Grants Commission (UGC), New Delhi (sanction letter No. F.10-11/93 (BP), dated August 12, 1993). Financial Assistance from the UGC is gratefully acknowledged.

It gives me immense pleasure to thank Dr. D.S. Claire, Principal, SGTB Khalsa College, University of Delhi, for his intellectual guidance, keen interest and important suggestions rendered during the tenure of my book writing scheme.

I feel complacent by placing on record my sincere thanks to Dr. S.S. Marwaha, Director, Punjab Council of Science and Technology, Chandigarh, for his perseverant help and cherishing attitude throughout the progress of the work embodied in this text.

I am oblidged to Dr. R.S. Jolly, Scientist, Institute of Microbial Technology, Chandigarh and Dr. Gajendra Singh, Lecturer, Department of Pharmaceutical Sciences, Guru Nanak Dev University, Amritsar, for their assistance in literature survey.

Thanks are due to Dr. E.T. Eliasson, Geothermal Project Manager, National Energy Authority of Iceland, Reykjavik; Mr. Jurgen Dischoff, Director, Asia and Pacific Centre for Transfer of Technology, New Delhi; and Mr. B.B. Baliga, Editor-in-Chief, Indian Science News Association, Calcutta, for providing suitable reference materials and the permission to reproduce the same.

G.S. Sodhi

Contents

Part I

Abiotic Origin

CHAPTER 1

Atoms and Molecules

Believe it or not, all naturally occurring elements which are the building blocks for all forms of matter, both the living and the nonliving, originated in the cores of stars. The study of the origin and build-up of these elements in the stars is called *nucleosynthesis*. The interstellar space was also the site where the first series of molecules were formed. Therefore, these molecules are referred to as *interstellar molecules*. To know more about nucleosynthesis and interstellar molecules, we have to know a great deal more about the origin of the universe.

Origin of the Universe

There are two theories pertaining to the origin of the universe: Static (or the steady state) theory and the evolutionary (or the big-bang) theory. The steady state theory (or the continuous creation hypothesis as is often called) propounds that the universe did not have a beginning in time and that matter is being created on a continuous basis. The big-bang theory, on the other hand, supports the idea that the universe had a beginning in time. It advocates that the universe was created at time t_0, about 10^{10} years ago, when a hyperdense sphere of matter packed into a volume roughly that of our solar system, with extremely high density and temperature, exploded in a 'big-bang' at or near the velocity of light.

How can these two patently divergent views be reconciled? Which of the two theories is more accurate? We do not know; perhaps we shall never know. However, the astronomical evidence is biased in favour of the big-bang hypothesis. We shall, therefore, choose the 'big-bang' as the starting point of our discussion on the origin of the universe.

The Three Eras

The big-bang model divides the history of universe into three eras—radiation era, matter era and the life era. Each era is subdivided into a number of epochs; each epoch corresponds to a specific period in the build-up of the universe. Fig. 1.1 shows the time scale delineation of the three eras and the corresponding epochs.

1. The radiation era

The hyperdense sphere of matter which caused the big-bang was a giant neutron ball consisting of 10^{78} to 10^{79} neutrons. This neutron ball is often referred to as *primordial matter* or *"Ylem"*. The temperature of the "Ylem" was about 10^{30} K. The first epoch, the *chaos epoch*, lasted for just 10^{-27} second, during which time the neutrons of the "Ylem" remained stable. Thereafter, the disintegration of neutrons into protons and mesons initiated, and lasted upto about 10^{-3} second. This stage is referred to as the *hardon epoch*. The neutrons, protons and mesons (collectively called hardons) collided with one another and were converted to high energy photons, thus creating a brilliant fireball of radiation. The temperature during the hardon epoch dropped to about 10^{12} K.

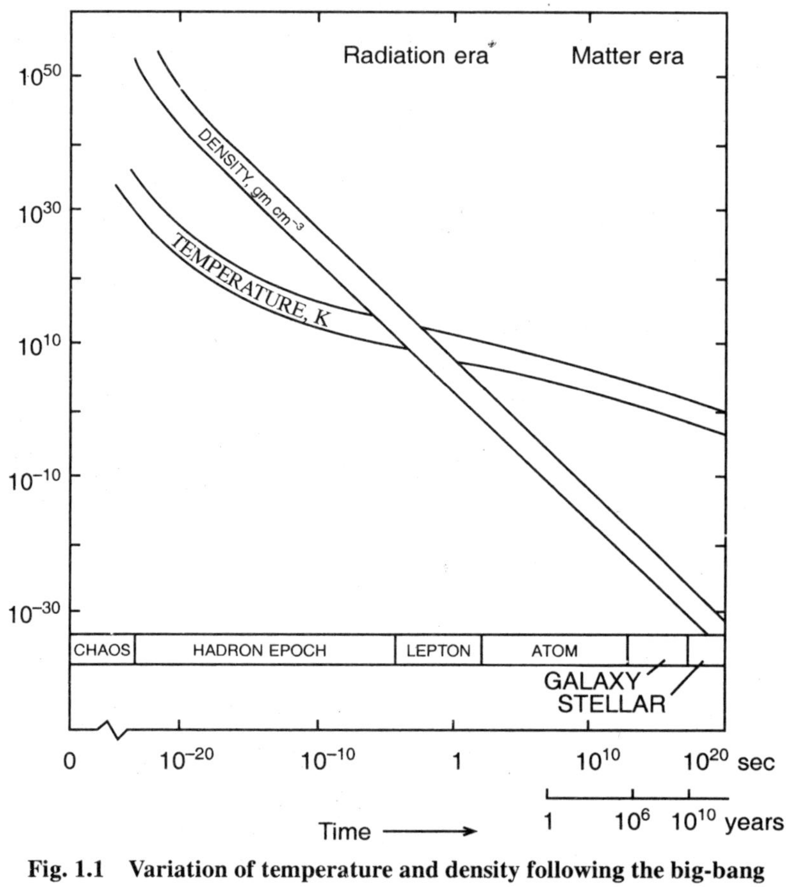

Fig. 1.1 Variation of temperature and density following the big-bang

As the universe expanded, its contents cooled. The average temperature during the third epoch, called the *lepton epoch*, dropped to about 10^{10} K. During this epoch, the lighter elementary particles, such as electrons, neutrinos and muons (collectively called leptons), began to emerge. However, like the hardons, these too were transformed into photons. The lepton epoch lasted upto a few hundred seconds after the big-bang.

During the first three epochs, the radiation density exceeded the matter density by a large amount. Photons of radiation outnumbered particles of matter. As soon as the elementary particles of matter had begun to coagulate, fierce radiation destroyed them. For this reason, the first three epochs are collectively referred to as the radiation era. Whatever matter existed was merely an inconspicuous precipitate suspended in a sea of dense, brilliant radiation.

2. The matter era

The fourth epoch, called the *atom epoch* extended in time from about a few hundred seconds to about a million years after the big-bang. Midway through this epoch, the average temperature had fallen to about 10^6 K. Towards the beginning of the atom epoch, radiation still dominated the matter and the universe remained flooded with photons. However, as the universe expanded the photon density decreased. The early dominance of radiation thus also diminished. Sometime between a few minutes and a million years after the big-bang, the charged elementary particles of matter were able to coagulate electromagnetically without being broken apart by radiation. Henceforth, matter would dominate radiation as the principal constituent of the universe.

The influence of radiation had grown so weak that it could no longer prohibit the combination of those leptons and hardons which had not been transformed into photons. As a result of this combination atoms began to appear. The first to be formed was hydrogen, since it required only a combination of a single electron to a single proton. Significant amounts of hydrogen atoms were synthesized in the early universe.

So long as the temperature of the universe exceeded 10^7 K, fusion of hydrogen atoms resulted in the formation of helium atoms. However, as the temperature dropped below the critical value of 10^7 K, the synthesis of helium, as well as of the higher atomic number elements, ceased. In contrast to the rapid cooling of the early universe, the dense interior of the stars were (and are) suitable for the generation of hotter temperatures and thus of heavier elements. The core of the stars were where the heavier elements were created—and where they are still being created. We shall discuss the plausible mechanisms for the synthesis of these heavy elements in the section on nucleosynthesis.

By the end of the atom epoch, matter was in firm control. During the fifth epoch, the *galaxy epoch,* gravity began to pull some of this matter together into enormous clumps. Galaxies were beginning to form; hot gases, intense radiation and turbulence of the clumps were conducive to galaxy formation. By the middle of the galaxy epoch, the average temperature of the universe had decreased to about 3000 K. The universe was becoming thinner, colder and darker.

The next epoch is the present *stellar epoch*. The dominant episode of this epoch was the formation of stars. As now, the universe has an average density of 10^{-30} g cm^{-3} and an average temperature of 3 K. Under these conditions, the formation of new galaxies is an impossibility. Nevertheless, stars are being formed within the existing galaxies.

The atom epoch, the galaxy epoch and the stellar epoch are collectively termed as the matter era. During this era the matter prevailed, while the radiation subsided. The epochs in the radiation era were spread over a few minutes; the epochs of the matter era were spread over millions of years. The reason being that when the universe was extremely hot, events occurred at a rapid pace; once it cooled, the pace of events slowed down. For example, the formation of stars and their life cycles cover millions and millions of years. During these eons heavy elements have been, and are being, synthesized at a slow pace.

3. The life era

The atoms formed in the universe combined to give molecules. Since these molecules were formed in space, they were termed interstellar molecules. Initially, the interstellar molecules were simple in composition. Gradually, more complex molecules were formed. The increasing complexity of the interstellar molecules with the passage of time is called *chemical evolution*. The chemical evolution finally led to the formation of amino acids and proteins—the molecular building blocks of life.

The life era covers not only the emergence of life on earth, but also the emergence of technologically intelligent life. Technology enables life to control matter, much as matter grew to dominance over radiation, billions of years ago. Matter is now losing its total dominance, at least at those isolated locations where technologically intelligent life persists. The transformation from matter era to life era was not instantaneous. It took millions of years for the matter to dominate over radiation; it took an even greater time for life to dominate over matter. And the transformation is not yet complete—at least when we strive for a more advanced, technologically intelligent life. Hence we name this epoch as the *future epoch*.

Nucleosynthesis

During the atom epoch, the combination of a proton and an electron produced a hydrogen atom. The heavier elements were produced by a series of nuclear reactions which occurred in the cores of stars. The build-up of the elements in the cores of stars is called nucleosynthesis. In order to discuss the role that stars played in the nucleosynthesis, we shall have to describe, qualitatively, the formation and structure of stars.

Life Cycle of a Star

A star is not a static heavenly body. It changes, evolves and dies. The younger stars, such as our sun, are called *Population I stars,* while the older ones are referred to as *Population II stars*. An earlier Population II star was obviously formed when the universe was rich in hydrogen and had small amounts of some lighter atoms like deuterium, helium and lithium. As the condensation of these atoms into the star proceeded, the gravitational potential energy of the atoms was converted into kinetic energy. As a result, a great diffuse mass of gas agglomerated at an accelerated pace, becoming progressively smaller, denser and hotter. Over the next millions of years, the star became compact and spherical and continued to consume hydrogen in various nuclear reactions. When hydrogen was exhausted in the core of the star, the star began to loss energy. To make up for the energy deficit, it contracted under the influence of gravitational forces and became hotter and denser. For the next millenia, the star became optically opaque and began to emit infrared radiation. It continued to lose energy and contract; the density and temperature continued to rise till the star exploded into a supernova, throwing much of its matter into the space and completing the material cycle.

Thus, throughout their life cycles, the stars lose energy into space as radiation. However, they compensate for the energy loss by gravitational contraction. Failure to maintain this balance triggers the supernova explosion and signals the death of the star. In our galaxy, supernovae occur about once in 250 years. One such supernova was recorded by Chinese astronomers in 1054 A.D. in the region called *Crab Nebula*. From the remnants of supernova, a Population I star is born and the cycle is repeated.

We have seen that various stages in the life history of a star are associated with different nuclear reactions. These reactions, in turn, give rise to the observed distribution of chemical elements in the cosmos.

Distribution of Elements

Table 1.1 lists the most abundant elements in the earth crust and of the universe. Table 1.2 presents the relative abundance data in an alternative but concise manner. The graphical representation of the data is shown in Fig. 1.2.

Table 1.1 Distribution of elements (%) in the universe and the earth crust

Universe		Earth Crust	
H	91	O	46.6
He	9.1	Si	27.2
O	0.067	Al	8.13
N	0.042	Fe	5.00
C	0.021	Ca	3.63
Ne	0.003	Na	2.83
Si	0.003	K	2.59
Mg	0.002	Mg	2.09
Fe	0.002		

Table 1.2 Relative abundance of elements in the universe

Elements		Fractional Abundance by Mass	
Hydrogen	: ^{1}H	0.73	(~ 98%)
	: ^{2}D	10^{-4}	
Helium	: ^{4}He	0.24	
	: ^{3}He	6×10^{-5}	
Li, Be, B	:	10^{-8}	
C, N, O, Ne		1.8×10^{-2}	(~ 2%)
Silicon group, Na-Sc:		2×10^{-3}	(~ 0.2%)
Iron group, A = 50–62:		2×10^{-4}	
Middle weight group, A = 63–100		10^{-6}	
Heavy weight group, A > 100		10^{-7}	

A = Mass number

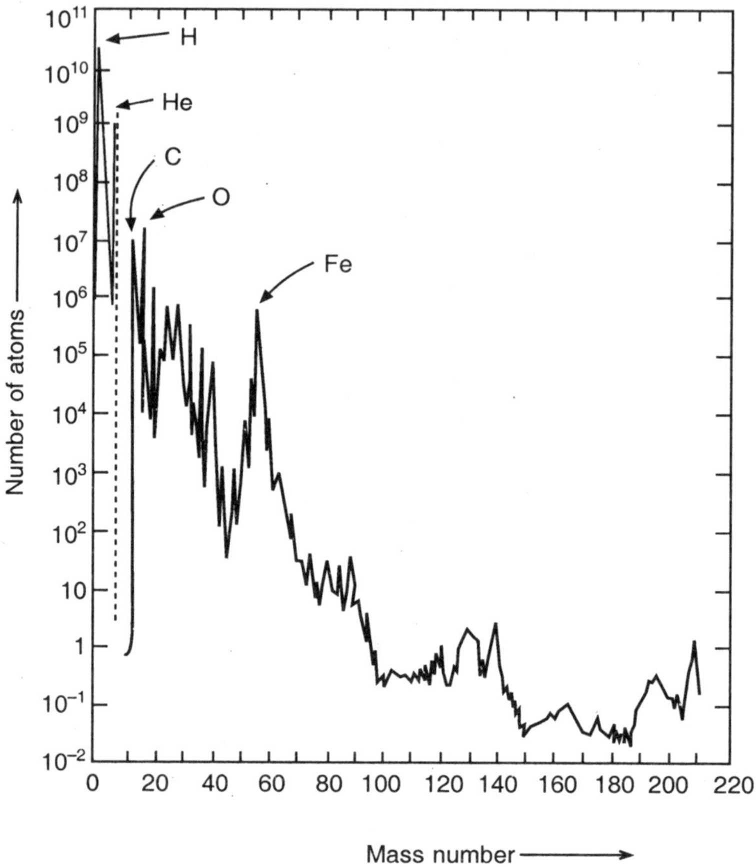

Fig. 1.2 Graphical representation of distribution of elements in the universe

The following features of the abundance data are noteworthy:

1. Hydrogen is the most abundant element in the universe.
2. Helium is the next most abundant element.
3. Lithium, beryllium and boron are particularly rare compared to the neighbouring elements.
4. Iron has a relatively high abundance.
5. Elements with even atomic numbers are more abundant than odd ones; this is called the *Oddo-Harkins rule.*

It is also pertinent to mention here that the elemental abundance varies from star to star. For example, the older Population II stars are deficient, by factors ranging from 100 to 1000, in heavy elements. The nucleosynthesis proceeds from low atomic number elements to high atomic number analogs. By the time the synthesis of high atomic number elements begins in Population II stars, the supernova occurs. Eventually, the scattered atoms become trapped by a gravitational field and condense into a Population I star. This time, however, the quantity of heavier elements in the star is quite large. Studies of stars, which are of recent origin, do indicate greater abundance of metals, particularly iron. It has also been observed that the element technetium is being formed in certain stars, called *S-type stars*. Technetium, however, does not occur naturally on earth because all of its fourteen isotopes are radioactive, with half lives of 2.6×10^6 years or less. These observations support the hypothesis, that:

(a) All the elements have been synthesized from primordial hydrogen.
(b) New nuclei are being synthesized all the time in stellar interiors.
(c) Over the eons, the stars have been ejecting matter back into the interstellar space.
(d) The universe is evolutionary; this, in turn, is considered as evidence for a big-bang cosmology.

Nuclear Reactions

We now come to the sequence of nuclear reactions which build-up, starting from primordial hydrogen, different elements in the stars. Various sets of nuclear reactions which are envisioned to synthesize the elements in stars are listed in Table 1.3. The details of nuclear reactions are described below.

Table 1.3 Nuclear reactions occurring in the stars

(I)	Exothermic Processes
	1. Hydrogen burning, including
	a. Proton-proton chain (PP-I, PP-II, PP-III)
	b. CN cycle
	c. CNO bicycle
	2. Helium burning
	3. Carbon burning
	4. Oxygen burning
	5. α-Process
	6. *e*-Process
(II)	Neutron Capture Processes
	7. *s*-Process
	8. *r*-Process
(III)	Miscellaneous Processes
	9. *p*-Process
	10. *x*-Process

(I) Exothermic Processes in Stellar Interiors

Various segments of exothermic processes are expected to occur successively in an evolutionary star as its core temperature and density rise progressively. The actual temperature at which the processes proceed depends on the density. The six processes (and their sub processes) falling under this category (Table 1.3) are the source of the elements upto iron group.

1. Hydrogen burning

In the early stages of the life history of a star, the gravitational attraction causes a slow contraction. As gravitational potential energy is converted into kinetic energy, the velocity of primordial hydrogen atoms increases and the temperature of the star rises along with the density of matter. When the temperature reaches 1×10^7 K to 3×10^7 K and density is about 10^2 g cm^{-3}, the activation energy barrier to fusion of hydrogen nuclei, as well as to the nuclear reactions between hydrogen and other lighter nuclei, is overcome, and hydrogen burning initiates. Hydrogen burning involves three main sequences of reactions.

(*a*) *Proton-proton chain:* These are further subdivided into three types: PP-I, represented by the reactions,

$$ {}^{1}_{1}\text{H} + {}^{1}_{1}\text{H} \rightarrow {}^{2}_{1}\text{H} + {}^{0}_{+1}\beta\,(\text{position}) + \gamma\ (\text{neutrino}) \quad (1.1)$$

$$ {}^{2}_{1}\text{H} + {}^{1}_{1}\text{H} \rightarrow {}^{3}_{2}\text{He} + \gamma\ (\gamma\text{-rays}) \quad (1.2)$$

$$ {}^{3}_{2}\text{He} + {}^{3}_{2}\text{He} \rightarrow {}^{4}_{2}\text{He} + 2\ {}^{1}_{1}\text{H} \quad (1.3)$$

$$ {}^{1}_{1}\text{H} + {}^{3}_{2}\text{He} \rightarrow {}^{4}_{2}\text{He} + {}^{0}_{+1}\beta + \nu + \gamma \quad (1.4)$$

PP-II, represented by the reactions,

$$ {}^{3}_{2}\text{He} + {}^{4}_{2}\text{He} \rightarrow {}^{7}_{4}\text{Be} + \gamma \quad (1.5)$$
(from 1.2)

$$ {}^{7}_{4}\text{Be} + {}^{0}_{-1}\beta\,(\text{electron}) \rightarrow {}^{7}_{3}\text{Li} + \nu \quad (1.6)$$

$$ {}^{7}_{3}\text{Li} + {}^{1}_{1}\text{H} \rightarrow {}^{8}_{4}\text{Be} + \gamma \quad (1.7)$$

$$ {}^{8}_{4}\text{Be} \rightarrow 2\ {}^{4}_{2}\text{Be} \quad (1.8)$$

PP-III, represented by the reactions,

$$ {}^{7}_{4}\text{Be} + {}^{1}_{1}\text{H} \rightarrow {}^{8}_{5}\text{B} + \gamma \quad (1.9)$$
(from 1.5)

$$ {}^{8}_{5}\text{B} \rightarrow {}^{8}_{4}\text{Be} + {}^{0}_{+1}\beta + \nu \quad (1.10)$$

$$ {}^{8}_{4}\text{Be} \rightarrow 2\ {}^{4}_{2}\text{He} \quad (1.11)$$

For the proton-proton chain reactions, the two positively charged hydrogen nuclei must be brought very close to each other and must overcome the activation energy barrier to fusion. This requires extremely high temperatures. On the other hand, when the two nuclei are very close to each other, the forces of repulsion may dominate. To counteract their mutual repulsion at close distances, the hydrogen atoms must possess high kinetic energy. Therefore, the two important conditions for the proton-proton reactions to materialize are high temperature and high kinetic energy.

It is also evident that the isotopes of lithium, beryllium and boron, which are produced in the reaction sequences PP-II and PP-III, are destroyed to produce helium. This accounts for the low abundance of lithium, beryllium and boron in the universe.

(*b*) *CN cycle:* The sequence of the CN cycle, also called *Bethe cycle,* is represented by the reactions,

$$^{12}_{6}C + ^{1}_{1}H \rightarrow \gamma + ^{13}_{7}N \rightarrow {}^{0}_{+1}\beta + \nu + ^{13}_{6}C \quad (1.12)$$

$$^{13}_{6}C + ^{1}_{1}H \rightarrow ^{14}_{7}N + \gamma \quad (1.13)$$

$$^{14}_{7}N + ^{1}_{1}H \rightarrow \gamma + ^{15}_{8}O \rightarrow {}^{0}_{+1}\beta + \nu + ^{15}_{7}N \quad (1.14)$$

$$^{15}_{7}N + ^{1}_{1}H \rightarrow ^{4}_{2}He + ^{12}_{6}C \quad (1.15)$$

$$4\,^{1}_{1}H \rightarrow ^{4}_{2}He + 2\,{}^{0}_{+1}\beta + 2\nu + 3\gamma \quad (1.16)$$

If the star contains a small amount of $^{12}_{6}C$, then CN cycle may be the major pathway for hydrogen burning. In this sequence of reactions, unstable nucleides, such as $^{13}_{7}N$ and $^{15}_{8}O$ decompose readily, but stable ones, such as $^{13}_{6}C$, $^{14}_{7}N$ and $^{15}_{7}N$ have long enough life times to undergo nuclear reactions with protons. Moreover, although $^{12}_{6}C$ is a reactant in reaction (1.12), it is regenerated in the reaction (1.15), thus serving as a catalyst.

(*c*) *CNO bicycle:* The sequence of the CNO bicycle is represented by the reactions,

$$^{15}_{7}N + ^{1}_{1}H \rightarrow ^{16}_{8}O + \gamma \quad (1.17)$$

$$^{16}_{8}O + ^{1}_{1}H \rightarrow ^{17}_{9}F + \gamma \quad (1.18)$$

$$^{17}_{9}F \rightarrow ^{17}_{8}O + {}^{0}_{+1}\beta + \nu \quad (1.19)$$

$$^{17}_{8}O + ^{1}_{1}H \rightarrow ^{14}_{7}N + ^{4}_{2}He \quad (1.20)$$

The $^{14}_{7}N$ generated in reaction (1.20) may then be consumed in the reaction (1.14) of the CN cycle. Hence the designation, bicycle. The CNO bicycle occurs 2000 times less frequently than the CN cycle.

Perspectives of hydrogen burning. Various segments (and subsegments) of hydrogen burning represent the most fundamental of the nuclear reactions leading to nucleosynthesis. The reactant, in each case, is the primordial hydrogen, the most abundant element in the universe. All the three pathways (viz. proton-proton chain, CN cycle and CNO bicycle) convert hydrogen to helium. Consequently, after hydrogen, helium is the next most abundant element in the universe. Elements with atomic numbers higher than helium are synthesized subsequently. However, the nuclear reactions involving higher atomic number species require extremely high activation energy, leading to a low abundance of the heavier nuclei. Therefore, we observe a rapid, exponential decrease in abundance with increasing atomic number.

As the hydrogen burning proceeds, two phenomena occur in the star. Firstly, as the reactant atoms are brought closer, a gravitational contraction of the star occurs. However, simultaneously there is a rise in temperature which causes the envelop of gaseous matter around the star to expand. The net effect is that the size of the star remains almost constant. Secondly, enormous amount of energy is released in the core of the star as a result of the hydrogen burning reactions. However, simultaneously radiation is lost from the outer, comparatively cooler, envelop of the star. The net effect is that the energy of the star remains constant. Since the size and the energy of the star remain constant, it means that during hydrogen burning the star remains stable. All the stars during their stable lifetimes are believed to be running on one or a combination of the three segments of hydrogen burning.

2. Helium burning

The denser, more massive helium nuclei, produced by hydrogen burning, tend to concentrate towards the center (core) of the star. The gravitational contraction initiates again, leading to a rise in temperature and density of the star. When a temperature of 10^8 K and a density of 10^5 g cm^{-3} are attained in the core of the star, the activation energy barrier to helium burning is exceeded and helium begins to be converted to carbon and oxygen by the following reactions:

$$^{3}_{2}\text{He} + ^{4}_{2}\text{He} \rightarrow ^{7}_{4}\text{Be} + \gamma \tag{1.21}$$

$$^{7}_{4}\text{Be} + ^{1}_{1}\text{H} \rightarrow ^{8}_{5}\text{B} + \gamma \tag{1.22}$$

$$^{8}_{5}\text{B} \rightarrow ^{8}_{4}\text{Be} + ^{0}_{+1}\beta + \gamma \tag{1.23}$$

$$^{8}_{4}\text{Be} + ^{4}_{2}\text{He} \rightarrow ^{12}_{6}\text{C} \tag{1.24}$$

$$^{12}_{6}\text{C} + ^{4}_{2}\text{He} \rightarrow ^{16}_{8}\text{O} \tag{1.25}$$

3. Carbon burning

Carbon burning is the next major thermonuclear epoch. It begins when the temperature and density of the star are of the order, 1×10^9 K and 5×10^5 g cm^{-3}, respectively. The pertinent primary reactions of carbon burning are as follows:

$$^{12}_{6}\text{C} + ^{12}_{6}\text{C} \rightarrow ^{20}_{10}\text{Ne} + ^{4}_{2}\text{He} \tag{1.26}$$

$$\rightarrow ^{23}_{11}\text{Na} + ^{1}_{1}\text{H} \tag{1.27}$$

$$\rightarrow ^{23}_{12}\text{Mg} + ^{1}_{0}\text{n} \tag{1.28}$$

$$\rightarrow ^{24}_{12}\text{Mg} + \gamma \tag{1.29}$$

We see that not only are neon, sodium and magnesium produced, but a fresh flux of α-particles, protons and neutrons becomes available for generating other nuclear species. In this way, isotopes of magnesium, aluminium and silicon may now be synthesized from secondary reactions.

4. Oxygen burning

Following the exhaustion of $^{12}_{6}$C, gravitational contraction once again takes over. It might be slowed down if *neon photodisintegration*,

$$^{20}_{10}\text{Ne} + \gamma \rightarrow ^{16}_{8}\text{O} + ^{4}_{2}\text{He} \tag{1.30}$$

were to occur, followed by the reaction,

$$^{20}_{10}\text{Ne} + ^{4}_{2}\text{He} \rightarrow ^{24}_{12}\text{Mg} + \gamma \qquad (1.31)$$

Nevertheless, the next main nuclear burning stage is reached around the temperature, 2×10^9 K (and density almost the same as in carbon burning) when oxygen burning commences. The primary reactions of oxygen burning are as follows:

$$^{16}_{8}\text{O} + ^{16}_{8}\text{O} \rightarrow ^{28}_{14}\text{Si} + ^{4}_{2}\text{He} \qquad (1.32)$$

$$\rightarrow ^{31}_{15}\text{P} + ^{1}_{1}\text{H} \qquad (1.33)$$

$$\rightarrow ^{31}_{16}\text{S} + ^{1}_{0}\text{n} \qquad (1.34)$$

$$\rightarrow ^{32}_{16}\text{Si} + \gamma \qquad (1.35)$$

The fresh supply of α-particles, protons and neutrons leads to the synthesis of chlorine and argon through secondary reactions.

5. α-Process

The coulombic repulsion between nuclei of atomic numbers 10 to 16 is sufficiently strong to discourage further burning by simple fusion. Therefore, further reactions take place by α-process. In this process the $^{28}_{14}\text{Si}$ nuclei, produced in oxygen burning (reaction 1.32), undergo photodisintegration with the emission of α-particles.

$$^{28}_{14}\text{Si} \rightarrow 7\ ^{4}_{2}\text{He} \qquad (1.36)$$

The emitted α-particles are then captured by another $^{28}_{14}\text{Si}$ nuclei.

$$^{28}_{14}\text{Si} + 7\ ^{4}_{2}\text{He} \rightarrow ^{56}_{28}\text{Ni} \qquad (1.37)$$

The overall reaction may be written as,

$$2\ ^{28}_{14}\text{Si} \rightarrow ^{56}_{28}\text{Ni} \qquad (1.38)$$

These reactions occur at a temperature around 3×10^9 K. The density of the star may be anywhere between 10^5 and 10^8 g cm^{-3}.

The $^{56}_{28}\text{Ni}$ nuclei produced in reaction (1.38) is unstable and disintegrates to $^{56}_{26}\text{Fe}$, $^{46}_{22}\text{Ti}$, $^{62}_{28}\text{Ni}$ and other nuclei in the similar atomic number range.

6. e-Process

The next major reaction sequence which takes place in the core of the star is the e-process. This process initiates with the so called silicon burning reaction,

$$^{28}_{14}\text{Si} + ^{4}_{2}\text{He} \rightleftarrows ^{32}_{16}\text{S} + \gamma \qquad (1.39)$$

followed by,

$$^{32}_{16}\text{S} + ^{4}_{2}\text{He} \rightleftarrows ^{36}_{18}\text{Ar} + \gamma \qquad (1.40)$$

$$^{36}_{18}Ar + ^{4}_{2}He \rightleftarrows ^{40}_{20}Ca + \gamma \text{ etc.} \tag{1.41}$$

These reactions occur when the temperature of the star is around 4×10^9 K and its density is of the order of 10^8 g cm^{-3}. Under these conditions, the (α, γ) reactions compete with the (γ, α) reactions and, therefore, reach equilibrium with them. Some of the nuclei so produced, undergo β-decay to give others such as, scandium, titanium vanadium, chromium, manganese and iron.

The e-process can build nuclei upto iron, which has the highest binding energy per nucleon (8.8 MeV). High binding energy means greater stability. The fact that $^{56}_{26}Fe$ is a very stable nucleus explains two important observations. Firstly, iron has a relatively high abundance in the universe. Secondly, since it is the most stable element, no more energy can be obtained from those nuclear reactions which involve $^{56}_{26}Fe$ as a reactant. The element processing in the stars, which use nuclear reaction energies as their prime source of power, take hydrogen to helium and helium ultimately to iron but no further. Thus, after the build-up of iron in the core of the star, it is a matter of short time before the supernova occurs.

(II) Neutron Capture Processes

If supernova occurs immediately after the synthesis of iron, how can we account for the presence of elements having atomic numbers greater than 26? The elements heavier than iron are synthesized by the neutron capture processes, since these reactions are not hindered by coulombic barriers. The temperature range in which these reactions occur is 2×10^9-4×10^9 K. The density of the star is in the range 10^5-10^8 g cm^{-3}.

Evidence for the occurrance of neutron capture comes from the fact that the abundance of an isotope resulting from these reactions will vary inversely with the neutron capture cross section of that isotope. Thus, if an isotope captures neutrons readily it will disappear fast in a nuclear flux and little of it will accumulate. If it does not capture neutrons easily, it would tend to accumulate. Indeed, it has been observed that the isotopes with low neutron cross sections, are the most abundant ones. This explains the relatively large abundance of magic number elements (A = 2, 8, 20, 50, 82, 126) which have a small neutron capture cross section.

There are two neutron capture processes: The slow neutron capture or the *s*-process and the rapid neutron capture or the *r*-process.

7. The s-process

The *s*-process occurs in Population I stars, initiating simultaneously at the stage of carbon burning. As the name implies, it is a slow process, taking 10^2-10^5 years to complete. It is responsible for the nucleosynthesis of elements heavier than iron and upto bismuth. The free neutrons are generated by (α, n) reactions, that is, those reactions in which α particles act as projectiles and neutrons are released. One such reaction is $^{13}_{6}C$ (α, n) $^{16}_{8}O$.

$$^{13}_{6}C + ^{4}_{2}He \rightarrow ^{16}_{8}O + ^{1}_{0}n \tag{1.42}$$

The $^{13}_{6}C$ nuclei is produced by reaction (1.12) of the CN cycle.

The slow neutron capture reactions may be represented by the following sequence, starting from element E which has an atomic number Z and mass number A.

$$^{A}_{z}E + ^{1}_{0}n \rightarrow {}^{A+1}_{z}E + \gamma \qquad (1.43)$$

$$^{A+1}_{z}E + ^{1}_{0}n \rightarrow {}^{A+2}_{z}E + \gamma \text{ etc.} \qquad (1.44)$$

The isotopes of element E are built up successively until one which is unstable to β–decay is produced. That isotope (let us assume it to be $^{A+2}_{z}E$) decays to produce the isobaric nucleide of element G, which serves as a target for further (n, γ) isotope-building reactions, until another β-unstable species is formed.

$$^{A+2}_{z}E \rightarrow {}^{A+2}_{z+1}G + {}^{0}_{-1}\beta + \nu \qquad (1.45)$$

$$^{A+2}_{z+1}G + {}^{0}_{1}n \rightarrow {}^{A+3}_{z+1}G + \gamma \text{ etc.} \qquad (1.46)$$

The process continues over the time period for which the star produces a reasonable flux of neutrons.

8. r-process

In the *s*-process it was assumed that the time scale for (n, γ) reactions is very long (10^2-10^5 years) compared to the time scale for β-decay (a few seconds). In the *r*-process, reverse is the case. By this process elements heavier than bismuth are produced by a chain of neutron captures which is so rapid (requiring just 10^{-2} to 10 seconds) that β-decay does not have the time to occur. The *r*-process can build-up very heavy radioactive nucleides, including those which have considerably long half-lives, such as uranium, thorium and plutonium.

The *r*-process can occur only if a very large neutron flux is available. The conditions which occur at the time of supernova can produce the vast quantities of neutrons required for the *r*-process. Thus, the elements heavier than bismuth are synthesized at the time of supernova explosion and are immediately dispersed off into the universe.

(III) Miscellaneous Processes

Under this category falls the *p*-process, which accounts for the nucleosynthesis of proton-rich heavy isotopes and the *x*-process (or unknown process) which accounts for the nucleosynthesis of certain light nuclei, such as lithium, beryllium and boron, which have extremely low abundance.

9. p-process

The relatively rare proton-rich and neutron-deficient isotopes of the heavy elements cannot be synthesized by neutron capture processes. The nucleosynthesis of these elements require proton capture by the nuclei formed through the *s*- and/ or the *r*-processes. These (p, γ) or (p, n) reactions obviously occur in the region which is proton-rich and neutron-deficient. The *p*-process nuclei are, therefore, produced during the supernova, in the proton-rich outer envelop of the stars. The density of the outer envelop is less than that of the core; it is of the order of 10^2 g cm^{-3}. The temperature in this region is also expected to be correspondingly lower than that of the inner region. However, due to the supernova, the temperature momentarily rises to 2×10^9-3×10^9 K and in this temperature range the *p*-process occurs. Some of the nuclei which are produced as a result of *p*-process are nitrogen, oxygen, fluorine and neon.

10. x-process

The theory of nucleosynthesis cannot explain the existence of three light elements: Lithium, beryllium and boron. The nuclei of these elements are extremely fragile and would rapidly disintegrate in the hot, dense and violent interior of most stars. Therefore, the three elements are relatively scarce (Table 1.2); yet they are not non-existent. For many years it was not known with certainty as to how these elements were produced. Thus, the process leading to their formation was designated as *x*-process. However, now it is believed that these elements were synthesized by the collision of cosmic rays with heavier elements. The cosmic rays primarily consist of high energy protons. These projectiles interacted with carbon, nitrogen and oxygen nuclei in the interstellar medium and fragmented them to produce lithium, beryllium and boron. The evidence for this theory comes from the observation that the concentration of lithium, beryllium and boron are high in those interstellar regions where the background concentration of cosmic rays is high.

Effects of Supernova on Surrounding Environment

A supernova is one of nature's grandest spectacles. During the first 10 seconds, it radiates 100 times more energy than the sun will radiate in 10 billion years. A supernova may result from the explosion of an ancient, burnt out star or the collapse of a comparatively younger star. In the latter case, spectra of light from the explosion reveal hydrogen left-over from the star's atmosphere. In contrast, the ancient stars, having burnt up all their hydrogen show no such emission lines.

Two significant supernovae have occurred in recent times—one on February 23, 1987 and the other on March 28, 1993. These were christened SN 1987A and SN 1993J, respectively. Both these supernovae resulted from the collapse of young supergiant stars.

A supernova leads to an enormous increase in the background cosmic radiation. Increased cosmic radiation has two important impacts on the surrounding environment.

1. Mass extinction of fauna

Cosmic radiation from supernova, if it reaches the earth, has the potential to cause mass extinction of fauna, without the simultaneous extinction of flora. The normal cosmic ray flux produces a radiation dose rate of about 0.03 roentgen year^{-1}. A 10,000 fold increase in the dose rate, as a result of supernova, would imply a dose of 300 roentgens year^{-1}, and 6000 roentgens in 20 years. The lethal dose for most animals is in the range 200-700 roentgens. It is, therefore, possible that if a supernova were to occur 10 to 15 light years from the earth, it could well cause a mass extinction of life. In fact, it has been suggested that a supernova explosion was responsible for the mass extinction that occurred about 60 million years ago in connection with the disappearance of dinosaurs. The lethal cosmic ray dose for most plants is well over 2000 roentgens. Thus, the plants would have been far less affected than animals by an increased dose of cosmic radiation from a supernova.

2. Depletion of ozone layer

Cosmic radiations in the atmosphere are known to produce free nitrogen atoms in excited state (N^*). These react readily with atmospheric oxygen to produce nitric oxide,

$$N^* + O_2 \rightarrow NO + O \tag{1.47}$$

which, in turn, depletes ozone:

$$NO + O_3 \rightarrow NO_2 + O_2 \tag{1.48}$$

The depletion of ozone layer in the stratosphere has hazardous consequences which are discussed in Chapter 19.

Interstellar Molecules

The interstellar molecules are so named because they are synthesized in interstellar clouds of dust and gas. Most of the interstellar regions of space consist of near perfect vacuum, populated by about one hydrogen atom per cubic centimeter. The interstellar clouds of dust and gas are, however, irregularly distributed in space. The dust in these clouds is not well characterized. Yet, there is evidence to suggest the presence of ice, silicates, graphite, macromolecular organic compounds and mixture of these ingredients. These dust particles were originally the constituents of stars, from where they were ejected into the interstellar medium by powerful solar winds. The gas is predominantly hydrogen, both in the atomic (H), as well as molecular (H_2) forms.

There are two types of clouds: *Diffuse clouds,* which contain hydrogen atoms and *Dense clouds,* which contain hydrogen molecules. Most of the simple molecules (diatomic) are synthesized in the diffuse clouds. Most of the complex molecules are synthesized in the dense clouds. A list of the molecules observed in the interstellar medium is presented in Table 1.4. An examination of the list of compounds reveals that both the inorganic and the organic molecules are found in space.

Synthesis of Interstellar Molecules

How are these chemically diverse and structurally complex molecules synthesized in space? The following two mechanisms account for their synthesis.

1. Synthesis of interstellar molecules on dust grains

The dust particles provide active sites on their surface, where an atom can be

Table 1.4 Common interstellar molecules

Inorganic		Organic		
H_2	Hydrogen	2-Atomic	CH	Methylidine
OH	Hydroxyl radical		CH^+	Methylidine ion
NO	Nitric Oxide		CN	Cyanide radical
NS	Nitrogen sulfide		CO	Carbon monoxide
SiO	Silicon monoxide		CS	Carbon monosulfide
SiS	Silicon sulfide			
SO	Sulfur monoxide			
H_2O	Water	3-Atomic	HC_2	Ethynyl radical
HNO	Hyponitrous acid		HCN	Hydrogen cyanide
H_2S	Hydrogen sulfide		HNC	Hydrogen isocyanide
SO_2	Sulfur dioxide		HCO^+	Formyl ion
			HCO	Formyl radical
			OCS	Carbonyl sulfide
NH_3	Ammonia	4-Atomic	C_2H_2	Acetylene
			C_2CN	Cyanoethynyl radical
			H_2CO	Formaldehyde
			HNCO	Isocyanic acid
			H_2CS	Thioformaldehyde
		5-Atomic	CH_4	Methane
			HC_4	Butadynyl radical
			H_2CNH	Methanimine
			H_2NCN	Cynamide
			HC_2CN	Cyanoacetylene
			HCO_2H	Formic acid
			H_2CCO	Ketene
		6-Atomic	CH_3CN	Methyl cyanide
			CH_3OH	Methyl alcohol
			$HCONH_2$	Formamide
		7-Atomic	CH_3C_2H	Methylacetylene
			CH_3NH_2	Methylamine
			H_2CCHCN	Vinyl cyanide
			$H(C_2)_2CN$	Cyanodiacetylene
			CH_3CHO	Acetaldehyde
		8-Atomic	HCO_2CH_3	Methyl formate
		9-Atomic	CH_3CH_2CN	Ethyl cyanide
			$H(C_2)_3CN$	Cyanotriacetylene
			CH_3CH_2OH	Ethyl alcohol
			$(CH_3)_2O$	Dimethyl ether
		11-Atomic	$H(C_2)_4CN$	Cyanotetracetylene

bound through physical adsorption, until another reactive atom approaches and reacts chemically with it. Thus, the dust grain localizes the atoms on its surface so that they may collide and react. In the absence of dust grains, and in view of the extremely low density of gaseous matter in the interstellar region, it is highly unlikely that two or more atoms would collide and produce an interstellar molecule.

Moreover, the dust grains act as a 'third body' and carry off the energy released as a result of molecule formation. For example, if two atoms of hydrogen combine to yield a molecule of hydrogen, energy equivalent to the bond energy of hydrogen molecule is released. Unless there is a 'third body' to carry off this energy, the hydrogen molecule would dissociate back to two hydrogen atoms, for the energy released during the formation of hydrogen molecule is the same as the energy required to break the hydrogen molecule.

Once the molecules are formed on dust grains, there must be some means of detaching them from the surface of dust particles. The molecules may be released from the dust grains by one or more of the following processes.

(a) The energy of chemical reaction that forms the molecule is taken away by the dust grain (third body). This energy, in form of heat, may partly be passed back to the newly formed molecule so as to evaporate it free from the dust grain.
(b) The impingement of a photon or an energetic particle on the surface of the dust grain may release the molecules.
(c) Ultraviolet light can also eject the newly formed interstellar molecule. However, this mode will work only in diffuse clouds. The dense clouds would not allow ultraviolet light to penetrate through them.
(d) Heat supplied by cosmic rays can be effective in evaporating interstellar molecules from the dust grains.
(e) The heating of dust grain by a nearby hot star can release the intersteller molecules from the dust grains.
(f) Shockwaves due to gravitational instabilities within a dust cloud can disintegrate the dust particles and release the interstellar molecules.
(g) Direct collision between dust grains can provide sufficient energy for evaporation of interstellar molecules.

Although the proposal that interstellar molecules form on dust grains has been widely accepted, the only molecule for which this mode of synthesis has been concretely established is hydrogen. Beyond this it is difficult to make predictions of chemical reactions on dust grains. In fact, as the size of the molecule increases, the efficiency with which it may be formed on grain surface decreases. This may be attributed to the following reasons.

(i) It is relatively easy to localize small hydrogen atoms on the surface of grain. However, for the synthesis of complex molecules, the precursors are larger atoms or molecules. These may not be localized so easily on the grain surface and hence would not get an opportunity to collide with each other.

(ii) The collision between the atoms or molecules is a necessary but not a sufficient condition for the reaction to occur. The reactants must possess sufficient energy to overcome the activation energy barrier required for their transformation into the products. Higher the temperature of the reaction medium, easier it is to cross this barrier. However, for most interstellar clouds the temperature varies in the range 3-100 K. At such low temperatures, the activation energy required to produce hydrogen molecule from two hydrogen atoms may be exceeded. However, the activation energy requirements to produce complex molecules necessiates higher temperatures, which are not available in interstellar regions.

(iii) Even if complex molecules are formed on the surface of the dust grains, it would be tedious to evaporate them. Hydrogen molecule requires relatively small energy for removal from dust particles. However, the larger molecules may be removed from dust grains by a macroscopic heating process, such as a nearby hot star, shockwaves or collision between dust grains (modes *e*, *f* and *g* listed earlier). These may not be always available. And when they are available, the energy may be supplied in such quantities that the larger interstellar molecules may dissociate into smaller analogs. For example, water molecules have been known to disintegrate to hydroxyl radicals.

How then can we account for the synthesis of larger and more complex interstellar molecules? Let us look into an alternative mode of their synthesis.

2. Synthesis of interstellar molecules by ion-molecule reactions

The reaction between an ion and a neutral molecule occurs with essentially zero activation energy. This means that there is no activation energy barrier for such reactions. Hence, these can occur even at low temperatures, that is, under the conditions prevalent in the interstellar medium. Most of the complex molecules are therefore built-up through ion-molecule reactions. Some such reactions are presented in Table 1.5.

The primary ionization reactions (1.49–1.51) are initiated by a flux of cosmic rays at energies of 100 MeV or more. These reactions produce mainly ionized hydrogen, helium, carbon and oxygen. Such reactive ionic species can enter into chain propagation sequences (1.52–1.63) that require little or no activation energy. The ions may recombine with electrons to form neutral molecules and thus end a reaction pathway. Alternately, the radicals may combine to give neutral molecules. As a result of these chain termination reactions (1.64–1.67), not all the products in the reaction scheme listed in Table 1.5 are identifiable. Many are just intermediates which, as a result of dissociative recombination or radical recombination reactions, are converted into relatively more stable analogs. For example, the species HCO may either undergo a proton transfer reaction (1.63), leading to carbon monoxide and methyl carbonium ion or a dissociative recombination reaction (1.67), leading to carbon monoxide and hydrogen radical.

It has been suggested that formaldehyde is produced by dissociative recombination of H_3CO^+ (reaction 1.66) which, in turn, is formed by radiative association of HCO^+ and H_2 (reaction 1.59). Alternately, formaldehyde may be formed by a radical recombination reaction (1.64) involving methyl and oxygen radicals.

Relative Abundance of Interstellar Molecules

The abundance of interstellar molecular species generally decreases as their complexity increases. However, there are striking additional variations in the relative abundance of various analogs, which depend upon the following factors.

1. Concentration of constituent atoms

Molecular hydrogen is the most abundant molecule in the interstellar medium. This is due to the large abundance of the constituent hydrogen atoms. For the same reason, the abundance of hydroxyl radical is significantly high.

Table 1.5 Some important ion-molecule reactions

Chain Initiation

Primary ionization

$$H_2 \xrightarrow{CR} 2H^+ + 2e^- \tag{1.49}$$

$$He \xrightarrow{CR} He^+ + e^- \tag{1.50}$$

$$C \xrightarrow{CR} C^+ + e^- \tag{1.51}$$

(CR = cosmic ray)

Chain Propagation

Secondary ionization

$$H_2^+ + H_2 \rightarrow H_3^+ + H \tag{1.52}$$

$$H_3^+ + O \rightarrow OH^+ + H_2 \tag{1.53}$$

$$H_3^+ + CO \rightarrow HCO^+ + H_2 \tag{1.54}$$

Hydrogenation

$$C^+ + H_2 \rightarrow CH^+ + H \tag{1.55}$$

$$CH^+ + H_2 \rightarrow CH_2^+ + H \tag{1.56}$$

$$CH_2^+ + H_2 \rightarrow CH_3^+ + H \tag{1.57}$$

Radiative association

$$C^+ + H_2 \rightarrow CH_2^+ + h\nu \tag{1.58}$$

$$HCO^+ + H_2 \rightarrow H_3CO^+ + h\nu \tag{1.59}$$

Insertion

$$CH_3^+ + CO \rightarrow CH_2CO^+ + H \tag{1.60}$$

$$C^+ + NH_3 \rightarrow CNH_2^+ + H \tag{1.61}$$

Proton transfer

$$CH_3^+ + CO \rightarrow HCO^+ + CH_2 \tag{1.62}$$

$$HCO^+ + CH_2 \rightarrow CH_3^+ + CO \tag{1.63}$$

Chain Termination

Radical recombination

$$CH_3 + O \rightarrow H_2CO + H \tag{1.64}$$

Dissociative recombination

$$CNH_2^+ + e^- \rightarrow HCN + H \tag{1.65}$$

$$H_3CO^+ + e^- \rightarrow H_2CO + H \tag{1.66}$$

$$HCO^+ + e^- \rightarrow CO + H \tag{1.67}$$

2. Bond dissociation energy

The relative abundance of interstellar molecules is directly related to its bond dissociation energy. The reason being that molecules having low bond dissociation energies are easily ruptured by ionizing radiation.

After hydrogen, the next most abundant molecule is carbon monoxide. This molecule has a bond order of three and thus possesses an enormously high bond dissociation energy. It, therefore, remains stable for almost 1000 years in the interstellar medium. On the other hand, complex molecules, such as, water, ammonia, methane and formaldehyde have low bond dissociation energies and possess a life time of only about 100 years. No doubt, the interstellar molecules are regenerated from the clouds many times during the life of the cloud; yet, those with a lower survival time (and lower bond dissociation energies) are less abundant. Accordingly, the bond dissociation energies of CS (761 KJ mol^{-1}) and CN (1000 KJ mol^{-1}) are quite high and so are their abundance.

3. Local environment

The abundance of an interstellar molecule in one interstellar medium is different from that in another. This gradation is attributed to the local conditions prevailing in a given interstellar region. For example, a star having more carbon than oxygen will contain most of the carbon as carbon monoxide. The residual carbon shall exist as CN and various hydrocarbons. On the other hand, if oxygen is more abundant than carbon, then most of the carbon shall still be contained as carbon monoxide, but instead of CN and hydrocarbons, the atmosphere of the star shall be rich in water.

A molecule with a low bond dissociation energy may survive if it is formed in a dense cloud. The reason being that ionizing radiation cannot penetrate the atmosphere of the dense cloud. Thus, the molecule will have a high degree of abundance. If the same molecule is formed in a diffuse cloud, it may disintegrate easily and hence its abundance may be quite low.

4. Mode of detection

Most of the interstellar molecules have been identified through microwave probes. Microwave spectroscopy can be used to identify only those molecules which have a permanent dipole moment. Thus, a hydrogen chloride molecule may be detected by microwave emission, but a chlorine molecule cannot be identified by this means. Nitrogen and oxygen molecules are expected to be present in the interstellar space in appreciable concentrations since these have significantly high bond dissociation energies and are, therefore, not liable to disintegration by ionizing radiation. Yet, these molecules have not been detected in the interstellar medium. The fact that they do not have a permanent dipole moment perhaps precludes their identification.

Importance of Interstellar Molecules

The molecules in the interstellar medium are difficult to identify, yet more difficult to characterize. Still considerable effort is expended on their investigation. What is the significance of discovering and studying the interstellar molecules? These discoveries bear directly on a number of important aspects:

1. The study of interstellar molecules provides a wealth of information about the composition of the universe.
2. The detection of interstellar molecules and analysis of their properties and abundance yields useful information on the constitution, location and velocities of the interstellar clouds.
3. Some of the interstellar molecules take away the heat of interstellar clouds and thus play an important role in the radiative cooling of the clouds.
4. The interstellar molecules absorb radiation from their surrounding environment and are transformed to excited states. A study of these molecules in their excited states provides information on the intensity of radiation in the surrounding environment.
5. It is speculated that the interstellar molecules made a significant contribution to the origin of life on earth. The fact that molecules as complex as cyanoacetylene or formaldehyde can exist in interstellar space is a convincing evidence that these, or molecules of similar complexity, could have formed during the time immediately following the origin of the earth. Aldehydes (e.g., H_2CO) and nitriles (e.g., CH_3CN) are the precursors of amino acids, while hydrogen cyanide is the precursor of the purine, adenine. Thus, interstellar molecules are the starting points of the chemical evolution, a process which culminated in the origin of life.

Inference

The highlights of this chapter may be summed up under the heading, *cosmic evolution*—the study of the seemingly endless changes in the composition and assembly of various components of the universe. It attempts to demonstrate that a clear thread links the emergence of radiation from the big-bang, the transformation of radiation to simple atoms, of atoms into galaxy and stars, of the stars into heavy elements, of heavy elements into molecules, of molecules into life, of life into intelligence, and of intelligence into culture and technology.

Cosmic evolution stipulates that complexity arises from simplicity. The concept of cosmic evolution seems straight forward: Light, radiation, atoms, galaxies, stars, heavy atoms, life, intelligence— an entire hierachy of material coagulation from radiation, to matter, to life. Yet, it is not so straight forward. We do not know the details; we do not know how, under turbulent conditions of the early universe, the galaxies were formed from matter comprising of simple atoms; we do not know anything about the destiny of the universe. Yet, we know that the universe is not static. It is changing with time; it is evolving.

References

Bethe, H.A. and Brown, G. (1985). How a supernova explodes, *Sci. Amer.,* **252**(5), 60–68.

Claisson , E.J. (1981) in, *'Life in the Universe'*, Billingham, J. (Ed.), The MIT Press, Cambridge, p. 1–16.

Lunine, J.I. (1995). Chemistry in the outer solar system, *Chem. Eng. News,* January 16, p. 40–52.

Malaney, R.A. and Fowler, W.A. (1988). The transformation of matter after the big-bang, *Amer. Sci.,* **76,** 472–477.

Metz, W.D. (1973). Interstellar molecules: New theory of formation from gases, *Science,* **182,** 466–468.

Peebles, P.J.E., Schramm, D.N., Turner, E.L. and Kron, R.G. (1994). The evolution of the universe, *Sci. Amer.,* **271**(4), 29–33.

Segan, C. (1972). Interstellar organic chemistry, *Nature (London),* **238,** 77–80.

Selbin, J. (1973). The origin of chemical elements, 1, *J. Chem. Ed.,* **50,** 306–310.

Selbin, J. (1973). The origin of chemical elements, 2, *J. Chem. Ed.,* **50,** 380–387.

Viola, V.E. and Mathews, C.J. (1987). The cosmic synthesis of lithium, beryllium and boron, *Sci. Amer.,* **256**(5), 34–41.

Watson, W.D. (1977). Interstellar chemistry, *Acc. Chem. Res.,* **10,** 221–226.

Weisskopf, V.F. (1983). The origin of the universe, *Amer. Sci.,* **71,** 473–480.

Wheeler, J.C. (1973). After the supernova, what? *Amer. Sci.,* **61,** 42–51.

Wheeler, J.C. and Nomoto, K.T. (1985). How stars explode, *Amer. Sci.,* **73,** 240–247.

Whitten, R.C., Cuzzi, J., Borucki, W.J. and Wolfe, J.H. (1976). Effect of nearby supernova explosions on atmospheric ozone, *Nature (London),* **263,** 398–400.

CHAPTER 2

Solid Earth

The interstellar atoms and molecules were finally condensed into the solar system, of which earth is one of the components. The upper 400 km of the solid earth is designated as *lithosphere*. It is this region which interacts with hydrosphere, atmosphere and biosphere.

Formation of the Earth

The interstellar atoms and molecules, as well as the cosmic dust particles collectively constituted the *solar nebula*. About 6×10^9 years ago, the constituents of the solar nebula aggregated under the influence of their mutual gravitational forces. As the solar nebula swirled together, the central part became a giant star—the sun. The outer portion separated into revolving rings of comparatively high density, with temperature decreasing radially outwards, as shown in Fig. 2.1. These rings attracted more cosmic particles of various sizes through gravitational forces. This process is called *accretion*.

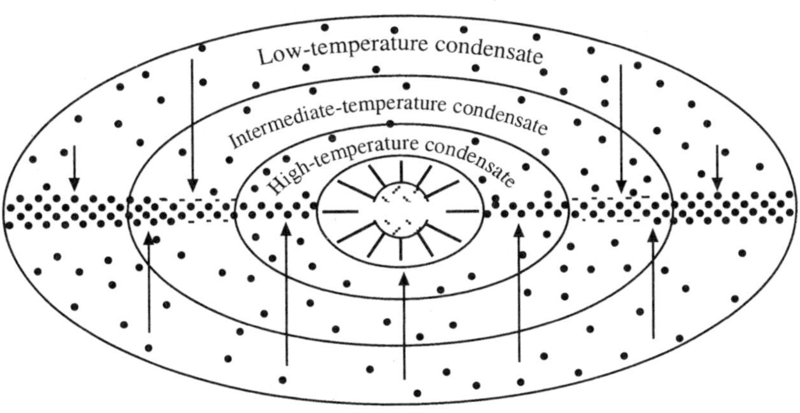

Fig. 2.1 Accretion of solar nebula

For accretion to commence, the relative velocities of dust grains had to be minimum so that they approached each other very slowly, thereby allowing surface chemical forces to cause them to adhere during gentle impacts. The accretion process initially resulted in small aggregations called *planetesimals*.

These planetesimals were initially moving in circular, planer orbits with low velocities. Other cosmic particles collided and fused with them. Thus, a hierarchy of planetesimals with different sizes developed. As the size of the planetesimals exceeded a certain critical value two effects emerged. Firstly, the planetesimal bodies developed gravitational fields of their own. As a result, the orbits of the other smaller planetesimals were perturbed. Such gravitational perturbations led to highly elliptical orbits of planetesimals. Secondly, the larger planetesimals would further sweep up mass at the expense of surrounding, smaller planetesimals. As a result, the larger planetesimal would grow more rapidly and would become a planet. Our earth is one of the planets—nearly spherical in shape with a radius of about 6400 km.

As the solar nebula underwent aggregation, followed by splitting into the sun and the planets, gravitational potential energy was converted into heat, and the temperature rose. The increase in temperature was supplemented by the radiogenic heat produced as a result of the spontaneous decay of the radioisotopes of uranium, potassium and thorium. Since the sun had a greater mass than the planets, its temperature rose to the highest level. In fact, it reached 2×10^7 K and the barrier to the hydrogen burning was crossed. The sun, therefore, started burning as a star. The rise in temperature of the less massive planets was correspondingly low. Therefore, the planets began to cool.

Thus, we see that all the constituents of the solar system—the sun, the planets, the meteorites, the asteroids—had a common origin. We therefore expect that the elemental composition of the earth to be almost the same as that of the universe. However, this is not so.

The Deficiency Factors

It has been observed that the terrestrial composition of certain elements is lower than their cosmic composition. It is useful to analyse this variation in composition in terms of deficiency factors of the elements. The deficiency factor is defined as the ratio of cosmic abundance to the terrestrial abundance of a particular element. Table 2.1 lists the deficiency factors for a few elements. The following factors account for the variation of terrestrial elemental composition *vis-a-vis* cosmic abundance.

1. According to the kinetic theory of gases, certain volatile substances may escape the earth's gravitational field and enter the universe. The cosmic abundance of such a substance would, therefore, be higher than the terrestrial abundance. Accordingly, their deficiency factors would be high.

The escape velocity is given by the expression,

$$V_{escape} = (2\ GM_b/r)^{1/2} \tag{2.1}$$

where G is the gravitational constant (6.6×10^{-11} m^3 s^{-2} kg^{-1}), M_b is the mass of the body whose gravitational field is to be escaped and r is the distance from the particle to the center of the body. At the surface of the earth, the values of M_b and r are 6.0×10^{24} kg and 6.4×10^6 m, respectively. From this data, the escape velocity is calculated to be 1.1×10^4 ms^{-1}.

It has been observed that those substances whose average molecular velocities are less than one-fifth of the escape velocity, would remain on the earth for an infinite time. Others would tend to escape to the universe. The average velocity of a molecule is expressed as,

$$V_{average} = (3RT/M)^{1/2} \tag{2.2}$$

where R is the gas constant, T the absolute temperature and M the molecular weight of the gas.

It is evident from equation (2.2) that lower the molecular weight (or atomic weight) of the substance, higher will be the average velocity. This means that low molecular weight species have been more prone to cross the escape velocity limit. Therefore, such molecules have comparatively large deficiency factors. Table 2.1 reveals that helium and neon are most deficient gases. Between the two, the latter is less deficient due to its higher atomic weight.

Table 2.1 Deficiency factors for selected elements

Element	Deficiency Factor
Hydrogen	6.55
Helium	14.17
Carbon	3.22
Nitrogen	52.5
Oxygen	0.79
Fluorine	0.96
Neon	10.46
Sodium	0.12
Magnesium	0.08
Aluminium	–0.04
Silicon	0.00
Phosphorus	–0.02
Sulfur	0.70
Chlorine	0.25
Argon	6.29
Iron	–0.56
Krypton	6.89
Xenon	7.03

2. If the volatile species form nonvolatile compounds, the deficiency factors are reduced. Thus, although the molecular weight of hydrogen equals the atomic weight of helium, the deficiency factor of the former is almost half that of the latter. Similarly, the atomic weight of neon (10) is only slightly less than the molecular weight of nitrogen (14), yet the deficiency factor for neon is almost double that of nitrogen. Both, hydrogen and nitrogen can interact with other elements to form nonvolatile compounds. Thus, lesser amounts of these gases escaped the earth's gravitational field. On the other hand, helium and neon are inert gases. They do not enter into compound formation. It was, therefore, easier for them to escape earth's gravitational field.

3. A number of gases, such as ammonia, methane, carbon dioxide, nitrogen and oxygen are quite deficient. These gases have comparatively high molecular

weights. Thus, their average velocities, according to equation (2.2), are expected to be correspondingly low, so that it is difficult to account for the deficiency of these species. However, equation (2.2) reveals that the average velocity of a molecule is also dependent upon the temperature. Higher the temperature, higher will be the average velocity. When the earth was initially formed, its average temperature was much higher than what it is now. Under the initial turbulent conditions, the aforementioned molecules would have attained average velocities approaching the escape velocity. Their average concentrations are, therefore, less than expected.

4. The deficiency factors of krypton and xenon are quite large (Table 2.1). These gases have substantially high atomic weights (84 and 131, respectively). For these species to escape the earth's gravitational field, the temperature of the earth at the time of its formation should have been of the order of 10^5 K. At such high temperatures, the molecules ordinarily dissociate. However, since molecules, such as methane, ammonia, carbon monoxide, etc., have survived the earth's formation, it is difficult to assume that the earth would have achieved a temperature as high as 10^5 K.

In order to explain deficiency of krypton and xenon, we once again refer to the theory of planetesimal bodies. The earth was formed by the condensation of a large number of planetesimals. Obviously, the mass (M_b) of these bodies was less than that of the earth. Therefore, according to equation (2.1), the velocity required for a constituent to escape from the gravitational influence of these planetesimal bodies would have been much less, compared to the V_{escape} of the earth's gravitational field. It is assumed that gases like krypton and xenon would have escaped the terrestrial sphere at the time of condensation of the planetesimals into the earth.

Zonal Structure of the Earth

Most of the information on the structure of the earth is derived from analysis of seismic (or earthquake) waves. An earthquake gives rise to different types of waves, of which two can penetrate the body of the earth. These are:

(1) The primary waves or *P* waves, which are transmitted in the direction of propagation of earthquake and possess small amplitude.

(2) The secondary waves or *S* waves, which are transmitted perpendicular to the direction of propagation of earthquake and possess large amplitude.

The velocities of *P* and *S* waves depend on the density of the material through which they pass. In the region of discontinuity, that is, where the density of the material changes, the *P* and *S* waves are subjected to reflection and refraction. This causes a change in their velocities. By comparing the velocities of *P* and *S* waves at different depths of the earth, it has been concluded that the internal terrestrial structure is heterogeneous and that two prominent discontinuities are present: The *Mohorovičic′ discontinuity* and the *Wiechert-Gutenberg* discontinuity. These are called the *first order discontinuities.* They divide the earth into the following three zones.

(a) *Crust,* from the surface of the earth to Mohorovičic′ discontinuity.

(b) *Mantle,* from the Mohorovičić discontinuity to the Wiechert-Gutenberg discontinuity.

(c) *Core,* from Wiechert-Gutenberg discontinuity to the center of the earth.

The zonal structure of the earth is depicted in Fig. 2.2. The first order discontinuities are shown by solid lines. The mantle is further subdivided into upper mantle and lower mantle by a transition zone. The transition zone also represents a discontinuity. However, since the variation of density between upper and lower mantles is not much, therefore, such a discontinuity is termed *second order discontinuity* and shown by broken lines in Fig. 2.2. Likewise, the core is also subdivided into outer (or upper) core, which is in liquid state and inner (or lower) core, which is in solid state, by a second order discontinuity.

Table 2.2 Physical characteristics of earth's zones

Zone	Thickness (km)	Volume ($\times 10^{27}$ cm^3)	Mean Density (g cm^{-3})	Mass ($\times 10^{27}$ g)	Mass (percent)
Crust	17	0.008	2.8	0.024	0.4
Mantle	2883	0.899	4.5	4.075	68.1
Core	3471	0.175	10.7	1.876	31.5
Whole Earth	6371	1.082	5.5	5.975	100.0

The crust and the upper mantle together constitute the *lithosphere.* Some physical characteristics of crust, mantle and core are presented in Table 2.2. The chemical composition of the three zones are described in sequel.

Composition of the Crust

From the stand point of environment of life on earth, it is the crust that is of prime significance, for it contains the continents and the oceans, the soil and the sediments, the water and the minerals. The average composition of the important elements in the earth crust is given in Table 2.3.

Table 2.3 Composition of the crust

Major Elements	Percent by Weight	Percent by Volume	Percent by Atoms
Oxygen	46.6	93.8	62.6
Silicon	27.7	0.9	21.2
Aluminium	8.1	0.5	6.5
Iron	5.0	0.4	1.9
Calcium	3.6	1.0	1.9
Sodium	2.8	1.3	2.6
Potassium	2.6	1.8	1.4
Magnesium	2.1	0.3	1.8

The eight elements listed in Table 2.3 make up nearly 99% of the total crustal composition by weight. Of these, oxygen is the most predominant. The earth's crust contains largely oxygen compounds, especially silicates. The silicates of aluminium, calcium, magnesium, sodium, potassium and iron are most abundant. Therefore, silicon is the second most abundant element in the earth's crust.

As is evident from Table 2.3, the oxygen composition exceeds 60% in terms of the number of atoms. If the volume of different atoms (or ions) is calculated, oxygen makes up more than 90% of the total volume. Thus, in brief, the crust of the earth is essentially a packing of oxygen anions in form of silicates.

If we were to extend Table 2.3, the ninth place would be occupied by titanium, which would be followed by hydrogen, phosphorus, barium and strontium. Surprisingly, some of the elements that play an important role in our economic life and have long been known to, and used by, man are actually quite rare. For example, zirconium is more abundant than copper; lead is almost as abundant as gallium; lanthanides (or the so called rare earths) are more abundant than mercury. Vanadium is more abundant than tin. Elements like titanium, zirconium or the rare earths, which are normally thought to be scarce, are in fact present in the crust in considerable amounts but are systematically dispersed throughout the crust and never occur at one place in high concentrations. These are, therefore, called *dispersed elements.* On the other hand, elements like zinc, copper or lead, which are normally thought to be abundant are in fact quite scarce, but have been concentrated into rich ores at a few scattered locations in the crust.

Composition of the Mantle

As is evident from Table 2.2, the mantle represents about 68% of the mass of earth. Figure 2.2 reveals that the mantle is subdivided into the upper, transition and the lower zones.

The *upper mantle* extends from the Mohorovičić discontinuity to about 400 km. This region contains predominantly three silicate materials: Olivine, $(Mg, Fe)_2SiO_4$; pyroxene, $CaO \cdot MgO \cdot 2SiO_2$ and garnet,

$(Ca, Mg, Fe, Mn)_3(Al, Fe, Cr)_2 \cdot Si_3O_{12}$.

The *transition zone* extends from a depth of about 400 km to about 1000 km. The three components of the upper mantle, viz., olivine, pyroxene and garnet, are also present in this region, but in chemically modified forms. At a depth of about 350 km, pyroxene forms a solid solution with garnet. At about 400 km, olivine undergoes a phase transition to β–olivine, β-$(Mg, Fe)_2 \cdot SiO_4$. The latter is transformed to γ-olivine, γ-$(Mg, Fe)_2 \cdot SiO_4$ at about 550 km. In the region 650–1000 km, the Mg_2SiO_4 component of γ-olivine disproportionates into magnesium silicate, $MgSiO_3$ and magnesium oxide, MgO. Magnesium silicate combines with alumina, Al_2O_3, to form a solid solution.

The *lower mantle,* which extends from a depth of about 1000 km to about 2900 km, consists of a mixture of oxides formed by the disproportionation of transition zone minerals under pressure. The main components are magnesium oxide, iron oxide and silica.

The overall composition of the mantle may be expressed either in terms of percentage of elements or in terms of percentage of oxides (Table 2.4).

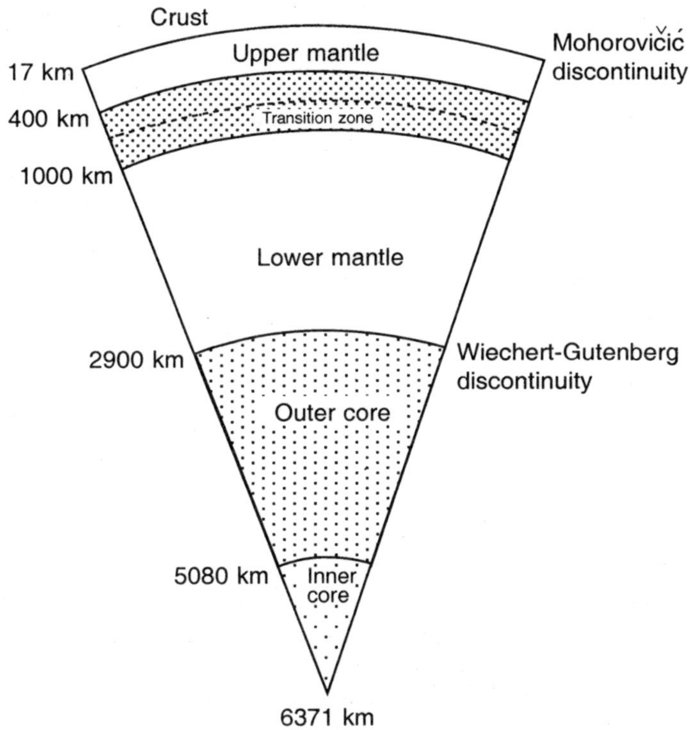

Fig. 2.2 Zonal structure of the earth

As in case of the crust, oxygen is the most predominant element in the mantle too. Most of the oxygen in the upper mantle and transition zone is present in the form of silicates. Therefore, silicon is the second most abundant element in the mantle. In the lower mantle, the oxygen is present as oxides. From Table 2.4, the following inference may be drawn as far as the oxide composition is concerned.

Table 2.4 Composition of the mantle

Major Elemental Composition		Major Oxide Composition	
Element	Percent by Weight	Oxide	Percent by Weight
Oxygen	44	SiO_2	48
Silicon	23	MgO	31
Magnesium	19	FeO	13
Iron	9.9	Al_2O_3	3
Calcium	1.7	CaO	2.3
Aluminium	1.6	Na_2O	1.1
Sodium	0.84	Cr_2O_3	0.55
Chromium	0.38	MnO	0.43
Manganese	0.33	P_2O_5	0.34
Phosphorus	0.14	K_2O	0.13
Potassium	0.11	TiO_2	0.13
Titanium	0.08		

1. More than 90% by weight of the mantle is represented by the system FeO-MgO-SiO_2 and no other oxide exceeds 4% in the mantle.
2. If the components, Na_2O-CaO-Al_2O_3 are added, then this accounts for more than 98% and no other oxide exceeds 0.6% in the mantle.

Composition of the Core

The outer core extends from the depth of about 2900 km to about 5080 km, while the inner core extends from about 5080 km to about 6371 km (Fig. 2.2). The inner core, which is in a solid state, contains nearly pure iron. The outer core, which is in a liquid state also contains predominantly iron. However, when the density of the outer core is compared with that of pure iron under the pressure-temperature conditions prevailing within the core, it is found to be 8–10% less dense than expected. This discrepency implies that the outer core contains about 5–15% of an element or elements of lower atomic weight than iron. One of the companions of iron in the outer core is nickel. In fact, the outer core is predominantly an iron-nickel alloy. Yet, the addition of nickel to iron does not completely overcome the discrepency in density value. A further addition of a still lighter element is necessary in order to explain the observed density measurements. This lighter element is a subject of considerable controversy and debate.

One school of thought considers sulfur to be the principal light element in the outer core. However, a formidable problem confronting this hypothesis is that sulfur is quite a volatile element. Its deficiency factor is about 0.7 (Table 2.1). It is, therefore, likely that sulfur would be depleted in the core, where the temperature is nearly 7500°C. Further, elements like chromium, manganese, sodium, rubedium, cesium and zinc, which are less volatile than sulfur and have lower deficiency factors (0.03-0.3) are not found in the core. It is, therefore, difficult to explain as to how sulfur is retained, while the less volatile elements are depleted in the core.

Another school of thought suggests that the density of the outer core can be explained if it is assumed to contain a mixture of about 60% liquid iron(II) oxide, FeO and 40% liquid Fe-Ni alloy. The Fe-Ni alloy has approximately the following composition: Fe, 94.5%; Ni, 5.5%.

Composition of the Earth as a Whole

We have seen that the core of the earth consists predominantly of iron, partly melted and partly solid. The mantle is predominantly composed of silicates. The crust, which contains only about 1% of the total mass of the earth also predominantly contains silicates, but in much more diverse composition that the mantle. The average composition of the earth as a whole is presented in Table 2.5.

Table 2.5 reveals that about 90% of the earth is made up of four elements: Iron, oxygen, silicon and magnesium. The only other elements that are present in amounts greater than 1% are nickel, calcium, aluminium and sulfur. The elements, sodium, potassium, chromium, cobalt, phosphorus, manganese and titanium are present in the 0.1 to 1.0% range.

Table 2.5 Composition of the earth as a whole

Element	Percent by Weight
Iron	34.63
Oxygen	29.53
Silicon	15.20
Magnesium	12.70
Nickel	2.39
Sulfur	1.93
Calcium	1.13
Aluminium	1.09
Sodium	0.57
Chromium	0.26
Manganese	0.22
Cobalt	0.13
Phosphorus	0.10
Potassium	0.07
Titanium	0.05

Differentiation of Elements

Various terrestrial elements can be broadly classified into the following four types:

1. *Siderophiles*—which tend to be associated with metallic iron.
2. *Chalcophiles*—which tend to be associated with sulfide minerals.
3. *Lithophiles*—which tend to be associated with oxide ion.
4. *Atmophiles*—which tend to occur as gaseous components of the atmosphere.

This four-fold geochemical classification of the elements is termed as the differentiation of the elements (Table 2.6). The relationship between geochemical classification of the elements and the periodic table is shown in Fig. 2.3. The differentiation of elements depends on a number of factors as explained below.

Table 2.6 Geochemical classification of elements

Siderophiles	Chalcophiles	Lithophiles	Atmophiles
Fe,[a], Co[a], Ni[a]	(Cu), Ag	Li, Na, K, Rb, Cs	(H), N, (O)
Ru, Rh, Pd	Zn, Cd, Hg	Be, Mg, Ca, Sr, Ba	He, Ne, Ar, Kr, Xe
Os, Ir, Pt	Ga, In, Tl	B, Al, Sc, Y, La-Lu	
Au, Re[b], Mo[b]	(Ge), (Sn), Pb	Si, Ti, Zr, Hf, Th	
Ge[b], Sn[a], W[b]	(As), (Sb), Bi	P, V, Nb, Ta	
C[c], Cu[a], Ga[a]	S, Se, Te	O, Cr, U	
Ge[a], As[b], Sb[b]	(Fe), Mo, (Os)	H, F, Cl, Br, I	
	(Ru), (Rh), (Pd)	(Fe), Mn, (Zn), (Ga)	

Elements in parenthesis show the indicated character to a lesser extent
(a) Chalcophile and lithophile in the earth's crust
(b) Chalcophile in the earth's crust
(c) Lithophile in the earth's crust

H		Atmophile: N Lithophile: Na															He
Li	Be	Chalcophile: Zn										B	C	N	O	F	Ne
Na	Mg	Siderophile Fe										Al	Si	P	S	Cl	Ar
K	Ca	Sc	Ti	V	Cr	Mn	Fe	Co	Ni	Cu	Zn	Ga	Ge	As	Se	Br	Kr
Rb	Sr	Y	Y	Nb	Mo		Ru	Rh	Pd	Ag	Cd	In	Sn	Sb	Te	I	Xe
Cs	Ba	La-Lu	Hf	Ta	W	Re	Os	Ir	Pt	Au	Hg	Tl	Pb	Bi			
			Th		U												

Fig. 2.3 Geoclassification of the elements

1. Chemical Affinity: The siderophiles, such as nickel, gold and platinum have a tendency to exist in metallic form. Moreover, they alloy readily with iron. Hence, they are associated with iron in the earth's core. The chalcophiles, such as copper and silver are distinguished by their great affinity for sulfide ions. In fact, they form sulfides which are highly insoluble in water as well in dilute inorganic acids. The lithophiles, such as the alkali and alkaline earth metals, tend to form oxides or silicates. The atmophiles, which include noble gases and nitrogen, are inert gaseous substances, that do not have any substantial affinity for iron, sulfide or oxide phases. Hence, they tend to concentrate in the atmosphere.

2. Electronic Configuration: The lithophiles have eight electrons in the outermost shell of their ions. In case of chalcophiles, the outermost shell accommodates eighteen electrons. Siderophiles are mostly the transition elements having incomplete outermost electronic shell. The atmophiles have eight electrons in the outermost shell, but in the elemental rather than ionic state.

3. Reduction Potentials: Lithophiles have highly negative reduction potentials. These elements are, therefore, easily converted into ionic states. Siderophiles have highly positive reduction potentials. These, therefore, exist in elemental states. The chalcophiles are intermediate in behaviour. They do not exist in elemental states, but are converted into ions with difficulty.

4. Heat of formation: It has been observed that if an element forms an oxide with a heat of formation greater than iron(II) oxide, then the element is a lithophile. Greater the difference between the heats of formation, greater is the lithophilic character of the element. On the other hand, chalcophiles or siderophiles have lower heats of formation of their oxides as compared to iron(II) oxide.

5. Density: Liquid iron migrated to the core because of its very high density. The less dense oxides, sulfides and silicates floated to the surface. However, density is not an important factor as far as the differentiation of the elements is concerned. Although uranium and thorium have high densities, yet they are concentrated in the earth's crust. This has been attributed to their tendency to enter into combination with oxide or silicate ions. On the other hand gold and platinum, which too have high densities, are found in the core not because they are heavy metals, but because they have a tendency to alloy with iron, rather than forming oxides or silicates.

6. Local Environment: The differentiation of elements depends upon the temperature, pressure and the chemical composition of the local environment. Therefore, some elements show affinity for more than one group. For example, chromium is a strongly lithophilic element in the earth's crust; but if oxygen is deficient, chromium is chalcophile, being present as daubreelite, $FeCr_2S_4$. Carbon and phosphorus, under oxidising conditions are siderophiles. Similarly, iron is siderophile in the core, but lithophile in the crust.

References

Allegre, C.J. and Schneider, S.H. (1994). The evolution of the earth, *Sci. Amer.,* **271**(4), 44–51.

Anderson, D.L. (1984). The earth as a planet: Paradigm and paradoxes, *Science,* **223,** 347–355.

Bassett, W.A. (1994). What is in the earth's core besides iron? *Science,* **226,** 1662–1663.

Bowring, S.A. and Housh, T. (1995). The earth's early evolution, *Science,* **269,** 1535–1540.

Burchfiel, B.C. (1983). The continental crust, *Sci. Amer.,* **249**(3), 114–129.

Harper, C.L. and Jacobsen, S.B. (1996). Noble gases and earth's accretion, *Science,* **273,** 1814–1818.

Jeanloz, R. (1983). The earth's core, *Sci. Amer.,* **249**(3), 56–65.

Mason, B. (1966), *Principles of Geochemistry',* 3rd Edn., John Wiley & Sons, New York, p. 28–66.

Mckenzie, D.P. (1983). The earth's mantle, *Sci. Amer.,* **249**(3), 66–78.

Ringwood, A.E. (1979) in, *'The Earth: Its Origin, Structure and Evolution,* McElhinny, M.W. (Ed.), Academic Press, London, p. 1–58.

Strahler, A.N. and Strahler, A.H. (1974), *'Introduction to Environmental Science',* Hamilton Publishing Co., California, p. 167–187.

CHAPTER 3

Hydrosphere

The total water present on the earth, in solid, liquid and gaseous phases, constitutes the hydrosphere. The water of the hydrosphere is present as ocean, snow and ice, fresh water bodies and water vapours. Although water is distributed in earth's reservoirs in reasonably fixed proportions, it does not remain locked in these reservoirs. Instead, it continuously circulates in a closed cycle, moving from one reservoir to another. The oceans are the largest reservoirs of the hydrosphere and, therefore, the most important ones on the global scale.

Characteristics of the Hydrosphere

The hydrosphere is a discontinuous shell of water at the surface of the earth. It comprises the oceans, with their connected rivers and streams, ground water, water vapours and snow and ice. Some physical characteristics of the hydrosphere are presented in Table 3.1. The hydrosphere may be split up into four components: The ocean, continental snow and ice, fresh water systems and water vapours. For every square centimeter of the earth's surface there is 273 liters of water. Its distribution in different components of the hydrosphere is given in Table 3.2. The mass of water in various components of the hydrosphere is presented in Table 3.3.

Table 3.1 Physical characteristics of the hydrosphere

Mean Thickness (km)	Volume ($\times 10^{27}$ cm^3)	Mean Density (g cm^{-3})	Mass ($\times 10^{27}$ g)	Area ($\times 10^{16}$ sq cm)
3.8	0.00137	1.03	0.00141	361

Table 3.2 Distribution of water for every square centimeter of the earth's surface

Component of Hydrosphere	Volume (l)
Sea water	268.450
Snow and ice	4.500
Fresh water	0.100
Water vapours	0.003
Total	273.053

Table 3.3 Mass of water in various components of hydrosphere

Component of Hydrosphere	Mass (g)
Sea water	1390×10^{21}
Snow and ice	22.83×10^{21}
Fresh water	0.51×10^{21}
Water vapours	0.000013×10^{21}

Almost 97% of the world's water is contained in the oceans. About 2% is locked up in the ice sheets. About 1% is represented by fresh water of the land. In vapour form, water is present only to the extent of 0.001%. Thus, the hydrosphere, which is a generalized term for the total water of the earth (in solid, liquid or gaseous state), is largely represented by the world ocean.

The Ocean

The term *world ocean* refers to the combined oceans and seas of the globe. The world ocean covers about 70.8% of the earth. Thus, the ocean is clearly of first magnitude. Yet, various marine systems show certain common characteristics.

Characteristics of the World Ocean

Some characteristics of the world ocean are described as follows.

1. Structure

The ocean has a three-layered structure (Fig. 3.1). The upper zone (or *surface zone*), represents about 2% of the volume of the ocean. It extends from the surface of the ocean to a depth of about 500 m. The intermediate zone extends from 500 to 2000 m depth. It is called *thermocline* and represents about 18% of the oceanic volume. The remaining 80% volume of the ocean constitutes the *deep zone* which extends from a depth of 2000 m to the sea floor. Near the North

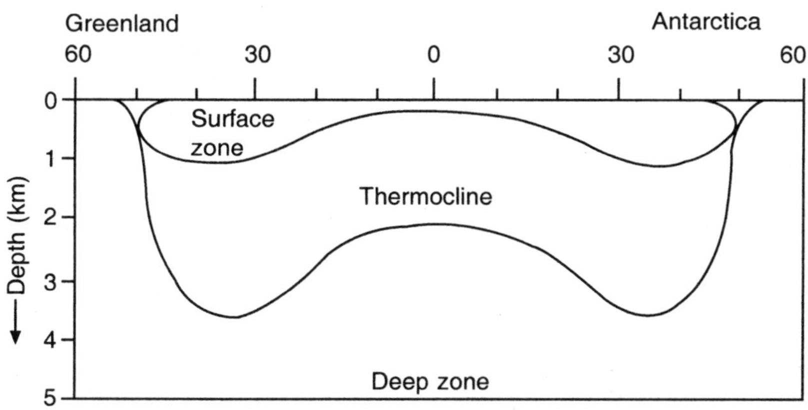

Fig. 3.1 Layered structure of the ocean

and South poles, however, the multi-layered structure disappears and is replaced by a single layer of cold water.

2. Temperature

The surface layer of the ocean is relatively warm. Due to the absorption of solar radiations, the temperature in the upper layer lies in the 20–25°C range. Within the thermocline the temperature rapidly declines. Below the thermocline is the deep layer of very cold water extending to the ocean floor. The temperature near the base of the deep layer is in the range 0–5°C. The variation in temperature with depth is shown in Fig. 3.2.

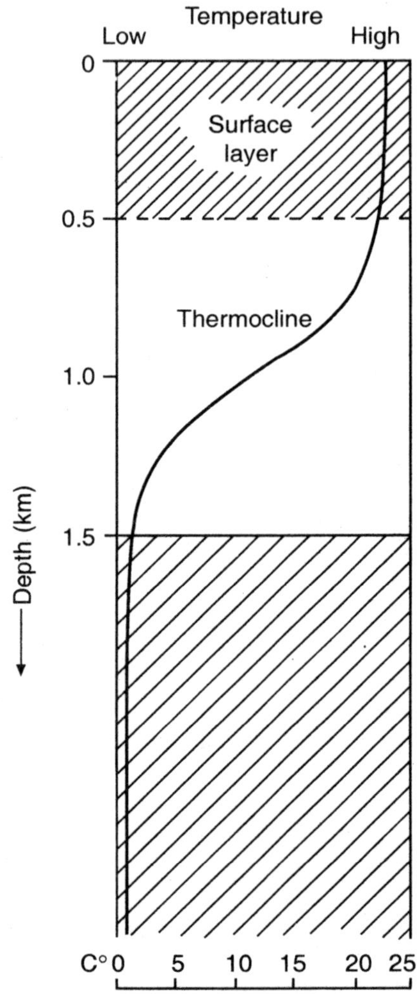

Fig. 3.2 Variation in temperature with depth of the ocean

3. Density

Sea water has a density ranging from 1.027 to 1.028 g cm^{-3}. Density depends on temperature: Warm water is less dense than cold water. The circulation of sea water depends on the change in densities within different layers of the ocean. For example, the warm surface water moves towards the poles, cools and sinks by displacing less dense water below. As a result vertical currents, called *convectional currents* are produced in the ocean. These currents cause the mixing between the surface and deep water. However, the mixing process is very slow, taking about 1000 years.

4. pH

The pH of the sea water is essentially constant at 8.1 ± 0.2, that is, the pH of the ocean is regulated by a buffer. The carbon dioxide equilibrium, involving carbonic acid, bicarbonate anions and carbonate anions was thought to be responsible for buffering action. This equilibrium is represented as follows:

$$CO_2 + H_2O \rightleftharpoons H_2CO_3 \rightleftharpoons HCO_3^- + H^+ \rightleftharpoons CO_3^{2-} + 2H^+ \quad (3.1)$$

If the alkali content of the ocean increases, the H$^+$ ions are removed from the system and equilibrium (3.1) shifts to the right. If the acid content of the ocean increases, the H$^+$ ions are added to the system and the equilibrium (3.1) shifts to the left, producing undissociated carbonic acid. Hence the pH varies over a very narrow range.

However, the buffering capacity of carbonate/bicarbonate anions in aqueous solution is very small. Since the oceans contain significantly high concentrations of various acids and bases, the carbon dioxide equilibrium (3.1) cannot possibly regulate the pH of the ocean. It has now been propounded that the silicate material of the sea sediments controls the pH. The silicate equilibrium may be represented as follows:

$$1.5Al_2Si_2O_5(OH)_{4(s)} + K^+ \rightleftharpoons KAl_3Si_3O_{10}(OH)_{2(s)} + H^+ + 1.5H_2O \quad (3.2)$$

This equilibrium has a very high buffering capacity. That is, even in the presence of high concentrations of acids or alkalis, the pH of the ocean is maintained at a constant value through the buffering action of silicates.

Composition of the Ocean

The unique characteristic of the world ocean is that although it contains a vast variety of chemicals, both organic as well as inorganic, its composition is essentially constant. This is not to suggest that the composition of sea water is static. However, the change in composition of the world ocean over relatively short periods is so small that it is almost immeasurable.

The composition of the ocean is generally expressed in terms of *salinity* (S). The latter is defined as the total amount in grams of solid material dissolved in one kilogram of sea water, when all the carbonate has been converted to oxide, all the iodide and bromide have been replaced by chloride, and all organic matter has been completely oxidised. Salinity is expressed as grams of salt per kilogram of sea water (g kg^{-1}), that is, as parts per thousand by weight (‰). Table 3.4 presents the contribution to salinity of various inorganic salts. The salinity of all the oceans or seas is about 35‰. However, small variations in salinity have been observed from place to place. In the equatorial regions, where the oceans receive abundant rainfall, salinity may drop to 34.5‰. Such is the case with Java Sea. Where there are desert belts below the ocean, evaporation raises the salinity to more than 35.5‰. Such is the case with Red Sea and Persian Gulf.

Table 3.4 Principal salts in the ocean

Salt	%
Sodium chloride	23.0
Magnesium chloride	5.0
Sodium sulfate	4.0
Calcium chloride	1.0
Potassium chloride	0.7
Others	1.3
Total	35.0

Another measure of marine composition is the *chlorinity* (Cl). It is defined as the total amount in grams, of chloride, bromide and iodide contained in one kilogram of sea water, assuming that bromide and iodide have been replaced by chloride. The average chlorinity of sea water is 19‰. The empirical relationship between salinity and chlorinity is expressed as,

$$S‰ = 1.80655\ Cl‰ \tag{3.3}$$

Yet another way of expressing the chemical composition of oceans is in terms of *chlorosity* (Cl_0). It is defined as the amount in grams of chloride, bromide and iodide in one liter of sea water, assuming that bromide and iodide have been replaced by chloride. At a particular temperature, chlorinity and chlorosity are related by the expression,

$$Cl\ ‰ = Cl_0/d \tag{3.4}$$

where d is the density of water at the specific temperature.

Some common ingredients of sea water are described below.

1. Electrolytes

The ions furnished by electrolytes in sea water have the highest concentration as compared to other chemical species. In fact, three ions, sodium, magnesium and chloride, account for about 90% of the stuff in ocean. Addition of potassium, calcium and sulfate ions accounts for nearly 93% of all the elements present in sea water. The concentrations of various ions in the ocean are presented in Table 3.5.

Table 3.5 Concentration of some important ions in the ocean

Ion	Concentration (%) (Chlorinity, 19‰)
Cl^-	55.05
Br^-	0.19
HCO_3^-	0.41
SO_4^{2-}	7.68
Na^+	30.61
K^+	1.10
Mg^{2+}	3.69
Ca^{2+}	1.16
Sr^{2+}	0.04

Some of the strong electrolytes in sea water include sodium chloride, sodium sulfate, magnesium chloride and potassium chloride (Table 3.4). It has been observed that the ratios of the concentration of major ions are nearly constant throughout all the seas and oceans. Table 3.6 depicts major constituent concentration to chlorinity ratios for some oceans and seas.

The concentration of calcium carbonate, an important electrolyte in sea water, varies with depth. Calcium carbonate is produced as calcite by planktonic organisms

Table 3.6 Major constituent concentration to chlorinity ratios for some marine systems

Ocean/Sea	$\frac{Na^+}{‰ Cl^-}$	$\frac{Mg^{2+}}{‰ Cl^-}$	$\frac{K^+}{‰ Cl^-}$	$\frac{SO_4^{2-}}{‰ Cl^-}$	$\frac{Br^-}{‰ Cl^-}$
Atlantic	0.5440 0.5567	0.0667	0.01953 to 0.0263	0.1393	0.00325 to 0.00380
North Pacific	0.5553	0.06632 to 0.06695	0.02096	0.1396 to 0.1397	0.00348
West Pacific	0.5497 to 0.5561	0.06627 to 0.06760	0.02125	0.1399	0.0033
Mediterranean	0.5310 to 0.5528	0.06785	0.02008	0.1396	0.0034 to 0.0038
Baltic	0.5536	0.06693	—	0.1414	0.00316 to 0.00344

as skeleton material. Since calcium carbonate is only sparingly soluble in water, the surface layer of the world ocean is always saturated with respect to calcite. However, in deep waters calcium carbonate solubilizes. Deep layer of the ocean contains high concentration of carbon dioxide. As a result, the sparingly soluble calcium carbonate is converted into soluble calcium bicarbonate.

$$CaCO_3 + H_2O + CO_2 \rightarrow Ca(HCO_3)_2 \tag{3.5}$$

Hence, the deep waters of the ocean are always undersaturated with respect to calcite. The depth at which the rate of dissolution of calcite equals the rate of supply from above is called *calcite compensation depth* (CCD).

The world ocean also contains weak electrolytes, such as boric acid, carbonic acid, phosphoric acid and siliconic acid. In addition, a number of trace elements are present as complex ions in sea water. Table 3.7 gives the compositions and concentrations of some of the complex ions furnished by trace elements in the ocean.

Table 3.7 Composition and concentration of complex ions of trace elements in the ocean

Element	Complex Ion	Concentration (mg l^{-1})
Uranium	$[UO_2(CO_3)_4]^{4-}$	0.003
Vanadium	$[VO_2(OH)_3]^{2-}$	0.002
Silver	$[AgCl_2]^-$, $[AgCl_3]^{2-}$	0.00004
Mercury	$[HgCl_3]^-$, $[HgCl_4]^{2-}$	0.00003
Gold	$[AuCl_4]^-$	0.000004

2. Gases

Sea water contains a number of dissolved gases, principally nitrogen, oxygen, carbon dioxide and noble gases. Trace amounts of ammonia, methane and hydrogen are also present. Table 3.8 depicts the average concentrations of some dissolved gases in the ocean.

Table 3.8 Concentration of the dissolved gases in the ocean

Gas	Concentration (ml l^{-1})
Oxygen	0–9
Nitrogen	8.4–14.5
Carbon dioxide	34–56
Argon	0.2–0.4
Helium and Neon	1.7×10^{-4}
Hydrogen sulfide	0–22

The dissolved gases owe their presence in the ocean to the fact that the atmosphere and the sea water are in contact with each other. Thus, at the air-sea interface, gases of the atmosphere freely enter and leave the sea surface. Moreover, the turbulent motion of the sea surface continually exposes new water to the gases of the atmosphere and encourages a free exchange of gases.

The surface waters are in equilibrium or near-equilibrium with nitrogen of the air. Although nitrogen is the most abundant gas in the atmosphere, it is relatively unimportant in the sea since it is not required by biological processes, except by nitrogen fixing bacteria. Therefore, its concentration in sea water does not undergo appreciable variation. The noble gases, argon, neon and helium, too are not utilized in any biological process.

However, oxygen and carbon dioxide play important roles in marine environment. The sea plants both photosynthesize and respire.

$$6CO_2 + 6H_2O + \text{Solar energy} \underset{\text{Respiration}}{\overset{\text{Photosynthesis}}{\rightleftharpoons}} C_6H_{12}O_6 + 6O_2 \quad (3.6)$$

The two processes occur simultaneously, but not at the same rate. If sufficient solar energy is available, the rate of photosynthesis exceeds that of respiration, reaction (3.6) proceeds to the right and oxygen is produced. When, however, sufficient sunlight is not available, the rate of respiration exceeds that of the photosynthesis, reaction (3.6) proceeds to the left causing a net removal of oxygen from sea water, a break down of sugars and a release of carbon dioxide. Photosynthesis predominantly occurs only near the sea surface, since sunlight cannot penetrate into the ocean depths. However, marine animals are found at all depths. During respiration they consume oxygen and produce carbon dioxide. Therefore, carbon dioxide is not only produced at all levels but is also absorbed from the atmosphere. Oxygen, on the other hand, is produced in the sea only in the upper sunlit layers or is absorbed by surface layer from the atmosphere.

As a result, oxygen is found in high concentrations at the surface and decreases with depth as respiration becomes more important than photosynthesis. A minima occurs in the 200–1000 m depth range where the biological demands are greatest and the supply rates are the smallest. In fact, it is in this zone that most of the organic matter falling from the surface undergoes decomposition, consuming oxygen. In deep waters the oxygen content holds to a uniform and moderate value down to the ocean floor. The source of oxygen in deep waters is the sinking of oxygen-rich surface waters. Though this is a very slow process, yet it replenishes sufficient oxygen to satisfy the demands of the sparse population of deep water marine animals. Moreover, most of the organic matter has already been decomposed before it reaches deep water. Therefore, the entire oxygen content, however small, is available for respiration. The variation in oxygen concentration with ocean's depth is depicted in Fig. 3.3.

If there are deep trenches or basins that are isolated by surrounding ridges—as in the Black Sea—the exchange of surface and deep water is restricted and the replenishment of oxygen is hindered. Under such conditions the consumption of available oxygen by biological processes exceeds the supply and near-anoxic conditions prevail. The stagnant anoxic regions are literally marine deserts. Yet, these deserts are not completely devoid of life. In the absence of oxygen as an energy source, the *sulfate bacterial reduction* process supplies the energy for life processes. Sulfate bacterial reduction is a process in which microorganisms use the oxygen of sulfate ions to oxidise organic matter, producing hydrogen sulfide as a by-product.

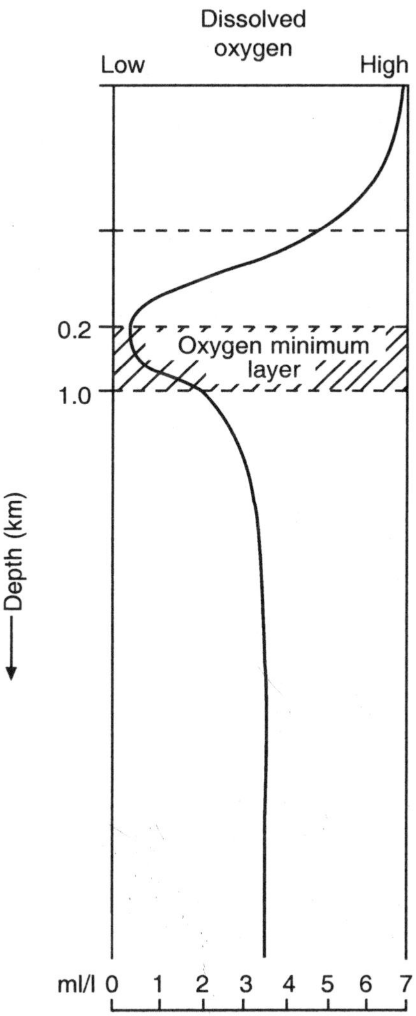

Fig. 3.3 Variation of oxygen concentration with depth of the ocean

$$SO_4^{2-} + 2C_{organic} + H_2O \rightarrow H_2S + 2HCO_3^- \tag{3.7}$$

Hence, in these oceans hydrogen sulfide is present to a significant extent.

3. Nutrients

Dissolved phosphates and nitrates are the most important nutrients in sea water. These are primary nutrients which act as fertilizers for phytoplanktons. In fact, phytoplanktons utilize phosphate and nitrate contents of sea water simultaneously. As a result, the N:P ratio of sea water stands constant at 16:1. Phosphates and nitrates are used not only by phytoplankton but also by other marine plants as they photosynthesize, grow and reproduce. Therefore, the nutrient supply is low in areas of increased photosynthesis, where oxygen is produced at the expense of carbon dioxide. When the plant cells die, they sink below the photosynthetic zone and decompose. Oxygen is consumed in the decomposition process and carbon dioxide is produced. The same process liberates nutrients and returns them in the water-soluble forms. In addition to phosphates and nitrates, dissolved silicates are also classified as nutrients, since these are required for building exoskeletons of the phytoplanktons.

4. Biomaterials

Biomaterials exist in sea water as carbohydrates, proteins, amino acids, lipids, enzymes and vitamins. These substances are as fundamental to life in the ocean as they are on land. They are deposited as excretory products of living organisms or as decomposition products of dead organisms.

Biomaterials are found in larger but variable concentrations in shallow, coastal waters and in smaller but constant concentrations in deep waters. Lipids are found in greater proportions since they are more persistent to decomposition than carbohydrates or proteins. Amino acids make an important contribution to the biomaterial budget of the sea. The important amino acids in the ocean include glutamic acid, lysine, glycine, aspartic acid, alanine, cystine and methionine. Saturated and unsaturated fatty acids, based on C_{16} to C_{18} skeletons, have also been found in sea water.

Biomaterials are important in several respects. Despite their low concentrations, they influence the life cycle of marine organisms and govern, in part, the succession and survival of species. They also determine how organisms may affect each other by production or consumption of substances that interfere with or are necessary for life processes. Moreover, they determine the extent of energy flow through the marine food chain.

5. Suspended matter

The quantity of suspended matter in the world ocean is small as compared to the dissolved constituents. The suspended matter may be of biotic or abiotic type. The biotic matter includes silicon dioxide, calcium carbonate and detritus. The latter is a highly variable biological debris, such as cell fragments, exoskeletons, faecal pellets and pollen grains. The abiotic particulates are mainly minerals like Kaolinite ($Al_2Si_2O_3(OH)_4$), talc ($3MgO \cdot 4SiO_2 \cdot H_2O$), quartz ($SiO_2$) and feldspar ($K_2O \cdot Al_2O_3 \cdot 6SiO_2$).

6. Radioisotopes

Radioisotopes of both, natural and artificial origin, are present in the world ocean. The most important natural nucleide present in sea water is potassium-40. It decays by β-emission and K-electron capture and accounts for more than 90% contribution of the oceanic radioisotopes. Next in quantity is rubedium-87, whose contribution is just 1%. The other natural radioisotopes include the long half-life nucleides, such as uranium-238, thorium-230 and radium-226 and the short half-life nucleides, such as carbon-14, silicon-32 and beryllium-10. The long half-life nucleides are the products of the uranium and thorium decay series and have been disintegrating since the origin of the earth. The short half-life nucleides are formed by cosmic radiations in the atmosphere and are then washed down to the oceans. The artificial nucleides include strontium-90, cesium-137, promethium-147, cobalt-60, manganese-54 and iron-55. These are added to the ocean either through underwater nuclear explosion tests or by human activities.

The Balance of Dissolved Material in the Ocean

The weathering of the earth's igneous rocks over geologic times has contributed different soluble substances to the ocean. These substances have been transported to the ocean by the rivers. It has been estimated that 600 grams of igneous rock has weathered for each kilogram of water in the ocean. It is thus expected that the concentration of dissolved substances in sea water would have increased through geologic times as more and more constituents of rock weathering are added. However, an analysis of sea water reveals that, for most elements, only a part of the material delivered to the sea has remained in solution. For example, silicon, aluminium and iron, the main constituents of the igneous rocks are present as minor elements in sea water, although their concentration in rivers is very high. Thus, there are processes at work which remove soluble substances from sea water.

On the other hand, certain volatile elements like chlorine and sulfur are present in far greater amounts than can have been supplied by weathering of rocks. Thus, there are processes, other than weathering, which contribute these elements to the ocean. The rate at which the various elements are added to the ocean is balanced by the rate of the return of these elements to the solid state as mineral deposits on sea floor. As a result, the chemical composition of the sea water has been essentially constant over eons. Table 3.9 is a balance sheet relating the potential supply of different elements to the ocean to the amount actually present in sea water.

The elements which tend to concentrate in sea water and for which there are no effective mechanisms of removal from solution are called *conservative elements.* The elements which can be removed from solution are called *nonconservative elements.* It may seem that conservative elements are simply accumulating in the ocean. However, this is not so. When the concentration of these elements increases enormously, one or the other removal process takes over. However, such removal processes are quite slow. For this reason, the residence times of conservative components are very large as compared to the nonconservative components. Residence time is given by the expression,

$$r = \frac{\text{Amount present in ocean}}{\text{Amount delivered per year}} \tag{3.8}$$

Table 3.10 compares the residence time of some conservative and non-conservative elements. As an example, consider the residence times of sodium (260×10^6 years), calcium (8×10^6 years) and silicon (0.008×10^6 years). The difference is quite apparent: Silicon is removed from sea water very effectively by phytoplankton. Hence, it is a nonconservative species. Sodium ions, on the other hand, can be removed only ineffectively and that too when their concentration

Table 3.9 Geochemical balance of some elements in the ocean

Element	Total Amount Supplied to Sea Water (g tonne^{-1})	Amount Present in Sea Water (g tonne^{-1})
Addition of Dissolved Constituents to the Ocean		
Chlorine	188.4	18980
Sulfur	312	884
Bromine	0.972	65
Boron	1.8	4.6
Loss of Dissolved Constituents from the Ocean		
Sodium	16980	10561
Magnesium	12540	1272
Calcium	21780	400
Potassium	15540	380
Carbon	192	28
Silicon	166320	4
Aluminium	48780	19
Manganese	600	0.01

in sea water reaches very high. Hence, sodium ions are conservative in nature. Calcium ions can be removed fairly effectively by organisms forming skeletons. Its rate of removal is almost compensated by its rate of replenishment. Hence, it is a borderline case.

Table 3.10 Residence time of some elements in the ocean

Element	Residence Time (Years × 10^6)
Sodium	260
Magnesium	45
Strontium	19
Calcium	8
Silicon	0.0080
Manganese	0.0014

Addition of Dissolved Constituents to the Ocean

Salts of the ocean have come from two sources through the geologic times. One is from the chemical products of rock weathering. The dissolved products are transported to the sea by streams and rivers. Contributions from rock weathering are principally sodium, magnesium, calcium and potassium. The second source of elements is from the earth's interior by a process called *outgassing,* in which water and a few volatile elements, such as sulfur, chlorine, bromine and boron, emerge from below the sea floor by underwater volcanism, hot springs and fumaroles (steam emissions). All the water of the oceans and the atmosphere is considered to have come from the earth's interior by outgassing.

Loss of Dissolved Constituents from the Ocean

Many of the minor and even some of the major constituents of the ocean can be extracted from solution and converted into insoluble materials by various chemical and biological processes.

1. Chemical processes: The following chemical processes account for the removal of elements from sea water solution.

(a) Ion exchange: Exchange of ions occurs between sea water and sediments. The extent of ion-exchange depends on the size and charge of the ions. Smaller the size and greater the charge on the ion, more strongly is it associated with exchanger sediments. Thus, the alkaline earth metal ions, such as magnesium and calcium are more easily removed from solution as compared to the alkali metal ions, such as sodium and potassium. Amongst the alkali metal ions, sodium and potassium are removed more effectively from solution by ion exchange than lithium. Although the ionic radius of lithium ion is smaller than that of sodium and potassium ions, yet lithium is more heavily hydrated. Its hydrated radius is, therefore, much greater than the remaining alkali metals. Hence, sodium and potassium ions are more easily associated with the sediments.

(b) Adsorption: Most of the trace metal ions in sea water are removed from solution by adsorption by sediments. Adsorption depends on the size and texture of the sedimentary materials. If the sediments are composed of small and fine particles, they adsorb more ions from solution than if the sedimentary particles are coarse and large. For this reason, the deep sea sediments, which have a fine texture, adsorb more trace metal ions than the near-shore sediments which are coarse in nature (Table 3.11).

Table 3.11 Distribution of trace metal ions in marine sediments

Trace Metal Ion	Concentration (ppm)	
	Near-Shore Sediments	Deep Sea Sediments
Vanadium	130	330
Copper	48	570
Lead	20	162
Nickel	55	293
Cobalt	13	116
Barium	750	2237
Gallium	19	20

(c) Halmyrolysis: Some elements are removed from solution by chemical reactions between their ions and solid material of the sediments. Such processes are called Halmyrolysis. For example, mercury ions disappear out of the water column by reacting with sulfhydryl group of the organic sediments. The driving force for such reactions is provided by the favourable interaction between a 'soft' acid (Hg^{2+}) and a soft base (–SH). The sandy sediments, which do not contain sulfhydryl groups in their structures, retain negligible concentration of mercury ions.

(d) Precipitation: Many dissolved elements in sea water are precipitated as insoluble substances which settle down, becoming a part of the sediments. For example, iron is precipitated as pyrites, FeS_2. Chromium is removed as hydroxide. Zinc, copper, lead and bismuth are also precipitated as insoluble salts. Precipitation depends on local conditions. Some ions are precipitated as soon as they enter the sea through river waters. Others are precipitated subsequently. Precipitation also depends on the concentration of and charge on the ion to be precipitated. Greater the concentration of the ion, greater are its chances to precipitate. The ions with high charges, such as aluminium and thorium, are more easily precipitated as compared to ions with smaller charges, such as sodium and potassium.

(e) Nodule formation: Manganese, iron and some trace metals are removed from sea water solution through the formation of ferromanganese nodules. A typical nodule contains, on an average, 22% manganese and 14.5% iron. The rest of it is composed of trace metals like cobalt, nickel, copper, titanium and zirconium. These nodules are economically alluring since manganese can be extracted from them. Manganese is concentrated in the nodules through the following chemical reactions.

$$Mn^{2+} + \tfrac{1}{2}O_2 + 2OH^- \rightarrow MnO_2 + H_2O \qquad (3.9)$$

$$5Mn^{2+} + \tfrac{1}{2}O_2 + 10OH^- \rightarrow 4MnO_2 \cdot Mn(OH)_2 \cdot 2H_2O + 2H_2O \qquad (3.10)$$

$$4MnO_2 \cdot Mn(OH)_2 \cdot 2H_2O + \tfrac{1}{2}O_2 \rightarrow 5MnO_2 + 3H_2O \qquad (3.11)$$

The MnO_2 phase is formed at a depth of 1000 to 1500 m, while the $4MnO_2 \cdot Mn(OH)_2 \cdot 2H_2O$ phase is formed at 4000 to 4500 m. Iron is present in the nodule as geothite, $FeO \cdot OH$.

Since the sea water is alkaline and contains dissolved oxygen, the hydrated oxides of manganese and iron are easily formed. Initially these are formed as charged colloidal particles. On the ocean floor they intercept oppositely charged ions. As a result, their own charge is neutralized and they precipitate as nodules. The growing nodule can concentrate trace elements through physical and chemical adsorption.

2. Biological processes: Biological activity is of great importance in controlling the concentrations of many of the elements in the sea, not only the abundant ones, such as calcium and carbon, but also rarer ones, such as copper and vanadium. For example, copper and iron are removed from the solution by marine organisms, since either of these is an important constituent of their blood. Thus, the concentration of copper and iron in sea water is extremely low. Likewise, marine plants remove large concentrations of soluble nutrients, such as phosphates and nitrates, for their growth. On death of marine animals and plants, these constituents re-enter the solution phase. Thus, a balance between consumption and liberation of nonconservative elements is created.

However, this balance is sometimes disturbed. It is possible that an element may enter the body of marine organisms and become bound as insoluble material. When the organism dies, this material sinks and is accumulated on the sea floor.

For many of the insoluble constituents, the chances of re-entering the solution phase are negligible and so they become a permanent part of the sediments. For example, calcium ions are removed from solution by marine organisms and converted into insoluble calcite ($CaCO_3$) for constructing protective shells. When these organisms die, the calcite tends to accumulate. The coral reefs which have formed over the eons are an accumulation of shell material from which calcium ions would never be reclaimed by the ocean. Strontium ions, though similar in character to the calcium ions, are not utilized in shell formation. Therefore, the concentration of strontium ions in sea water is four times that of calcium ions.

A significant amount of phosphorus content of sea water also becomes a permanent part of the sea sediments, from where it may be extracted as phosphatic nodules. We have seen that ferromanganese nodules are formed by chemical processes. The phosphatic nodules are formed from biogeneous materials. The remnants of dead phytoplankton contain appreciable amounts of phosphate. On reaching the sea floor the dead phytoplankton is incorporated into the sediments. The sediments also contain significant amounts of calcium phosphate. The latter, in combination with phosphate content of dead phytoplankton, forms the phosphatic nodules. The concentration of phosphorus in these nodules, measured as P_4O_{10}, may reach as high as 32%. The remaining components of phosphatic nodules are calcium oxide, silicon dioxide, ferric oxide and magnesium oxide.

El Nino

El Nino refers to an anomalous warming of sea water in the eastern region of equatorial Pacific Ocean. Ordinarily trade winds and ocean currents move from the colder eastern region of the equatorial Pacific Ocean to the warmer western sector. This movement is controlled by a geographical phenomena called *Southern Oscillations*. The latter is a trans-pacific linkage of atmospheric pressure systems. When the atmospheric pressure rises in the east, it falls in the west and *vice-versa*. Quantitatively, Southern Oscillation index, Q, is given by the expression,

$$Q = P_E - P_W \tag{3.12}$$

where P_E is the pressure in the eastern sector and P_W the pressure in the western sector. Under normal circumstances P_E is greater than P_W and, therefore, Oscillation index is positive. Thus, the ocean's surface water moves from eastern to western region. As a result, the surface water of eastern ocean is continuously replaced by colder water from the depth. Due to the consistent upwelling of water to the surface, the thermocline (boundary between the warm upper surface and the cold deep water) in the eastern Pacific is relatively shallow. Due to the consistent accumulation of eastern waters in the western Pacific, the thermocline in the latter region is correspondingly deep.

Occurrence

However, it has been observed that in a typical year, during the month of October, the trade winds collapse. As a result, the Southern Oscillation index, Q, begins to fall. Ultimately P_E becomes less than P_W and Q becomes negative. Under such conditions, the warm water piled up in the western Pacific flows back towards the east. The warm water from the west reaches the coast of Peru and Ecquador in the east by December end. Due to its proximity to Christmas, the local fishermen have named the anomalous flow of water as El Nino—the Spanish term for Christ child.

On an average, El Nino occurs about once every four years, but the interval between successive events has been as short as two years and as long as ten. Although El Nino generally lasts uptil March/April, it may sometimes persist for a year. Such relatively intense El Nino's were observed in 1953, 1957–58, 1965, 1972–73, 1976–77, 1982–83, 1987 and 1994. During the 1982–83 event the sea surface off Peru warmed by more than 7°C.

Effects

El-Nino is known to perturb marine ecology and global climate.

1. Marine ecology: El Nino raises the sea surface temperatures in two ways: Firstly, by physically bringing warm waters from the west and secondly, by depressing the thermocline. When thermocline is depressed, the upwelling of cold water from the deep zone of the ocean is prevented. The lowering of thermocline is associated with loss of marine biological productivity. The biodiversity in aquatic systems depends on the amount of sunlight that penetrates through it and the amounts of nutrients that are present in it.

The intensity of sunlight decreases exponentially as a function of depth. This means that the sunlight can penetrate only a few meters below the sea surface. On the other hand, most of the nutrients, such as phosphates, nitrates and silicates, are distributed below the thermocline. When the thermocline is shallow, the primary sea organisms, that is, phytoplanktons, can capture both the sunlight and the nutrients. Enhanced primary production is reflected in richer biodiversity.

Thus, under normal conditions, the regions of the ocean, such as the eastern Pacific, having a thermocline near the surface, are biologically richer than the western Pacific. However, during El Nino the sea level rises—resulting in reduced penetration of sunlight through the marine waters, and the thermocline depresses—resulting in reduced upwelling of nutrients. In this manner the supply of both sunlight and nutrients is reduced as El Nino strengthens in intensity. The strength of primary producers, phytoplanktons, falls and the strength of secondary producers, fish, sea birds, marine mammals etc., is proportionally reduced, crippling the local economy.

During El Nino, fishes like sardine and peruvian anchovy are reduced in number. Lack of nutrients affects the reproductive fitness of adult anchovy. Since larval survival is especially dependent on the availability of phytoplanktons, the reproductive success of the species is impaired.

The reproductive ability of sea birds is also affected, since the scanty food available is just sufficient for survival rather than reproduction. For example, during the 1982–83 El Nino it was observed that the population of blue-faced booby (*Sula dectylatra*) on Christmas Island (2°N, 157°W) decreased. The number of great frigate birds (*Fregato minor*) also decreased. Further, there were cases of nest abandonment by these birds.

The marine mammals provide the clearest evidence for the food stress that accompanies El Nino. During the 1982–83 El Nino the galapages fur seal and California sea lion suffered high mortality. Most of their young ones died. The female species were unable to provide enough milk to prevent pups from starving.

2. Global climate: We have stated that EI Nino causes the sea surface temperatures to rise. This encourages more evaporation. Evaporation of water requires heat energy. Later, when water vapours condense and rain falls, the latent heat stored in the clouds is released, warming the air. Thus consistent El Ninos produce global warming.

El Ninos also change atmospheric circulation on a global scale. After the 1982–83 El Nino, droughts plagued the western Pacific rim countries. Dry atmosphere caused forest fires in eastern Borneo. Regions of Australia experienced severe dust storms. Meanwhile, the central Pacific regions received enormously higher quantum of rainfall. Tropical storms hit Hawaii and Tahiti. Iceland's bird population were decimated as their nests (in sand bars) were flooded by heavy rains.

Origin of the Ocean

The origin of the world ocean may be traced back to the early history of earth, although at that time its volume was less than and its composition was different from what it is today. At the time of the origin of ocean, the earth's atmosphere was reducing rather than oxidising in nature. It was dominated by nitrogen and carbon dioxide, but devoid of oxygen. As a result, the ocean too was a reducing ecosystem.

In consequence—

1. Organic matter was not susceptible to oxidation and was embedded in the sediments.
2. Iron content of the ocean was present in the ferrous state (Fe^{2+}) rather than ferric state (Fe^{3+}).
3. Sulfur of the ocean was present as sulfide ions (S^{2-}), rather than sulfate ions (SO_4^{2-}).
4. The partial pressure of carbon dioxide was very high and the pH of the sea water was lower than its present value of 8.3 ± 0.2.

It is speculated that under these reducing conditions, sea surface temperatures were as high as 50°C. Sometimes about 2.5×10^9 years ago photosynthesizing algae appeared. These built up the store of atmospheric oxygen at the cost of carbon dioxide. More and more of carbon dioxide became fixed in organic matter and its partial pressure progressively decreased. Due to the corresponding increase in the partial pressure of oxygen, ferrous content of the ocean was slowly oxidised to ferric state. Further, the sulfide ions were oxidised to sulfate ions. The organic matter buried in the sediments also began to be oxidised. Simultaneously, the earth started cooling and the sea surface temperatures dropped.

These conditions prevailed till about 1×10^9 years ago. At this stage the sea water reached a composition which is not very different from its present composition. This is borne out by the fact that the characteristics of sea sediments deposited since 1×10^9 years ago have remained essentially the same. The fossil remains of some marine organisms, such as echinoderms, found in old sediments, also support the hypothesis of unaltered sea water composition. The basic physiology of these organisms would not have tolerated a substantial departure from the present day composition and temperature of the ocean, suggesting that there have been no major changes.

Snow and Ice

Snow and ice are present in the atmosphere, in lakes, rivers and oceans, on land and also beneath the earth's surface. In fact, snow and ice are significant components of the world's hydrological system. Sometimes their appearance is brief and local, such as a sudden snow flurry or snowfall in the lower reaches of the Himalayas. In other places snow and ice may dominate the earth's surface for a whole geologic period. The ice caps of polar regions, for example, appear to be a permanent feature of the world.

Characteristics

Ice is colourless to white or pale blue and varies from transparent to translucent. It is characterized by crystals of hexagonal symmetry. The size of the crystals, however, depends on the local environment. For example, crystals in a glacier may vary from 1 mm to 1 m in length.

Density relationships between ice and liquid water are anomalous. For most substances the solid state has a greater density than the liquid state. However, a reverse relationship exists between ice and water. The density of water at 4°C and 760 mm pressure is 1.0 g cm^{-3}. The density of the ice is variable, depending upon local conditions, but is always less than 1 g cm^{-3} (Table 3.12).

Table 3.12 Comparison between densities of water and various forms of ice

Water/Ice	Density (g cm^{-3})
Water (0°C; 760 mm)	1.00
Sea ice	0.91–0.93
Lake ice	0.89–0.90
Glacier ice	0.84–0.85
Snowflakes	0.004–0.33

The anomalous density of ice serves a useful purpose. If the density of ice would have been greater than that of liquid water, ice would have sunk to the bottom of lakes, rivers or oceans. Since solar radiations do not penetrate deep into water, very little of this ice would have melted seasonally. There would thus be a progressive accumulation of ice from depth upwards in the earth's aquatic systems.

Zones of Occurrence

There are three principal zones of occurrence of snow and ice in the hydrosphere.

1. Snow and ice which occurs and originates directly in the atmosphere from freezing of previously condensed droplets. Examples include snowflakes and hail. The existence of ice in the atmosphere is very brief.
2. Snow and ice which occurs on the earth's surface. Examples include frost, glacier ice, as well as sea, lake and river ice, which results from the freezing of water already present at the surface.
3. Ice which occurs beneath the earth's surface. Example includes interstitial ice in soil. Depending on location, such ice may be seasonal or relatively permanent.

Environmental Implications

Snow and ice represent both a valuable resource and a menacing hazard. On the positive side, snow is the principal source of water, not only in many of the world's mountainous areas, but also in the densely settled plains and valleys adjacent to them. Moreover, in the critical dry years, snow and ice are more dependable than other sources of water. On the negative side, the most detrimental effect of snowfall lies in the disruption of communication systems. Economically, it is quite expensive to combat snow. Snowfall also obscures visibility. Further, glaciers are known to cause rock erosion. This is particularly true where the glaciers are thick and fast moving.

The Fresh Water Systems

The fresh water systems include rivers, streams, lakes and springs. The total water content of these systems is called *terrestrial waters.* The main source of terrestrial waters is rain, although thermal springs beneath the earth's surface also contribute to the fresh water systems. As is evident from Table 3.3, the amount of fresh water on earth is insignificant as compared to that of the world ocean. Yet, the fresh water systems, especially the rivers, are important geochemically because they are responsible for most of the weathering and erosion of land masses.

Since water has immense solvent power, a number of substances become dissolved in it during weathering and erosion. These substances are carried by rivers to the world ocean. It has been estimated that the world's rivers contribute about 2.73×10^{12} kg of dissolved material to the ocean each year. However, a comparison of Table 3.5, which gives the average concentrations of ions in the world ocean with Table 3.13, which gives the average concentrations of ions in the rivers reveals an important anomaly.

In the world ocean, the concentrations of ions vary in the order: $Na^+ > Mg^{2+} > Ca^{2+}$ and $Cl^- > SO_4^{2-} > HCO_3^-$. In the rivers, the trends are: $Ca^{2+} > Na^+ > Mg^{2+}$ and $HCO_3^- > SO_4^{2-} > Cl^-$. These opposing trends are baffling, since the river-supplied compounds are the primary source of sea salts. However, this anomaly may be explained as follows.

When water causes weathering and erosion of land masses, the most soluble

Table 3.13 Concentration of some important ions in river waters

Ion	Concentration (% by weight)
Cl^-	6.5
NO_3^-	0.8
HCO_3^-	48.6
SO_4^{2-}	9.3
Na^+	5.3
K^+	2.0
Mg^{2+}	3.4
Ca^{2+}	12.5

substances are leached into solution phase. During the earlier geologic times the rivers would have leached the highly soluble salts like sodium chloride and potassium chloride from the earth's crust. After these highly soluble substances would have been removed, slightly less soluble substances like magnesium chloride, would have solubilized. Today, the rivers are leaching the least soluble substances like calcium sulfate and, therefore, such substances are present in high abundance in fresh water bodies. However, since the dissolution of these less soluble substances has begun quite late, their concentrations in the world ocean are still quite low. The reversal in the chemical composition of river waters compared to that of sea water also suggests that once soluble salts are delivered to the sea they are not returned to the elevated land except in minor quantities on the global scale.

It has also been observed that the concentrations of the oxides of iron, aluminium and silicon are higher in the river waters than in the world ocean. These oxides are the weathered products of the igneous rocks. The weathering process has continued since the earliest geological times. However, the fact that the world ocean contains only small amounts of the weathered products of igneous rocks proves, that during the earlier history of earth, the composition of the crustal rocks was different from what it is now.

Water Vapours

Water vapours are inducted in the atmosphere through evaporation of water from the earth's surface. Net transfer of water molecules into air occurs only if there is a vapour pressure gradient between the evaporating surface and the atmosphere. This implies that when the relative humidity is 100%, evaporation is nil. Evaporation from a moist surface involves a change of state from liquid to vapour. Therefore, an external heat source is a must to affect evaporation. This may be solar radiation or heat content of atmosphere and/or surface from which evaporation takes place. However, solar radiation is the principal energy source of evaporation.

The rate of evaporation depends primarily on the temperature of the air or of the surface. At higher temperatures more water molecules can leave the surface due to their greater kinetic energy. In addition, the wind speed also controls the evaporation rate, since air movement carries fresh, unsaturated air to the evaporating surface. Salinity depresses the evaporation rate in proporation to the solution concentration. Hence, for sea water the rate is about 2-3% lower than for fresh water. However, since the amount of water in oceans is astronomically greater than in rivers and lakes, the net evaporation from sea is far greater than from fresh water systems. The oceans provide 84% of the water vapours annually and the fresh water sources, 16%.

Water loss may occur not only from aquatic bodies but also from plants. Evaporation of water from plants is called *transpiration*. This phenomenon occurs when the vapour pressure in air is less than that in the leaf cells. About 95% of transpiration occurs during day time because water vapour is transpired through small pores in leaves called *stomata,* which open in response to stimulation by sunlight. Transpiration is a passive process so far as the plant is concerned, but it performs a vital function in affecting the internal transport of nutrients and in cooling leaf surface.

References

Barnes, S.S. (1967). Minor element composition of ferromanganese nodules, *Science,* **157,** 63–65.

Bowen, H.J.M. (1979), *'Environmental Chemistry of the Elements',* Academic Press, London, p. 13–29.

Broecker, W.S. (1983). The Ocean, *Sci. Amer.,* **249**(3), 146–160.

Deacom, C.E.R. (1965). Chemistry of the sea, 1/Inorganic, *Chem. Brit.,* **1,** 48–53.

Drever, J.I. (1982), *'The Geochemistry of Natural Waters',* Prentice-Hall, Inc., New Jersey, p. 229–249.

Duxbury, A.C. (1971), *'The Earth and its Oceans',* Addison-Wesley Publishing Co., Massachusetts, p. 114–130.

Jin, F.F., Neelin, J.D. and Ghil, M. (1994). El Nino on the devil's staircase: Annual subharmonic steps to chaos, *Science,* **264,** 70–72.

Marcus, M.G. (1969) in *'Water Earth and Man',* Chorley, R.J. (Ed.), Methuen & Co. Ltd., London, p. 359–367.

Mason, B. (1966), *'Principles of Geochemistry',* 3rd Edn., Johy Wiley & Sons, New York, p. 192–207.

Park, K., Williams, W.T., Prescott, J.M. and Hood, D.W. (1962). Amino acids in deep sea water, *Science,* **138,** 531–532.

Roony, J. (1969) in, *'Water-Earth and Man',* Chorley, R.J. (Ed.), Methuen & Co. Ltd., London, p. 389–401.

Sullén, L.G. (1967). The ocean as a chemical system, *Science,* **156,** 1189–1197.

Strahler, A.N. and Strahler, A.H. (1974), *'Introduction to Environmental Science',* Hamilton Publishing Co., California, p. 29–37.

Wuethrich, B. (1995). El Nino goes critical, *New Scientist,* February 4, p. 32–35.

CHAPTER 4

Atmosphere

The atmosphere is defined as a thick envelop, containing a mixture of gases and small amounts of finely divided liquid and solid aerosols, that surrounds the earth and permeates its crust and oceans. The atmosphere plays a dominant role in sustaining life on earth, for it provides oxygen which nearly all organisms need in their respiration. The constituents of the atmosphere are instrumental in causing weathering and erosion of rocks. The dynamic nature of the atmosphere serves to distribute the energy of solar radiations received by the earth. The resulting weather and climatic patterns profoundly influence the activities of man.

Origin of the Atmosphere

The theory of the origin of atmosphere begins with an idea of how the earth itself originated. As has been discussed in Chapter 2 (p. 25), the earth was formed by accretion of solid materials that condensed from the solar nebula. The atmosphere which existed then was entirely different from today's atmosphere. The present atmosphere evolved in three stages (Fig. 4.1).

Stage-I

Stage-I lasted from the time of accretion of the earth to about 3.5×10^9 years ago. During its later stages of development, the earth was able to gravitationally attract and hold gases that were normal constituents of the original cosmic dust cloud. These gases were mainly hydrogen, methane, ammonia, water vapours and noble gases. These constituted the first or the primitive atmosphere of the earth. The primitive atmosphere was a reducing or a hydrogen-rich atmosphere. There was no free oxygen at that stage.

The accretion of the earth was an exothermic process. Some of the constituents of the primitive atmosphere absorbed the heat released as a result of accretion. These constituents acquired sufficient kinetic energy to disperse out of the earth's atmosphere. The components which oozed out of the earth's primitive atmosphere were mainly the noble gases. It is, therefore, not surprising that the present abundance of noble gases on earth is a minute fraction of their cosmic abundance. Hence the name rare gases. Other gaseous components were more reactive. They entered into chemical combination with the earth's solid materials and, for the time being, were able to resist depletion. Subsequently, however, the earth was to yield up these ingredients through volcanism.

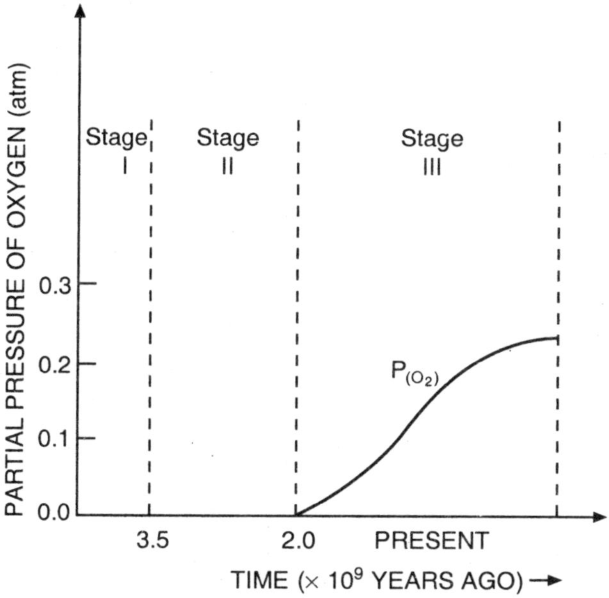

Fig. 4.1 Three stages of atmospheric evolution and development of oxygen partial pressure

Stage-II

Stage-II of the atmospheric evolution initiated at about 3.5×10^9 years ago and ended at about 2×10^9 years ago. During this period the earth differentiated into three distinct zones: Crust, the uppermost zone; mantle, the middle zone; and core, the deepest zone. During this period itself, the heat released due to culmination of accretion process, as well as the decay of radioisotopes, caused volcanoes to occur. Large quantities of magma, saturated with dissolved gases, moved towards the surface of the earth. These gases constituted the second atmosphere of the earth.

During the earlier era of stage-II, the iron rich core of the earth was not completely formed. The iron was largely contained in the mantle—the middle zone. The gases which were ejected out from the interior of the earth had to pass over iron. Due to the reducing nature of iron, the evolving gases were in a reduced state. Nitrogen, hydrogen, hydrogen sulfide, carbon monoxide and water vapours were the main constituents given out from the earth's interior.

During the later era of stage-II, iron present in the mantle melted and drifted towards the core. Now the ejected gases did not interact with iron and were, therefore, in a more oxidised state. These included carbon dioxide, sulfur dioxide and water vapours.

Any free oxygen which escaped out of the earth's interior during stage-II combined with hydrogen, residual methane or ammonia. Thus, even during stage-II, the atmosphere was a reducing, non-oxygen one.

Stage-III

The third and present day atmosphere began to evolve about 2×10^9 years ago.

At about this time, oxygen production began to exceed oxygen consumption. Stage-III, therefore, represents a transition from the oxygen-poor to an oxygen-rich atmosphere.

Two phenomena contributed to the induction of oxygen in the final stage of atmospheric evolution. The first was the photochemical dissociation of water molecules. This process occurred in the upper atmosphere, when water molecules absorbed ultraviolet rays from the sun and were split into hydrogen and oxygen. Photochemical dissociation of water, however, added only minor amounts of oxygen to the atmosphere. The major chunk of oxygen was inducted as a result of the second phenomena—photosynthesis. This process began after the advent of life on earth. In this process the green plants captured carbon dioxide and water vapours in the presence of chlorophyll and sunlight, and converted them into carbohydrates, the plant foodstuff, and oxygen.

Slow Build-Up of Atmospheric Oxygen

As shown in Fig. 4.1, although oxygen began to accumulate in the atmosphere about 2×10^9 years ago, its build-up was quite gradual. Until about 0.6×10^9 years ago, the concentration of aerial oxygen was just one-tenth of what it is now. Thereafter it rose steeply.

It is easy to explain the slow rate of accumulation of atmospheric oxygen from 2×10^9 to 1.2×10^9 years ago. During this time, ferrous iron, carried into the oceans by surface water, consumed a large portion of oxygen released at the beginning of stage-III. The soluble ferrous iron was converted into insoluble ferric oxide, which was subsequently incorporated into sediments. Due to the presence of ferric oxide, these sediments acquired red colour and are called *red beds*. The formation of red beds culminated by about 1.2×10^9 years ago.

Once the formation of red beds ceased, the atmospheric oxygen concentration was supposed to rise steeply. However, the steep rise in oxygen level began only 0.6×10^9 years ago. What retarded the oxygen build-up between 1.2×10^9 to 0.6×10^9 years ago? For a very long time this question remained unanswered. It was only recently that the mystery was unveiled.

It has been hypothesized that at the beginning of stage-III of atmospheric evolution, photosynthesis was carried through in the ocean by phytoplankton (we shall learn in Chapter 6 that life evolved in the world ocean). These organisms occupied the sea surface. Other organisms at sea surface were the heterotrophic bacteria. These bacteria fed on dead phytoplankton and, in presence of oxygen, used to break down the cellular matter of phytoplankton into simple hydrocarbons that were rich in carbon-13 isotope. So a large part of the oxygen, produced by photosynthesis, was consumed by heterotrophic bacteria. The evidence for this hypothesis is provided by the observation that sediments as old as 1×10^9 years are rich in carbon-13 containing hydrocarbons.

About 0.6×10^9 years ago, the first multicellular marine organisms appeared. They consumed the photosynthetic phytoplankton that would previously have died and floated in water where heterotrophic bacteria could reach them. After being eaten by phytoplankton-grazing organisms, the cells' remains quickly sunk to the ocean floor. They were thus quickly rendered out of reach of heterotrophic bacteria which could survive only on sea surface. With a decrease of phytoplankton levels, the heterotrophic bacteria also registered a decline in population. With fewer bacteria to mop-up oxygen, the gas started to build-up in the atmosphere at a much faster rate.

Composition of the Atmosphere

The mass of the present atmosphere is estimated to be 5.6×10^{14} tonnes. The constituents which make up this load may be classified into three broad categories: Gases, water vapours and aerosols.

1. Gases

The atmosphere is made up of relatively stable mixture of gases. The proportions of gases, excluding water vapours, are nearly uniform over the earth's surface and show little variations upto an altitude of about 80 km. The composition of the dry atmosphere by volume is presented in Table 4.1

Table 4.1 Composition of the dry atmosphere

Component	Volume (%)
Nitrogen	78.084
Oxygen	20.946
Argon	0.934
Carbon dioxide	0.033
Helium	0.00052
Krypton	0.00001

Almost 99% of the atmosphere is composed of two gases—nitrogen and oxygen. The noble gases, hydrogen, sulfur oxides and methane are present in trace amounts. Most of the argon originates from the decay of radioactive potassium. Helium, a trace component of the atmosphere, continuously escapes from the top of the atmosphere to interplanetary space. However, it is also being continually replenished by the radioactive decay of uranium and thorium isotopes. Although helium and argon are produced at comparable rates, helium being lighter is depleted from the atmosphere more easily than argon. Under natural conditions, carbon dioxide is released to the atmosphere through cellular respiration of living organisms, and also by decomposition of dead organic matter. Human activities add carbon dioxide to the atmosphere by burning of fossil fuels.

The mixture of gases constituting the atmosphere has no sharp upper boundary. Hence the atmosphere does not have a specific thickness or height. The upper layers of the air compress the layers near the earth's surface. Due to this, the bulk of the atmosphere lies close to the earth. As shown in Fig. 4.2, the air pressure decreases with increasing elevation. The air peters out with increasing height. Thus, the greatest density of air, 1.2×10^{-3} g cm^{-3}, is observed at sea level.

2. Water Vapours

Strictly speaking, water vapours should be treated along with the gaseous components of the atmosphere. However, they merit a separate consideration for two reasons. Firstly, unlike the gases whose concentrations in the lower atmosphere are nearly constant, the amount of water vapours is variable. Secondly (and more importantly) the amount of water vapours over a particular region affects the climatic and weather patterns of that region.

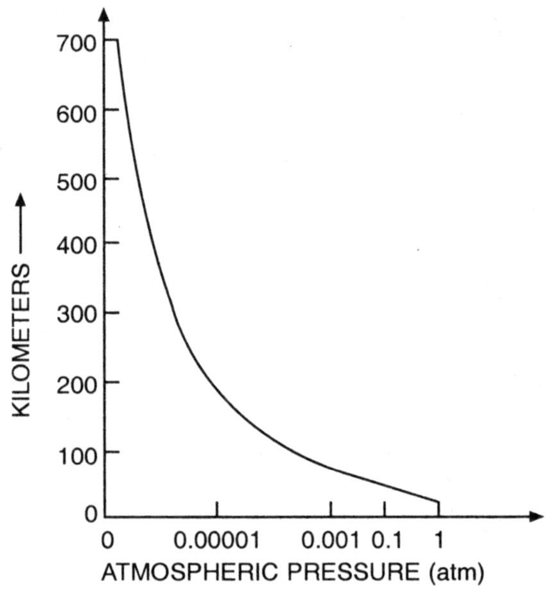

Fig. 4.2 Variation of atmospheric pressure with height

Water vapours occur in the atmosphere in minor and variable amounts. Their concentration varies from place to place and from time to time at the same place. In general, the amount of water vapours over a desert region is negligible, while that over humid tropics may be as high as 4%. The global average of water vapours is about 1%.

The amount of water vapours over a given region also varies with height. As shown in Table 4.2, the water vapour composition decreases sharply with elevation. Thus, mountainous regions have a lower moisture content than the plains.

About 90% of the atmospheric moisture lies below an altitude of 5 km. This seems quite surprising for a water molecule is quite light, having a molecular weight of just 18, and is therefore expected to rise to great heights. However, the atmospheric water vapours are close to their critical temperature, that is, the temperature at which they change from gaseous to liquid state. Therefore, as the vapours are carried upwards, they cool to their critical temperature, condense into cloud particles and return to the earth's surface by precipitation (rain or snow).

Water vapours exert a profound influence on local climate. The solar radiation which strikes the earth's surface during the day is reflected back as heat energy to the outer space after sunset. Water molecules possess the potential to absorb the reflected heat energy. In humid regions, where the moisture content is very high, the heat energy is absorbed by water molecules and is prevented from

Table 4.2 Variation of water vapour concentration with elevation

Height Above Sea Level (km)	Water Vapours (%)
0	1.3
1	1.0
2.	0.69
3	0.49
4	0.37
5	0.27
6	0.15
7	0.09
8	0.05

being transmitted into outer space. Therefore, in these regions the atmosphere remains relatively warm during the night. On the other hand, in desert and mountainous regions the water vapours are virtually absent and the earth's heat rapidly escapes to the outer space. Hence, the cool nights in these regions.

Atmospheric water vapours also influence the weather changes. When water vapours are formed from ice or liquid water, energy in form of latent heat has to be expended. This latent heat remains stored in water molecules as potential energy. When the atmospheric vapours condense by precipitation, the latent heat is released. The released latent heat is the source of energy for violent storms like hurricanes, tornadoes and thunderstorms.

3. Aerosols

Another important constituent of atmosphere are the aerosols or suspended particles. The atmosphere is never free from suspended particles. In urban and industrial areas, the concentration of aerosols may be as high as 10^5 particles per cubic centimeter of air.

Suspended particles too influence the weather conditions. They act as a screen that retards the incoming solar radiations—a factor that reduces the atmospheric temperature. They also act as a blanket and retard the outgoing, reflected radiations—a factor that increases the temperature. Thus, the local temperature conditions depend, to quite some extent, on the concentration of suspended particles. Dust particles also act as *cloud condensation nuclei*, that is, they trigger cloud formation and thus influence the rainfall pattern over local regions. These, as well as other environmental impacts of suspended matter are discussed in Chapter 20.

Structure of the Atmosphere

Based on the manner in which the temperature varies with increasing altitude, the atmosphere has been divided into the following four distinct zones (Fig. 4.3).

1. Troposphere

The troposphere is the lowest zone of the atmosphere, reaching to a height of 16 km over the equator and gradually decreasing towards the poles to about 8 km. The depth of troposphere also varies with season; it is deeper in the summers and shallower in the winters.

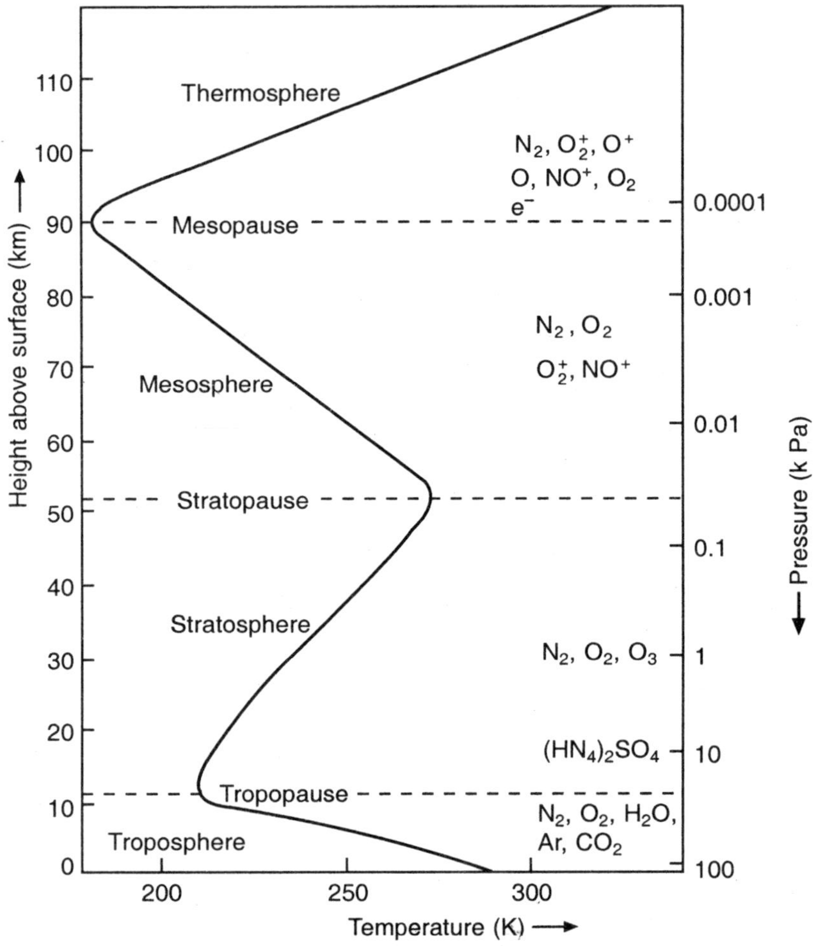

Fig. 4.3 Zonal structure of the atmosphere

By far, the largest percentage of the mass of total atmosphere is found in the troposphere. Essentially all the water vapours and dust particles are concentrated in the troposphere. The troposphere is characterized by vigorous and turbulent movement of air currents.

Some of the gaseous components of the troposphere, like water vapours, carbon dioxide, nitrous oxide and methane, absorb the heat energy reflected by the earth's surface and hence increase the average atmospheric temperatures. These gases are called *greenhouse gases.* We shall learn more about these gases in Chapter 18. Like other components of the atmosphere, the concentration of greenhouse gases decreases with height. Hence at high altitudes, the heat energy reflected by the earth's surface is not retained through absorption, but is transmitted to outer space. For this reason the atmospheric temperature, within the troposphere, decreases with increasing altitude, at a rate of about 6.4°C per km. This decrease in temperature is called *normal environmental lapse rate.*

2. Stratosphere

The next atmospheric zone, stratosphere, extends from about 16 km to about 56 km above the sea level. The boundary between troposphere and stratosphere is called *tropopause.* The stratosphere is characterized by the absence of turbulent conditions or changing weather patterns. Therefore, materials once injected into the stratosphere tend to remain there for years together. A good example of this is the presence of ammonium sulfate aerosols at an altitude of about 20 km.

The most important feature of the stratosphere is the presence of a thin layer of ozone. The incoming solar ultraviolet rays are absorbed by this layer and converted into heat. It is for this reason that within the stratosphere the temperature increases with elevation. Had the ozone layer not been present in the stratosphere, the ultraviolet rays would have reached the earth's surface and proved to be deleterious to life forms. We shall learn more about the role of stratospheric ozone in Chapter 19.

3. Mesosphere

Mesosphere extends from about 56 km above the sea level to about 90 km. The boundary between stratosphere and mesosphere is called *stratopause.* Mesosphere contains very low concentrations of oxygen and nitrogen. These gaseous molecules have symmetrical structures and, therefore, they possess little capacity to absorb solar radiation. Hence, within the mesosphere, the temperature decreases with height.

4. Thermosphere

The uppermost zone of the atmosphere, thermosphere, extends above 90 km altitude. The boundary between mesosphere and thermosphere is called *mesopause.* Thermosphere contains extremely minute concentrations of oxygen ions (O^+), molecular oxygen ions (O_2^+) and molecular nitrogen ions (N_2^+). These ions can efficiently absorb invisible, short wave solar radiation. As a result, the temperature within the thermosphere increases with elevation.

Temperature Inversion

We have noted that within the troposphere there is a decrease in temperature with increasing altitude. However, at times, and over specific pockets, the temperature may increase with elevation in the lower layers of the troposphere. This phenomenon is called temperature inversion. There are two types of inversions: Static and dynamic.

Static temperature inversion occurs when the earth's surface becomes colder than the air above and, in turn, chills the lower layers of the air. Inversions of this type are common at night in midlatitudes, especially during winters. *Dynamic temperature inversion* occurs when two large bodies of air at different temperatures collide and the colder, denser layer slides below the warmer, lighter layer. This type of inversion generally occurs in industrial towns located in valleys. At such places the collisions between the hot air from chimney stacks and the cold mountainous air are quite common.

Whatever be its mechanism of occurrence, temperature inversion is considered to be an environmental hazard. Under normal circumstances, the warmer air currents continuously travel upwards, penetrating and eventually displacing the upper, colder layers of air. The obnoxious air pollutants are carried upwards by the warm air currents and are ultimately blown out to the outer space. But for this natural phenomenon, the pollutants would have concentrated at breathing levels and their adverse effects would have been several times more than are actually felt.

When an inversion occurs, the cold layer of air becomes trapped below the warm layer. The cold currents cannot penetrate the overlying warm currents. Nor can the pollutants penetrate out of the lower atmosphere. Thus, the unwanted air contaminants are retained at the breathing levels. Though the pollutants cannot move up vertically, they are easily dispersed horizontally. Their effect is thus felt at great distances.

Not much can be done to prevent an inversion from occurring or to break up an inversion once it has occurred. One has to wait for the sun's rays to warm up and break the inversion layer. This may take a few days. During this time the original contaminants present in the cold layers of air may interact with each other to form secondary pollutants, which are generally more dangerous than the original, primary pollutants.

The secondary pollutants often accumulate over an area in form of thick smog. The smog formation is very common in the Los Angeles city, where the local topography and industrial infrastructure prompts frequent temperature inversions. Due to this the smog of secondary pollutant was named *Los Angeles Smog*. Subsequently, however, it came to be called *photochemical smog* because many reactions leading to the formation of secondary pollutants were catalysed by sunlight. We shall learn more about photochemical smog in Chapter 17.

Heat Balance of the Earth

The heat balance of the earth is defined as the ratio of energy received by the earth from the sun to the energy the earth loses by reflection and by radiation in form of heat. This ratio is essentially unity. That is, the earth loses as much heat energy as it receives. As a result, a near steady state of the average global temperature has been maintained.

Thus, over the eons the earth has neither progressively cooled nor warmed. Although there have been imbalances in local regions, the average temperature of the earth has remained more or less constant over a span of millions of years.

The atmosphere plays a significant role in maintaining the heat balance of the earth. It is through the atmosphere that the solar radiations reach the earth's surface. It is also through the atmosphere that the earth radiates back the heat energy or reflects back the solar radiations.

On an average 33% of the solar radiations received from the sun are reflected back, primarily by clouds and partially by snow and ice sheets on the earth's surface. The reflecting capacity of the earth is called *albedo*. The remaining 67% of the solar energy is absorbed. Of this, one-fourth is absorbed by the atmosphere, three-fourth by the earth's surface. Ultimately both the atmosphere and the earth's surface radiate back the absorbed energy as heat. The amount of heat energy radiated into the outer space almost equals 67%. The heat gain and loss factors are thus balanced over the global scale.

Although there is a heat balance for the planet as a whole, all parts of the earth and its atmosphere do not maintain a strict balance between the incoming solar radiations and outgoing terrestrial radiations. The heat balance varies greatly from one region of the earth to another. This serves an important purpose. The imbalance between the heat surplus areas at low latitudes and the heat deficiency at polar latitudes causes the atmospheric and oceanic circulations to occur. This, in turn, controls the climatic and weather patterns.

It is feared that human activities are likely to disturb the delicate heat balance of the earth. Industrial operations have induced large amounts of greenhouse gases, like carbon dioxide, ozone, nitrous oxide, methane and water vapours into the atmosphere. These gases tend to increase the average global temperature. The industrial operations have also injected particulate load into the atmosphere. The particulates tend to counter the warming effect of greenhouse gases.

The heating and cooling effects of pollutants discharged by human activities are not exactly balanced. At present, the warming effect of greenhouse gases predominates over the cooling effect of suspended matter. Over the last decade the average global temperature has steadily risen. This phenomenon is likely to disturb the natural heat balance of the earth and prove to be deleterious to the environment.

References

Aldhous, P. (1995). How ancient sewage breathed life into the earth, *New Scientist,* July 8, p. 16.

Cloud, P.E. (1968). Atmospheric and hydrospheric evolution on the primitive earth, *Science,* **160,** 729–736.

Davidson, C.F. (1965). Geochemical aspects of atmospheric evolution, *Proc. Nat. Acad. Sci., U.S.,* **53,** 1194–1205.

Donn, W.L. (1972), *'The Earth: Our Physical Environment',* John Wiley & Sons, Inc., New York, p. 386–405.

Fischer, A.G. (1965). Fossils, early life and atmospheric history, *Proc. Nat. Acad, Sci., U.S.,* **53,** 1205–1215.

Holum, J.R. (1977), *'Topics and Terms in Environmental Problems',* John Wiley & Sons, New York, p. 37–41.

Ingersoll, A.P. (1983). The atmosphere, *Sci. Amer.,* **249**(3), 162–174.

Kasting, J.F. (1993). Earth's early atmosphere, *Science,* **259,** 920–926.

Levin, H.L. (1978), *'The Earth Through Time',* W.B. Saunders Co., Philadelphia, p. 188–191.

Lounsbury, J.F. and Ogden, L. (1979), *'Earth Science',* 3rd Edn., Harper & Row, New York, p. 46–73.

Mason, B. (1966), *'Principles of Geochemistry',* 3rd Edn., John Wiley & Sons, Inc., New York, p. 208–222.

Wayne, R.P. (1988). Origin and evolution of the atmosphere, *Chem. Brit.,* **24,** 225–230.

CHAPTER 5

Biosphere

Biosphere is the thin shell of the earth, including the oceans and the atmosphere, where the living organisms exist. It was the last abiotic entity of our planet to come into existence. Its environment laid the foundation for the biotic origin. Biosphere thus represents an interface between the nonliving and the living. Living organisms and their nonliving environment are inseparably interrelated and interact upon each other. That the abiotic environment controls the activities of living organisms is quite obvious. Less obvious, however, is the fact that the biotic community also influences its abiotic surroundings in many ways. The difference in the composition of soil in inhabited areas and in moorlands is a case in point. Without living organisms the earth, like any other planet, would have been a relatively unchanging mass, composed of a fewer kinds of materials. Thus, it is the interplay between the living and the nonliving which determines the characteristics of the biosphere.

Characteristics of the Biosphere

Biosphere is defined as the thin layer of soil, rock, water and air that surrounds the earth, along with the living organisms for which it provides support. The biosphere extends from approximately six kilometers above the sea level to about ten kilometers below the sea level. It is endowed with certain special conditions which are necessary to support life. These conditions are:

1. A supply of air is available.
2. A supply of water is available.
3. An area where the interface between the earth's crust, hydrosphere and atmosphere exists is available.
4. A suitable temperature range exists.
5. A supply of usable energy is available.
6. A supply of nutrients is available.

The elemental composition of the materials comprising the biosphere is given in Table 5.1. The elements are broadly classified into two categories: Macronutrients and micronutrients. *Macronutrients* are required in substantial quantities by living organisms. Carbon, hydrogen and oxygen are the principal components of the organic living matter. Nitrogen is contained in amino acids and proteins. Phosphorus, calcium and silicon are used in skeletal structure of vertebrates.

Sodium and potassium play a role in regulating osmotic pressure in living cells. Magnesium is an important constituent of enzymes and chlorophyll. Sulfur is a component of essential amino acids, such as cysteine and methionine. *Micronutrients* are the elements which are required by living organisms in very small quantities, although they are just as vital as the macronutrients. Micronutrients are also called *trace elements*. Examples of trace elements are iron, molybdenum and manganese. Most of these elements play a specialized role in biochemical reactions, such as photosynthesis and metabolism.

Table 5.1 Elemental composition of the biosphere

Element	Concentration (moles hectare^{-1})
	Macronutrients
Hydrogen	13100
Carbon	6560
Oxygen	6540
Nitrogen	71.6
Calcium	18.9
Potassium	11.7
Silicon	8.6
Magnesium	8.2
Sulfur	4.4
Phosphorus	3.3
Sodium	1.6
	Micronutrients
Aluminium	4.1
Chlorine	2.8
Iron	1.4
Manganese	0.76

Energy for the biosphere is provided through the action of photosynthesis. In this process the green plants use solar energy, extract carbon dioxide and water from the environment and covert them into carbohydrates. The carbon is then said to be 'fixed' by the plants. The energy stored in the fixed carbon flows through the biotic community.

The biotic community is thus sustained by the special conditions prevailing in the biosphere. The biotic community also modifies the biosphere in the direction that either enhances or lessens its life supporting capacity. Therefore, both the biotic community and the abiotic environment exert a profound influence on each other; both are necessary for maintenance of life as we have it on the earth.

A community of organisms interacting with each other and with their nonliving surroundings is called an *ecosystem*. The ecosystems are the subdivisions of the biosphere. In fact, the total of all the world's ecosystems constitute the biosphere.

Ecosystems may be classified as small or large. The example of a small ecosystem is an aquarium, while that of a large ecosystem is a tropical forest.

Ecosystems may also be classified as simple or complex. A community of lichens growing on a rock in the Himalayas is an example of a simple ecosystem. A coral reef in the ocean is a complex ecosystem.

Each ecosystem has two parts: Living and nonliving; biotic and abiotic. The study of the interrelationship between the living species and their nonliving environment is called *ecology*. Ecosystems are dynamic entities. Changes in any component of an ecosystem, whether natural or as a result of human perturbation has far fetched repercussions. Human actions can enhance the life supporting characteristics of the ecosystems; they can degrade them as well.

Biogeochemical Cycles

In order to maintain the biosphere, and thus life, chemical elements must be recycled so that after use they become available again. The cycling of elements makes nutrients continuously available to living organisms. The cycling also prevents the accumulation of elements in forms, quantities and locations that are deleterious to life. During the course of each cycle the interchange of nutrients occurs among both the biotic community and the abiotic environment. The more or less circular paths of the nutrients passing back and forth between organisms and environment are known as biogeochemical cycles or *nutrient cycles.* Bio refers to biotic community and geo to abiotic environment. Thus the biogeochemical cycles connect the living and the nonliving components of the biosphere.

Any area or location, such as the atmosphere, the ocean, the soil or the earth's crust, where a material is concentrated is called a *biogeochemical pool.* There are two types of biogeochemical pools: *Active pools,* where materials are in forms and places easily accessible to life processes and *storage pools,* where materials are more or less inaccessible to life. For example, atmospheric carbon dioxide is circulated rapidly amongst the plant community by way of photosynthesis and respiration. On the other hand, it takes many thousand years for a coral reef to uplift and decompose into carbon dioxide. Therefore, in the carbon cycle, atmosphere is an active pool, while the coral reef is a storage pool.

The quantity of material passing from one pool to another per unit time and per unit area or volume of the ecosystem is called *flux rate*. Pathways connecting the active pools have high flux rates; these pathways are usually controlled by life processes. Pathways connecting storage pools have low flux rates; these pathways are usually controlled by physical processes.

The biogeochemical cycles are broadly classified into two types: Gaseous and sedimentary. Elements that form gaseous compounds at temperatures usually encountered in the biosphere are said to participate in gaseous cycles. Biogeochemical cycles of carbon, oxygen, nitrogen and sulfur are gaseous cycles. Even the water or hydrologic cycle is a gaseous cycle. For such cycles, atmosphere is a significant pool. Elements which rarely form gaseous compounds at normal temperatures and for which atmosphere is not a significant pool, follow the sedimentary cycles. Phosphorus cycle is an example of a sedimentary cycle.

As indicated in Table 5.1, the elements hydrogen, carbon, oxygen and nitrogen are of prime importance to the living organisms. The biogeochemical cycles of these elements are discussed in sequel. Since hydrogen usually enters and leaves the living systems in form of water, this element has been analysed in the context of hydrologic cycle. The other nutrient cycles considered in the present text are those of sulfur and phosphorus. This is not to imply that the remaining elements are not important to the biosphere. They are— and all have their natural biogeochemical cycles. However, the technological advancement of the present day society has put maximum load on the biogeochemical cycles of water, carbon, oxygen, nitrogen, sulfur and phosphorus. Any stress on these cycles is likely to rebound on human beings. It is, therefore, worthwhile to study these cycles as they occur in nature and to examine how they are perturbed by human intervention.

The Water Cycle

Of all the biogeochemical cycles, the water cycle is the most important in the sense that it carries along with it the materials moving through other nutrient cycles. Water also serves as a reservoir for many dissolved and suspended materials. Moreover, water is a vital component of living cells and many biochemical reactions take place in aqueous medium.

The water cycle is depicted in Fig. 5.1. About 97% of the world's water is present in active pools like atmosphere, streams, lakes and oceans. The remaining 3% is inaccessible, for it is in storage pools like icecaps, glaciers and ground water. The ground water reservoirs are so deeply seated beneath the earth that very little interchange occurs with actively circulating water.

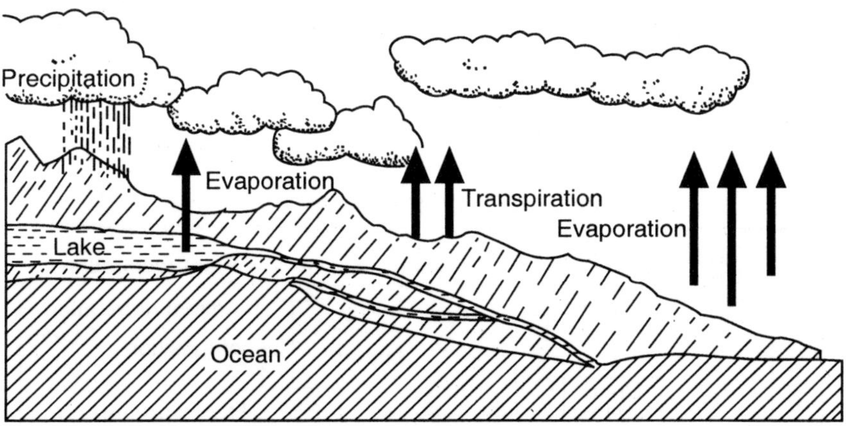

Fig. 5.1 The water cycle

The circulation of water over the land is brought about by rivers. The rivers also serve as conduits for dissolved and suspended materials. The rivers ultimately carry the water to the oceans. The continuous loss of water from land is made up by evaporation of marine water. Evaporation also takes place on land, from lakes, streams and soil. Water is also lost from the stomata of plants through transpiration.

The evaporated water becomes a constituent of the atmosphere. The water vapours on cooling condense into clouds and ultimately the water returns to the earth by precipitation (rain or snow). Of course, precipitation occurs over both land and ocean, but the topography of land is such that it receives greater amount of precipitation than the oceans. This compensates for the loss of water which the rivers empty into the oceans.

Human Perturbations

With the advent of industrial revolution, there has been a sharp increase in the use of fossil fuels as energy resources. Burning of fossil fuels raises the atmospheric concentration of carbon dioxide and produces a special type of environmental hazard called greenhouse effect. We shall discuss the greenhouse effect in details in Chapter 18. At present suffice it to state that the greenhouse effect raises the average global temperature. Over the last few decades the average global temperature has steadily risen and this is likely to disturb the water cycle. Due to the greenhouse effect the icecaps and glaciers have started melting and the sea level has started rising. If the present trend continues, the low lying inhabited areas would become more prone to floods; the low lying agricultural land would become more susceptible to submersion.

The Carbon Cycle

All living organisms are composed of carbon compounds of one form or another. It is, therefore, not surprising that the biosphere contains a complex mixture of carbon compounds, in a continuous state of creation, transformation and decomposition. The carbon cycle is shown in Fig. 5.2. The active pools of carbon include atmosphere and the sea surface. The storage pools are the carbonate sediments and fossil fuels.

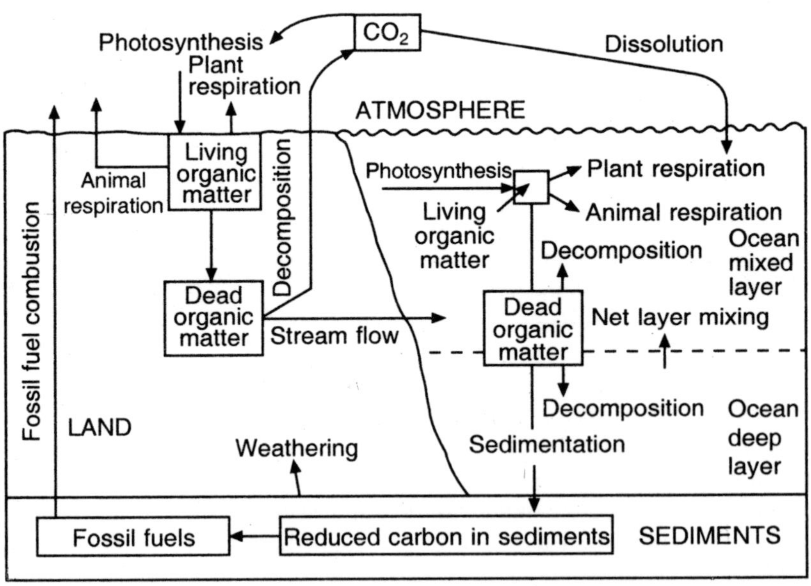

Fig. 5.2 The carbon cycle

In the atmosphere carbon is present chiefly as carbon dioxide. The sources of atmospheric carbon dioxide are cellular respiration by plants and animals, decaying of organic matter, burning of fossil fuels and weathering of carbon-containing sediments. The carbon dioxide from the atmosphere is converted by terrestrial plants by photosynthesis into organic compounds. Photosynthesis affects the atmospheric carbon dioxide levels both diurnally and seasonally. Diurnally, the carbon dioxide concentration falls during daytime because uptake through photosynthesis is possible only in presence of sunlight. At sunset, photosynthesis ceases but respiration by plants continues, leading to a rise in carbon dioxide concentration. Seasonally, the lowest levels of carbon dioxide are observed during spring when vegetation blooms and photosynthesis is at its peak.

The atmospheric and oceanic pools of carbon are in equilibrium with each other. Therefore, some of the atmospheric carbon dioxide enters the ocean. Here it is fixed through photosynthesis by phytoplanktons.

The plants on land and phytoplanktons in the sea provide food for a host of organisms. The carbon originally fixed by photosynthesis, passes from one organism to the other in form of organic molecules of precise architecture and rich diversity. When these organisms die, the microorganisms of decay act upon them and convert the organic compounds present in the dead body into carbon dioxide, which is released to the atmosphere.

Some of the dead organic matter in the sea is, however, not completely decomposed. On land too a part of the organic debris is left undecomposed. It flows through the rivers into the ocean, supplementing the undecayed organic matter already present there. The undecomposed debris settles down into the sea sediments through sedimentation. A part of the dissolved carbon dioxide in the ocean reacts with calcium and magnesium ions and is converted into insoluble rock material, which also settles down into the sediments. At present, greater quantities of carbon are stored as fossil fuels and rock material as compared to carbon in active circulation.

Human Perturbations

As more and more of the fossil fuels are being exploited, carbon from the storage sedimentary pool is increasingly being transferred into the active atmospheric pool. This has led to abnormally high concentrations of carbon dioxide in the atmosphere and the concomitant threat of greenhouse effect. The latter is an environmental adversity leading to a rise in average global temperature, fall in soil fertility, rising sea levels and melting of polar ice. The impacts of greenhouse effect are discussed in Chapter 18.

The Oxygen Cycle

Oxygen not only supports life, but it arises from life. It constitutes roughly 21% of the atmospheric gases. The oxygen cycle is depicted in Fig. 5.3. Its active pools are atmosphere and oceans. Its storage pools are crustal and sedimentary rocks.

Molecular oxygen was not present in the primordial atmosphere and hydrosphere. The principal evidence for this is the absence of oxidised minerals,

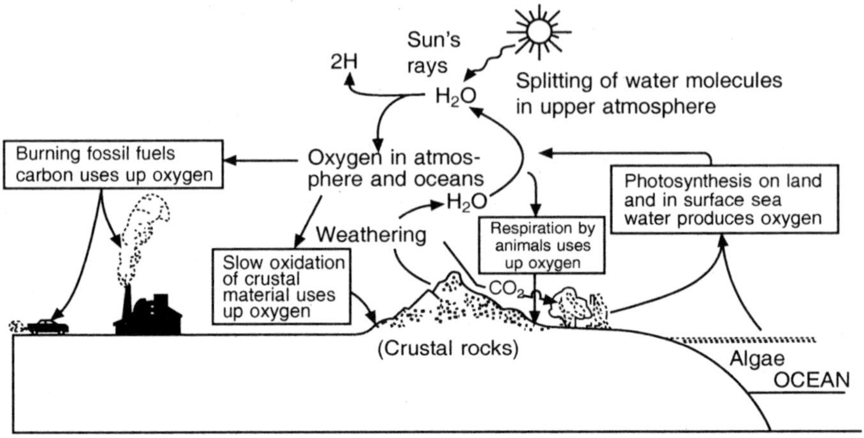

Fig. 5.3 The oxygen cycle

such as ferric oxide, in the early sedimentary rocks. It was the photosynthesis by green plants which inducted oxygen in the atmosphere. Minor amounts were also added through photodissociation of water vapours in the upper atmosphere.

$$2H_2O + h\nu \rightarrow 2H_2 + O_2 \tag{5.1}$$

During cellular respiration, organisms consume oxygen and release carbon dioxide; during photosynthesis, carbon dioxide is consumed and oxygen is released. Thus, respiration and photosynthesis alone are sufficient to complete the oxygen cycle. However, oxygen also reacts with crustal materials, giving oxide and carbonate rich sediments. On weathering, these sediments do not directly release oxygen. Instead they degrade into carbon dioxide and water. The former is converted into oxygen through photosynthesis; the latter through photodissociation.

Human Perturbations

In recent years, man's activities tend to reduce the forest cover. Deforestation reduces the extent of photosynthesis and hence the amount of aerial oxygen. However, the oxygen cycle continues to be in a near steady state condition. It is hypothesized that sulfate reducing bacteria, occurring in anaerobic environment, compensate for the oxygen loss. These bacteria use the sulfate ions as an oxygen source for oxidising organic matter.

$$SO_4^{2-} + 2C \rightarrow 2CO_2 + S^{2-} \tag{5.2}$$

Although reaction (5.2) does not release free oxygen, it does carry through the decomposition of organic matter with consequent release of carbon dioxide. The latter is transported to a site where it is utilized in photosynthesis, leading to release of molecular oxygen.

The Nitrogen Cycle

All living organisms require nitrogen for synthesizing proteins and other biomolecules like purines and pyrimidines. Therefore, nitrogen has to be cycled through various pools in the environment (Fig. 5.4). The inorganic nitrogen in the soil and the organic nitrogen in the decaying matter are the active pools of this element. The sediments and the atmosphere are the storage pools.

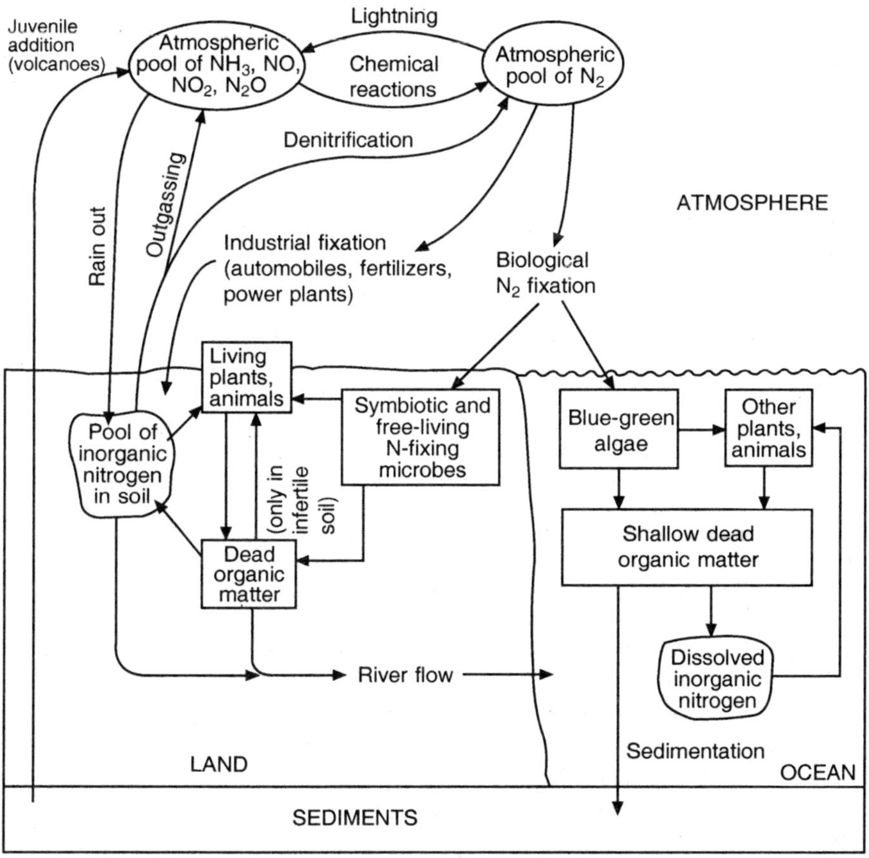

Fig. 5.4 The nitrogen cycle

Although atmosphere contains 79% of molecular nitrogen, most organisms cannot use it for life processes. The reason being that the nitrogen molecule is quite inert because of the presence of a triple bond between the constituent atoms ($N \equiv N$). To break the triple bond requires substantially high amounts of energy.

It is, therefore, necessary that the inert, molecular nitrogen be 'fixed' into forms that can be taken up by plants and animals. Fixation of nitrogen can take place through three processes: Atmospheric fixation, industrial fixation and biological fixation.

Atmospheric fixation involves conversion of molecular nitrogen into nitrogen oxides and ammonia, by the action of lightning or cosmic radiations. These natural phenomena provide the activation energy for transforming molecular nitrogen into combined forms. The nitrogen compounds are washed down by rainfall and on interacting with soil, are converted into nitrites and nitrates—the plant nutrients.

The most important *industrial fixation* method is the Haber's process. In this process nitrogen and hydrogen are reacted together in 1: 3 molar ratio at high temperatures and pressures, in presence of a catalyst. The initial product, ammonia, is subsequently converted into fertilizers like ammonium nitrate and urea. In soil, the fertilizers are converted into nitrites and nitrates.

Biological fixation of nitrogen is carried out by nitrifying microorganisms. Bacteria of the genus *Rhizobium* which live as symbionts in nodules on the roots of legume are able to convert molecular nitrogen into ammonia.

$$2N_2 + 6H_2O \rightarrow 4NH_3 + 3O_2 \tag{5.3}$$

The energy required by reaction (5.3) is extracted by bacteria from the plant. The ammonia is then transformed into amino acid.

$$2NH_3 + 2H_2O + 4CO_2 \rightarrow \underset{\text{Glycine}}{2CH_2NH_2COOH} + 3O_2 \tag{5.4}$$

Alternately, bacteria of genus *Nitrosomonas* convert ammonium ions into nitrite ions,

$$2NH_4^+ + 3O_2 \rightarrow 2NO_2^- + 4H^+ + 2H_2O \tag{5.5}$$

which are subsequently oxidised to nitrate ions.

$$2NO_2^- + O_2 \rightarrow 2NO_3^- \tag{5.6}$$

In aquatic environment, biological fixation is carried out by blue-green algae.

In addition to these three processes, plants growing in infertile ecosystems can directly take up organic nitrogen from the soil.

The nutrients obtained by fixation of atmospheric nitrogen are directly consumed by plants. These nutrients build-up the nitrogen-based biomolecules in the plant tissue. The nitrogen compounds then enter the body of animals which feed on plants. When the living organisms die they are acted upon by microorganisms of decay. A part of nitrogen biomolecules is converted into inorganic nutrients, which are recycled through the living organisms. The remaining nitrogen-based compounds enter the sediments through the process of sedimentation.

From the sediments, nitrogen is ejected into the atmosphere through volcanic action. However, this step occurs extremely slowly. Were it the only phenomenon to induct nitrogen back into the atmosphere, it could have taken several thousand years to complete the nitrogen cycle. And during this time, the atmosphere would have been depleted of nitrogen, while the soil would have become extremely toxic due to the accumulation of excessive nitrates. However, the nitrogen cycle is completed relatively rapidly by the action of denitrifying bacteria, such as *Pseudomonas denitrificans*. These microorganisms convert the nitrate content of soil into molecular nitrogen,

$$4NO_3^- + 2H_2O \rightarrow 2N_2 + 5O_2 + 4OH^- \tag{5.7}$$

which is added to its atmospheric pool.

Human Perturbations

The growing demands for agricultural produce have led to an ever increasing production of industrially fixed nitrogen in form of fertilizers. The agricultural runoff, rich in fertilizers, sometimes enters a closed water body, such as a lake. Due to the presence of fertilizers, excessive algal blooms crop up in the lake. They deplete the dissolved oxygen and choke the other aquatic flora and fauna. The lake becomes stratified. This phenomenon is called eutrophication. We will learn more about eutrophication in Chapter 29.

An additional perturbation of nitrogen cycle has resulted due to the combustion of fossil fuels. Nitrogen oxides and ammonia, the products of combustion, interact with soot, another product of combustion, to produce soluble nitrogen-containing organic compounds. The concentration of these dissolved nitrogen compounds is steadily increasing in the world's ocean, putting a stress on the nitrogen cycle.

Yet another component of stress in the nitrogen cycle arises when nitrogen oxides, released from fossil fuel combustion, are precipitated as acid rain. The obnoxious effects of acid rain are discussed in Chapter 14.

The Sulfur Cycle

Sulfur is a constituent of two of the twenty essential amino acids. It is incorporated in several proteins and is, therefore, a vital component of the living organisms. The biogeochemical cycle of sulfur is shown in Fig. 5.5. The atmosphere is an active pool for sulfur compounds; the sedimentary rocks are a storage pool.

Sulfur enters the atmosphere as sulfur dioxide, hydrogen sulfide and sulfate aerosols from four sources: (1) Juvenile addition due to volcanic activity; (2) burning of fossil fuels; (3) outgassing from land and marine environments; and (4) blowing of sea salt sprays bearing sulfate aerosols. Hydrogen sulfide in the atmosphere is oxidised to sulfur dioxide by ozone. The reaction is catalysed by particulates present in the atmosphere.

$$H_2S + O_3 \rightarrow H_2O + SO_2 \tag{5.8}$$

A part of sulfur dioxide is further oxidised to sulfur trioxide, which is subsequently hydrolysed to sulfuric acid,

$$SO_2 + O + M \rightarrow SO_3 + M \tag{5.9}$$

$$SO_3 + H_2O \rightarrow H_2SO_4 \tag{5.10}$$

where M is a molecule of oxygen or nitrogen which carriers off excess energy.

Sulfur dioxide and sulfate aerosols are washed down by rainfall. They are taken up by living organisms, both on land and in ocean. When the organisms die, the microorganisms of decay extract sulfur from their body. Under aerobic conditions, sulfate ions are produced; under anaerobic conditions, hydrogen sulfide is produced. A part of sulfate ions is reconsumed by living organisms. The remaining sulfate ions are converted by microorganisms, first to hydrogen sulfide and then to dimethyl sulfide (DMS, CH_3SCH_3). Dimethyl sulfide is easily oxidised to sulfur dioxide, which escapes into the atmosphere.

Some of the hydrogen sulfide interacts with iron content of earth, producing iron sulfide, which becomes embedded in sedimentary rocks. This reaction also

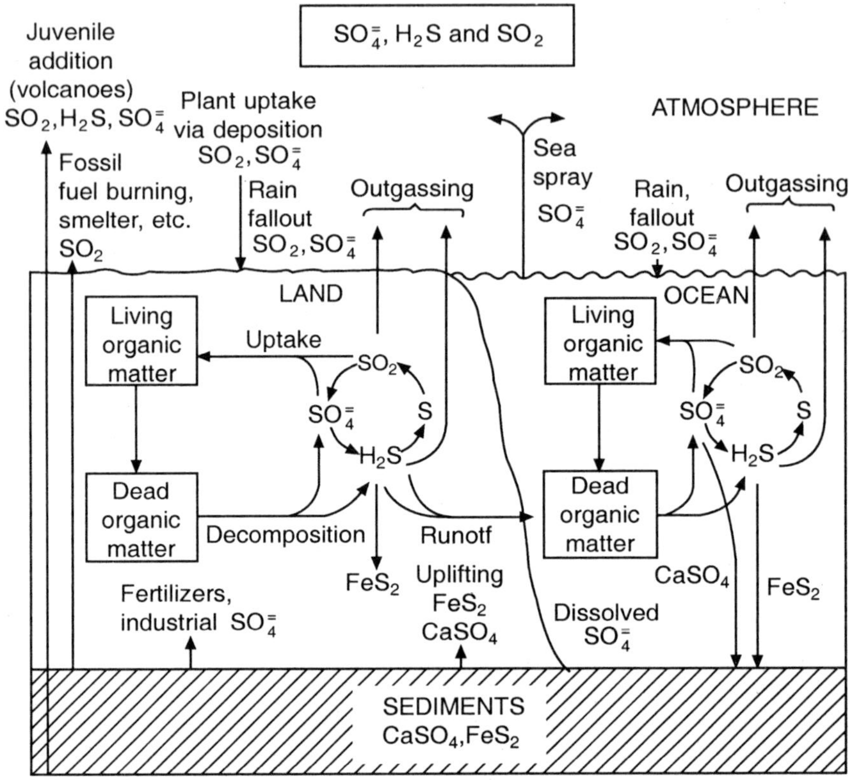

Fig. 5.5 The sulfur cycle

occurs in the ocean. In addition, a part of sulfate ions in the marine environment interact with calcium ions, producing calcium sulfate, which joins the sediments.

From the sediments, sulfur compounds are ejected extremely slowly due to uplifting, weathering or volcanic activity. Sulfur compounds may also be industrially extracted from the sediments to synthesize fertilizers, fungicides and other commercial compounds. Smelting of ores and burning of fossil fuels also release sulfur compounds from the sediments.

Human Perturbations

The combustion of fossil fuels and metallurgical operations have released abnormally large amounts of sulfur compounds, especially sulfur dioxide, and placed a burden on the natural sulfur cycle. Higher the concentration of sulfur dioxide in the atmosphere, greater is the threat of acid rain. The ill effects of acid rain are detailed in Chapter 14.

The Phosphorus Cycle

Although phosphorus content in living matter is very small, yet it plays an indispensable role. It is a constituent of nucleic acids and cell membranes. In addition, the bones and teeth of animals contain significant quantities of phosphorus. Fig. 5.6 depicts the biogeochemical cycle of phosphorus. The decaying organic matter is an active pool of phosphorus; the sediments are a storage pool. In fact, almost all phosphorus is incorporated into the sediments. Phosphorus from this pool can become available to living organisms through the incredibly slow process of uplifting and weathering or through human intervention by mining.

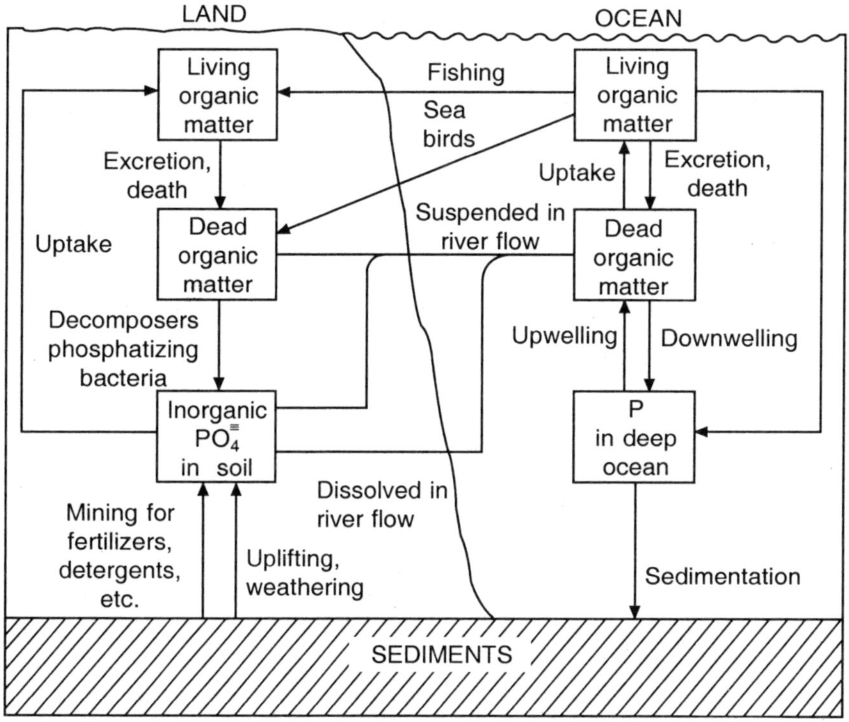

Fig. 5.6 The phosphorus cycle

The phosphorus released from the sediments enters the soil as phosphate ions. These ions are taken up by living organisms on land for use in various biological processes. When the living organisms die, the microorganisms of decay transfer the phosphorus content from the body of dead animals to the soil. It may then be recycled to the living organisms. A part of phosphorus, however, flows through the rivers into the ocean, where it is consumed by marine life.

Once phosphorus enters the ocean, there are only two processes which transfer it to the terrestrial environment. One is through sea birds and fishes that harvest phosphorus passing through marine food chains and then return it to land in their excrement. The other is that phosphorus becomes a part of sea sediments and through slow geological uplifting of the sediments reaches land. This phenomenon is very slow, taking millions of years.

Human Perturbations

The need for higher agricultural yields has prompted manufacture of phosphorus-based fertilizers. This, in turn, has led to excessive mining of phosphate sediments. The fertilizers are sprayed in the fields in quantities abnormally higher than optimum dose. The excess, unused fertilizers become a part of agricultural runoff and are ultimately drained into water bodies. In closed water systems like lakes, these fertilizers cause unwanted algal blooms to grow. The other flora and fauna in the lake chokes due to stratification. The lake dies due to eutrophication. The mechanism of eutrophication and its harmful consequences are discussed in Chapter 29.

Oxygen-Phosphorus Cycle Interlink

Oxygen can carry through vital natural processes only if its atmospheric concentration is moderate. Too little, and oxygen breathing organisms would all suffocate. Too much, and firestorms would sweep the earth. The atmospheric oxygen content is regulated through an interlink between biogeochemical cycles of oxygen and phosphorus.

The amount of oxygen in the atmosphere is always roughly in balance with the oxygen content of the oceans— if one of these reservoirs of the gas becomes depleted, oxygen diffuses into it from the other. Most of the oxygen in the atmosphere is produced by photosynthetic algae floating near the surface of the sea. If oxygen is produced in excess, it is consumed in weathering of rock material. The weathering process, in turn, releases phosphorus which eventually finds its way into the sea, where it serves as an essential nutrient for the marine algae. Passing through the marine food chain, phosphorus is subsequently incorporated into the sea sediments by bacteria.

During episodes of mountain formation, the amount of weathering increases and earth's oxygen content decreases. Left unchecked, this process would asphyxiate the planet. However, no sooner than oxygen concentration begins to fall, phosphorus is released from the sediments. When it reaches the sea surface, the phosphorus causes a bloom of algae that pump out oxygen, replenishing the gas lost through weathering.

If, on the other hand, an algae bloom were to occur when little or no weathering is taking place, the earth's oxygen level would rise to dangerous levels. However, in such an event, the sea dwelling bacteria take phosphorus out of water and deposit it in the sediments. The surface algae get starved and the rise in oxygen level is arrested.

Soil

For terrestrial ecosystems, soil serves as a medium of entry of the nutrients into living beings. The plants absorb their nutrients from the soil through their roots. With these they synthesize essential biomolecules upon which they, as well as other living organisms depend.

Soil is defined as a thin layer on the earth's surface formed by the interaction of crustal rocks, sunlight, atmosphere and living organisms. Soil without living organisms is called *mineral substrate.* The soil on planets other than earth is mineral substrate, which would have been converted into true soil had life been present on them. Thus true soils are product of life; in turn they support life.

Soil Formation

Soil is formed through a combination of physical and biological processes.

Physical Processes

Soil building always initiates with physical processes. These processes involve weathering of rock material. There are two types of weathering: Mechanical and chemical.

Mechanical Weathering involves the fragmentation of rocks by geological agents. In regions where atmospheric temperatures are high, the rocks tend to fracture because of uneven expansion. In extremely cold regions, water present in the crevices of rocks expands as it freezes, causing the rock to break down. A glacier causes collision between rocks and breaks them into smaller pieces. Moving water also results in friction between rocks, causing mechanical weathering. The presence of smooth rocks and pebbles in a stream in a mountanous region is an evidence that moving water has caused collision between rocks, removing their sharp edges.

Chemical Weathering is brought about by chemical reagents. Aerial oxygen brings about the oxidation of the constituents of rocks; carbon dioxide brings about their carbonation; water hydrolyses them. As a result of these chemical reactions, a part of rock material becomes soluble. Under the influence of rainfall the soluble components dissolve. The residual rock is much smaller in size than it was before the advent of chemical weathering.

Biological Processes

Biological processes begin when the living organisms gain a foothold in the rock weathered by physical processes. Roots of plants secrete carbonic acid which dissolves rock materials much more rapidly than does chemical weathering. Plants and small animals contribute their deciduous parts to rock fragments which later become soil. Decaying organic matter, called *humus*, becomes mixed with the top layer of rock particles and constitutes an important ingredient of the soil. The presence of humus in the soil makes possible the growth of a varied flora and fauna that would not have existed otherwise. Humus also increases the acidity of soil so that inorganic nutrients, which are more soluble under acidic conditions, become easily available to plants.

Microorganisms decompose a part of the organic matter and add inorganic nutrients to the soil. Burrowing animals like earthworms also contribute to the biological processes of soil formation. These animals, through their movement in soil, cause a mixing of organic and inorganic nutrients. They also bring nutrients into those layers of soil where plant roots are concentrated.

Soil Characteristics

Soil represents an extremely complex matrix of minerals derived from the parent rock, organic matter and substances carried by geologic agents and biotic community. The characteristics of the soil are determined by its texture and structure.

Soil texture depends upon the size of the constituent rock particles within the soil. The rock particles may be grouped into four categories on the basis of their diameter (Table 5.2).

Table 5.2 Size range of particles in the soil

Particles	Diameter (mm)
Gravel	>1.00
Sand	1.00–0.05
Silt	0.05–0.002
Clay	<0.002

When large particles like gravel and sand aggregate, tiny spaces appear between them. These spaces allow the passage of air through the soil. Thus soils containing large particles have the advantage of being adequately aerated. The disadvantage of such soils is that water drains rapidly through the spaces, carrying valuable nutrients beyond the reach of plant roots.

When extremely small clay particles aggregate, no spaces are left out. The soils containing a close packed arrangement of finely divided clay particles have the advantage that nutrients and moisture are retained for longer periods. The disadvantage of such soils is that they are poorly aerated.

An ideal soil is the one which combines the good aeration properties of large particles, with nutrient and moisture retaining ability of small particles. Such a soil is called *loam.* It contains 40% sand and gravel, 40% silt and 20% clay.

Soil Structure refers to the manner in which particles of different diameters aggregate with each other to form clumps. Sand and gravel particles, being large in size, do not clump. Therefore, sandy soils lack structure. On the other hand, clay particles, being small in size, tend to stick together in close packing, forming large clumps. A good soil, such as loam, containing a balanced composition of large and small particles, forms clumps of moderate size. These soils would crumble easily and create passage for water, air and nutrients to reach the plant roots.

Soil Profile

The term soil profile refers to a series of distinct horizontal layers of different chemical compositions and physical characteristics. Each layer is called a *horizon.* The pattern of various horizons is shown in Fig. 5.7. The layered structure of soil results from the action of living organisms and climatic influences of the region.

The uppermost layer of the soil is called *topsoil* or *horizon A.* This horizon is rich in inorganic nutrients, as well as humus. The presence of humus imparts a dark colour to this layer. Most of the living organisms are found in the topsoil. In mountainous regions the thickness of the topsoil is much less than in the plains.

Beneath the topsoil is the *subsoil.* This layer is also called *horizon B.* Some of the inorganic nutrients present in horizon A are transported to horizon B by water flowing from topsoil to deeper layers. This process is called *leaching.* The amount of leaching primarily depends on the available rainfall. Greater the rainfall, greater is the extent of leaching. Therefore, the areas which receive scanty rainfull have poorly developed subsoil. Horizon B contains scarce organic matter and is, therefore, lighter in colour as compared to horizon A.

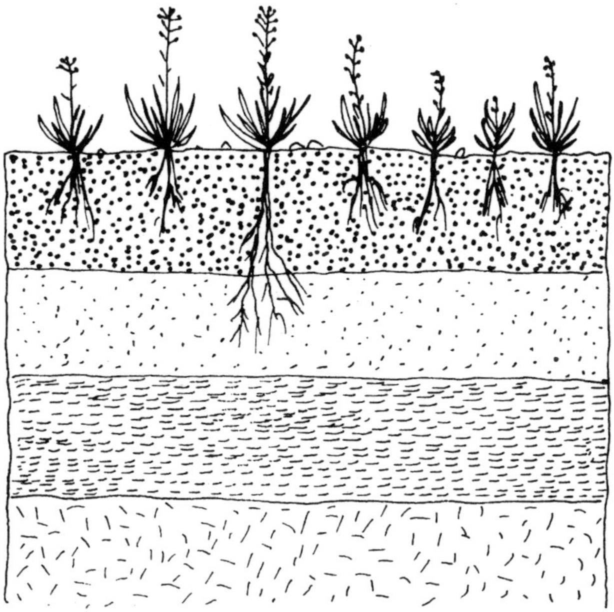

Fig. 5.7 The soil profile

Beneath the subsoil is a layer of weathered parent material. It is called *horizon C.* It represents the initial stages of fragmentation of rocks. The parent rock material occupies the lowest layer or *horizon D.* The chemical nature of the parent material has a great influence on the properties of soil derived from it. It also influences the soil's rate of water absorption and retention.

Soil Classification

Depending on the thickness of horizon A and horizon B, soils may be classified into two broad types: Chernozem and podzol.

Chernozem soils are also called grassland soils. Grassland areas usually receive limited amount of rainfall. Therefore, there is a low degree of nutrient leaching from topsoil to subsoil. As a result, chernozems have a deep horizon A and a thin horizon B (Fig. 5.8). Since horizon B lacks nutrients, it supports little root growth.

Podzol soils are also called forest soils. These are found in areas which receive abundant rainfall. Therefore, plenty of nutrients flow down from horizon A to horizon B. Clay particles, being fine in texture, also pass on to the horizon B. Due to this, horizon A in podzols is comparatively shallow, while horizon B is quite thick (Fig. 5.9). The plant roots are found in both the horizons.

Soil Erosion

Soil erosion refers to the weathering away of soil and its transportation by water or wind. During the last four decades, nearly one-third of the world's aerable land has been lost by erosion and continues to be lost at a rate of about ten

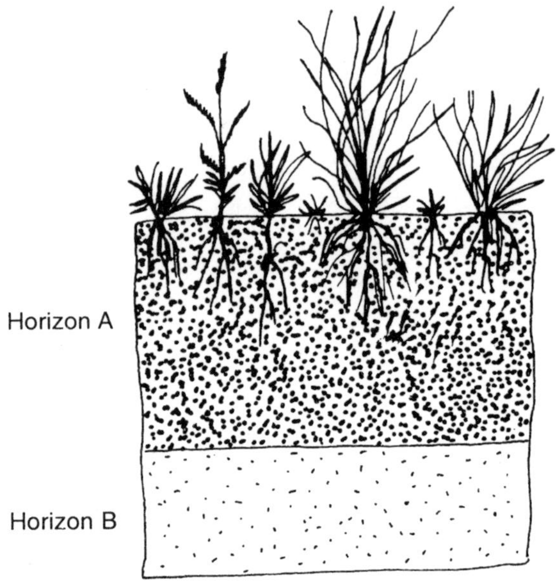

Fig. 5.8 The chernozem soil

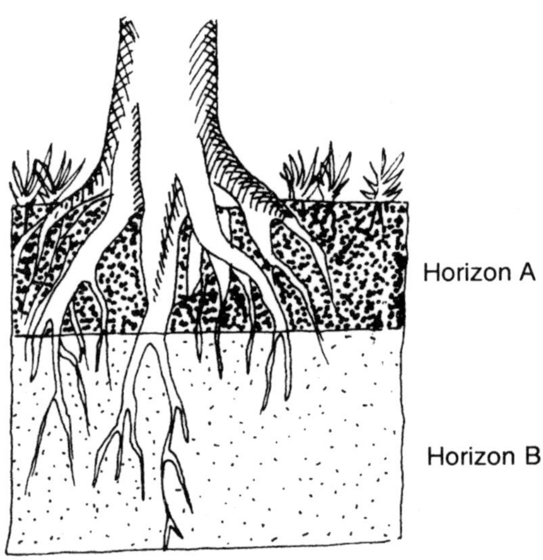

Fig. 5.9 The podzol soil

million hectares per year. Therefore, soil erosion constitutes a major environmental threat to the sustainability and productive capacity of biosphere.

Causes

Erosion is caused by the force of water and wind. There are several factors which influence the rate of erosion. It has been observed that:

1. Loss of vegetative cover and deforestation increase the rate of erosion.

2. Cultivation of slopes of mountainous regions increases the rate of erosion.
3. The presence of living organisms and decaying organic matter in the soil reduces the rate of erosion by intercepting and dissipating rain water and wind energy.
4. Soils with fine texture and low organic content are more susceptible to erosion.

Consequences

Erosion depletes the inorganic nutrients like nitrogen, phosphorus, potassium and calcium present in the soil. It also uproots the organisms living in the soil, as a result of which the organic nutrients too are lost. Moreover, when erosion occurs, the amount of water runoff increases, so that less water enters the soil matrix. All these factors lead to a reduction in soil productivity. The plant growth is stunted and crop yield declines. Because of erosion related loss of productivity, and population growth, the *per capita* food supply has been reduced over the past ten years and continues to fall.

Mitigation

Soil formation is an extremely slow phenomenon. It takes between 200 and 1000 years to form one inch of topsoil. It is, therefore, imperative that appropriate soil conservation practices must be followed so as to reduce the rate of erosion. Various soil conservation techniques reduce the erosion rates by maintaining a protective vegetative cover over the soil. Some important erosion mitigating schemes are now briefly outlined.

Rotation of crop ensures that some part of the land is continually covered with vegetation. *Contour farming* involves tillage at right angle to the slope of land. This creates a series of ridges that slow down the flow of water and prevent erosion. This conservation method is useful on gentle slopes. On steep slopes, *strip farming* reduces the extent of erosion. This practice involves alternating strips of closely sown crop. Alternating bands of wheat and soya beans are commonly sown. The closeness of the strips retards the flow of water (and erosion). On very steep land, the most practical method of reducing erosion is *terracing*. Leveled areas are constructed at right angle to the slope to retain water and prevent erosion. Another way of soil conservation involves construction of *waterways*. These are protected channels which allow the water to flow off the cultivated land. Soil erosion caused by wind can be reduced by installing *windbrakes.* This practice involves planting of trees or other plants that protect the bare soil from the full force of the wind. Windbrakes reduce the speed of wind and hence reduce the quantity of soil that it can carry away.

Inference

The entire biotic community of the earth is not only sustained by the biosphere but has also shaped its evolution. The integration of living beings everywhere on and near the earth's surface has resulted in the interaction of the lithosphere, the hydrosphere and the atmosphere with the biosphere. Rock weathering and soil formation has been carried through by biological processes. The oxygen in the atmosphere has evolved overwhelmingly from photosynthesis by green plants. The plants on which animals feed depend upon the nitrogen fixing activities of bacteria. It is through the living organisms that the nutrients are cycled biogeochemically. Thus, in the long history of biosphere, the most important event was the origin of life. In the next section we shall examine as to how life originated on the earth.

References

Belwiche, C.C. (1970). The nitrogen cycle, *Sci. Amer.,* **223**(3), 136–146.

Bergeron, L. (1996). How the seabed saves the world, *New Scientist,* February 3, p. 15.

Bolin, B. (1970). The carbon cycle, *Sci. Amer.,* **223**(3), 124–132.

Chapin, F.S. (1995). New cog in the nitrogen cycle, *Nature (London),* **377,** 199–200.

Clapham, W.B. (1983), *'Natural Ecosystems',* 2nd Edn., Macmillan Publishing Co., Inc., New York, p. 3–19.

Cloud, P. and Gibor, A. (1970). The oxygen cycle, *Sci. Amer.,* **223**(3), 110–123.

Cloud, P. (1983). The biosphere, *Sci. Amer.,* **249**(3), 176–189.

Cornell, S., Rendell, A. and Jickells, T. (1995). Atmospheric inputs of dissolved organic nitrogen to the oceans, *Nature (London),* **376,** 243–246.

Ehrlich, P.R. and Roughgarden, J. (1987), *'The Science of Ecology',* Macmillan Publishing Co., New York, p. 553–571.

Enger, E.D. and Smith, B. (1992), *'Environmental Science',* 5th Edn., Wm. C. Brown Publishers, Dubuque, p. 284–307.

Kimball, J.W. (1975), *'Man and Environment',* Addison-Wesley Publishing Co., California, p. 220–221.

Money, D.C. (1976), *'Climate, Soils and Vegetation',* University Tutorial Press Ltd., London, p. 140–153.

Penman, H.L. (1970). The water cycle, *Sci. Amer.,* **223**(3), 98–108.

Pimental, D., Harvey, C., Resosudarmo, P., Sinclair, K., Kurz, D., McNair, M., Crist, S., Shpritz, L., Fitton, L., Saffouri, R. and Blair, R. (1995). Environmental and economic costs of soil erosion and conservation benefits, *Science,* **267,** 1117–1123.

Safrany, D.R. (1974). Nitrogen fixation, *Sci. Amer.,* **231**(4), 64–80.

Strahler, A.N. and Strahler, A.H. (1974), *'Introduction to Environmental Science',* Hamilton Publishing Co., California, p. 534–558.

Tiessen, H., Cuevos, E. and Chacon, P. (1994). The role of soil organic matter is sustaining soil fertility, *Nature (London),* **371,** 783–785.

Part II

Biotic Origin

CHAPTER 6

Life

The question of the origin of life has engaged the minds of men since they first contemplated the nature of their existence on the earth. Yet, for centuries scientific thought did not attempt to answer this question. There were two main reasons for the neglect of this important area. Firstly, there was a lack of information on the chemical environment of the prebiotic earth. Secondly, there was a prevailing opinion that experimentation in this field of science would be an invasion of the precincts of religion. Thus, everyone continued to believe that 'life can come only from life'. It was in the early 1920s that a young Russian biochemist, A.I. Oparin, expressed the idea that life on earth has arisen from a 'pool' of organic compounds. It was from within this 'pool' that organic molecules reacted, interacted, aggregated and transformed into the constituents of a living cell. As the knowledge of the environment of prebiotic earth advanced, it became possible to formulate reaction schemes which would have transformed simple organic molecules from the 'pool' to the complex molecules of life.

Chemical Evolution and the Prebiotic Environment

Formation of the earth from a dust cloud about 4.5×10^9 years ago marked the culmination of cosmic evolution. At that time the only chemical species present on the earth and in its atmosphere were the different elements and a variety of interstellar molecules (Chapter 1). Life in the form of unicellular organisms (protozoa) appeared sometimes about 3.1×10^9 years ago. The intermediate period ($4.5–3.1 \times 10^9$ years) was the era of *chemical evolution*. During this era the atoms of various elements and interstellar molecules interacted to yield simple molecules. The simple molecules underwent different kinds of aggregations and were eventually transformed to a size and complexity that could contain and sustain living processes, leading to life as we now know. Thus, the term chemical evolution is defined as a set of chemical reactions that took place on the primitive, prebiotic earth, about 4.5 to 3.1 billion years ago, leading to the appearance of the first living cell.

How did chemical evolution begin? To answer this question we must examine the nature of the atmosphere of the primitive earth. The atmosphere would, in turn, have controlled the chemistry of oceans and surface rocks.

The primitive earth's atmosphere was dominated by hydrogen. Therefore, it was of reducing nature. Its other constituents were methane, ammonia and water vapours. Two strong evidence support the reducing nature of the earth's early atmosphere. Firstly, the cosmic dust from which the earth is believed to have been formed contained a large excess of hydrogen. Therefore, there is no reason to doubt that the earth's atmosphere in those times was also dominated by hydrogen. Secondly, the meteorites which reach the earth from the outer atmosphere have been observed to contain chemical species in a reduced form. An analyses of the meteorites reveal that their iron content is present mostly in elemental form or as ferrous sulfide; carbon is present in elemental form or as iron carbide; and phosphorus is present as phosphides.

At present the atmosphere is oxidising in character. Oxygen is not only an important constituent, but also an essential part of the atmosphere. Without oxygen no life would be possible. Yet, life would not have evolved if the primitive atmosphere would have been oxidising in nature. Had oxygen been present then, the molecules of life would have combusted as soon as they were formed.

At present the direct or indirect source of free energy for living organisms is the sunlight, which is utilized by the photosynthetic organisms. However, before the life forms evolved, photosynthesis was not possible. Therefore, other sources of free energy must have been used. Table 6.1 lists the major sources of energy in the terrestrial region. It is believed that large amount of energy from ultraviolet light had made a significant contribution to the synthesis of organic compounds. The constituents of the primitive atmosphere—hydrogen, methane, ammonia and water vapours were activated by ultraviolet light of wavelength less than 200 nm. These species underwent photochemical interactions in the upper reaches of the atmosphere to yield other simple organic molecules which, in due course, were leached from the upper atmosphere to the oceans. Subsequent chemical evolution to complex molecules of life occurred in the oceans.

Table 6.1 Terrestrial energy sources

Energy Source	Energy (cal year^{-1} × 10^{-19})
Total radiation from sun	132,000
Ultraviolet light $\lambda < 250$ nm	300
$\lambda < 200$ nm	45
$\lambda < 150$ nm	1.8
Electric discharges	2.1
Radioactivity	0.4
Volcanoes	0.07
Cosmic rays	0.0008

Following factors favoured the progress of chemical evolution in the oceans.

1. In the prebiotic environment there was no ozone shield in the stratosphere. At present the stratospheric ozone shield protects the organisms from harmful ultraviolet radiations with wavelength in the 280–320 nm range (Chapter 19). These radiations are known to destroy DNA. Since in ancient times there was no ozone layer, the harmful ultraviolet radiations would have eliminated all life

forms, had the organisms originated on land. However, in the depths of oceans, water protected the molecules of life from malicious ultraviolet radiations.

2. The ocean was an effective medium for the mixing together of different classes of organic compounds formed at separate places and at separate times.

3. The sediments of the oceans acted as heterogeneous catalysts for the synthesis of complex organic compounds. For example, peptides were synthesized from condensation reactions of amino acids. Such reactions proceed with a loss of water molecule from two molecules of amino acids. In a highly dilute medium, which the ocean would have provided, the thermodynamically favoured reaction would have been hydrolysis and not condensation. Yet, the sediments of ocean adsorbed the different amino acids and caused their condensation reactions to occur on the catalytic surface.

Over the eons, the molecules in the marine environment continued to form aggregates. Some aggregates possessed a favourable constitution and organisation and these grew in number and complexity. Sometime, somewhere or may be several times and at several places, such an aggregate reached a state which we call *alive.* Thus, the oceans were the birthplace of life on our planet.

Over the eons, important changes occurred in the atmosphere too. Hydrogen was once the dominant constituent of the atmosphere. However, being the lightest of all gases, it continuously diffused out of the atmosphere. Its partial pressure in the primitive atmosphere was 10^{-3} atmosphere. At present it is about 10^{-6} atmosphere. We have stated that apart from hydrogen, the other constituents of ancient atmosphere were methane, ammonia and water vapours. Methane and ammonia are stable only under high concentration of hydrogen. As the partial pressure of hydrogen decreased these were converted into carbon dioxide and nitrogen, respectively. Simultaneously, ultraviolet radiations photolyzed water molecules, producing oxygen.

$$H_2O + h\nu \rightarrow H + OH \tag{6.1}$$

$$OH + h\nu \rightarrow O + H \tag{6.2}$$

$$O + O \rightarrow O_2 \tag{6.3}$$

However, only a small amount of oxygen was produced as a result of photochemical dissociation of water. Photolysis of water was self-limiting because when oxygen had accumulated to about 1/10,000 that of present atmospheric level, the oxygen itself absorbed the very wavelengths that photolyzed water. The major share of oxygen was liberated from water by photosynthetic organisms.

The escape of hydrogen from the atmosphere and the accumulation of oxygen affected a change from the primitive, reduced atmosphere to the present, oxidised one. As the concentration of oxygen increased, nitrogen content of the earth converted to nitrates, sulfur to sulfates and elemental and ferrous iron to ferric forms. Moreover, a part of oxygen in the upper atmosphere was converted into ozone, which formed a shield in the stratosphere.

$$O_2 + h\nu \rightarrow O + O \tag{6.4}$$

$$O + O_2 \rightarrow O_3 \tag{6.5}$$

Now that there was an ozone layer, the harmful ultraviolet radiations were filtered out and life forms transitioned from the ocean to the land. Hence, the entry of oxygen into the atmosphere not only permitted the development of respiration, but also allowed the life to emerge from water on to land.

Figure 6.1 shows the transition from hydrogen to the present day life. Hydrogen became the dominant component of the universe after the big-bang. In due course it was transformed to other elements through nucleosynthesis. From these elements a number of interstellar molecules were formed. This part of the transition is called *cosmic evolution*. We have discussed cosmic evolution in Chapter 1. Chemical evolution initiated with the appearance of simple molecules like methane, ammonia, water, etc. through organogenic synthesis. The conversion of simple molecules, first to biomonomers, and then to biopolymers marked the culmination of chemical evolution. The first living organisms progressed to the present day life through *biological evolution* or *Darwinian evolution*. The biological evolution is more concerned with ecology than with environmental chemistry and is, therefore, beyond the scope of the present text.

Stages of Chemical Evolution

We shall now examine the steps involved in the chemical evolution.

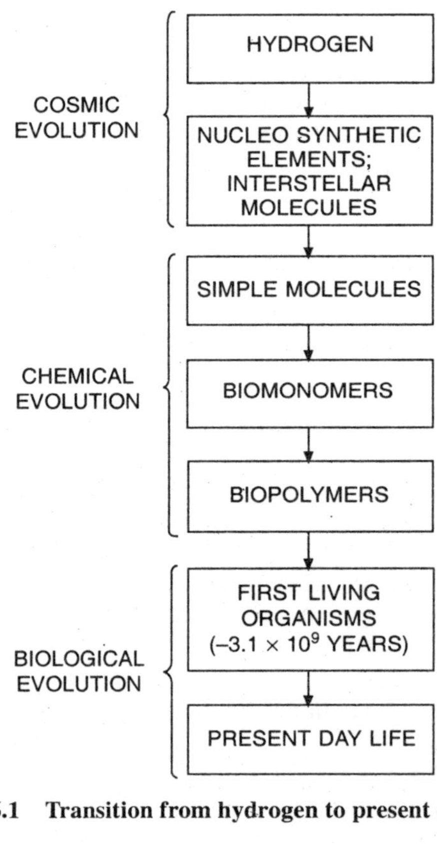

Fig. 6.1 Transition from hydrogen to present day life

From Atoms to Simple Molecules

The atoms formed as a result of nucleosynthesis combined to give simple molecules like hydrogen, methane, ammonia and water. These were the constituents of the ancient atmosphere. As the atmosphere transitioned towards oxidation, carbon monoxide, carbon dioxide and nitrogen were also formed. These molecules then interacted with each other to give a host of chemical species like hydrogen cyanide, formaldehyde, formic acid and acetic acid. The reaction of organic acids with ferrous sulfide produced hydrogen sulfide and with phosphide minerals produced phosphine. Most of the reactions leading to the synthesis of simple molecules were catalysed by ultraviolet radiation, electric discharge or thermal energy from volcanoes.

In due course, a substantial number of primeval molecules were synthesized in the atmosphere (Fig. 6.2). They were subsequently washed down to the oceans, where they served as the 'building blocks' for the structures of cell constituents.

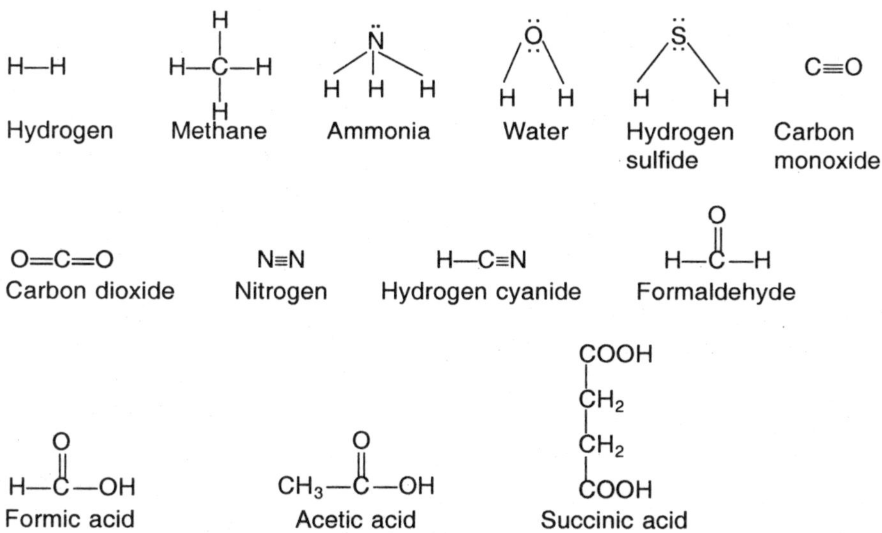

Fig. 6.2 Simple primeval molecules

From Simple Molecules to Biomonomers

The biomonomers important from the point of view of the origin of life are amino acids, nucleic acid components, nucleosides, nucleotides and porphyrins. We shall now examine the mechanisms of their synthesis.

1. Amino acids

If a mixture of hydrogen, methane, ammonia and water (the constituents of the earth's ancient atmosphere) is subjected to an electric discharge, a number of amino acids such as glycine, α-alanine, aspartic acid, α-aminobutyric acid, etc. are produced. Since in the primitive atmosphere electric discharges were of common occurrence, it is likely that naturally occurring amino acids were formed through the catalytic action of electric discharges. It is believed that action of electric discharge on the constituents of the primitive atmosphere first produced aldehydes and hydrogen cyanide, which subsequently reacted to give α-amino acids (Fig. 6.3a). In fact, all the twenty amino acids which polymerize to form natural proteins are α-amino acids. These amino acids are called *proteogenic amino acids*. It seems that when life got started on the earth, the α-amino acids were the commonest types available for protein construction. Those organisms which could metabolize α-amino acids would be the ones most likely to survive.

Yet another mechanism for the formation of amino acids proposes that the action of electric discharge on methane-ammonia mixture would have resulted in the formation of α-aminonitriles, which subsequently would have got hydrolysed to α-amino acids (Fig. 6.3b). The interaction of amino acids with hydrogen sulfide would have produced sulfur-containing homologues. Fig. 6.4 presents the structures of proteogenic amino acids.

$$\text{(a) } RCHO \xrightarrow{NH_3,\, HCN} \underset{\text{Cyanohydrin}}{RCH(NH_2)CN} \xrightarrow{H_2O} \underset{\alpha\text{–Amino acid}}{RCH(NH_2)COOH}$$

$$\text{(b) } CH_4 + NH_3 \longrightarrow \underset{\alpha\text{–Aminonitrile}}{H_2NCH_2CN} \xrightarrow{H_2O} \underset{\text{Glycine}}{H_2NCH_2COOH}$$

Fig. 6.3 Prebiotic synthesis of amino acids

2. Nucleic acid components

The components of nucleic acids are purines, pyrimidines, sugars, nucleosides and nucleotides.

(i) *Purines.* Two purines are present in the nucleic acid structures. These are, adenine and guanine. Their structures are shown in Fig. 6.5.

Of the two purines, adenine is more important. It is a constituent of both DNA and RNA. It is also present in many important cofactors, such as adenosine diphosphate and adenosine triphosphate. Moreover, of all the biologically important purines and pyrimidines, adenine has the greatest reasonace energy. This not only confers more stability to adenine, but also makes its synthesis more likely.

The mechanisms for the formation of adenine and guanine are shown in Fig. 6.6. Hydrogen cyanide in aqueous medium polymerizes to give its tetramer. A part of hydrogen cyanide reacts with ammonia to yield formamidine. The tetramer and formamidine combine with each other to give 4-aminoimidazole-5-carboxamidine. The latter on reaction with formamidine yields adenine. 4-Aminoimidazole-5-carboxamidine may also be hydrolysed to give 4-aminoimidazole-5-carboxamide which, on reaction with formamidine produces guanine.

(ii) *Pyrimidines.* Three pyrimidines are present in the nucleic acid structures. These are, uracil, cytosine and thymine (Fig. 6.7). While the probable mechanisms for the formation of uracil and cytosine under prebiotic conditions are known, that of thymine is not yet understood.

Uracil was supposedly produced by the reaction of malic acid and urea in presence of polyphosphoric acid, at a temperature of 100–140°C (Fig. 6.8a). Malic acid could have been produced by the reaction of acetic acid and formaldehyde. Urea would have been synthesized from carbon dioxide and

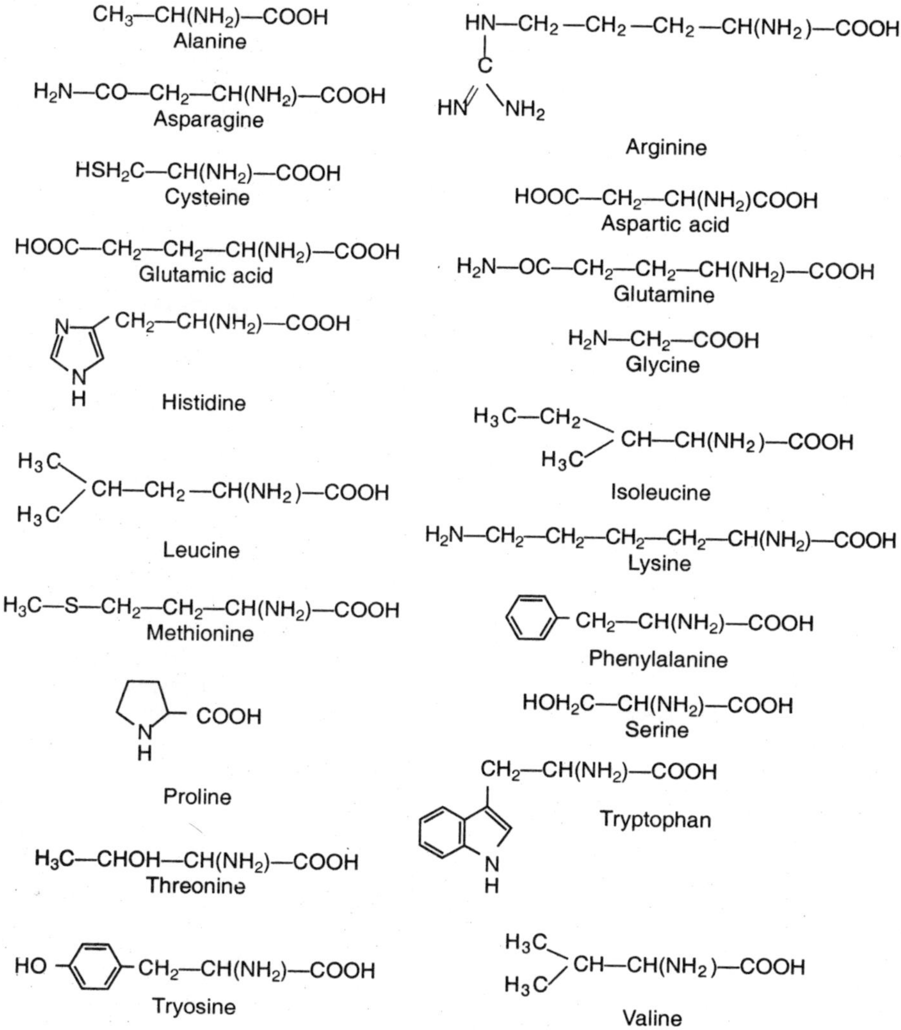

Fig. 6.4 Proteogenic amino acide

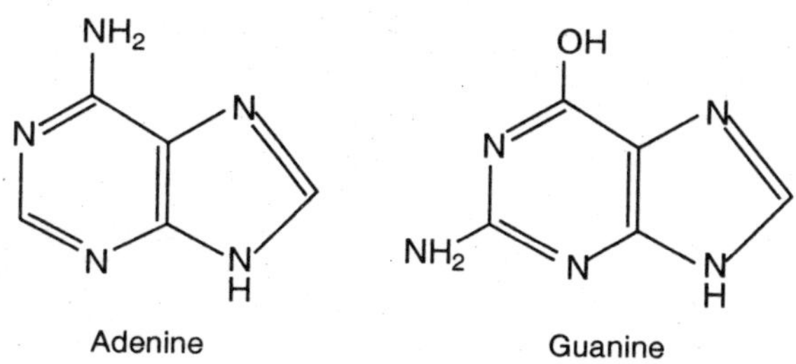

Fig. 6.5 Purine components of nucleic acids

ammonia. Phosphates were present both in marine environment, as well as in rocks (see the section on nucleotides and Fig. 6.13). Temperatures in the range 100–140°C would have been possible in the selected volcanic regions of the earth. However, we have hypothesized that chemical evolution occurred in the

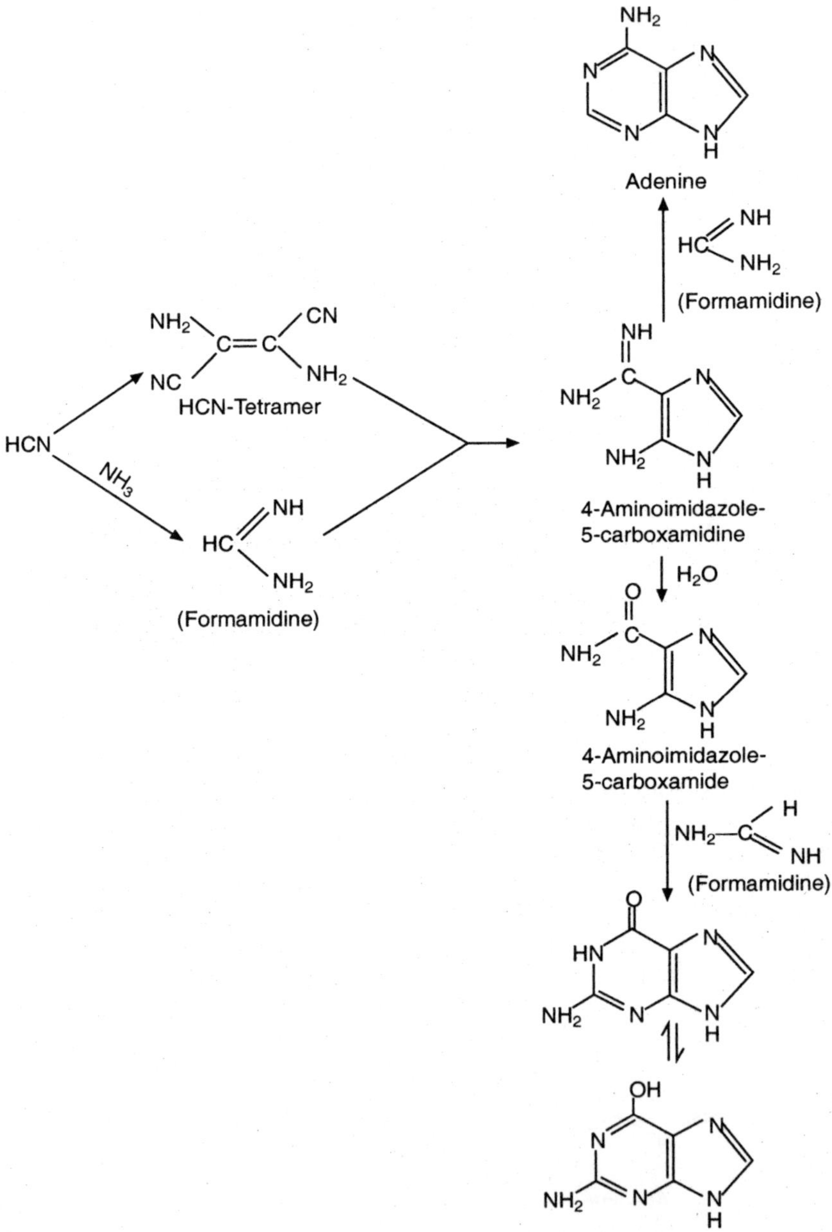

Fig. 6.6 Prebiotic synthesis of purines

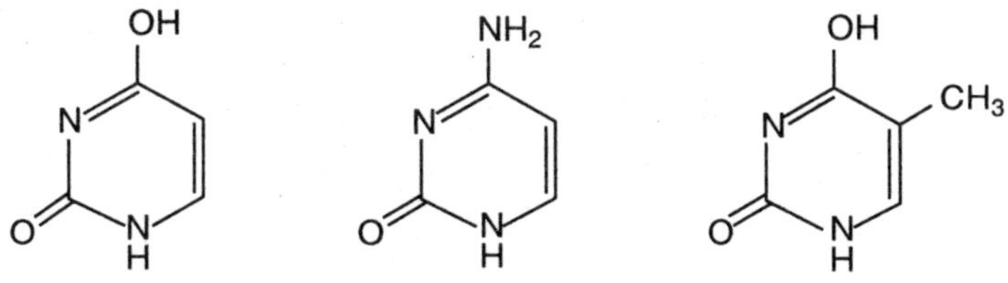

Fig. 6.7 Pyrimidine components of nucleic acids

ocean. It is difficult to explain such high temperatures in the marine systems. It is, therefore, speculated that the reactants required for the synthesis of uracil were washed out from the ocean to the volcanic cinder cones. After the reaction occurred, the resulting product, that is, uracil, was washed back to the ocean.

Cytosine was formed by subjecting cyanovinylurea to high pH values. Cyanovinylurea was produced by the reaction of cyanoacetylene and cyanate ions (Fig. 6.8b). Cyanoacetylene, in turn, would have been produced by sparking a mixture of methane and nitrogen. Cyanate ions are believed to have been produced by the hydrolysis of cyanogen in neutral or mildly alkaline medium. Another synthetic route to cytosine and uracil involves the reaction between urea and cyanoacetaldehyde in enolic form. Concentrated urea solutions required for this reaction were formed in evaporating lagoons. The reaction initially yielded cytosine, which was subsequently hydrolysed to uracil (Fig. 6.8c).

(iii) *Sugars.* The irradiation of a mixture of methane, ammonia and water vapours produce a set of pentose and hexose carbohydrates. Since we are here concerned with nucleic acid constituents, we would focus our attention on pentoses, for both D-ribose and D-2-deoxyribose belong to this class of compounds (Fig. 6.9).

The initial product of irradiating methane-ammonia-water mixture with electric discharge or ionizing radiation is formaldehyde. Two molecues of formaldehyde condense to give glycolaldehyde, which undergoes aldol-type condensations with additional formaldehyde molecules to yield pentoses (Fig. 6.10a). The formation of glycolaldehyde, however, is possible only if the carbon atom of one molecule of formaldehyde becomes a nucleophile, through the loss of a proton, and then attacks the carbon atom of the second molecule of formaldehyde. Under normal conditions this is not possible. However, in aqueous environment, formaldehyde exists as a monohydrate ($HCHO \cdot H_2O$). This species ionizes through a loss of proton, as shown in Fig. 6.10b.

The nucleophile may be stabilized through coordination to a divalent metal ion. In such circumstances, the equilibrium reaction (Fig. 6.10b), shifts to the right. Certain clay materials like illite contain divalent cations, such as magnesium(II) and iron(II). These, therefore, would have stabilized the nucleophile of formaldehyde, leading to the formation of glycolaldehyde and finally sugars. This reaction speaks of the importance of sea sediments in chemical evolution.

(iv) *Nucleosides.* Nucleosides are formed by the condensation of purine or pyrimidine bases with sugars (either D-ribose or D-2-deoxyribose). Prior to condensation, the nucleic acid bases and sugars would have been concentrated through adsorption on sea sediments.

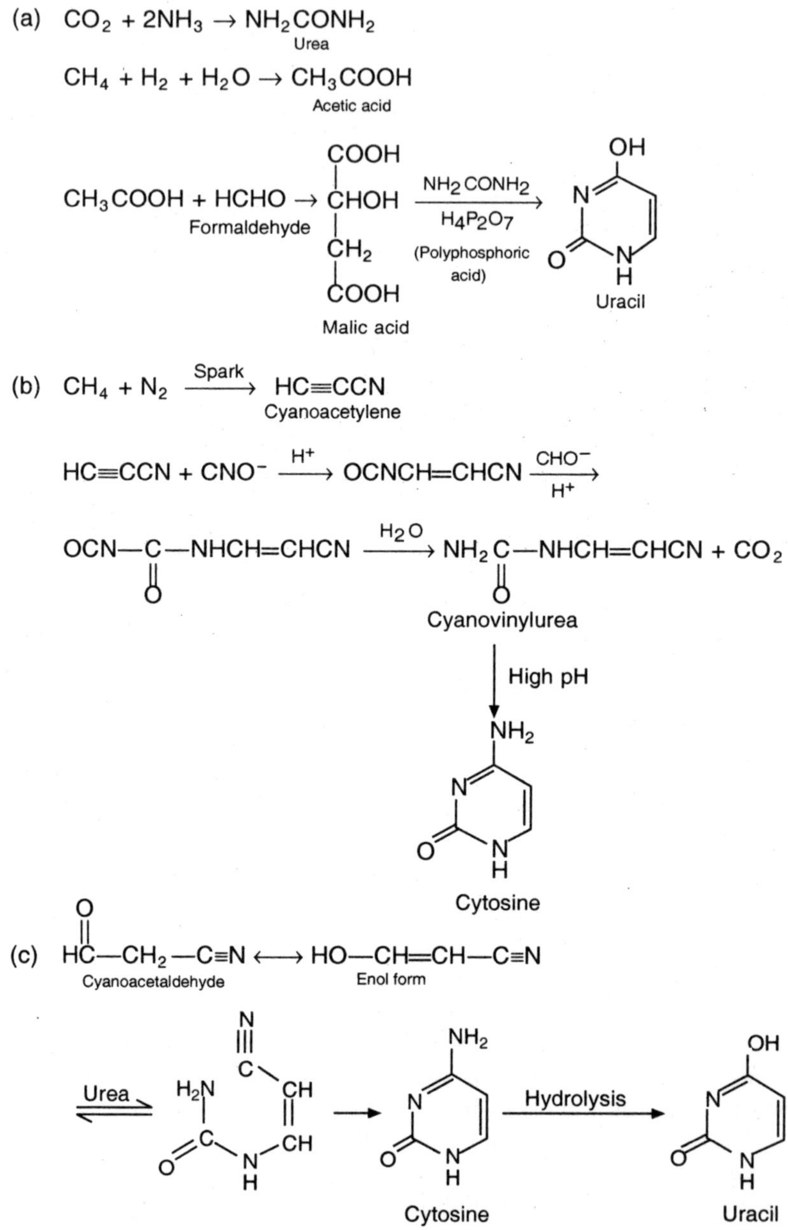

Fig. 6.8 Prebiotic synthesis of pyrimidines

As a representative case, the condensation reaction between adenine (purine base) and D-ribose is shown in Fig. 6.11. The resulting nucleoside is called adenosine.

Figure 6.11 is a simplistic representation of a nucleoside formation. The actual mechanism of this reaction is still not known.

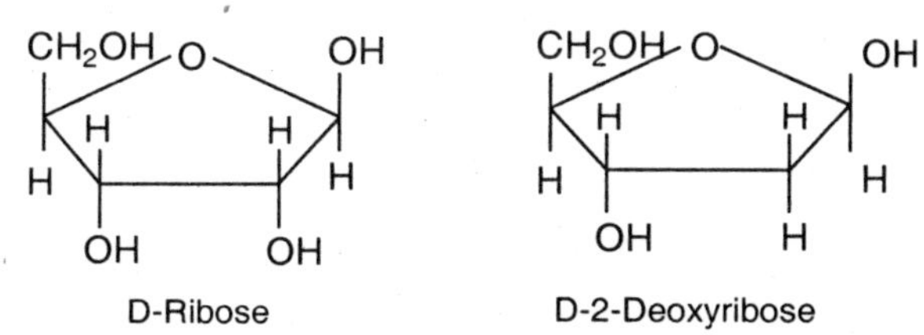

Fig. 6.9 Sugar components of nucleic acids

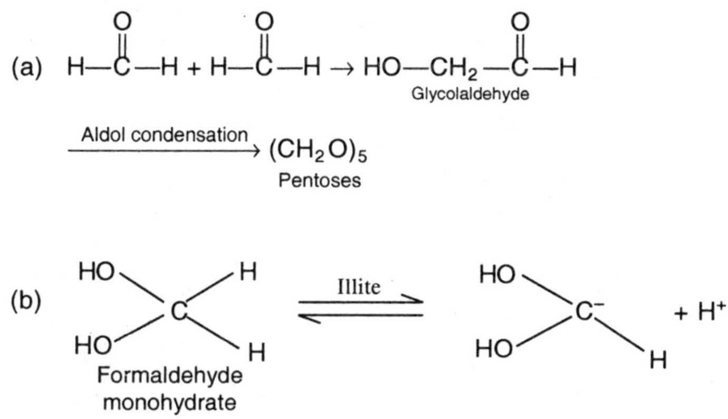

Fig. 6.10 Prebiotic synthesis of sugars

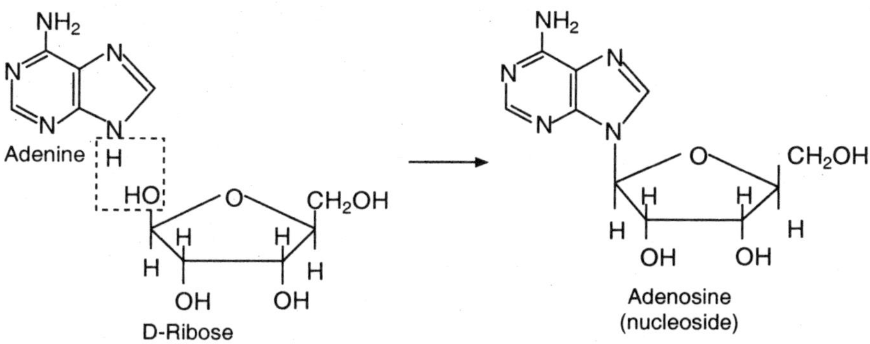

Fig. 6.11 Prebiotic synthesis of nucleosides

(v) *Nucleotides.* Nucleotides are nucleoside phosphates. They are formed by the condensation of nucleosides with a suitable phosphorylating agent in the presence of a condensing reagent like hydrogen cyanide or cyanogen. In Fig. 6.12 the phosphorylating agent is shown to be orthophosphoric acid. This is an over simplification. Under experimental conditions, orthophosphoric acid does not produce nucleotides from nucleosides. However, condensed phosphates, such as pyrophosphoric acid, are good phosphorylating agents. It is believed that condensed phosphates were formed by the reaction of cyanate ions with the phosphate content of hydroxyl apatite, $Ca_{10}(PO_4)_6(OH)_2$. The latter is the earth's most common phosphorus-containing material. On reaction with cyanate ions it produces carbamyl phosphate, as shown in Fig. 6.13. Carbamyl phosphate is not very stable, yet it may react immediately with another phosphate ion to produce

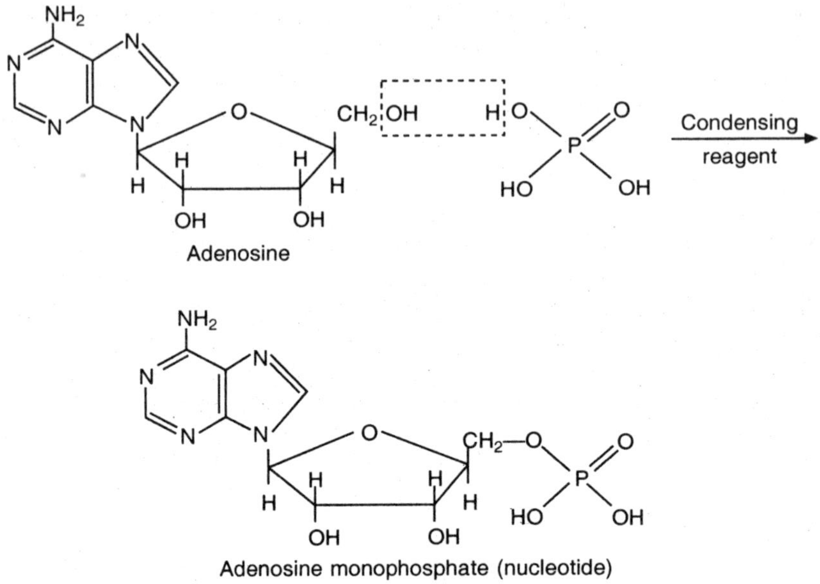

Fig. 6.12 Prebiotic synthesis of nucleotides.

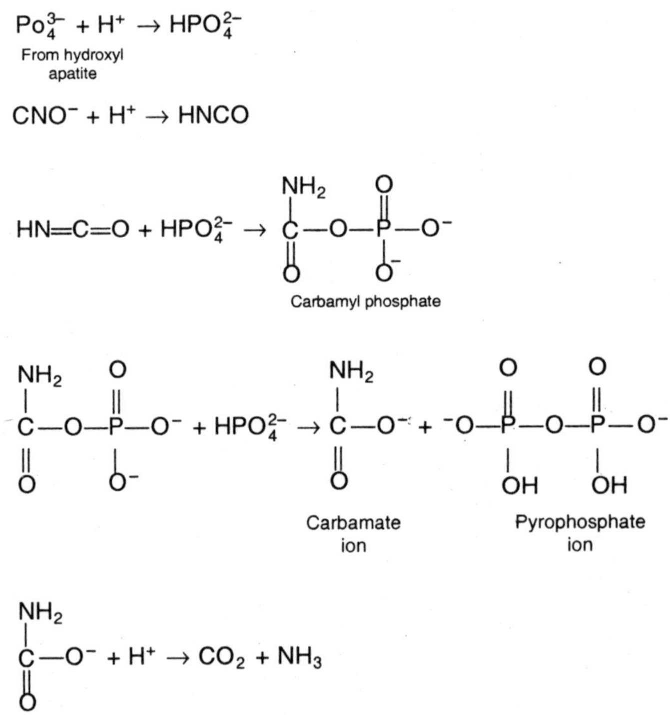

Fig. 6.13 Prebiotic synthesis of pyrophosphate

carbamate ions and pyrophosphate ions. The carbamate ions then decompose into carbon dioxide and ammonia.

3. Porphyrins

The synthesis of porphyrins is considered to be a necessary step for the origin of life. Chlorophyll and heme are, respectively, magnesium and iron chelates of substituted porphyrins. Since chlorophyll is an essential component of photosynthesis, porphyrins had to pre-exist the first algae. The synthesis of porphyrins initiated with the reaction of succinic acid (present as simple, primeval molecule) and glycine. The resulting product, adipic acid, on decarboxylation yielded δ-aminolevulinic acid. Two molecules of δ-aminolevulinic acid condensed to yield a pyrrole skeleton, which under the action of ultraviolet radiations resulted in the formation of a porphyrin unit (Fig. 6.14). Once porphyrin was formed, it acted as an autocatalyst and accelerated its own synthesis from succinic acid and glycine.

From Biomonomers to Biopolymers

With the passage of time the primitive ocean came to possess a wide variety of biomonomers. Some of these were in the dissolved state, while others were adsorbed on clay materials. The biomonomers underwent condensation reactions to produce biopolymers. In chemical evolution, we are concerned with three biopolymers: Proteins, polysaccharides and nucleic acids. We shall now discuss their synthesis.

1. Proteins

In view of the fundamental and ubiquitous role of proteins in contemporary metabolism, it is probable that an accumulation of substantial amounts of polypeptides on the prebiotic earth was required for subsequent development of living systems. The polypeptides, in turn, were produced through condensation polymerization of amino acids. In such reactions it is necessary to remove a water molecule between two monomers. However, removal of water has to be achieved in presence of water itself, since chemical evolution proceeded in the oceans. Ordinarily, condensation reactions do not occur in aqueous medium. Yet, in presence of molecules like hydrogen cyanide, such reactions can be carried through even in aqueous environment. Hydrogen cyanide molecule contains a carbon-nitrogen triple bond. Such multiple bonds act as stable, high energy sources for absorption of water. They do not react rapidly with water; they react preferably with a source of water. The condensing monomer amino acids are the source of water. As shown in Fig. 6.15, hydrogen cyanide reacts with the water liberated from the condensation of two molecules of amino acids to give formamide. The condensation reaction produces dipeptide, which subsequently polymerizes to proteins.

2. Polysaccharides

As shown in Fig. 6.16, polysaccharides are formed by condensation of monosaccharides. An energy source like hydrogen cyanide is essential to carry out such condensation polymerization reactions.

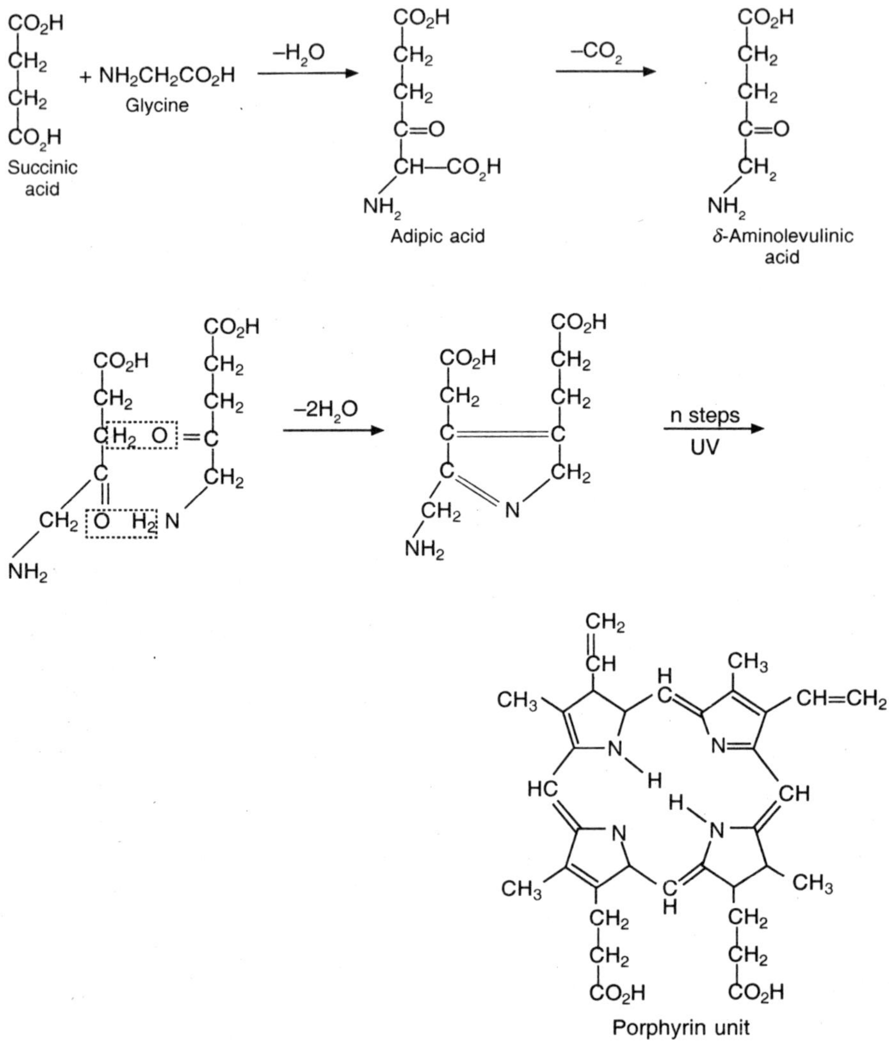

Fig. 6.14 Prebiotic synthesis of porphyrins

3. Nucleic acids

Nucleic acids are polynucleotides. These too are formed by condensation polymerization reactions. The monomer reactants are the nucleotides. As in case of proteins and polysaccharides, the formation of nucleic acids requires an energy source like hydrogen cyanide. The formation of nucleic acids is presented in Fig. 6.17. It is believed that the synthesis of polypeptides and polynucleotides occurred in almost the same phase of chemical evolution and that each tended to catalyse the formation of the other.

Inference

In this chapter we have discussed the formation of various biomonomers and biopolymers which are important components of chemical evolution. Needless

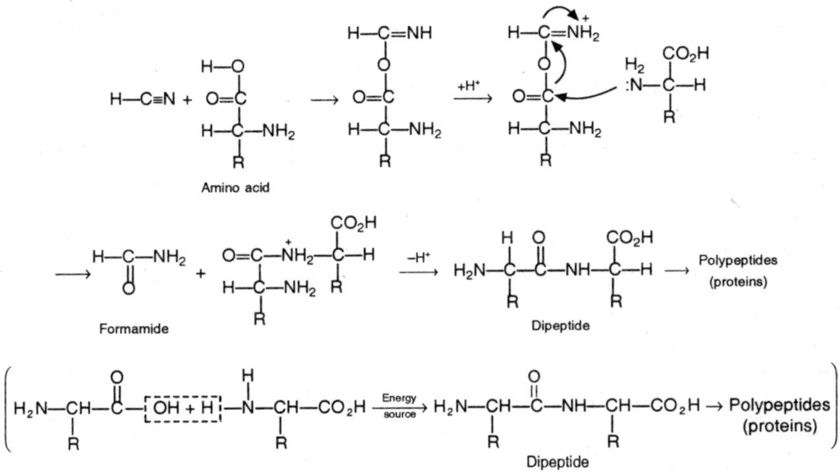

Fig. 6.15 Prebiotic synthesis of proteins

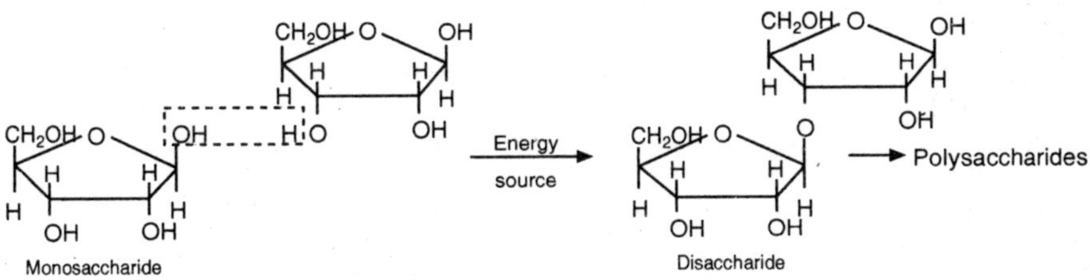

Fig. 6.16 Prebiotic synthesis of polysaccharides

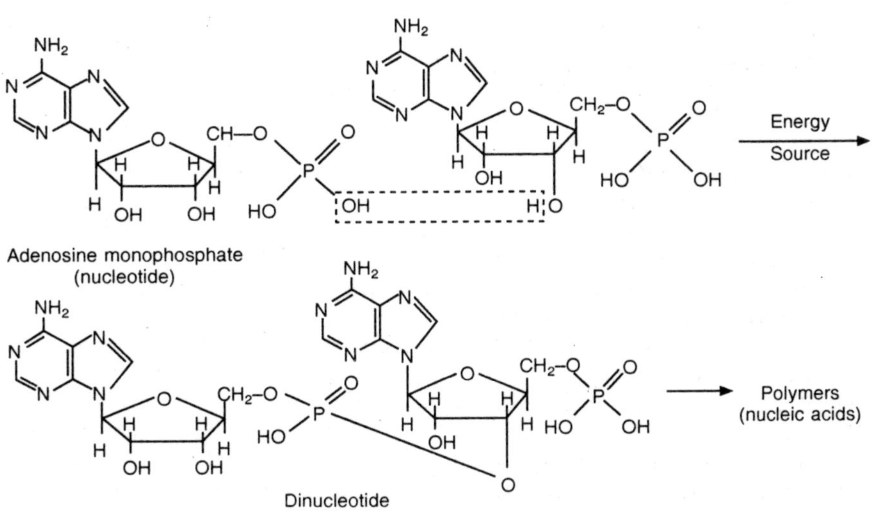

Fig. 6.17 Prebiotic synthesis of nucleic acids.

to mention, the reaction schemes formulated in each case are only plausible mechanisms. There was no one to verify these pathways at the time they originally occurred. Yet, experiments performed under laboratory conditions, simulating the prebiotic environment, support the pathways presented in these schemes. Further, these are not the only pathways to the synthesis of biomolecules. A particular analogue may have been synthesized by more than one reaction sequence. It is, however, neither possible nor important to list all the mechanistic pathways leading to chemical evolution. The idea in this chapter was to bring home the importance of synthetic pathways to biomolecules. The emergençe of life is the inevitable outcome of the associational forces that acted on these biomolecules and organised them as constituents of the living cell. The synthesis of biomonomers and biopolymers are but halting steps on the long road to the understanding of the genesis of the living cell. How these biomolecules progressed towards the degree of organisation that we see in the cellular environment? This shall be the subject matter of the next chapter.

References

Briggs, D.E.G. and Eglinton, G. (1994). Chemical traces of ancient life, *Chem. Brit.*, **30,** 907–912.

Calvin, M. (1959). Evolution of enzymes and the photosynthetic apparatus, *Science,* **130,** 1170–1174.

Calvin, M. (1969). Chemical evolution, *Chem. Brit.,* **5,** 22–28 (1969).

Calvin, M. (1975). Chemical evolution, *Amer. Sci.,* **63,** 169–177.

Ferris, J.P., Sanchez, R.A. and Orgel, L.E. (1968). Studies in prebiotic synthesis. III. Synthesis of pyrimidines from cyanoacetylene and cyanate, *J. Mol. Biol.,* **33,** 693–704.

Fox, S.W. and Harada, K. (1961). Synthesis of uracil under conditions of a thermal model of prebiological chemistry, *Science,* **133,** 1923–1924.

Friedmann, N. and Miller S.L. (1969). Phenylalanine and tryosine synthesis under primitive earth conditions, *Science,* **166,** 766–767.

Gabel N.W. and Ponnamperuma, C. (1967). Model for origin of monosaccharides, *Nature (London),* **216,** 453–455.

Lemmon, R.M. (1970). Chemical evolution, *Chem. Rev.,* **70,** 95–109.

Lohrmann, R. and Orgel, L.E. (1968). Prebiotic synthesis: Phosphorylation in aqueous solution, *Science,* **161,** 64–66.

Miller, S.L. (1953). A production of amino acids under possible primitive earth conditions, *Science,* **117,** 528–529.

Miller, S.L. and Parris, M. (1964). Synthesis of pyrophosphate under primitive earth conditions, *Nature (London),* **204,** 1248–1250.

Miller, S.L. and Urey, H.C. (1970). Organic compound synthesis on the primitive earth, *Science,* **130,** 245–251.

Orgel, L.E. (1994). The origin of the life on the earth, *Sci. Amer.,* **271**(4), 53–61.

Ponnamperuma, C., Sagan, C. and Mariner, R. (1963). Synthesis of adenosine triphosphate under possible primitive earth conditions, *Nature (London),* **199,** 222–226.

Ponnarmperuma, C. (1964). Chemical evolution and the origin of life, *Nature (London),* **201,** 337–340.

Robertson, M.P. and Miller, S.L. (1995). An efficient prebiotic synthesis of cytosine and uracil, *Nature (London),* **375,** 772–774.
Sanchez, R.A., Ferris, J.P. and Orgel, L.E. (1966). Cyanoacetylene in prebtiotic synthesis, *Science,* **154,** 784–785.
Sanchez, R.A., Ferris, J.P. and Orgel, L.E. (1967). Studies in prebiotic synthesis. II. Synthesis of purine precursors and amino acids from aqueous hydrogen cyanide, *J. Mol. Biol.,* **30,** 223–253.

CHAPTER 7

Cellular Environment

The origin of life is to be considered a transition from lifeless to living matter, from the inanimate to the animate realm of nature. The phenomenon of chemical evolution, outlined in the previous chapter, was a necessary preamble to the emergence of life. It initiated during the period when earth contained only simple inorganic molecules and culminated in the formation of biopolymers, mainly proteins and nucleic acids, in the hydrosphere. However, these biopolymers themselves were lifeless. The characteristics of the living organisms—metabolism, growth and reproduction—were not to be found in these macromolecules. Yet, when the biopolymers organised themselves in a stable, three dimensional pattern of a cell, they could perform all the vital functions of life. The origin of cells serves to explain their fitness in terms of the environment into which they developed. The structure of the cell explains how these building blocks of plant and animal life perform the vital functions that distinguish the living matter from the nonliving one.

Origin of the Living Cell

There is every reason to believe that a certain period of the earth's history was marked by complete sterility, that is, absence of organisms. This was the era of chemical evolution which, during its last phase, produced two important sets of macromolecules—proteins and nucleic acids. Proteins containing nucleic acids are the only constituents of organisms which are known to possess the capacity to metabolize, grow and reproduce. Proteins were formed from peptides and nucleic acids from nucleotides. Since chemical evolution occurred in the oceans, the precursors to proteins and nucleic acids were dispersed and uniformly distributed in the marine environment. They were inseparable from the dispersing medium, that is, the sea water. However, the living cell contains the proteins and nucleic acids in a highly organised, stable and three dimensional pattern. Hence, there must have occurred a pheonomenon which would have brought the proteins and nucleic acids from the dispersed state to a highly organised state. This phenomenon was based upon the colloidal properties of the biopolymers.

In the prebiotic environment, the peptides underwent condensation polymerization to produce proteins. Likewise, the nucleotides polymerized to nucleic acids. As the polymerization progressed, the design of these biopolymers became more complicated and their properties were altered. With the elongation of the chain structure and increase in the molecular weights of these polymers, their solubility in water diminished. As a result, the proteins and nucleic acids formed colloidal sols in water.

The colloidal sols are classified into two types: Hydrophobic—the solvent hating and hydrophilic—the solvent loving. The hydrophilic colloids are more stable since the surrounding sheath of water prevents their breakdown. The proteins incorporate polar groups like $-NH_2$ and $-COO^-$ in their structures. Likewise, the purine and pyrimidine constituents of nucleic acids have polar substituents like –NH, $-NH_2$, –OH and C=O. The water molecules are also polar in nature. Therefore, there was a strong interaction between the biopolymers and the surrounding water, due to which the proteins and nucleic acids acquired hydrophilic nature. These macromolecules were swathed with a more or less thick membrane of water, which kept the individual particles in a stable state. As a result of this stability the colloidal solution separated into two distinct layers: The dispersed phase, rich in colloidal macromolecules and the dispersed medium, the sea water. The dispersed phase was in form of a semiliquid gel and was designated as *coacervate*. The phenomenon by which coazervates were formed is called *coacervation.*

The coacervate did not settle down as suspension, but remained in form of minute microscopic droplets floating in the dispersed medium. Though the coacervates were in equilibrium with the surroundings, yet they were sharply demarcated from the medium by a clearly discernable surface. These droplets fused with each other (Fig. 7.1), but they never mixed with the dispersed medium. The fusion of coacervate droplets was one of the most powerful method of concentrating polymerized biomolecules from very dilute solutions. Before the advent of coacervation, the biopolymers were uniformly distributed in sea water. The formation of coacervates brought them together. Thus, the organic matter became concentrated at different regions of the marine waters and, at the same time, sharp division occurred between the coacervate and the medium. Previously, the macromolecules had no real boundary separating them from the disperse medium, but with the onset of coacervation, the biopolymers became separated from the aqueous phase by a real, sharp boundary.

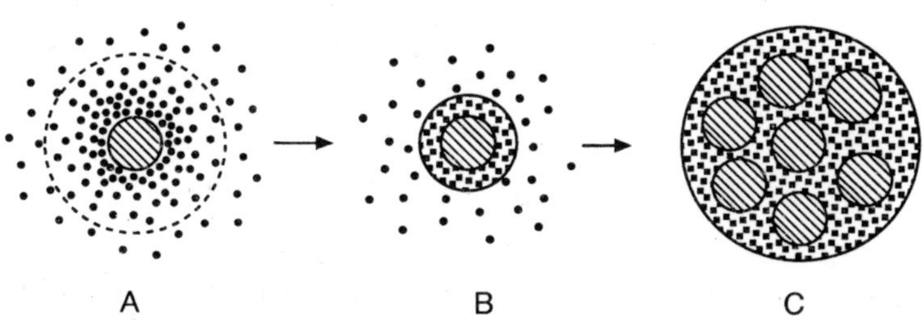

Fig. 7.1 The phenomenon of coacervation. A. Colloidal particle with diffuse hydration layer; B. Hydration layer becomes delimited; C. Formation of coacervate

The following example imitates the phenomenon of coacervation: Dilute solutions of gelatin and gum arabic on mixing become turbid at optimum pH and temperature. This is due to the fact that the molecules of gelatin and gum arabic are initially distributed throughout the volume of the solution but subsequently unite into integral molecular 'clusters'. When the dimensions of these clusters reach a particular size, they separate from the solution in the form of droplets visible under ordinary microscope. The droplets float freely in the medium, which now becomes deficient in these polymers.

As has been stated, a sharp boundary segregated the coacervate drops and the surrounding medium. Yet, there existed an equilibrium between the coacervate and the medium. Moreover, the coacervate was not suspended in pure water, but in sea water which contained several other inorganic and organic compounds. There was a continuous exchange of matter between the coacervate and the surrounding medium. Therefore, the equilibrium between the coacervate and the medium acquired a dynamic character.

In fact, dynamic equlibrium is a characteristic property of a living cell. The latter, during the course of its existence, is in a state of continuous interaction with the suróunding environment. The living system exists only as long as the dynamic equilibrium lasts. As soon as this equlibrium breaks down, the cellular system is destroyed. Its prolonged existence and the consistency of its form, therefore, reflects not a static character but a dynamic nature of the cellular system.

In a primitive form such exchange of matter must have been operative in the coacervate and would have been brought about in two ways. Firstly, the coacervate would have simply adsorbed materials selectively from the external medium. For example, if methylene blue dye is added to the mixture of gelatin and gum arabic, it is observed that the dye becomes adsorbed on the colloid. The aqueous medium, accordingly, becomes lighter in colour. Secondly, the adsorbed organic and inorganic compounds would have undergone chemical changes within the coacervate. The unwanted reaction products would have been discharged to the medium.

The exchange of matter between the coacervate and the souroundings is depicted in Fig. 7.2. The rectangle represents the coacervate drop, X and Y the external medium, A, the matter entering the drop and B, the reaction product capable of diffusing into the external medium. If the biochemical reaction A $\rightarrow$ B occurred more rapidly than the reverse reaction, the concentration of A in the coacervate decreased, while that of B increased. The equilibrium shifted so as to allow more of A to enter the drop and more of B to diffuse out.

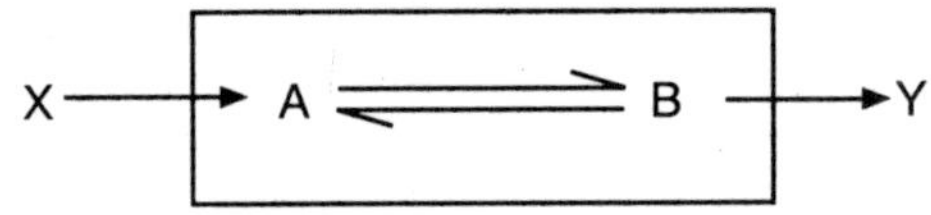

Fig. 7.2 Exchange of matter between coacervate and the surroundings

On its own, the reaction A $\rightleftharpoons$ B would have occurred very slowly. The most effective way of increasing its rate was to introduce a catalyst in the system. The catalysts which enhance the speed of biochemical reactions of the type A $\rightleftharpoons$ B are called *enzymes*. All enzymes are proteins. However, all proteins are not enzymes. The non-enzymatic proteins serve as structural matter.

The increase in speed of a given biochemical reaction, brought about by enzymatic proteins within the coacervate droplets, made the colloidal system more dynamic and hence more stable. Initially, the enzymes were adsorbed by the coacervate from the external environment. In due course, the enzymes were depleted from the medium. However, by that time some of the coacervate droplets developed the ability to synthesize enzymes by the reaction of suitable compounds extracted from the environment. Others did not or could not develop this ability. At that stage the forces of natural selection came into play. Those coacervate systems which could synthesize enzymes survived. Others were wiped off the face of the earth.

The inclusion of enzymes in the coacervate ensured that the material substrate of life was constantly synthesized and decomposed. This represented *metabolism.* As the substrate material was added from the surrounding medium to the coacervate, the size and weight of the colloidal system increased. This represented *growth.* The growth of the coacervate could not proceed indefinitely. When the size and weight crossed a threshold limit, the external mechanical stress or the forces of surface tension broke the coacervate into daughter droplets. This represented *reproduction.* The daughter droplets possessed greater surface area and lower mass than the parent coacervate. Therefore, they were endowed with a better ability to grow. Moreover, the daughter droplets had the same organisational pattern as the parent droplet since they represented the same original system.

Merely by acquiring the ability to metabolize, grow and reproduce, the coacervate drops were not able to sustain life. Sustainence and preservation of life required a highly ordered arrangement of biomolecules as is found in the living cell. Before the formation of coacervate there was a random distribution of biopolymers in the primitive oceans. However, in the process of coacervate formation, organic substances not only became concentrated at definite points in space but, to a certain degree, also acquired specific orientation with respect to each other. This orientation was governed by the spatial alignment and mutual relationship of the molecules. Thus, the structure of coacervate during growth proceeded to a more perfect order.

With the introduction of catalytic activity, the orderliness became still more perfect. The enzyme catalysts function by a lock and key mechanism. According to this mechanism, a substrate must fit in the enzyme structure, just as a key fits in the lock (Fig. 7.3). If the substrate happened to be a polynucleotide, it must have acquired a suitable orientation so as to couple with the enzymatic protein. Once the polynucleotide acquired a specific orientation, its synthesis within the coacervate would have been catalysed by the enzymes. The polynucleotide, in turn, would have played a role in protein synthesis. It would have directed the amino acids to polymerize in a particular sequence so that the resulting protein structure would surround it as a protective sheath. This type of cooperative interaction is also found in a living cell. The proteins catalyse the combination of polynucleotides to nucleic acids. The nucleic acids serve as a template for the formation of protein structures. In fact, nucleic acids cannot replicate without proteins and proteins cannot be made without nucleic acids.

The proteins are concerned with metabolism and growth. The nucleic acids

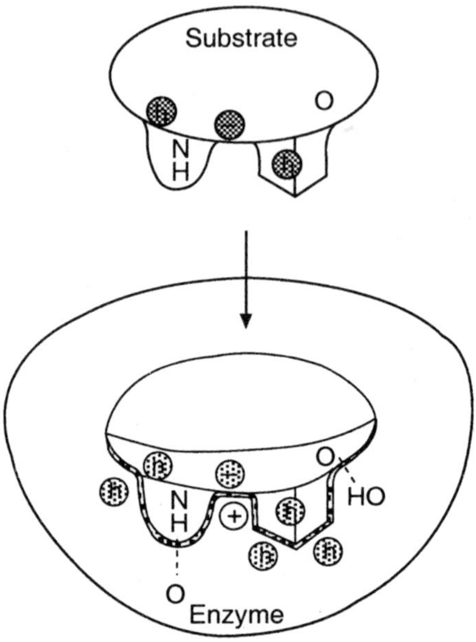

Fig. 7.3 Lock and key mechanism of enzyme catalysis

are concerned with reproduction. To ask which came first, proteins or nucleic acids, is just like asking in layman's term as to which came first, chicken or egg. The answer is that neither the chicken, nor the egg came first. They evolved together from the lower forms of life.

The lower forms of life were *heterotrophic,* that is, they obtained their basic constituents from the environment instead of synthesizing them. Just as initially the coacervate droplets obtained enzymes from external medium, but subsequently started synthesizing them, the heterotrophic organisms too were able to synthesize porphyrin molecules which paved the way for photosynthesis—and higher forms of life. The higher forms of life were *autotrophic.* These were capable of synthesizing their food from inorganic compounds. The transition from heterotrophs to autotrophs is called *heterotrophic hypothesis.* The synthesis of porphyrin structures represented a step further towards better organisation of matter.

Another aspect of higher organisation of biomolecules needs to be mentioned. Many biomolecules possess asymmetric carbon atoms and hence are optically active. They owe the capacity to rotate the plane of polarized light to the right, in which case they are called dextrorotatory (D-) or to the left, in which case they are called laevorotatory (L-). When synthesized artificially, such molecules form racemic mixtures containing equal amounts of D- and L- forms. Racemic mixtures are optically inactive. They do not rotate the plane of polarized light.

The living organisms, however, invariably incorporate either D- or L- forms of biomolecules, but never both. For example, all natural amino acids are laevorotatory and all natural sugars are dextrorotatory. The living organisms derive the following advantages by incorporating only one configuration of biomolecules.

1. The biomolecules acquire a favourable geometry with one configuration

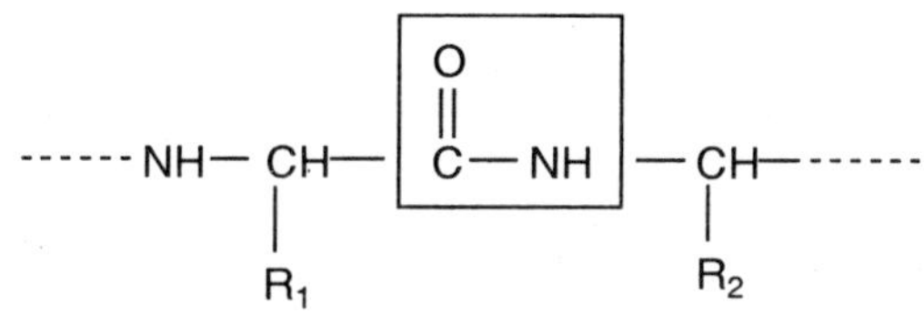

Fig. 7.6 The peptide bond

example actin, are responsible for the active movement in living organisms. *Transport proteins,* like haemoglobin, carry oxygen in the vertebrate blood.

The behaviour and properties of proteins are determined by their structures. Theoretically, there is no limit to polypeptide chain length and all permutations and combinations of the twenty constituent amino acids are possible. However, all twenty amino acids are not present in every protein. Nevertheless, structural variations are offered by attachment of non-protein prosthetic groups and by attainment of different levels of polymerization of amino acids. Hence, the possible diversity of structures and functions of proteins is almost limitless. There are about 40,000 proteins in humans.

Proteins are neither rigid nor motionless. Within a protein molecule, the atoms of the backbone, as well as of the side chains are in a state of thermal motion, making the structure flexible. The dynamics of protein conformation range over a broad spectrum, from picoseconds to longer than milliseconds, from the rotation about the individual bonds to the bending of the backbone. The dynamic properties of proteins are crucial to their biological functions. For example, enzymes would not be able to show their catalytic properties, had they possessed a rigid, non-flexible structure.

Proteins are broadly classified into two types: Unconjugated proteins and conjugated proteins.

(a) Unconjugated Proteins

The unconjugated proteins, also called the simple proteins, are those which are not conjugated to a non-protein moiety. They are subdivided into *globular proteins* and *fibrous proteins.* Globular proteins are soluble in water and dilute salt solutions. They consist of one or more peptide chains held together by hydrogen bonds. Examples include albumin and globulin. Fibrous proteins are insoluble in water and dilute salt solutions. They are stable to acids and alkalis and possess greatly extended molecular structures with very high degree of molecular asymmetry. Examples include flagellin and pilin.

(b) Conjugated Proteins

Conjugated proteins are those which, in addition to amino acids, contain a non-protein prosthetic group. The prosthetic group is essential for performance of a specific function. It is attached to the protein moiety by a covalent, coordinate or polar bond. A prosthetic group may be a carbohydrate, a lipid, a nucleic acid or a metal ion. The proteins containing these prosthetic groups are called, glyco-, lipo-, nucleo- and metallo-proteins, respectively. Haemoglobin is an example of metalloprotein.

2. Nucleic Acids

Nucleic acids are the polymers containing repeating nucleotide residues. Nucleotides, in turn, are the phosphate esters of nucleosides. Each nucleotide consists of three structural components—a purine or a pyrimidine base, a sugar moiety, and a phosphate ester group. A typical nucleotide is shown in Fig. 7.7.

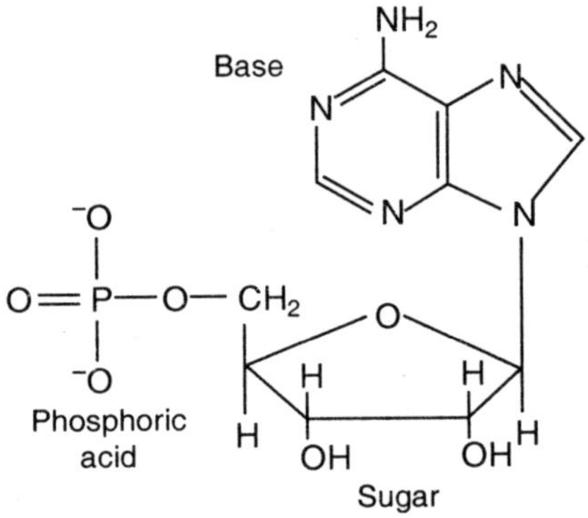

Fig. 7.7 A nucleotide

During polymerization, each nucleotide unit is linked to its neighbour by the phosphate group, which forms an ester linkage with position 3′ of one sugar and 5′ of the neighbouring sugar. This 3′, 5′ linkage results in a linear chain of phosphate-linked sugar residues. A base (purine or pyrimidine) is attached to C–1′ of each sugar by an N-glycosidic linkage.

Nucleic acids are acidic in nature. They possess a molecular weight of 10^4 to 10^9. There are two types of nucleic acids: Deoxyribonucleic acid (DNA), which contains 2-deoxy-D-ribose and ribonucleic acid (RNA), which contains 2-D-ribose. The two nucleic acids perform different biological functions. DNA stores genetic information. It is replicated during cell division, so that each daughter cell receives DNA that is identical in structure and informational content to the mother cell. RNA is intimately involved in protein biosynthesis and is primarily responsible for translating the information of the DNA into the primary structure of specific proteins.

(a) Deoxyribonucleic Acid

The mononucleotide repeating unit of DNA consists of phosphorylated D-2-deoxyribose which is glycosidically linked to either of the four bases: Adenine, guanine, cytosine or thymine (these are abbreviated A, G, C or T, respectively). The mononucleotides are linked by 3′, 5′–diester bridges in an unbranched polynucleotide backbone, as shown in Fig. 7.8. The base content of DNA varies from organism to organism; yet the amount of adenine is always equal to that of thymine, while the amount of guanine is same as that of cytosine.

The DNA molecule has a double helix structure (Fig. 7.5). It consists of two complementary, but not identical strands made up of sugar-phosphate chains from which the purine and pyrimidine bases protrude at regular intervals into the

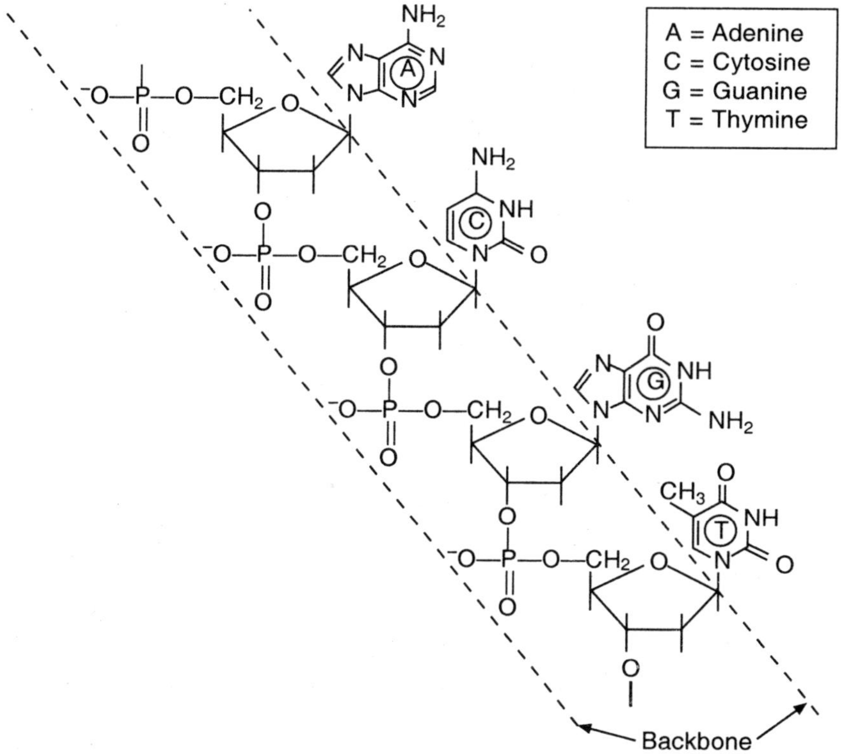

Fig. 7.8 DNA chain of nucleotides

interior of the helix. Thus, the backbone of the DNA consists of phosphate groups and sugar molecules and the inward pointing side chains consist of purine and pyrimidine bases. The two strands are held together by hydrogen bonds which unite the bases on the opposite strands. In order for the strands to fit together on the helix, a purine on one strand faces a pyrimidine on the opposite strand. The hydrogen bonding can exist only between adenine and thymine or between guanine and cytosine. Therefore, the sequence of bases along one strand determines the sequence of bases on the other. The genetic information is encoded in the sequence of bases in the DNA.

The double helix is not symmetrical. It has a broad and a narrow groove between the chains, which provide steric orientation for the process of replication and transcription.

(b) Ribonucleic Acid

The mononucleotide unit of RNA consists of ribose, phosphorylated at C-3 and linked by N-glycoside bond to one of the four bases: Adenine, guanine, cytosine or uracil (these are abbreviated A, G, C or U, respectively). The polymerized nucleotides form linear chains via 3′, 5′-phosphodiester bonds (Fig. 7.9). RNA possesses a single α-helical structure.

As has been stated, RNA is involved in the biosynthesis of proteins. Protein biosynthesis is a cyclic, energy requiring, multistage process, in which the amino

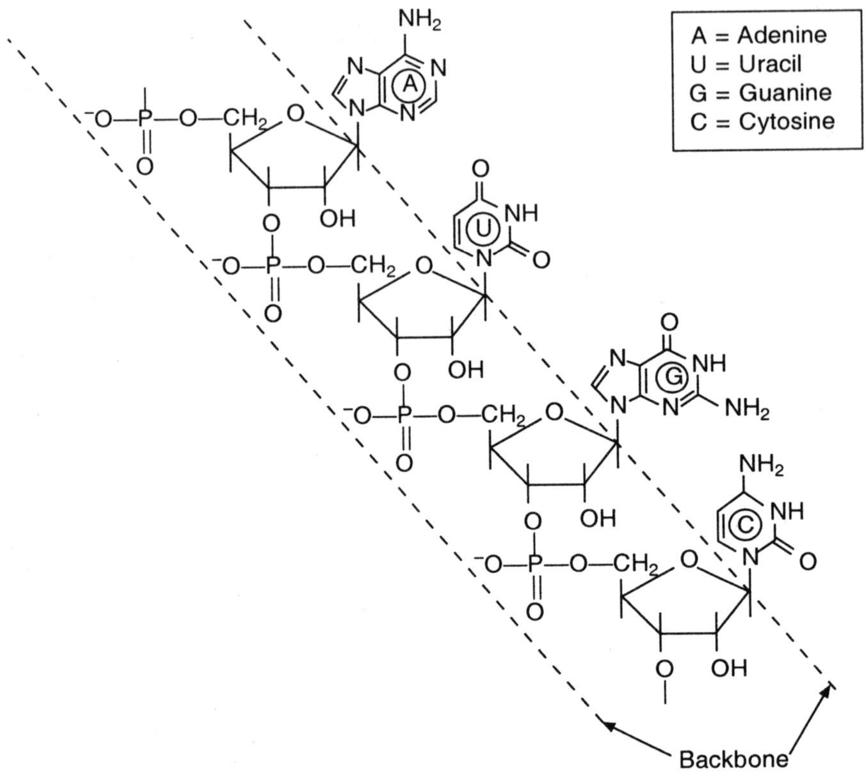

Fig. 7.9 RNA chain of nucleotides

acids are polymerized in a genetically determined sequence to form proteins. The genetic sequence is determined by the DNA and has to be translated to synthesize specific proteins by the RNA. There are three types of RNAs which carry out protein synthesis: Transfer-RNA (t-RNA), ribosomal-RNA (r-RNA). and messanger-RNA (m-RNA). The t-RNA combines with the amino acids, esterifying them and preparing them for polymerization. The r-RNA represents the site where protein biosynthesis takes place. Before the biosynthesis commences, the genetic information of the DNA is carried by m-RNA to r-RNA.

3. Lipids

Lipids are the constituents of the cell membranes. The most important lipids present in the cell membranes are the phospholipids, which are the triglycerides based on phosphatidic acid skeleton.

Lecithin is an example of phospholipids (Fig. 7.10). Such compounds incorporate both ionic groups (phosphates) and covalent groups (hydrocarbons). The former are hydrophilic in nature, while the latter are hydrophobic. For this reason, phospholipids are partially soluble in both water and fats. Therefore, they serve an important role in the cell in that they hold the water-soluble and the fat-soluble constituents together.

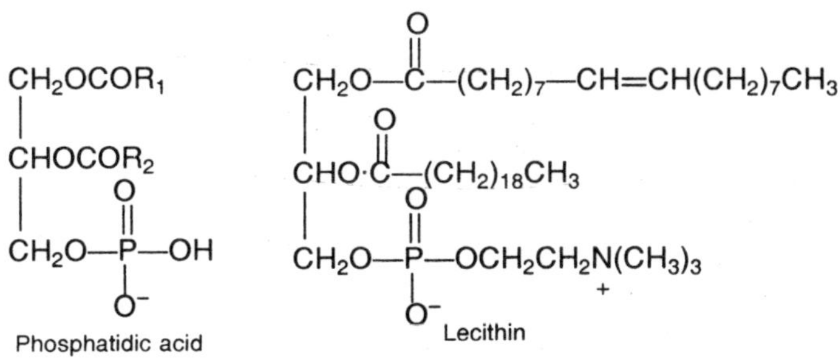

Fig. 7.10 Phospholipids

4. Carbohydrates

Carbohydrates are based on the empirical formula CH_2O. The simple monomeric carbohydrates are termed *monosaccharides* (Fig. 7.11). The latter are subdivided into pentoses, if the molecule has five carbon atoms and hexoses, if the molecule has six carbon atoms. D-Ribose, present in the RNA and D-2-deoxyribose, present in the DNA are the examples of pentoses. D-Glucose, a hexose, plays an important role in cellular metabolism. However, within the cell, glucose is stored in a polymerized, insoluble form—as glycogen in animal cells and as starch in plant cells. Such polymerized carbohydrates are called *polysaccharides.* There are enzymes in the cells that are able to digest these insoluble food reserves, mobilizing them for use when the need arises.

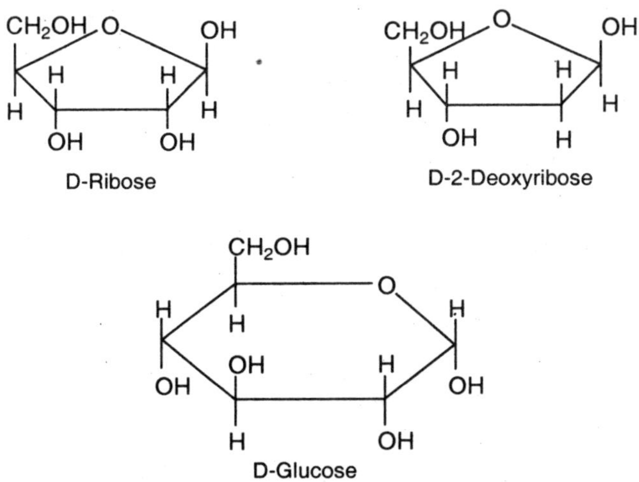

Fig. 7.11 Monosaccharides

Apart from their nutrient value, the polysaccharides contribute to the structure of the cell. The cell wall (of the plant cells) is composed of *lipopolysaccharides,* the heteropolymers based on repeating pentose and lipid structures. Similarly, the conjugates of proteins and polysaccharides, called *glycoproteins,* are the constituents of cell membrane.

5. Electrolytes

Electrolytes are required in the nutrient media for all cells. The importance of electrolytes can be judged from the fact that when cells are immersed in distilled water, they lose their salt content and die. Further, in human beings, loss of sodium ions from the cells, following excessive perspiration, leads to the leakage of potassium ions into the extracellular fluid. This induces muscle cramps. An intake of sodium chloride causes potassium ions to re-enter the cells and reduces the cramps.

The important cations present in the cells are sodium, potassium, magnesium and calcium. The important anionic constituents of the cells are chloride, phosphate and bicarbonate. In addition, certain ions, such as iron, copper, fluoride and iodide, are present in trace amounts and are essential for enzymatic activity.

Structure of the Cell

Cellular organisms are classified as *eukaryotes* and *prokaryotes.* Eukaryotic cells are generally larger than prokaryotic cells. All complex, multicellular organisms, including all animals and plants as well as many unicellular species are eukaryotes, whereas all bacteria are prokaryotes. Further, eukaryotic cells have a nucleus, while prokaryotic cells do not have a nucleus. It is thus obvious that eukaryotic cells are more complex in their structure as compared to the prokaryotic cells. We shall, therefore, present an outline of the eukaryotic cell structure, a graphical representation of which is depicted in Fig. 7.12. The cell consists of three important components: The cell membrane, the cytoplasm and the nucleus.

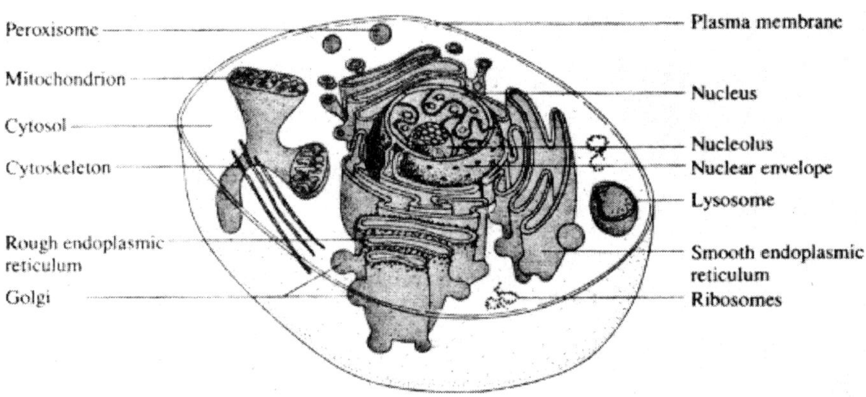

Fig. 7.12 Graphical respresentation of a eukaryotic cell

The Cell Membrane

The cell is encased by a plasma membrane, called the cell membrane. It is composed of a double layer of lipids, proteins and glycoproteins. The plant cells have a cell wall in addition to the cell membrane. The cell wall is composed of lipopolysaccharides.

The Cytoplasm

Beyond the cell membrane is the cytoplasm. It is composed mainly of proteins and RNA. The liquid part of cytoplasm is called *cytosol*. The latter consists of water and water-soluble constituents, including proteins, salts, minerals and cellular nutrients.

Along the inner surface of cell membrane, the cytoplasm contains a set of microtubules and microfilaments called *cytoskeleton*. The cytoskeleton provides stability to the cell membrane and hence maintains the cellular structure. Cytoskeleton also plays a prominant role in bringing about morphological changes that accompany cell division.

The cytoplasm also contains a number of fine structures called *organelles*, that appear as mere granules. The principal organelle is the *mitochondria*, whose function is to produce adenosine-5′-triphosphate, ATP (Fig. 7.13). The energy of the cell is stored in the ATP and, as and when required, made available to the energy consuming processes. Synthesis of ATP is catalysed by enzymes that are embedded within the mitochondria.

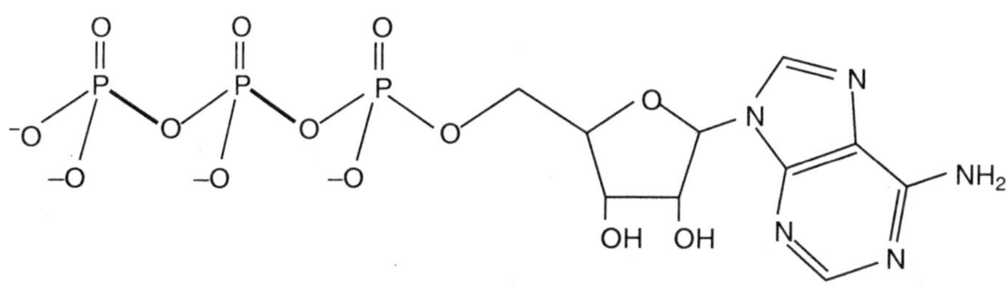

Fig. 7.13 Adenosine 5′-triphosphate

Another group of organelles within the cytoplasm are the *endoplasmic reticulums*. These are engaged in protein and steroid-hormone biosynthesis. Those which are involved in protein biosynthesis contain a large number of ribosomes. The ribosomes impart to the organelle a rough texture and hence it is called *rough endoplasmic reticulum*. Those endoplasmic reticulums which are involved in steroid-hormone biosynthesis contain a fewer ribosomes and do not appear to be rough. They are, therefore, termed *smooth endoplasmic reticulums*.

The cytoplasm contains two other organelles, *lysosomes* and *peroxisomes* which decompose foreign substances that invade the cell. These organelles contain biodegradative enzymes which eliminate the cellular debris. The lysosomes have hydrolytic enzymes and degrade the unwanted substances by hydrolysis. Peroxisomes contain oxidative enzymes and get rid of harmful substances by oxidising them to harmless substances.

The *Golgi complexes* or *Golgi apparatus* are the organelles which serve as the site for the combination of proteins and polysaccharides, producing glycoproteins. The proteins synthesized in the endoplasmic reticulum pass to the Golgi complexes, where the locally synthesized polysaccharides become conjugated to them.

The Nucleus

The nucleus is a complex structure contained in the cytoplasm. It houses the genetic material and represents the chief site for the storage, replication and expression of genetic information. The nucleus is surrounded by a double phospholipid membrane called the *nuclear envelope.* The nuclear envelope contains pores called *annuli,* which open to the cytoplasm and allow various chemical constituents to move in either direction.

Every cell, sooner or later, must undergo division, that is, a mother cell grows and gives rise to two daughter cells. The crucial event in this process is the equal division of bodies in the nucleus called *chromosomes.* The chromosomes are the agents of heredity. During their self-replication, they convey to the daughter cells all the capacities of the mother cell. The principal constituent of the chromosomes is the DNA, bound to proteins called histones. The genetic information is encoded in the DNA. The sequence of nucleotide base pairing in the chromosomal DNA is called a *gene.*

Each nucleus contains a number of comparatments called *nucleoli.* The nucleolus is the active center for the synthesis of ribosomes. The ribosomes then traverse the nuclear envelope, through the annuli, and enter the cytoplasm, where they play a central role in protein biosynthesis.

How a Cell Functions

In subsequent chapters we will learn that many of the hazardous effects of environmental pollution arise from the disruption of cellular functions. Therefore, before we conclude this chapter, we would briefly outline the functioning of a typical cell (Fig. 7.14).

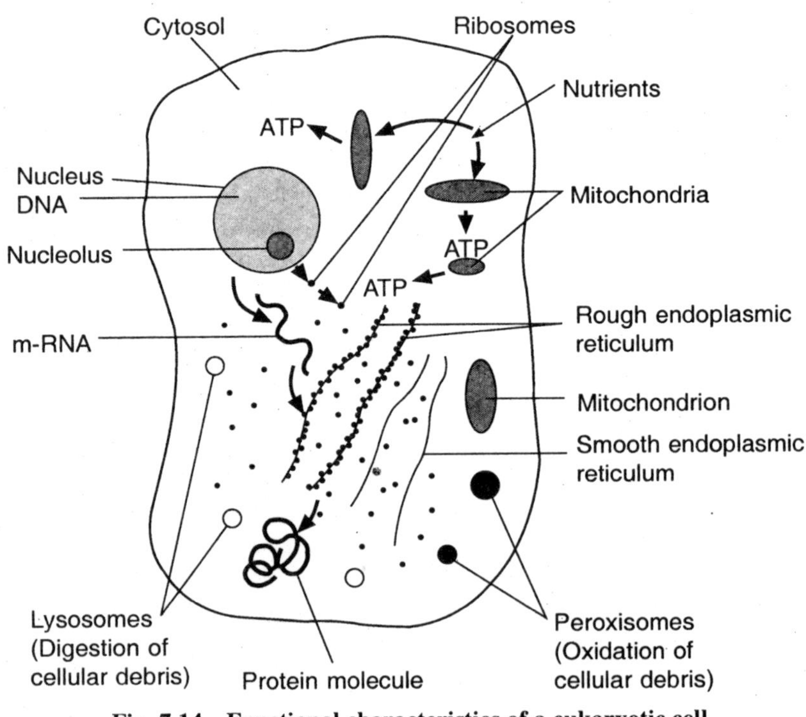

Fig. 7.14 Functional characteristics of a eukaryotic cell

All cellular functions require energy. Therefore, one of the most fundamental biochemical function of any cell is the production of energy in forms that are useful for sustaining the cell. Sustenance of the cell requires continuous biosynthesis of cellular constituents. The energy for these biological processes is derived from the nutrients that enter the cell.

Within the cell the nutrients are metabolized by the mitochondria and the energy so released is stored as chemical free energy in the molecules of adenosine-5′-triphosphate. The clevage of the phosphoanhydride bonds (shown as dark lines in Fig. 7.13) of the ATP converts the chemical free energy into biological useful forms.

Using this energy, DNA is synthesized in the nucleus and RNA in nucleoli. The r-RNAs migrate to the cytoplasm and become associated with endoplasmic reticulum. There they act as the sites for protein biosynthesis. The m-RNA, produced by transcription from DNA in the nucleus, carries the genetic code to the ribosomes. The genetic code is translated to a molecule of protein by the polymerization of amino acids, which are brought to ribosomes by t-RNA. The unwanted cellular debris is degraded by lysosomes and peroxisomes.

References

Abeles, R.H., Frey, P.A. and Jencks, W.P. (1992), *'Biochemistry'*, Jones and Bartlett Publishers, Boston, p. 2–11.

Alberts, B., Bray, D., Lewis, J., Raff, M., Roberts, K. and Watson, J.D. (1994), *'Molecular Biology of the Cell'*, Garland Publishing, Inc., New York, p. 3–40.

Bada, J.L. (1995). Origin of homochirality, *Nature (London)*, **374,** 594–595.

Brachet, J. (1961). The living cell, *Sci. Amer.*, **205**(3), 51–61.

de Duve, C. (1996). The birth of complex cells, *Sci. Amer.*, **274**(4), 38–45.

Dickerson, R.E. (1978). Chemical evolution and the origin of life, *Sci. Amer.*, **239**(3), 62–78.

Ferris, J.P. (1995). A theory of macromolecular and cellular origins, *Nature (London)*, **205,** 328–340.

Giese, A.C. (1979), *'Cell Physiology'*, 5th Edn., W.B. Saunders Co., Philadelphia, p. 7–34.

Gogarten, J.P. (1995). The early evolution of cellular life, *Trends Ecol. Evol.*, **10,** 147–151.

Ingle, M.R. (1985), *'The Eukaryotic Cell'*, Basil Blackwell, Oxford, p. 41–49.

Oparin, A.I. (1968), *'Genesis and Evolutionary Development of Life'*, Academic Press, New York, p. 101–126.

Pain, R.H. (1983). Dynamic proteins, *Nature (London)*, **305,** 581–582.

Ponnamperuma, C. and Mac Dermott, A.J. (1994). Cosmic asymmetry: The meaning of life, *Chem. Brit.*, **30,** 487–490.

Watson, J.D. and Crick, F.H.C. (1953). A structure for deoxyribose nucleic acid, *Nature (London)*, **171,** 737–738.

CHAPTER 8

Cellular Alterations

The adult human body consists of a complex community of about 100 billion cells. In a healthy organism each cell performs its functions perfectly—in a way that is integrated with the common cause. However, when living cells interact with certain hazardous environmental agents—physical, chemical and biological—their normal constitution is altered. The cells then stray away from their appointed functions. When alterations occur in the chromosomes, the process is called mutagenesis. Of particular importance are the mutations occurring in the germ cells, since these are passed on to the next generation and are retained in the population. When alterations occur in the developing embryo, the process is called teratogenesis. When alterations induce the cells to divide at an abnormally high rate, the process is called carcinogenesis. We shall discuss these malfunctions separately.

Mutagenesis

A mutation is an alteration of the base sequence in cellular DNA. The alteration is capable of being replicated in the genetic material of the organism, leading to inherited differences. An agent which causes these abnormalities is called a *mutagen.* The train of events which culminate in a mutation is called mutagenesis.

The genetic information of a cell is carried by DNA which resides in the chromosomes. As has been stated, the DNA double helix consists of two strands of polydeoxynucleotides, which are held together by hydrogen bonds between the complementary bases (Chapter 7, p. 112). These bases are the purines, adenine (A) and guanine (G) and the pyrimidines, thymine (T) and cytosine (C). In the DNA they are arranged in four base pairs: A-T, T-A, G-C and C-G.

The information contained in the DNA is determined by the precise sequence of the four base pairs and is preserved by its precise duplication. An alteration in the base sequence causes an alteration in the genetic information. The ensuing mutation causes changes in several aspects of development, of structure and of functions of the organism.

Types of Mutations

Mutations are classified according to the type of structural changes that they impose on DNA. There is a continuum in size of the mutations.

At the lower end, with small structural alterations, are the *point mutations*. These involve changes in a single base pair. Point mutations are subdivided into two classes, known as transitions and traversions. In a transition, a purine is replaced by a purine and, at the same time, a pyrimidine is replaced by a pyrimidine. For example, A-T to G-C or G-C to A-T. In a transversion, a purine is replaced by a pyrimidine and a pyrimidine by a purine. For example, A-T → T-A or A-T → C-G.

Mutations can also result from addition or deletion of one or more base pairs. Such mutations are of intermediate size and are called *frameshift* or *sign mutations*. These mutations frequently occur in the DNA sequence that encodes polypeptide synthesis.

At the upper end of the size spectrum are the *chromosomal aberrations*. These involve alterations in the quantity of chromosomes in the cell nuclei or a rearrangement of chromosomal material without altering its total amount. A number of physical agents like intense radiation and chemical agents like methylmercury ion break the chromosomes. This generally leads to cell death. However, in most cells, broken chromosomes can heal by joining two broken segments. During rejoining, the original chromosome undergoes reformation or rearrangement. As a result of these changes mutations arise. For this reason, all known chromosome breaking agents are classified as mutagens.

Mechanisms of Mutagenesis

A mutation which occurs in the absence of a mutagen is called *spontaneous mutation*. If it results from exposure to a mutagenic agent, it is called *induced mutation*. In this text of environmental chemistry we are more concerned with induced mutations.

Physical agents like X-rays and ultraviolet radiation, as well as a host of chemical agents produce changes in the DNA framework. The structures of some common chemical mutagens are depicted in Fig. 8.1. All mutagens do not act by the same mechanism. Structural alterations leading to mutagenesis may arise in a number of ways.

1. Incorporation of Base Pair Analogs into DNA

Mutations may arise because of the incorporation of a base analog into DNA. For example, 5-bromouracil can be incorporated into DNA in place of thymine. As has been stated, thymine in the DNA pairs-up with adenine. However, 5-bromouracil prefers to become attached to guanine. Hence in presence of 5-bromouracil, the normal A-T interactions give way to abnormal 5-bromouracil-guanine interactions.

2. Deamination of DNA

Sodium nitrite, a common food additive (Chapter 28, p. 395), when exposed to low pH, produces nitrous acid. The latter deaminates the bases that make up the DNA molecule. For example, nitrous acid causes deamination of cytosine to uracil and adenine to hypoxanthin. The conversion of adenine to hypoxanthin induces A-T → G-C transitions since hypoxanthin pairs-up preferentially with

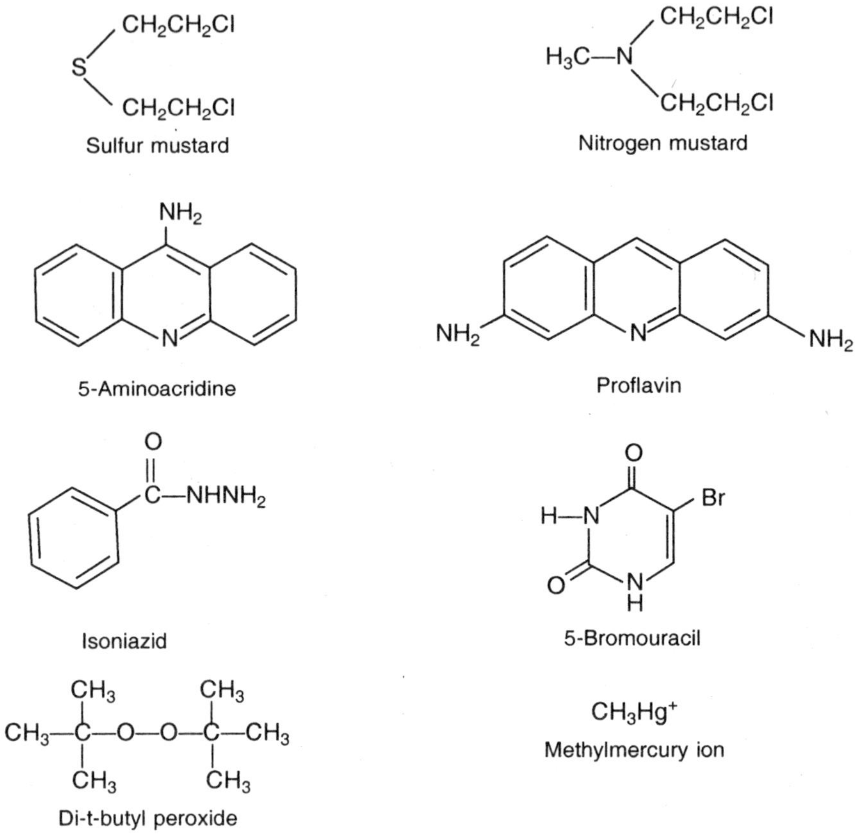

Fig. 8.1 Chemical mutagens

cytosine. Similarly, the deamination of cytosine to uracil causes G-C → A-T type transitions since uracil forms base pairs much more readily with adenine than guanine.

3. Alkylation of DNA

Alkylating agents constitute the largest class of potential mutagens. They were amongst the first chemicals recognized as mutagens. Examples include nitrogen mustard and sulfur mustard. These chemicals bring about the alkylation of the DNA's base components. Guanine is most easily alkylated. Upon alkylation it prefers to pair-up with thymine, rather than cytosine, producing transition type of mutations. Alkylated guanine also prefers to pair-up with another guanine segment present in the same strand of the DNA. In this process, the G-C link is replaced by alkylated guanine-guanine link, producing transversion type of mutations.

4. Free Radical Attack on DNA

Ionizing radiation, ultraviolet light, X-rays, peroxides (for example, hydrogen peroxide and di-t-butyl peroxide) and hydrazine derivatives (for example, isoniazid) produce reactive free radicals in the body. These radicals then attack the DNA molecules and open many chemical bonds. Reactive pyrimidine and purine analogs are produced. When the bonds are reformed, the normal pairing sequence is not necessarily maintained. For example, a reactive pyrimidine moiety would prefer to recombine with another pyrimidine, rather than pairing up with a purine. Thus DNA exposed to free radical attack contains pyrimidine dimers instead of normal purine-pyrimidine links.

5. Intercalation of DNA

A number of chemical compounds bind physically with DNA, inducing structural alterations. During these interactions no new chemical bonds are formed. For example 5-aminoacridine and proflavin become physically intercalated between adjacent base pairs of the DNA double helix. As a result of intercalation, the length of the DNA molecule increases. Moreover, a certain amount of uncoiling or unwinding of the helix occurs to provide space for the intercalating agent. These structural changes produce frameshift mutations.

6. Interaction of DNA with Metals

Transition metals like iron and copper catalyse the decomposition of peroxides or hydrazines into free radicals and thereby enhance the rate of mutation. Moreover, heavy metals like mercury tend to bind to the purines in the DNA, inducing chromosomal aberrations.

Teratogenesis

A *teratogen* is a physical, chemical or biological agent that causes birth defects. *Teratology* is the study of the causes and development of disabling and lethal birth defects and congenital malformations that are present at or soon after birth. The cellular changes leading to congenital defects are called teratogenesis. The defect may be functional, such as lack of a limb, histological, such as lack of a tissue or biochemical, such as lack of an enzyme.

Teratogens act by causing damage to cells or by despoiling interaction between tissues of developing embryos. In contrast, mutagens produce changes in genetic materials. While most mutagens are also teratogens, the reverse is not necessarily true. Whether a given agent is teratogenic depends on its physical and chemical characteristics, the developmental stages at which it is appropriated, the dose, the genotype of the mother and interaction with other environmental factors.

For a very long time it was advocated that uterus was an impregnable barrier protecting the embryo from environmental insults. Any congenital malformation was considered to have arisen because of genetic and not environmental factors. However, in 1941, an Australian ophthalmologist, Gregg, noted that a large number of children with congenital cataracts were born to women who were infected with rubella (German measles) during pregnancy. Deafness and heart defects were soon recognized as additional teratogenic effects of maternal rubella. A biological agent, the rubella virus, was identified as the teratogen. During this period, the adverse effects of a physical teratogenic agent, radiation, also came to light. Exposure of pregnant women to high doses of radiation often resulted in mental retardation amongst the offsprings.

Despite rubella and radiation, the comfortable certainity that the uterus provided an essentially impermeable barrier to toxins continued to persist. This complacency was shattered, once and for all, when a chemical teratogen, thalidomide, caused a grave tragedy.

Thalidomide Episode

Thalidomide was a sedative and analgesic that was heavily advertised in Germany as being totally safe and non-toxic. Its purported harmlessness made it particularly attractive for the relief of nausea, tension and sleeplessness during pregnancy. It was marketed 'over the counter', that is, no prescription was required for its use. Unfortunately, thalidomide was an inadequately tested drug. Its screening covered only adult animal species and not pregnant animal species.

Between 1959 and 1962 approximately 8000 infants with severe malformations were born to women who had consumed thalidomide during early stages of pregnancy. In some cases the whole limb was missing, with the hand or foot attached directly to the body. This condition is referred to as *phocomelia* or seal limbs. In other cases children were born with paralysed faces and deformed ears. Internal malformations like defective hearts were noted in some of the children. About one-third of the offsprings died in early infancy, primarily due to internal malformations.

The drug was soon taken off the market. But by then irrepairable damage had been done. Nevertheless, thalidomide episode exposed the vulnerability of the human fetus to environmental hazards. An emphasis was now placed on the reproductive testing of drugs, pesticides and other environmental chemicals to which pregnant women might be exposed. As a result, the teratogenic effects of many other chemical agents became apparent. The structures of some important chemical teratogens are displayed in Fig. 8.2.

Classification of Teratogens

As has been stated, teratogenesis results by the action of physical, chemical or biological agents. These agents can cause adversities only if they reach the developing tissue of the embryo, crossing placenta—a membranous tissue which nourishes the fetus through umblical cord. Some common teratogenic agents, along with their adverse effects are listed in Table 8.1.

1. Physical teratogens

The most important physical teratogen is the radiation which, because of its high penetration power, reaches the embryo and disrupts the cellular mechanism. *Hypothermia,* the condition when the body temperature becomes abnormally low, freezes the biochemical reactions of the growing embryo cells and hence acts as a teratogen. Repeated maternal fever during pregnancy causes *hyperthermia,* a condition when the body temperature becomes abnormally high. As a result, the biochemical reactions of the growing embryo cells become too rapid, causing malformations. *Anoxia,* a condition in which reduced levels of oxygen penetrate the embryo, prevents normal growth of fetus and hence acts as a teratogen.

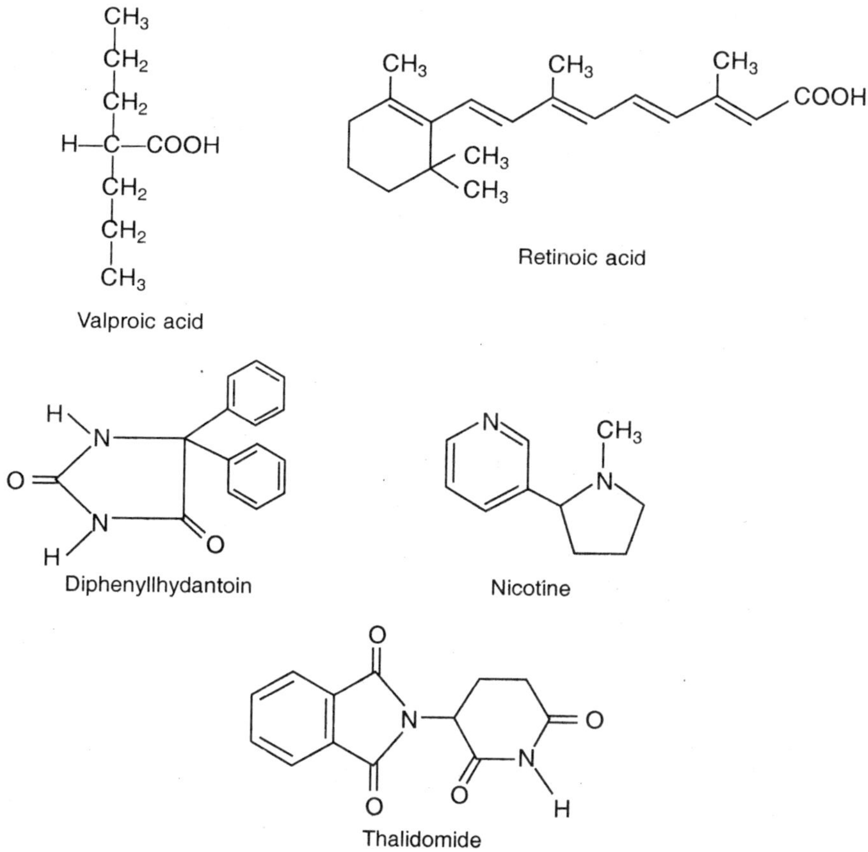

Fig. 8.2 Chemical teratogens

2. Chemical teratogens

A chemical agent acts as a teratogen only if it passes through placenta and reaches the fetus. A fetus does not possess the degradative ability to metabolize environmental chemicals. Hence, even those reagents which are not noxious to adults, because they are rapidly metabolized, may well affect the fetus.

3. Biological teratogens

Biological teratogenic agents include viruses and bacteria. These cross the placenta and decrease the pregnancy progress by interfering with the biochemical cellular reactions.

Mechanisms of Teratogenesis

The natural sequence of development is altered by the teratogens through the following mechanisms.

1. Selective cell death

Most of the teratogens act by disrupting the cell division. Since the cells of the

Table 8.1 Some common teratogenic agents and their effects

Agent	Reason of/for Exposure	Malformation(s)
Physical Teratogens		
Anoxia	Cigarette smoking	Skeletal malformation
Hyperthermia	Repeated maternal fever	Microphthalmia; Seizures
Radiation	Repeated X-rays; Radioactive fallout	Microcephaly (small head)
Chemical Teratogens		
Alcohol	Addiction	Psychomotor retardation; Hyperactivity; Distractibility
Diphenylhydantoin	Anti-epilepsy drug	Wide spaced eyes; Short nose; Deformed fingertips and nails; Cleft lip; Heart defects
Nicotine	Cigarette smoking	Skeletal defects
Retinoic acid	Vitamin A precursor	Deformed ears and brain
Valproic acid	Anti-epilepsy drug	Neural tube defect; Heart malformation
Biological Teratogens		
Herpes virus	Infection	Brain damage
Rubella virus	Infection	Cataract
Syphilis	Infection	Blindness

embryo multiply very rapidly, interference with cell division results in selective cell deaths. The fall in the number of cells leads to the failure of inductive sequences. For example, if ureter does not grow, the kidney does not develop. If the cell division in the eyes does not proceed to an optimum level, the retinal tissue too does not develop completely.

2. Altered biosynthesis

The DNA monitors the protein biosynthesis in the cell. Teratogenic agents interfere with the ability of DNA to give signal for the requisite biosynthesis. Altered biosynthesis prevents the differentiation of tissues in the embryo, causing structural or functional malformations.

3. Energy inhibition

The embryonic cells must possess sufficient energy to grow and multiply, to draw nutrients from the mother and to metabolize, as far as possible, the unwanted substances. A teratogen may prevent the cells from receiving the required amount of energy. For example, under anoxic conditions, the embryo cells receive lower concentrations of oxygen and hence produce lesser amount of energy. The differentiation of organs slows down and malformations result.

Factors Affecting Teratogenesis Mechanisms

The conditions leading to teratogenesis depend on the following factors.

1. Developmental stage

Susceptibility to teratogenic attack varies with the development stage of the embryo. The gestation may be roughly divided into three time periods: *Preimplantation,* which begins at conception and lasts for about a fortnight; *organogenesis,* which lasts from approximately sixteenth to seventy fifth day; and *fetal period,* which extends beyond the seventy fifth day.

During preimplantation, the embryo consists of relatively few cells, most of which are *totipotent,* that is, if they are moved to a different site within the embryo, they will become what their new position dictates. Since the cells are uncommitted, they do not undergo any serious damage by way of teratogenesis. The fetal stage is marked by growth and maturation. The differentiation of the organs is nearly complete. The developing fetus can, to some extent, metabolize external agents and, therefore, this stage too is less sensitive to teratogenesis.

As shown in Fig. 8.3, the organogenesis period is most sensitive to teratogenic effects. For example, it has been observed that the risk of teratogenic defects attributed to rubella is 90% if the infection occurs within the first 75 days of pregnancy; it falls to 25% thereafter. Likewise, the chances of congenital cataract is 50% if the mother becomes infected with rubella during 15th to 30th day of pregnancy; it falls to 5% after 120 th day.

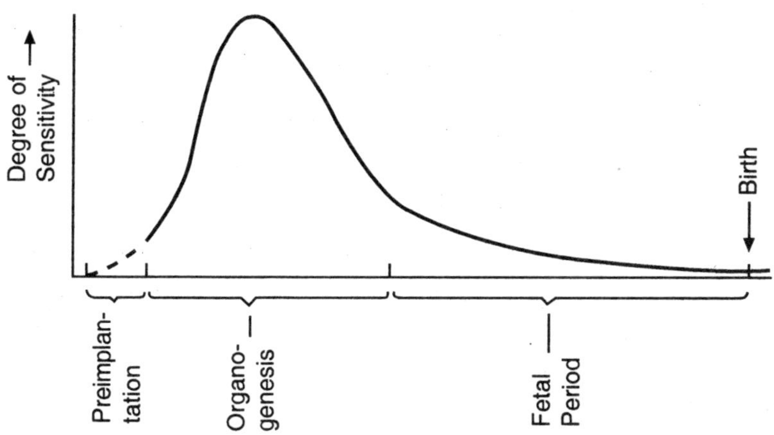

Fig. 8.3 Vulnerability of teratogenic effect with respect to developmental stage of the embryo

Organogenesis is characterized by rapid growth and differentiation of most organ systems. Extremely rapid cell division leaves the embryo particularly susceptible to exogenous agents. Generally, cytotoxicity or lack of neutrients causes cell deaths. However, within the embryo, the cell division is so fast that the cells continue to survive, though they get damaged. The result is a deformed child.

Within the organogenesis period, each organ has its own critical time of vulnerability. The relationship between the time of exposure to thalidomide and the malformation of various organs is shown in Table 8.2.

Table 8.2 Relationship between the time of exposure to thalidomide and the malformation

Organ Deformity	Period of Ingesting Thalidomide (days)
Heart defect	27–29
Deformation of thumb	38–43
Deformation of arms	38–44
Phocomelia of arms	38–46
Dislocation of hips	38–48
Deformities of ears	39–43
Deformities of legs	41–45
Phocomelia of legs	42–48

2. Hereditary predisposition

The susceptibility of an organism to a teratogenic agent depends on the fetal genotype. Thus, a teratogenic agent, cortisone, produces cleft palate in A/J strain of mouse to an extent of 100%, while an equal dose produces the deformity in only 20% of the C57BL/6 strain. This shows that there exists a genetically determined difference in developmental patterns.

Not all the pregnant women who ingested thalidomide during vulnerable period of organogenesis produced malformed infants. The gene-type of some was strong enough to ward off the ill effects. Can this be considered a blessing? For the individual, yes; for the society, no. The more conspicuous the ill effect, the more easily is it detected and more rapidly is the teratogen phased out. Had thalidomide produced no effect other than reducing the children's IQ by a few points, even the most alert physician would have failed to recognize the cause and effect relationship. The society would have continued to be flooded with children whose mental rating would have been below the natural levels.

3. Clinical prehistory

There are several maternal conditions that can increase the probability of malformations. As has been stated earlier, the incidence of rubella during pregnancy can lead to congenital cataract. Likewise, herpes simplex infection during pregnancy can cause brain damage in the child. The frequency of heart and neural tube defects is high in the offsprings of a diabetic mother. Mothers suffering from phenylketonuria give birth to children with heart defects and mental retardation.

4. Synergism

It is too often the case that a woman takes extensive medication from the beginning of pregnancy until the birth of the child. However, it must be noted that drugs taken in combination may produce malformations even though the drugs taken singly would have no or little teratogenic effect. This condition is called synergism. For example, diphenylhydantoin, an anti-epilepsy drug, is known to cause cleft lip and congenital heart disease. Its teratogenic effect increases if phenobarbital, another anti-epilepsy drug, is simultaneously taken.

5. Dosage

In contrast to mutagenesis and carcinogenesis, teratogenesis is a threshold phenomena. The *teratogenic threshold* is defined as the dose at which the agent induces malformations. A dosage below the threshold causes no defects. A really vicious teratogen is one that has a threshold for maternal toxicity well above the threshold for fetal toxicity. That is, it would cause an ailment in mother only when a very high dose is ingested, but would cause a deformity in the child at a very low dose. Under these circumstances, the cautious respect given to a toxin is absent. The agent is considered harmless because it does not affect the adults and, therefore, hardly any preventive measures are adopted. Thalidomide and diphenylhydantoin are examples of the drugs which have low adult toxicities but are otherwise potent teratogens.

Prevention of Teratogenesis

The parents of a child with a birth defect usually have a feeling of guilt. For example, the general attitude of the women who took to thalidomide may be summed up in the statement, "If only I had possessed more will power, my baby would have been normal".

Most of the teratogenic effects are related to environmental exposure and are, therefore, avoidable. What is required is a more cautious approach towards diet, medication and life style. However, the preventive measures should start early—earlier than conception. Most of the organs are vulnerable to teratogenic attack about a month after conception—a time when the mother is just becoming convinced that she is pregnant. She then decides to be more cautious; but by then it is too late.

1. Cautionary diet

It has long been realized that the nutrition of the mother is reflected in the child and that an adequate diet during pregnancy is important. With relevance to birth defects it has been noted that the fetus is less susceptible to teratogens if the mother is well nourished and healthy. Teratogens manifest more severe effects in malnourished women. Inadequate proteins and minerals in the diet can cause death of fetus, spontaneous abortion or retarted growth. Excessive intake of vitamin A should be avoided by pregnant women. It is known to cause deformities in ears, brain and heart of the offspring.

2. Cautionary medication

A pregnant women should take all possible preventions against viral and bacterial infections. This means not only getting vaccinated against diseases like rubella, but also avoiding contact with infected persons. It was observed that nurses who attended to the infants with rubella syndrome themselves produced affected children; the virus in the rubella syndrome children had persisted for months after birth and then infected the nurses.

All drugs and antibiotics should be avoided unless they are absolutely necessary. It must be remembered that with regard to teratogenesis, the basic frustration is expressed by *Karnofsky's law* which states that every chemical, if given in the right dose, to the right species, at the right (read wrong) time, is teratogenic. Streptomycin taken during pregnancy causes fetal damage to ear, producing deafness; tetracyclin causes cataract in offsprings; penicillin increases the incidence of miscarriage.

Women with epilepsy present a paradoxy. On one hand all anti-epileptic drugs are teratogenic and, on the other, these drugs have to be taken on a long term basis; they cannot be taken off even during pregnancy. Therefore, management of an epileptic woman who wants children is difficult. The best approach is to be sure she knows about the teratogenic risk before she gets pregnant. If she does conceive, she should be kept on the lowest possible dose, preferably of just one drug. The blood level of the drug should be continuously monitored during pregnancy and should not be allowed to overshoot.

3. Cautionary lifestyle

Cigarette smoke is a potent teratogen. It causes teratogenesis both by physical and chemical mechanisms. As a physical agent it prevents oxygen from penetrating to the fetus and causes anoxia. The growth of the fetus gets retarted. As a chemical agent it inducts noxious chemicals like nicotine and carbon monoxide to the fetus causing heart defects and skeletal deformities. Smoking is harmful in yet another way. It increases the susceptibility of fetus to attack from other teratogenic agents.

More than 50% of the children born to chronic alcoholic mothers face the risk of teratogenesis. During pregnancy, even 10 g of alcohol a day produces detectable deformities. The malformations found in children of addicted mothers include reduced birth weight, short nose, microcephaly and mental retardation, the last being the most common. After Down's syndrome and spina bifida, alcohol is the third largest contributor to mental retardation—and the only one that is preventable.

Carcinogenesis

Cancer is defined as an abnormal mass of tissue, the growth of which exceeds and is uncoordinated with that of normal tissues, and persists in the same excessive manner after the cessation of the stimuli which evoked the change. The abnormal tissues are called *tumors* or *neoplasms,* meaning new forms. The cells that constitute the tumors are called *neoplastic.* The sequence of events that results in a cancer is called carcinogenesis.

Cancer is not a single disease. It is a generic term which includes more than a hundred forms of ailments. Normally, cancers are described by the organs where they originate, for example, stomach cancer or lung cancer. However, they may also be classified on the basis of the tissue from which they originate. Thus, cancers which start from epithelial tissue—the internal or external lining of body surface—are called *carcinomas.* Skin cancer is a form of carcinoma. Tumors arising from the body's structural tissues are known as *sarcomas.* Bone cancer is a form of sarcoma. The classification of cancers on the basis of tissues is presented in Table 8.3.

How Cancer Arises

Normal, healthy cells tend to undergo *proliferation,* that is, they grow in number

Table 8.3 Classification of tumors

Tissue	Tumor
Epithelial	Carcinoma
Connective	Sarcoma
Blood cells	Leukemia
Lymphatic system	Lymphoma

through cell division. When a normal cell divides, the resulting daughter cells are identical to each other. This means that they have the same amount and type of DNA in them. Therefore, before a cell can divide in normal way, the amount of DNA in it has to be duplicated.

Like the normal cells, the cancer cells too proliferate. However, unlike the normal cells, the distribution of DNA between the two daughter cells is not equal because of some biological error, which arises most frequently due to environmental factors. The result is that the daughter cells receive chromosomes which are unidentical, either in number, or in structure. Such cells are said to be mutated.

Further proliferation of the altered or mutated cells is controlled by genes that are carried in the chromosomal DNA. A gene specifies the sequence of amino acids that must be linked together to make a particular protein. Two classes of genes play a key role in the uncontrolled cell proliferation seen in human cancers: The *proto-oncogenes* and the *tumor suppressor genes.* Under normal conditions, the former encourage the cells to divide and grow, while the latter inhibit cellular proliferation. In mutated cells, however, both play an undesired role. Proto-oncogenes contribute to cancer by synthesizing growth-stimulating proteins. Tumor suppressor genes contribute to cancer by becoming inactive and depriving the cells of crucial brakes that prevent unwanted growth.

Even when the proto-oncogenes become overactive and the tumor suppressor genes become inactive, a back-up system takes over to ward off cancer growth. The normal cells in the immediate vicinity of the mutated cell begin to synthesize additional quantities of a special protein, called *p 53 protein.* One of the functions of *p 53 protein* is to trigger the mutated cell to undergo *apoptosis*—also called cell suicide. Destruction of a mutated cell is bad for the cell itself but good for the body as a whole.

Most often apoptosis acts as a safeguard against cancer. However, at times the mutated cells evade suicide by inactivating *p 53 protein.* Once they are successful in this venture, they become 'immortal'. They are the ones which subsequently grow into a tumor.

How Cancer Spreads

A tumor which remains at the place of its origin, well demarcated from its surrounding tissue, is called *benign.* A tumor which tends to infiltrate and spread to neighbouring or distant tissues in the body is called *malignant.* The invasion of cancer cells to various sites in the body is called *metastasis.*

The difference between benign and malignant cancer is the crux of the disease.

A surgeon can remove a benign tumor easily but a cancer that has metastasized usually reaches so many organs that cure by surgery alone becomes impossible. Matastasis is what makes cancer lethal—and dreaded.

To metastasize successfully, cancer cells have to detach from their original location, invade a blood or lymphatic vessel, travel to a distant site and establish a new cellular colony. Normal cells, apart from blood cells, cannot float away from their tissue and establish themselves somewhere else. The reason being that normal cells adhere both to each other and to an insoluble meshwork of protein (filling the space between them) called *extracellular matrix.*

Cell-cell adhesion is affected chiefly by a substance called *E-cadherin*. This substance dissolves away in cancer cells. Adhesion of cells to extracellular matrix allows the cells to survive and proliferate. Unattached cells stop growing and undergo apoptosis because a nuclear protein, called cyclin E-CDK2 complex, that regulates the growth and division of cells, becomes inactive. Cancer cells get around this requirement. The cyclin E-CDK2 complex in such cells stays active whether the cells are attached or not. The reason being that the proto-oncogenes relay a false message to the nucleus that the cell is properly attached when it is not, thereby preventing the cells to die through apoptosis.

The detached, surviving cancer cells pass into the blood stream. They often get trapped in the vascular bed, that is, the network of blood capillaries. The first vascular bed encountered by blood leaving most organs is in the lungs. Accordingly, the lungs are the most common sites of metastasis. However, circulation in blood is not the only cause of metastasis. The constituents of the cancer cells show affinity for specific tissues. This explains the predilection of the cancer cells to migrate selectively. For example, the prostate cancer cells usually metastasize to the bones.

The latent period between a cell becoming mutant and metastasis runs into decades. For this reason the signs and symptoms of cancer generally become evident in the old age. Yet, in some cases the disease strikes the young. This is explained by the inheritance from one or the other parent of a mutant cancer causing gene. Colon cancer is an example of familial cancer syndrome.

What Causes Cancer

A single, genetically disturbed cell has the potential to promote cancerous growth and metastasis. The question is how, in the first instance, the cell becomes genetically altered. It has been estimated that upto 80% of the cancer cases arise because of environmental carcinogens—the cancer causing agents. Here the term environment is used in a broad sense to include diet, lifestyle and exposure to radiation, chemicals or viruses. Apart from virus, the other environmental stresses generate free radicals in the body. The free radicals interact with gene's DNA molecules. The molecules become disrupted and the genes mutated. Viruses promote cancer by accelerating the rate of cell division. Some common types of cancers, along with the associated risk factors are listed in Table 8.4.

1. Diet

Our food contains many chemicals that are possible carcinogens or that can be

Table 8.4 Some common cancers

Type	Risk Factors: Occupational	Risk Factors: General	Warning Signs
Bladder	Workers dealing with rubber, leather, textile and metal works	Cigarette smoking	Blood in urine; Pain during urination; Frequent urination
Breast	—	Early onset of menstruation; Never having children; Having first child after age 30; High fat diet	A painless lump in breast; Change in colour, shape or texture of breast often accompanied by discharge
Colorectal	—	Physical inactivity; Low fibre diet	Blood in stools; Stomach discomfort; Weight loss
Kidney	Coke-oven workers; Asbestos workers	Cigarette smoking; Drinking	Blood in urine; Lump in the area of kidney; Back pain; Sometimes high blood pressure
Leukemia	Exposure to ionizing radiation and aromatic hydrocarbons	Exposure to radio-activity from a damaged nuclear reactor; Genetic abnormalities	Fatigue; Paleness; Weight loss; Repeated infections; Nosebleeds
Lung	Workers dealing with asbestos, arsenic; Miners	Cigarette smoking	Persistent cought; Sputum streaked with blood; Wheezy breathing; Chest or shoulder pain; Swelling in face or neck; Recurring pneumonia or bronchitis
Melanoma	Farmers; Sailors	Excessive childhood exposure to sunlight	A change in the size, colour, texture or shape of a mole; Appearance of a new, abnormal mole; Spontaneous bleeding from a mole
Non-Hodgkins lymphoma	Workers dealing with herbicides	Organ transplants; Virus	Enlarged lymph nodes; Generalized itching; Fever; Night sweat; Weight loss
Ovarian	—	Never having been pregnant; First pregnancy after age 30;	Enlargement of abdomen; Digestive discomfort
Pancreatic	—	Cigarette smoking; Chronic pancreatitis; High fat diet	None, until disease is advanced
Prostate	—	High fat diet	Weak and interrupted urine flow; Painful urination; Back or pelvic pain
Uterine	—	Cigarette smoking; Hypertension and obesity; Never having been pregnant	Abnormal uterine bleeding

changed into carcinogens by digestive enzymes in the stomach and intestine. Barbecued and smoked meats and sausages are implicated in causing cancer of esophagus, the tube connecting mouth and stomach. High incidence of this disease has been noted amongst Iranian and Afghani tribes whose main diet is smoked meat. Carcinogenic chemicals from soot and smoke settle down on smoked food and are subsequently ingested.

High fat diet, particularly red meat, is known to increase the risk of colon and rectum cancers. Red meat has also been implicated in prostate cancer. Excessive intake of salt is associated with stomach cancer. Those who consume a great deal of salted fish show excessive rates of nasopharynx cancer (the upper part of pharynx which reaches the nasal passage). Drinking beverages while they are very hot, including mate, a south American tea-like drink, has been shown to increase the risk of esophagus cancer.

2. Lifestyle

Cigarette smoking causes cancer of lungs, upper respiratory tract, esophagus, bladder and pancreas. Smoking is also implicated in chronic myelocytic leukemia. Whether smoking will result in malignancy depends on several factors, including the frequency of smoking, the cigarettes' tar content and most important—the duration of the habit. Taking up smoking at a young age increases the risk.

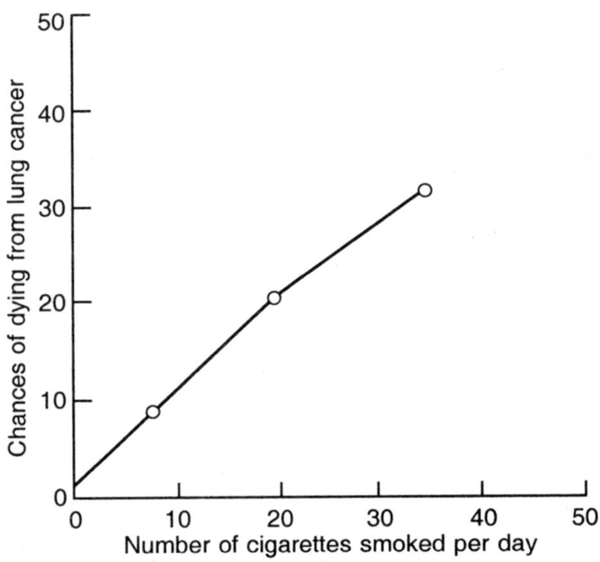

Fig. 8.4 Relationship between smoking and risk of lung cancer

Compared to non-smokers, smokers are twice as likely to be afflicted with cancer of bladder, but eight times more likely to contact cancer of the lungs. The relationship between smoking and risk of lung cancer is depicted in Fig. 8.4. Passive smoking or inhalation of tobacco in the environment also causes lung cancer, though the risk is substantially less. In people who quit smoking, the relative risk of lung cancer begins to fall and by 20 years comes to the level of non-smokers (Fig. 8.5).

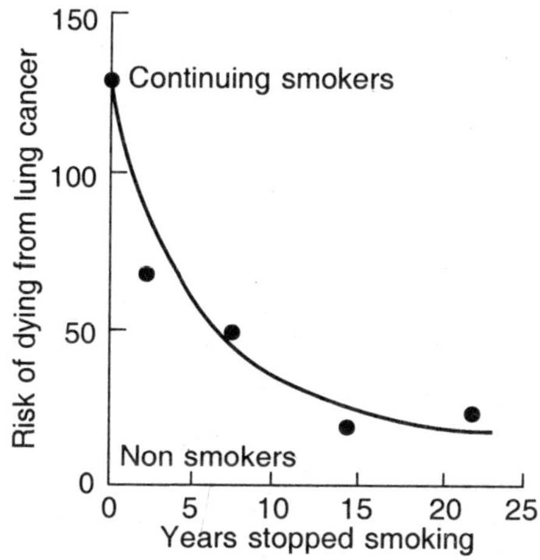

Fig. 8.5 Incidence of lung cancer in smoking quitters.

It has recently been found that babies whose mothers smoke during pregnancy are born with a carcinogen in their bodies. The compound, 4-methylnitrosoamino-1-(3-pyridyl)-1-butanone, has been detected in the urine of newborns whose mothers smoke. This compound is a potent carcinogen and is found only in tobacco smoke. It has not been detected in newborns whose mothers do not smoke. However, in case of smoking mothers, the level of the nitrosamine derivative in newborns is a tenth of those found in adult smokes. The fetus absorbs this compound through the placenta. This continues throughout pregnancy as the fetus grows and develops.

Consumption of large quantities of alcoholic beverages, particularly by smokers, increases the risk of cancer of the upper respiratory and digestive tracts. Alcoholic cirrhosis leads to liver cancer. Amongst the women, the chances of breast cancer increase with alcoholic consumption.

Excessive childhood growth, due to overeating and insufficient exercise pushes girls into menstruating when they are young, and early menstruation is a major risk for breast cancer. Late first pregnancy and late menopause too tend to increase the breast cancer risk. Obesity in adult life causes cancer of endometrium amongst the women.

3. Exposure to radiation

Radiation, both from natural sources and man-made sources, has been implicated in promoting cancer. For example, ultraviolet component of sunlight is a pervasive form of radiation which can damage DNA and cause skin cancer. Thus, agricultural workers in sunny climate are known to suffer from melanoma. Another natural source of radiation is radon, a radioactive, inert gas that is emitted from the earth in some regions. It can seep into buildings, especially in the basement areas. Its concentration in mines is especially high. Prolonged inhalation of the gas increases the quantum of radiation in the body and causes lung cancer.

Exposure to radiation emanating from a damaged nuclear reactor, as in the Chernobyl episode (Chapter 13, p. 210) or from a nuclear bomb, as in case of Hiroshima and Nagasaki disasters, causes leukemia in children and thyroid cancer in adults. Radiofrequency electromagnetic radiation emitted by cellular telephones and microwave ovens are suspected to cause brain cancer, although an empirical cause and effect relationship is yet to be established.

4. Exposure to chemicals

Exposure to a number of chemicals carries a high risk of cancer. Both the inorganic (Table 8.5) and the organic (Table 8.6, Fig. 8.6) chemicals have been implicated in this regard. Prominent industrial chemicals known to cause cancers in humans include arsenic, asbestos, benzidine, 2-napthyl amine and vinyl chloride. Incomplete combustion of fossil fuels and tabacco yields a wide range of polycyclic (e.g., benzo (α) pyrene) and heterocyclic (e.g., nicotine) compounds, which are known to cause cancer. Polycyclic hydrocarbons are also found in broiled foods, as well as in smoke of incinerators, fire-places and auto-exhaust.

Table 8.5 Inorganic Carcinogens

Chemical	Use/Source	Cancer Type
Arsenic	Insecticide, herbicide sprays	Lung, skin
Asbestos	Brake-lining; fire proofing	Mesothelioma
Cadmium	Electroplating; vinyl stabilizers; pigments	Prostate
Chromium	Chrome plating	Lung
Nickel	Metallurgy, refining	Lung, nasal
Radon	Mines, underground structures	Lung

Table 8.6 Organic carcinogens

Chemical	Use/Source	Cancer Type
Benzene	Solvent	Marrow
Benzidine	Manufacture of dyes, polymers	Bladder
Benzo (α) pyrene	Cigarette and coal smoke	Lung
Bis(chloromethyl) ether	Plasticizer	Lung
Creatinine	Cigarette smoking	Lung
2-Napthyl amine	Optical bleach	Bladder
Vinyl chloride	Polyvinyl chloride manufacture	Liver

5. Exposure to viruses

While the other environmental factors promote cancer by breaking apart the helices of DNA, viruses act by an entirely different mechanism. Viruses are the simplest forms of life. They consist of a collection of genes, protected by a protein coat. Since they do not possess other structures found in a cell, they do not have the mechanism to build proteins. Therefore, they cannot reproduce on their own. To do this they have to invade a host cell.

When a virus enters a human body, it attaches itself to the cell membrane. It then transfers its genes into the DNA of the host cell. Instead of duplicating itself, the host cell starts producing the copies of viral genetic material. The virus

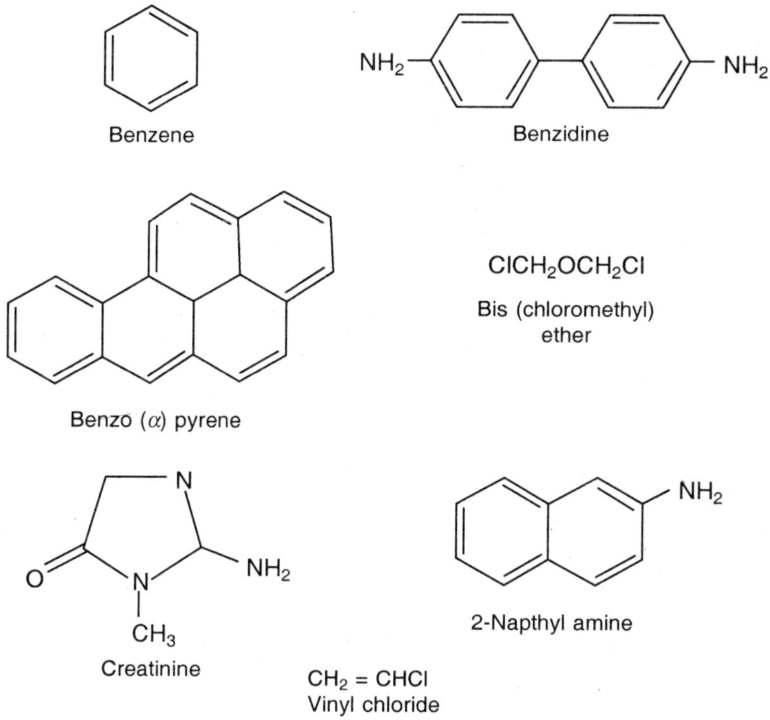

Fig. 8.6 Chemical carcinogens

genes also instruct the host cell to produce their protective protein coat, so that after a while a complete, new collection of virus emerges. These move on to infect other cells. Cancer of liver is caused by hepatitis B and C viruses.

How to Minimize Cancer Risk

A cancer death can be avoided through prevention or through detection of the disease early enough to treat it successfully. As in the case of other ailments, prevention of cancer is better than its detection or cure. The following facts justify this assertion.

1. The 'war on cancer', which aimed at searching for a treatment, has met with limited success and, therefore, should be balanced by a more extensive effort in prevention.
2. Since almost 80% cases of cancer arise because of environmental factors, it follows that malignant diseases are largely preventable. After all, if outside factors can increase cancer risk, avoiding them would decrease it.
3. Prevention avoids the shock and pain of being diagnosed and treated for cancer.
4. Many cancer prevention methods, like regular exercising and balanced diet, reduce the risk of cardiovascular diseases as well.
5. Prevention of the disease is more cost-effective than its cure.
6. Failure of prevention still leaves treatment as a last resort.

The most important strategy to prevent the onset of a number of cancers is not to smoke and, if it is too late, to give up this practice. Excessive alcoholic consumption too should be given up. The routine diet should be high in vegetables, fruits and legumes, such as peas and beans. It should be low in red meat, saturated fats, salt and sugar. Carbohydrates should be consumed largely as wheat bread. Cooking fats should come mainly from plants and should be unsaturated. Olive oil is especially beneficial. So are multivitamin supplements.

It has recently been reported that a key ingredient in soya beans thrawts cancer. Genistein, a plant oestrogen, plays a pivotal role in suppressing the growth of cancer cells. Asian diets high in soya have been linked with low incidence of cancers, particularly breast, colon and prostate cancers. The link has been reinforced by evidence that when Asians migrate to Europe and abandon the high-soya diet, then risk of developing these cancers increases.

Overweight should be avoided through physical activity. Exercising vigorously for 20 minutes a day reduces the incidence of colon cancer and prevents heart disease as well. Regular physical activity during childhood and adolescence avoids an early onset of menstrual cycle and prevents cancers of the reproductive system.

Melanoma may be prevented by avoiding being in the sun between 11 AM and 3 PM and by more widespread use of sunscreens. Likewise, strict control measures at the workplace can shrink the proportion of fatal cases caused by occupational exposure. For example, it is now mandatory for asbestos workers and miners to wear masks. Those employed in nuclear power plants wear lead-coated overalls that are impermeable to radiation.

Hepatitis B and C viruses are known to cause hepatocellular carcinoma, a form of liver cancer. The recently introduced vaccines against the hepatitis B virus is expected to reduce the incidence of liver cancer. The vaccine has, however, not gained popularity firstly, because its cost is high and secondly, because it is understood to be an antidote only against liver infection. Most people are not aware of its long term potential to prevent liver cancer.

To conclude, it must be emphasized that with regard to prevention of cancer, it is never too soon to intervene—but it is often too late.

References

Coghian, A. (1998). Cancer killer, *New Scientist,* March 14, p. 14.

Drake, J.W., Glickman, B.W. and Repley, L.S. (1983). Updating the theory of mutation, *Amer. Sci.,* **71,** 621–630.

Francis, B.M. (1994), *'Toxic Substances in the Environment',* John Wiley & Sons, Inc., New York, p. 199–293.

Fraser, F.C. and Nora, J.J. (1986), *'Genetics of Man',* 2nd Edn., Lea & Febiger, Philadelphia, p. 219–231.

Freese, E. (1971) in, *'Chemical Mutagens',* Vol. 1, Hollaender, A. (Ed.), Plenum Press, New York, p. 1–56.

Guterman, L. (1998). Pregnant Smokes have run out of excuses, *New Scientist,* August 29, p. 6.

Lajtha, L.G. (1970) in, *'What we know about Cancer'*, Harris, R.J.C. (Ed.), George Allen & Unwin Ltd., London, p. 34–54.

Perera, F.P. (1997). Environment and cancer: Who are susceptible? *Science,* **278,** 1068–1079.

Ruoslahti, E. (1996). How cancer spreads, *Sci. Amer.,* **275**(3), 42–47.

Smedley, H., Sikora, K. and Stepney, R. (1985), *'Cancer—What it is and how its Treated'*, Basil Blackwell, New York, p. 1–29.

Tomlinson, I. (1993). Mutations, evolutionary theory and cancer, *Trends Ecol., Evol.,* **8,** 107–110.

Trichopoulos, D., Li, F.P. and Hunter, D.J. (1996). What causes cancer? *Sci. Amer.,* **275**(3), 50–57.

Weinberg, R.A. (1996). How cancer arises, *Sci. Amer.,* **275**(3), 32–40.

Willett, W.C., Colditz, G.A. and Mueller, N.E. (1996). Strategies for minimizing cancer risk, *Sci. Amer.,* **275**(3), 58–63.

Zubay, G. (1987), *'Genetics'*, The Benjamin Cummings Publishing Co., Inc., California, p. 308–360.

Part III

Energy

CHAPTER 9

Energy Flow

We have seen in the last section that life originated as a result of complex organisation of matter in the living cell. This complex organisation of the cellular environment had to be maintained against the random disorganising forces of nature. In order to sustain their extraordinary structural and functional complexity, the living organisms require a continuous input of energy. It is, therefore, imperative that energy is essential to all life forms. In fact, energy is the driving force of all activities in the universe. Nothing moves, changes, grows or decays without the expenditure of energy. Energy flows downhill from a high-potential, high-temperature source in form of the sun to the low-potential, cosmic cold of the outer space. Earth intercepts some of this energy. A part of the captured energy is used for the performance of useful work. The remaining is re-radiated into the outer space. This energy flow is necessary for the sustenance of all natural and human-developed systems on earth.

A Chemist's Perspective

Energy is defined as the capacity to perform work. Work (W) is done when a force (F) moves a body over a distance (1) according to the relation,

$$W = F \cdot 1 \tag{9.1}$$

Energy is commonly expressed as Kcal, where one Kcal is the quantity of energy that, when completely converted into heat, will warm 1 kg of water through 1°C.

Energy occurs in many forms: Heat energy is the energy of random motion of atoms and/or molecules that constitute the matter; mechanical energy is the energy of organised motion of matter; electrical energy is the energy of charged particles acting on each other; chemical energy is the energy of chemical bonds in the molecules; nuclear energy is the energy that binds the nucleons within the atomic nuclei; and light energy is the energy of electromagnetic radiation.

Mechanical energy exists in two forms: Kinetic energy and potential energy. *Kinetic energy* (E_K) is the energy due to the motion of a body. Mathematically,

$$E_k = \tfrac{1}{2} mv^2 \tag{9.2}$$

where m is the mass of the body, v is its velocity. *Potential energy* (E_P) is the energy due to the position of a body. An object at rest possesses potential energy by virtue of its height (h) above some reference level. Mathematically,

$$E_p = mgh \tag{9.3}$$

where g is the acceleration due to gravity.

Water stored in a reservoir behind a hydroelectric dam is an example of potential energy. If the water is allowed to flow, the potential energy is converted into kinetic energy. The flowing water can be used to turn the blades of a turbine, in which case the kinetic energy is quantitatively converted into electrical energy.

Laws of Thermodynamics

The conversion of one form of energy into another is governed by the laws of thermodynamics. The first law of thermodynamics states that energy can neither be created nor be destroyed, but one form of energy can be converted into another form. This means that if one form of energy is changed to another form, the same total quantity of energy remains after the transformation.

Though the quantity of energy remains the same before and after a specific transformation, its quality is altered. During each transformation a part of the energy becomes unfit for doing useful work. When water flows over a hydroelectric dam and falls to the pool below, its mechanical energy is converted into heat. However, this heat cannot be tapped to perform useful work. Likewise, when a barrel of oil is burned, the heat energy so produced cannot be converted into work with 100% efficiency. That is, a part of heat escapes out and cannot be utilized to perform useful work.

Why can't heat energy be completely converted into work? Because some of the heat energy is taken up by individual molecules making up the system. In case of the water flowing over a hydroelectric dam, the water molecules take up heat energy when the water strikes the pool. In case of combustion of oil, the products of combustion, that is, carbon dioxide and water molecules, take up some of the heat energy. This heat becomes unavailable for the performance of useful work. All it can do is that it increases the kinetic energy of the individual molecules which absorb it. These molecules are randomized to a greater extent than they previously were. The disorderliness of the system increases.

The extent of disorderliness or randomness of a system is measured in terms of entropy (S). The greater the disorder, the greater is the entropy. The extent of useful work performed by a system is measured in terms of Gibbs free energy (G). The latter is related to the entropy by the equation,

$$G = T - TS \tag{9.4}$$

where T is the absolute temperature and H is the enthalpy or the heat content of the system. Equation (9.4) shows that, at a fixed temperature, the ability of the system to perform useful work decreases with increasing entropy.

Entropy of the universe increases during the course of every spontaneous process. Water flowing down a hydroelectric dam is an example of a spontaneous process. So is the combustion of oil. As we have seen, the increase in entropy in these processes restricts the conversion of heat into useful work. This is the basis of the second law of thermodynamics which states that heat cannot be converted into work with 100% efficiency.

Efficiency

Efficiency of a system is defined as the ratio of useful work delivered to the total energy expended, when both are measured in the same energy units. Experience has shown that heat spontaneously flows from a body at higher temperature (T_1) to one at lower temperature (T_2). The maximum efficiency (E_{max}) of the system is given by,

$$E_{max} = \frac{T_1 - T_2}{T_1} \qquad (9.5)$$

E_{max} is called Carnot efficiency. Due to the entropy factor, Carnot efficiency is always less than 100%. For example, for a fossil fuel based power plant, the Carnot efficiency for conversion of heat energy into electrical energy is 65–80%. In practice, the efficiency is still lower, of the order of 35–40%. This is because maximum Carnot efficiency can be achieved only under ideal conditions, one of which is that the transformation must occur infinitely slowly. Humans are obviously concerned with processes that operate on a finite time scale. The faster a given process is carried out, the lower its efficiency becomes. The trade-off between the rate of energy conversion and the theoretical maximum efficiency with which it is converted to work has important implications not only for industrial processes, but for biological systems as well.

A Biologist's Perspective

Living organisms maintain their organised state by capturing high-quality, low-entropy energy from the environment, using it to grow, respire and reproduce, and then releasing the residual energy back to the environment in form of low-quality, high-entropy waste heat. In order to maintain a state of physiological and chemical equilibrium, a state known as *homeostasis,* each organism must invest energy. However, one-time investment does not serve any purpose. Each organism is constantly subjected to catabolic or entropic processes. These processes tend to break apart the highly ordered structure of cells. An organism must also bear additional stress due to hostile environmental conditions or disease. All these factors shift the homeostatic state towards disorder. To achieve physiological and chemical order, the organism must keep investing energy as long as it survives.

Energy Investment by Organisms

The acquisition of energy by a living organism is governed by the laws of thermodynamics. For example, if a plant captures 100 units of energy from sunlight, and converts 5 units into chemical energy tied up as food, the remaining 95 units are dissipated to the environment. Thus, in accordance with the first law of thermodynamics, the total energy is conserved. When a herbivore feeds on that plant, it is able to capture only a small fraction of stored energy. The remaining is discarded as waste heat. Thus, in accordance with the second law of thermodynamics the energy transfers in natural ecosystems are inefficient.

The energy of sunlight is essential to all forms of life. It is taken up directly or indirectly by the entire community of plants, animals and microorganisms. However, green plants alone have the potential to harness the energy of sunlight directly through the process of photosynthesis. This phenomenon enables the green plants to capture carbon dioxide and water from the environment and convert them, by reactions catalysed by sunlight and chlorophyll, into glucose—the plant foodstuff. Organisms such as the photosynthesizing plants which can manufacture their own foodstuff by directly utilizing sunlight are called *autotrophs*. These plants store the energy of sunlight as chemical potential energy.

The amount of chemical potential energy stored by the plant community over a given period of time is called *net productivity.* Some of the net productivity continues to be stored over the eons in the plants which become buried under the earth's crust. Such plants are ultimately converted into fossil fuels. Some of the net productivity is consumed by the organisms of decay, chiefly fungi and bacteria. The remaining net productivity is harvested by herbivores, such as a grazing antelope, which walks from plant to plant, appropriating the sunlight-derived energy. The herbivores, in turn, provide food for the carnivores. The herbivores and carnivores, which cannot manufacture their foodstuff by directly utilizing sunlight, are called *heterotrophs.*

Food Chain

Food chain is a sequence of organisms in which each uses the next lower member of the sequence as a food source and is eaten by the one above it. A food chain is thus a diagramatic representation of the food relationship within an ecosystem (Fig. 9.1).

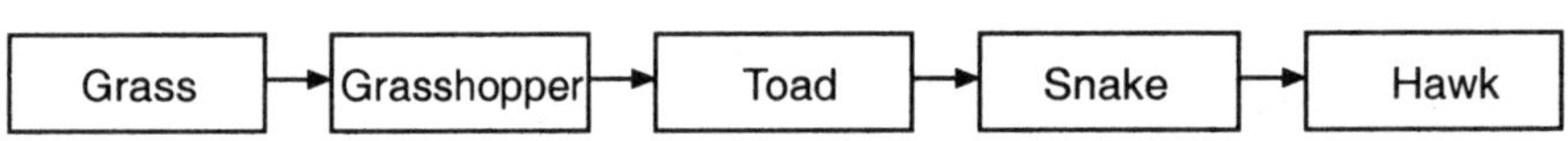

Fig. 9.1 A segregated food chain

All food chains start with an autotrophic organism, that is, a photosynthesizing green plant. As has been stated, these organisms can produce their own food from inorganic raw materials drawn from the environment. Hence the autotrophs are also sometimes called *producers.* The heterotrophs, which are dependent for food consumption on autotrophs, are called *consumers.* Each level of consumption in a food chain is called a *trophic level.* For example, a grasshopper that feeds directly on grass occupies the first trophic level. A toad which feeds upon the grasshopper occupies the second trophic level. The toad, in turn, may be consumed by a snake; the snake occupies the third trophic level. A hawk, which may eat the snake, occupies the fourth trophic level.

The transference of energy from one trophic level to another is governed by the second law of thermodynamics. That is, the energy transfer is never 100% efficient. Thus, when a grasshopper feeds on grass, only a part of the energy stored in the autotroph is transferred to the grasshopper. The remaining escapes out as waste heat. This process continues throughout the food chain. At each link a substantial portion of sun's energy, originally trapped by the autotrophic organism, is dissipated back into the environment.

Thus, the total amount of energy stored in the body of a given population is dependent on its trophic level in the food chain. There is a decrease in the total available energy at each higher level. This decrease in energy is illustrated in a biotic pyramid (Fig. 9.2). In the biotic pyramid the largest amount of food energy is found in the lowest layer. Thus, the autotrophs constitute the base of the pyramid. The carnivorous hawk occupies the narrow apex.

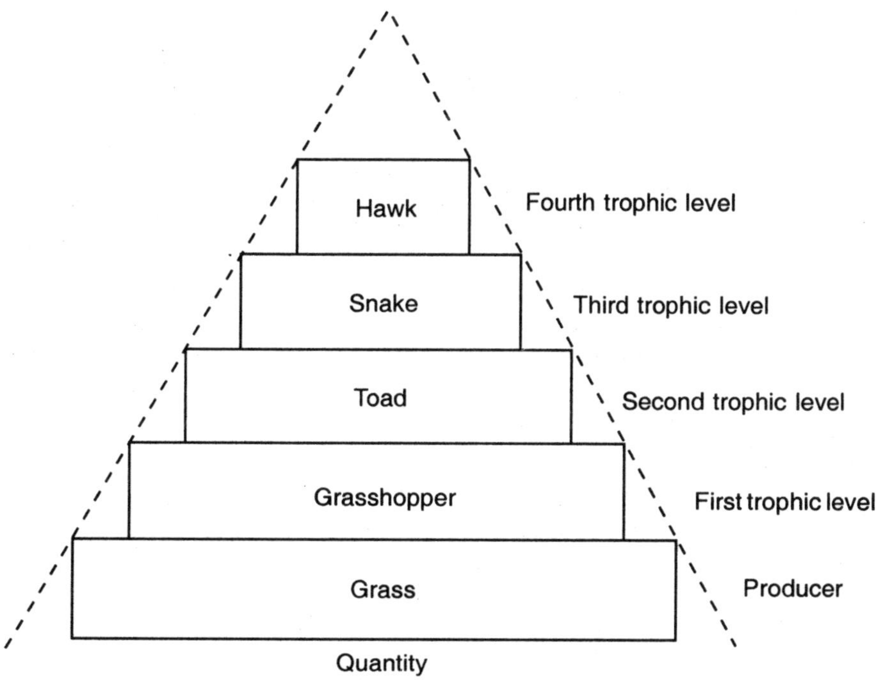

Fig. 9.2 A biotic pyramid

The example of the food chain cited above, grass to grasshopper to toad to snake to hawk (Fig. 9.1), is a segregated case. Although food chains may be artificially separated out, as has been done in this case, yet most food relationships in an ecosystem are far more complicated. Most of the heterotrophs consume a varied diet and, in turn, provide food for a variety of creatures that prey on them. For example, a toad will consume not only grasshoppers but mosquitoes and caterpillars as well. Similarly, the toads are eaten not only by snakes but by vultures too. Therefore, various food chains are interlinked. The interwoven food chains are called *food webs* (Fig. 9.3).

Final Transformation

We have seen that the energy of sunlight trapped in the autotroph is transferred through various trophic levels of the food chain. When an organism dies, some of the energy it had captured while it was alive, is retained in the body. The final

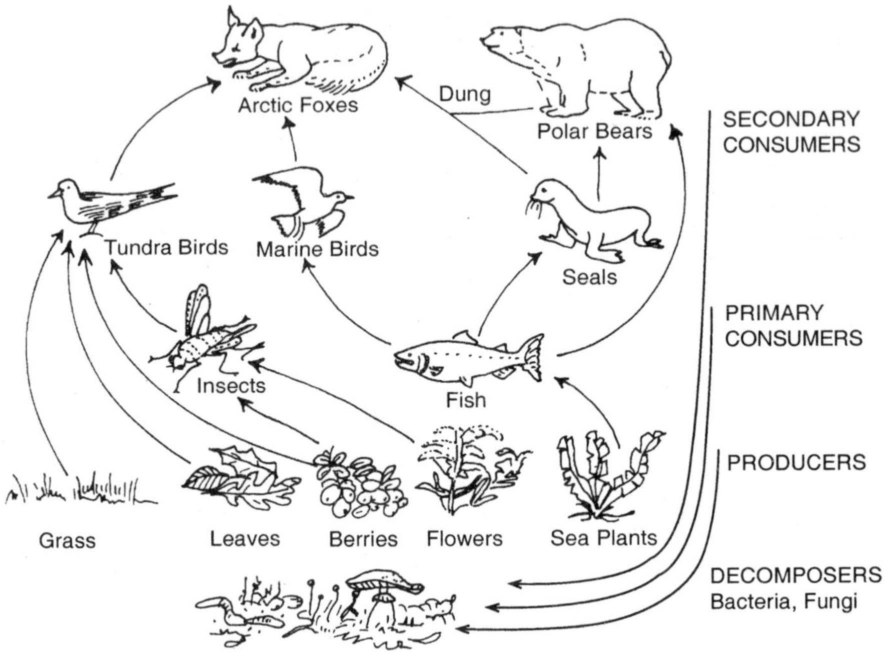

Fig. 9.3 A food web

transformation of this energy is brought about by microorganisms of decay. These are certain varieties of fungi and bacteria, which are found in abundance in natural ecosystems. They extract the residual energy stored in the body of dead organism. They use a very small fraction of the extracted energy to sustain themselves and radiate the remaining to the environment as waste heat. Simultaneously, they degrade the organic constituents of the organism's body into inorganic products, such as carbon dioxide, water and ammonia—raw materials of a living cell—and pass them out into the environment. The inorganic chemicals then enter a new generation of plants or animals.

Thus, the flow of energy through the biosphere proceeds from the sun to the producers, then to the consumers, and lastly to the microorganisms of decay. It is through the microorganisms of decay—working at each trophic level of the food chain—that the final transformation of energy is achieved.

Energy Resources

For most organisms the energy required for sustenance is derived from food. Food energy is transferred from one trophic level to another along a food chain. The transference of energy is inefficient—a limitation imposed by the second law of thermodynamics. Theoretically, the second law of thermodynamics also holds true for human beings, but the complexity of our use of energy relieves us of some simple limitations imposed on other organisms. For us energy is more than food energy. Unlike other organisms, we not only require food energy but also energy to drive machines, to produce heat, to generate electricity, to transport limited materials, etc. Therefore, we have to look for those resources which are easily accessible and from which energy can be economically extracted and utilized for the welfare of humanity.

Those energy resources whose geographical location is known with certainty and from which energy is recoverable by the application of existing technologies are called *reserves*. The Raniganj coal fields in West Bengal are fossil fuel reserves. The unexplored resources from which energy may possibly be extracted in the future, using the existing technologies, are called *inferred resources*. The hot springs in the Chanderkhani Pass in Himachal Pradesh are examples of inferred geothermal resources. The term *hypothetical resources* applies to those sources from which the extraction of energy is not economically viable. Nuclear energy from a fusion reaction is, at present, a hypothetical resource.

Energy resources like coal and petroleum, which are commonly used, are called *conventional resources*. Those like oil shales, which are not currently in commercial use, are called *unconventional resources*. Some energy resources like coal and petroleum are not replenished as rapidly as they are used. These are called *nonrenewable resources*. Other resources like solar energy are inexhaustible. They are referred to as *renewable resources*.

Energy resources may be broadly classified into four categories: Fossil fuels, terrestrial, solar and nuclear (Fig. 9.4). These are discussed in Chapters 10, 11, 12 and 13, respectively. The various subcategories of each energy resource, along with their environmental implications are also described in the respective chapters.

Energy Conservation

Over the last few decades several national governments have recognized that rapid economic growth is the only way to alleviate poverty. In order to achieve economic goals, the energy resources were used luxuriantly. No concern was shown for the fast depleting, nonrenewable energy resources. No thought was given to develop technologies for exploiting renewable energy resources. If the exploitation of energy resources continues unabated, the annual global energy demand by the middle of this century would be about four times the present level. Therefore, conservation of energy has become an absolute necessity.

Energy conservation is defined as an activity to prevent the physical waste of a nonrenewable energy source. Its purpose is to affect economy, overcome shortages of essential commodities, enhance living standards and to stretch the supplies of limited energy resources. Some of the measures which can negate energy crisis and boost energy conservation are—

1. Using highly efficient methods for extraction of fossil fuels.
2. Substituting renewable energy resources for nonrenewable ones.
3. Converting primary fuels into usable forms of energy, such as heat and electricity.
4. Encouraging urban passenger traffic from automobile to mass transit systems.
5. Introducing more efficient industrial processes and equipment.
6. Using super high voltage electrical transmission lines.

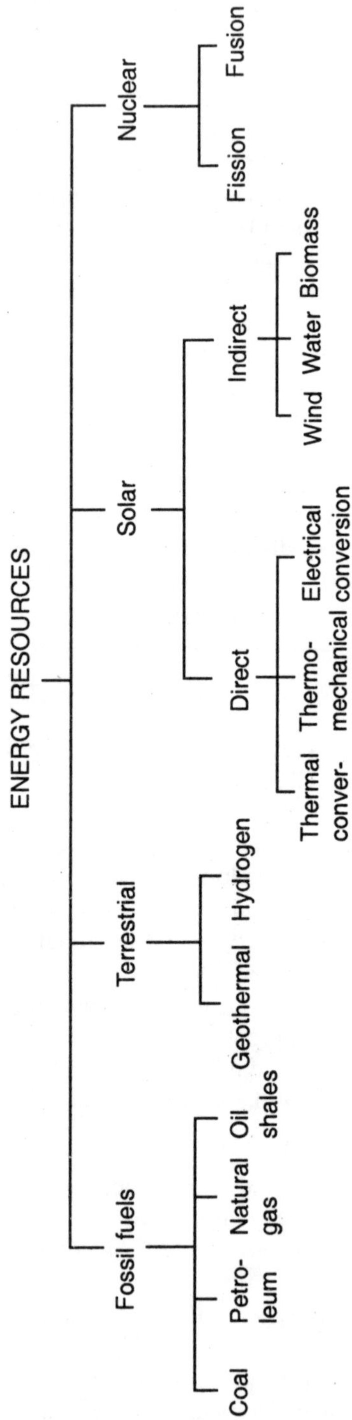

Fig. 9.4 Energy resources

On the face of it, these measures seem to be simple and attractive. Yet, there are a number of constraints which limit their implementation. For example, high voltage transmission lines are, no doubt, energy efficient but they scar the landscape. Radical changes in the pattern of passenger transportation are shunned by those whose lifestyles are affected. Though nuclear power plants are more energy efficient than fossil fuel based electricity generating units, the Chernobyl-type disasters mould the public opinion against nuclear energy utilization. Energy-efficient appliances are usually costly. People tend to buy cheaper products, without realising that in the long run they would be paying more in terms of energy costs.

The depleting reserves and rising prices of oil have prompted several governments to look for substitutes. There are two alternatives to oil: Coal and solar energy. However, coal utilization causes serious environmental damage and the technologies for solar energy utilization are not yet geared-up to replace oil. The most promising approach to conserve oil appears to be a utilization of mixed resources. That is, to use a combination of coal and solar energy resources so as to 'buy time' for developing other viable alternatives like geothermal or nuclear sources.

It is thus obvious that conservation of energy cannot be achieved in an isolated manner. Measures ensuring conservation of environment and development of alternative energy sources must be carried out simultaneously. Further, experience has shown that pursuing extreme policies of energy conservation, environmental protection or development of alternative energy sources have not yielded the desired results. The wisest course is to follow a balanced approach, incorporating an array of opinions and flexible strategies.

References

Bakthavatsalam, V. (1995). Renewable energy technologies, *Asia Pacific Tech. Monitor,* **12**(3), 9–14.

Berg, C.A. (1973). Energy conservation through effective utilization, *Science,* **181,** 128–138.

Cabana, G. and Rasmussen, J.B. (1994). Modelling food chain structure and contaminant bioaccumulation using stable nitrogen isotopes, *Nature (London),* **372,** 255–257.

Carter, A.P. (1974). Applications of input-output analysis to energy problems, *Science,* **184,** 325–329.

Dasmann, R.F. (1976). *'Environmental Conservation',* 4th Edn., John Wiley & Sons, Inc., New York , p. 310–338.

Ehrilich, P.R. and Roughgarden, J. (1987), *'The Science of Ecology',* Macmillan Publishing Co., New York, p. 521–552.

Gates, D.M. (1985), *'Energy and Environment',* Sinauer Associates, Inc., Sunderland, p. 48–58.

Häfele, W. (1974). A systems approach to energy, *Amer. Sci.,* **62,** 438–447.

Hall, C.A.S., Cleveland, C.J. and Kaufmann, R. (1986), *'Energy and Resource Quality',* John Wiley & Sons, New York, p. 3–23.

Hammond, A.L. (1972). Conservation of energy: The potential for more efficient use, *Science,* **178,** 1079–1081.

Lincoln, G.A. (1973). Energy conservation, *Science,* **180,** 155–162.

McNaughton, S.J. and Wolf, L.L. (1979), *'General Ecology'*, 2nd Edn., Holt, Rinehart and Winston, New York, p. 128–148.

Ramade, F. (1981), *'Ecology of Natural Resources'*, John Wiley & Sons, New York, p. 20–25.

Starr, C., Searl, M.F. and Alpert, S. (1992). Energy sources: A realistic outlook, *Science*, **256,** 981–987.

CHAPTER 10

Fossil Fuels

Fossil fuels are the energy resources which originated from the organisms that lived millions of years ago. After their death they became covered with layers of sediments and were ultimately subjected to high temperatures and pressures in the depths of the earth's crust. In the absence of oxygen their decomposition proceeded only partially. Under these conditions, the once living tissue became fossilized. When the fossilized materials are extracted and burnt, the energy stored in their chemical bonds is released as heat. Hence, the term fossil fuels. Under the category of fossil fuels there are three principal resources—coal, petroleum and natural gas—along with a minor resource—oil shales. Although fossil fuels are widely distributed, there are two problems associated with them. Firstly, their supplies are fast dwindling. They cannot be replaced except by natural processes which are extremely slow. Secondly, their extensive use is harmful to the environment. No other energy source causes as large a damage to the human environment as do fossil fuels. In this chapter we shall endeavour to analyse the merits and demerits of fossil fuels.

Coal

Coal is a black or brown-black nonmetallic solid formed by the alteration of fossilized remains of organic plant matter. It consists chiefly of carbon, hydrogen, nitrogen and oxygen and varying amounts of moisture and other noncombustible minerals. It is used as a fuel for direct heating and for generating electric power, as a raw material for coke and as a source of industrial chemicals. In recent years the importance of coal as a future energy source has gained momentum. This is attributed partly to the rising prices of oil and gas and partly to the social non-acceptability of nuclear energy resources.

Origin of Coal

Coal was formed from the debris of giant tree ferns and other vegetation that grew in coastal and inland swamps and bogs about 300 million years ago. When the plants died they were covered with shallow layers of sediments and water. Under these conditions the decay of vegetation took place anaerobically and extremely slowly. Before the decomposition could proceed to completion, the dead organic matter became covered with fresh debris of plants and soil. Gradually, the remnants became buried to a considerable depth. They were subjected to high pressures, stemming not only from the weight of the material above them, but also from the movement of the earth's crust. They were also heated by molten igneous rocks. Under the conditions of high temperatures and pressures, the fossilized remains were subjected to physical and chemical alterations that ultimately led to the origin of coal.

Classificiation of Coal

The chemical and physical characteristics of coal are determined essentially by the duration of time the vegetation remained buried and the magnitude of pressure to which it was subjected. The coals that have undergone the greatest degree of compression possess a higher C:H ratio and lower amounts of moisture and volatile contents. Such coals have higher heating values per kilogram.

Based on the duration and depth of the fossilized materials, coals are classified into four ranks: Lignite, subbituminous, bituminous and anthracite. Lignite is the lowest ranking coal, while anthracite is the highest. As shown in Table 10.1, the change from lignite to anthracite is a change towards increasing heating value, decreasing volatile matter and increasing economic utility of the coal.

Table 10.1 Classification of coals

Rank	Analysis %				Heating Value ($Kcal\ kg^{-1}$)
	Carbon	Oxygen	Volatile Matter	Moisture	
Lignite	60–70	18	45	35	5500–7000
Subbituminous	75–83	15	40	17	7000–8000
Bituminous	78–90	10	20	6	8000–8500
Anthracite	92–98	3	5	1	8500–8700

Lignite represents an early stage of transformation of vegetation into coal. Due to its high hydrogen content as compared to other varieties, lignite is an attractive source for coal gasification technology, an energy transformation technique (see next section). The next higher rank, subbituminous, possesses large ash and volatile matter content and, therefore, burns with dense smoke. Subbituminous coals rapidly deteriorate on exposure to air. The next higher rank, bituminous, has the most widespread geological distribution. It is also the most extensively used variety of coal. Most power plants employ bituminous coal. Anthracite, the highest ranking coal, is produced under unique geologic conditions that subject the coal deposits to unusually high temperatures and pressures due to large scale folding and faulting movements of the earth.

Environmental Impacts of Coal

Compared to other fossil fuels, coal utilization poses greatest threat to the environment. In fact, every phase of coal utility—extraction, processing, transportation and cosumption—affects our air, water and land resources.

1. Air pollution

The major problem associated with coal burning centers around the emission of noxious gases. Oxides of sulfur and nitrogen released as a result of coal combustion are washed down to the earth in form of acid rain. The environmental problems associated with acid rain are highlighted in Chapter 14, p. 224. The release of carbon dioxide from coal combustion produces greenhouse effect, the deleterious effects of which are detailed in Chapter 18, p. 260. Coal combustion also releases particulates and fly ash to the atmosphere. The adverse effects of particulate matter are described in Chapter 20, p. 297.

2. Water pollution

In humid mining regions, acid mine drainage is a serious problem associated with extraction of coal. Rain water infiltrates in the mining regions and reacts with sulfide minerals like iron pyrites to produce sulfuric acid, which contaminates the ground water resources of the area.

3. Occupational hazards

Coal miners suffer more respiratory diseases and impairments and, therefore, have a higher occupational morbidity rate than their counterparts in other industries. Shortness of breath and persistent cough increase with a worker's years underground. Miners lungs and bronchial track are the primary sites of illness. Minute particles of coal dust become lodged in the respiratory system and lead to or exacerbate bronchitis and *coal workers pneumoconiosis.* The latter disease occurs when coal particles are deposited in the terminal respiratory bronchioles and alveoli, where they are engulfed by alveolar macrophages. Tissue necrasis may then occur leading to increase in the incidence of heart failure.

Control measures

In order to curtail the environmental problems associated with coal utilization two broad methods are available. These are: Coal treatment technologies and coal gasification technology.

Coal treatment technologies. The aim of the coal treatment methods is to reduce the emission of sulfur oxides and nitrogen oxides to the atmosphere. Methods which reduce the levels of sulfur oxides and nitrogen oxides from coal combustion are outlined in Chapters 14 (p. 222) and 15 (p. 233).

Coal gasification technology. Coal gasification promises a clean, alternative source of fossil fuel energy. It aims at converting solid coal into methane, also called *synthetic natural gas* or SNG. Methane has a high energy content of about 37 megajoule per cubic meter at standard temperature and pressure. It burns without causing pollution.

Coal gasifiers incorporate a mixture of coal, steam and oxygen as feed. Air may be substituted for oxygen. However, the disadvantage of using air is that the product gas is contaminated with noncombustible nitrogen. Hence, its heating value is lowered. The gasifiers operate at temperatures of 900–1000°C.

Coal first combines with steam to produce water gas

$$C_{coal} + H_2O \rightarrow H_2 + CO \qquad (10.1)$$

Water gas has a low energy content of about 11 megajoule per cubic meter and hence a low heating value. A part of the carbon monoxide formed in reaction (10.1) is treated with steam to produce more hydrogen.

$$CO + H_2O \rightarrow H_2 + CO_2 \quad (10.2)$$

Reaction (10.2) is called *shift reaction.* The additional hydrogen so produced combines with the remaining carbon monoxide to produce synthetic natural gas.

$$3H_2 + CO \rightarrow CH_4 + H_2O \quad (10.3)$$

Reaction (10.3) is called *methanation reaction.* Apart from methane, several other lower alkanes are also produced. However, as the concentration of methane increases, the heating value of the product gases becomes closer and closer to the natural gas.

The conversion of coal into methane may be improved by using a coal gasification catalyst. Potassium carbonate and sodium carbonate are the most important active catalysts for the coal conversion process. Using a catalyst concentration of 10–20%, commercially acceptable gasification rates are achieved at 700°C, compared to 900–1000°C required without a catalyst.

Petroleum

Petroleum, from the Greek word *rock oil,* refers to fluid fossil hydrocarbons that are available as fuel either in form of oil or natural gas. Petroleum freed of natural gas is referred to as *crude oil.* Crude oil is a mixture of a wide variety of hydrocarbons, both aliphatic and aromatic.

Origin of Petroleum

The formation of petroleum began with the death and subsequent deposition of organic matter in shallow marine environment. For this reason the major oil-bearing regions are located in coastal environments.

It is theorized that in a calm, shallow sea, dead plants and animals fell to the bottom and were covered by silt. Steady accumulation of mud above the organic matter produced anoxic conditions. Under these conditions the organic matter was acted upon by anaerobic microbes, which transformed it into methane and an insoluble, tar like residue called *kerogen.* The latter was the main precursor to petroleum in the marine sediments.

As the sedimentation continued at a steady pace, the organic matter was subjected to high pressures. Simultaneously, the molten igneous rocks raised the temperature of the decaying matter. Increasing temperature and pressure now replaced microbial activity as the main agent of transformation. Thus, the conversion of kerogen into crude oil was brought about largely due to the conditions of high temperature and pressure. At a depth of 1500–4500 m and a temperature range of 60–150°C, kerogen was thermally transformed into a mixture of hydrocarbons. It has recently been postulated that the carbon-sulfur bonds in kerogen first cleaved. The sulfur radicals, so generated, brought about the cleavage of carbon-carbon bonds. As a result, the kerogen got depolymerized. The lighter hydrocarbons constituted the natural gas; the heavier ones constituted the crude oil. During this transformation, most of the compounds containing sulfur, oxygen and nitrogen were driven off.

The crude oil and its associated natural gas were then forced by the high pressure into channels or porous rocks around the pockets containing them. They migrated several kilometers till they reached an impervious strata. Here they became trapped in the form of reservoirs. The petroleum reserves may be tapped by drilling through the impervious rock formations.

Composition and Refining of Petroleum

As has already been stated, petroleum is a mixture of aliphatic and aromatic hydrocarbons. The hydrocarbons are separated into various fractions (Table 10.2) by the technique of fractional distillation (Fig. 10.1).

Table 10.2 Some important fractions of petroleum

Fraction	Approximate Composition	Boiling Range (°C)
Gaseous	C_1–C_4	–160–30
Petroleum ether	C_5–C_7	30–70
Gasoline	C_5–C_{10}	40–200
Kerosene oil	C_{10}–C_{16}	175–290
Gas oil	C_{15}–C_{18}	250–320
Lubricating oil	C_{17}–C_{30}	320–400

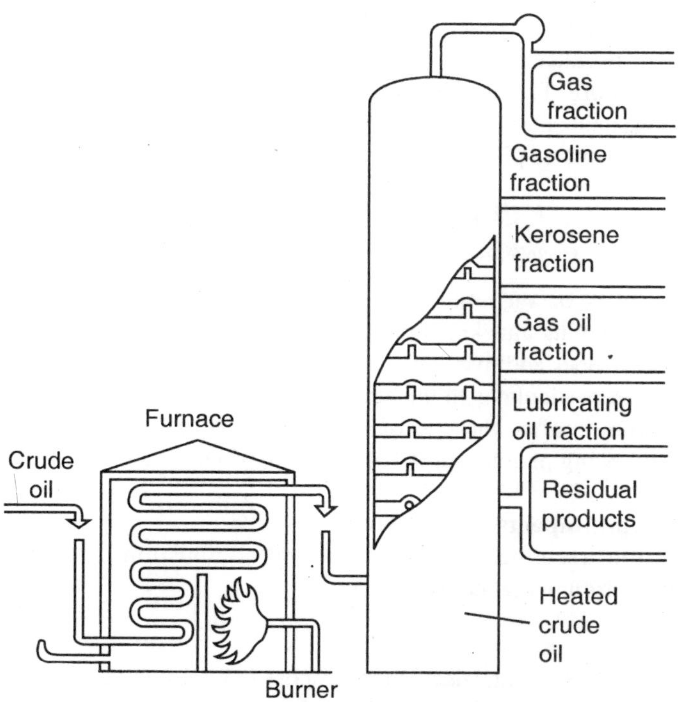

Fig. 10.1 Fractional distillation of petroleum

The most important fraction of petroleum from a commercial viewpoint is the gasoline (C_5–C_{10} hydrocarbons). A more efficient utility of petroleum, therefore, demands that the non-gasoline fractions be converted into gasoline. Thus, the hydrocarbon fractions having less than five carbon atoms are subjected to *polymerization,*

$$mC_nH_{2n} \rightarrow C_{m+n}H_{2\,(m+n)} \tag{10.4}$$

while the hydrocarbon fractions having more than ten carbon atoms are subjected to *cracking.*

$$C_{m+n}H_{2\,(m+n)+2} \rightarrow C_mH_{2m} + C_nH_{2n+2} \tag{10.5}$$

In an automobile engine, a mixture of air and gasoline is compressed and ignited by a spark. The mixture burns and moves the piston, thereby producing mechanical work. However, some hydrocarbons have a tendency to ignite spontaneously. Such pre-ignitions produce a knocking sound and reduce power output and efficiency. They also accelerate engine wear. Obviously, such hydrocarbons are undesirable components of the gasoline fraction. It has been noted that the straight chain hydrocarbons produce greater knocking than the branched chain ones. Aromatic hydrocarbons have the best anti-knocking characteristics.

The tendency of different gasolines to produce engine knocking is measured in terms of *octane rating.* On this scale n-heptane is given a rating of zero, while isooctane or 2, 2, 4-trimethylpentane is given a rating of 100. A fraction whose tendency to produce knocking in a test engine is equal to a mixture containing 9% n-heptane and 91% isooctane is rated at 91-octane.

In order to produce a high octane rating fraction, the gasoline is subjected to *catalytic reforming.* This involves the conversion of straight chain hydrocarbons into branched chain or aromatic hydrocarbons using a catalyst such as molybdenum trioxide on alumina or platinum on clay.

$$\underset{\text{Cyclohexane}}{C_6H_{12}} \xrightarrow{MoO3/Al2O3} \underset{\text{Benzene}}{C_6H_6} + 3H_2 \tag{10.6}$$

Alternately, the octane rating may be enhanced by adding very small quantities of compounds having anti-knocking properties. Tetraethyl lead has been used as a common anti-knock additive to gasoline. However, leaded gasoline on combustion discharges lead particulates to the atmosphere. In view of the adverse effects of lead on human health, the use of tetraethyl lead as a gasoline additive is progressively being phased out (Chapter 22, p. 331).

Environmental Impacts of Petroleum

The extraction, refining, transportation and combustion of petroleum have significant environmental consequences.

1. Extraction

The degree of environmental degradation imposed by petroleum extraction depends, in large part, on whether the drilling site is on land or below water. Impact of terrestrial petroleum wells is little. Such oil rigs do not require significant quantities of land from agricultural or natural production sites. In many cases agriculture continues around the immediate area of the rig. Oil spilled on land is absorbed by the ground and causes only a local mess. However, petroleum extraction from marine wells has severe impacts. Oil spilled in water is carried afar by ocean currents. It then produces a host of environmental problems which are discussed in the next section.

2. Refining

Gaseous effluents from petroleum refineries include carbon monoxide, sulfur oxides, nitrogen oxides, particulates and hydrocarbons. The harmful consequences of these pollutants are discussed in Part IV of the text, which deals with air pollutants. Petroleum refineries also produce toxic inorganic compounds like acids and alkalis, as well as organic compounds like phenols and cresols. These ultimately enter the aquatic systems and cause water pollution. The harmful consequences of these pollutants are outlines in Chapter 21, p. 307.

3. Transportation

Construction of pipeline for the transportation of petroleum products through the ecologically fragile areas threatens the ecology of these regions. Transportation through seafaring vessels is all the more dangerous. If these vessels meet an accident, the oil spills out in the open sea, causing severe damage to the marine life. The consequences of oil spills are highlighted in the next section.

4. Combustion

As in the case of coal, combustion of petroleum produces mainly carbon dioxide and unburnt hydrocarbons. However, unlike coal, sulfur oxides and nitrogen oxides are not produced. Thus, the environmental impacts of burning petroleum are less than those of coal combustion.

Oil Spills

From an environmental point of view, oil spills are the most feared consequences of petroleum utilization. An oil spill refers to an accidental dumping of oil or other petroleum products into ocean and its coastal water bays and harbours. A wide spectrum of biotic community suffers the harmful impacts of oil spills.

Causes of Oil Spills

Two sources—exploitation of offshore fields and transportation of crude oil by sea—have led to a multiplicity of massive accidental spills over the past few decades. Drilling wastes discharged from sea rigs contaminate the surrounding waters. Sometimes, blowouts occur on offshore drilling platforms, in which case the oil spillage is much more spread out.

However, major spills occur when a seafaring oil tanker meets an accident. Grounding, hull failure and collisions are the most common types of accidents associated with oil tankers. Hose failure and pumping errors account for most of the oil spills in ports and harbours.

Fate of the Spilled Oil

Crude oil has a lower density than water. Hence on spillage, the oil floats on the sea surface, imparting it a dark colour. For this reason, oil spills are also referred to as *black tides*. The black tides spread away from the original source, generally at a speed of 1–4 km h^{-1}. Timely human intervention may contain and clean-up a major part of the spilled oil. However, if left to nature, the removal of oil from water takes a very long time. During this duration, the oil undergoes one or more of the following transformations.

1. It may disperse and dissolve into the water column. So long the oil remains afloat, only the biotic community prevailing on the sea surface is affected. However, once the oil goes deep into the water column, a far greater quantum of marine life is likely to be affected.

2. It may be oxidised by chemical and biotic pathways. Complete oxidation of oil into carbon dioxide and water totally wipes out the black tides. However, complete oxidation of oil takes a very long time. Over a short period, the hydrocarbons are partially oxidised to peroxides or phenols. Such oxidation reactions are catalysed by ultraviolet component of natural light. Peroxides and phenols are more toxic than the original hydrocarbons comprising the crude oil. Moreover, they are water soluble and, therefore, easily enter the food chain, causing deleterious consequences.

3. It may be incorporated into the bottom sediments. Although oil is less dense than water and is, therefore, expected to remain afloat, yet a part of it may end up in the bottom sediments. Oil droplets are rapidly adsorbed by suspended clay particles which carry them to the sea sediments. Oil in the sediments is very persistent and degrades very slowly over years or even decades.

Consequences of Oil Spills

The spilled oil is harmful to the marine life in several ways.

1. Reduced photosynthesis

Black tides reduce the transmission of sunlight across the water surface. As a result the photosynthesis by aquatic algae slows down and the oxygen level of water falls. In consequence, all forms of marine life are affected.

2. Sea birds

Marine and coastal birds are extremely vulnerable to oil spills. It has been observed that the population of auks, puffins and razorbills register a decline after every oil spill incident. Guillemors are especially vulnerable to oil pollution because they dive in response to danger, rather than fly away.

Sea birds may directly ingest the oily material from the sea water. Most components of the oil elicit toxic effects. They also induce behavioural changes, so that nesting, mating and migratory activities are seriously disrupted. Moreover, the hydrocarbons present in the oil are fat-soluble and, therefore, become concentrated in the tissue of the birds. The toxic compounds then bioaccumulate in the food chain. Further, black tides dissolve the natural oils and waxes that insulate and protect the birds from loss of body heat. Oil also damages the feathers.

3. Fish

The larval and egg stages of fish are sensitive to the toxic effects of oil. The larvae and eggs of many fish float at or near the surface for 4–5 months while they develop. An oil spill occurring in this period, therefore, causes the fish population to decline. Further, oil components confuse the predators by blocking their odour and taste receptors. Migratory fish are believed to depend partly on their sense of smell to find their way into particular streams. The migratory behaviour of fish is, therefore, impaired in the event of an oil spill.

4. Benthic community

The benthic community refers to those marine animals that reside on the sea floor. These are affected by the oil residues which become a part of the sediments. Mortality rates of crabs, zoanthids, hydrocorals and snails increase after an oil spill. Behavioural disorders, such as locomoter impairment and abnormal burrow construction, have also been observed among these species.

The most significant biological impact of oil has been on the burrowing brittle star. This animal is a dominant member of sea floor community. Its burrowing helps oxygenate the sediments, so encouraging other organisms to grow. The average density of brittle star is 100 individuals per square meter. In the vicinity of sea rigs or in the event of black tides, the population of brittle star is completely wiped out.

5. Marine vegetation

Seagrass meadows, which are found in the shallow coastal regions, are instantly killed after an oil spill. Seagrass provides food, habitat and refuge for diverse population of invertebrates and fish. Mangrove trees, which line up along the sea shore, are also killed when the black tide is washed ashore.

Mitigation of Oil Spills

When crude oil is discharged to the environment, its hydrocarbon components are attacked by microorganisms and converted into non-toxic inorganic compounds. This phenomena is called biodegradation. However, biodegradation is a very slow process and by the time it is complete, ecological degradation begins to manifest itself. Moreover, in case of oil spills, the hydrocarbons spread out in the open sea at a much faster rate than the rate of their consumption by biodegradation.

The harmful consequences of oil spills can be controlled only if the natural phenomena of biodegradation of hydrocarbons is somehow accelerated. A relatively new technology called *bioremediation* seeks to increase the rate of natural biodegradation of pollutants by adding microorganisms with specific enzymes to a contaminated site and/or establishing conditions that favour microbial decomposition by indigenous mircoorganisms. It is a safe and cost-effective method to restore the contaminated sites in general and oil-contaminated sea water in particular.

Hydrocarbon-degrading microorganisms are ubiquitous in most ecosystems and over 30 different genera of oil-degrading bacteria and fungi have been identified. However, in the oceans the population density of these microorganisms is low because marine waters contain very small concentration of nitrogen and phosphorus nutrients required by the microorganisms. Therefore, natural degradation of oil occurs very slowly. Bioremediation stimulates biodegradation by adding nitrogen and phosphorus fertilizers to the site of oil spillage. In some cases, the oil-degrading microorganisms are added to the contaminated site along with the fertilizer nutrients. As the colonies of microorganisms multiply, the crude oil is rapidly decomposed into harmless compounds.

Case Histories of Some Important Oil Spills

The history of transportation of oil by sea is marked by tragic consequences in form of great black tides—in 1967 it was the Torrey Canyon, in 1978 the Amoco Cadiz, in 1989 the Exxon Valdez, in 1996 the Sea Empress and in 1997 the Nakhodka.

1. Torrey Canyon

This Liberian tanker, loaded with 1.19 lakh tonnes of crude oil from Kuwait, ran aground in the early hours of March 18, 1967 near the Sicily Islands. Several cisterns burst due to impact and oil immediately started spilling into the sea. Ten days later, the ship broke into two, releasing the rest of the cargo. All attempts at redeeming the situation proved futile. Even the endeavour to set afire the residual oil by aerial attacks failed.

2. Amoco Cadiz

The French tanker, Amoco Cadiz, carrying 2.33 lakh tonnes of light oil and 4000 tonnes of heavy oil ran aground off the coast of Brittany (France) on March 16, 1978. Over the next twelve days the entire cargo was spilled into the sea. Rough meteorological conditions on one hand interfered with remedial measures and on the other rapidly dispersed the oil in the sea water. A vast brown coloured emulsion—called *chocolate foam*—spread over the sea. Long stretches of Brittany shoreline became contaminated. However, natural biodegradation occurred rapidly. It is believed that the microbial population in the region were adapted to metabolize hydrocarbons, because they were frequently exposed to oil releases from ballast tankers.

3. Exxon Valdez

The American tanker, Exxon Valdez, carrying a cargo of 40,000 tonnes of crude oil wrecked on the rocks near Alaska coast (USA) on March 24, 1989. The oil was rich in aromatic hydrocarbons. Over the next ten days the entire cargo poured out in sea water. Several birds and sea animals were killed. However, bioremediation rapidly and dramatically improved the clean-up. The success of this operation gave bioremediation a strong political support.

4. Sea Empress

Sea Empress, a British carrier loaded with 1.3 lakh tonnes of oil, ran aground near Milford Haven (Wales) on February 15, 1996. About 6000 tonnes of oil was immediately lost to the sea. The oil spread out to about 190 km of coastline.

Initial attempts to pump the remaining oil to another vessel failed since the pump rooms of Sea Empress were flooded. Subsequently air had to be pumped to increase the buoyancy and carry the ship away from the shore where the tides were too strong. All the while oil continued to leak. Pumping of the remaining oil ultimately began on February 24, 1996.

*5. **Nakhodka***

The Russian tanker, Nakhodka, loaded with 19000 tonnes of oil broke into two near the Oki Island in Japan on January 2, 1997. Rough weather hampered the clean-up operations. All the while thick fuel oil continued to leak. One of the containers sank to a depth of 1900 m and could not be salvaged. Although local people and volunteers worked frantically to clean-up the mess, large slicks approached Wakasa Bay. Several nuclear power plants are located on the shoreline of the Bay. They are provided with oil fences which prevent the suction of oil when sea water is drawn for cooling purposes. However, the oil slicks from Nakhodka were so large that at one time it was feared that the power plants might have to be shut down.

Natural Gas

We have seen that petroleum is formed when large amount of organic material is deposited in anoxic environments and subsequently altered by microbial and geologic activities. The fraction of petroleum that at ordinary temperatures and pressures is in a gaseous state is called natural gas.

Classification and Composition

At some sites, natural gas overlies the crude oil. Since it is in contact with oil, it is termed *associated natural gas* or *wet gas*. Associated natural gas is collected from oil wells. At other sites, however, natural gas oozes from gas wells which are quite remote from the oil-bearing strata. In such cases, the gas migrates through porous rocks away from oil. It is not in physical contact with oil and is, therefore, termed *non-associated natural gas* or *dry gas*.

The average composition of natural gas is depicted in Table 10.3. Methane is the most common constituent of natural gas. Decreasing amounts of ethane, propane and butane are also present, along with varying proportions of carbon dioxide, nitrogen and oxygen. Non-associated natural gas has a higher percentage of methane

Table 10.3 Composition of natural gas

Component	Volume (%)
Methane	70–90
Ethane Propane Butane	0–20
Carbon dioxide	0–8
Nitrogen	0–3
Oxygen	0–0.2

than the associated natural gas. The heating value of the former is, therefore, higher.

Non-associated natural gas has another advantage in that it tends to flow to the surface at higher pressure which makes its transportation in pipeline easier than associated natural gas, which often needs artificial boosting just to get it to the surface. Moreover, associated gas must be separated from the crude oil—a step that is not necessary for non-associated gas. All these factors make non-associated natural gas commercially more important.

Advantages of Natural Gas

Natural gas is the most desirable fossil fuel from an environmental point of view. It mixes easily with air, making complete combustion possible. Thus, no unburnt or partially burnt hydrocarbons are emitted when natural gas is employed as a fuel. Since natural gas is free of sulfur, the oxides of sulfur are not emitted during combustion. The level of nitrogen oxides emitted from natural gas combustion is also very low. Additionally, natural gas requires no local storage facilities like other fossil fuels and, therefore, is readily available to consumers through the pipeline.

In addition to it being a versatile fossil fuel, natural gas has industrial applications as well. Ethane is generally removed from natural gas and converted into ethylene for the plastic and petrochemical industry. Propane is separated and marketed as *bottled gas* in areas lacking pipeline. Butane is removed to make synthetic gasoline or is added to gasoline to increase its volatility.

Oil Shales

Large quantities of energy are stored in carbon-rich sedimentary rocks known as oil shales. The organic matter in these deposits is essentially in form of kerogen.

Origin of Oil Shales

Oil shales originated because of accumulation of dead organic matter at shallow depths in stagnant streams, lagoons or bogs. Due to anaerobic microbial activity the organic matter was converted into kerogen. This polymeric mineral was also an intermediate in petroleum formation. However, in case of petroleum it was progressively buried to a depth of 1–3 km below the sediments. As a result it was subjected to high temperatures and pressures, which ultimately converted it into oil. In case of oil shales, kerogen is not buried very deep. Therefore, its transformation to oil is not complete. Hence, oil shales originated due to insufficient geologic conditions.

Recovery of Oil from Shales

The recovery of oil from shales involves two general steps: Mining, during which the shale is extracted from the environment; and retorting, during which the extracted shale is subjected to pyrolysis. The pyrolysis of kerogen occurs at about 480°C. It yields a highly viscous crude oil, gas and water.

On an average, one tonne of oil shales yields 10–40 gallons of viscous crude oil. However, only those shales yielding 25 or more gallons per tonne of processed rock are considered commercially viable. The crude oil extracted from these resources is nearly similar in composition to the heavy oil pumped from petroleum wells. The only real difference is that oil from shales is more viscous and, therefore, more difficult to recover.

Environmental Impacts of Oil Shales

Several environmental problems are associated with the extraction of oil from shales.

1. Water pollution

Oil shales are alkaline in nature. When the shale rock is extracted and crushed, alkaline salts are leached into the aquatic bodies. After the retorting process, the waste shale is deposited above ground, rather than underground in the mines, since the latter process entails a higher cost. Spent shale is heavily watered to remove salts. The runoff from a bare shale is, therefore, extremely salt laden. Since salinity reduces the usability of irrigation water, oil shale development is a potential threat to agriculture. Moreover, for each barrel of oil produced, about three barrels of water are needed for washing away the salts. In some parts of the world water is a costly commodity.

2. Air pollution

An oil shale industry has the potential to pollute the air with dust from mining, crushing and disposal operations. The harmful impacts of particulates are outlined in Chapter 20. The retorting of crushed shale releases sulfur oxides to the atmosphere. The hazards posed by sulfur oxides are detailed in Chapter 14.

Normally, the retorting process uses an external heating system which operates at about 480°C. However, in some cases, a part of the shale is burnt to produce heat which, in turn, is used for retorting. This step, no doubt, lowers the heating cost, but in such cases the heating temperatures shoot upto 700–1100°C. At such high temperatures, the carbonate minerals in the rocks begin to calcine, releasing massive quantities of carbon dioxide to the atmosphere. Carbon dioxide is a greenhouse gas, the hazardous consequences of which are listed in Chapter 18.

Control Measures

Some of the environmental problems associated with the recovery of oil from shales may be obviated if *in situ* retorting process is used. In this process, the release of oil by cracking and burning the shale is carried out underground. The production rate is nearly doubled and the spent shale is left underground. Thus, the problem of waste disposal is averted. The water used in the disposal of spent shale above ground is also saved. Lesser quantities of dust and sulfur oxides are released to the atmosphere.

However, the technology required for *in situ* conversion of shale to oil is yet to be perfected. In order that a flame front should burn evenly through a large underground region, a huge void must be created through the shale. This is uneconomical since natural shale is quite impermeable.

Shale oil has a low heating value. Its extraction requires massive and costly mining operations which result in extensive environmental hazards. All these factors suggest that the future production of oil from shales shall depend on the price of competitive sources of energy and an assessment of environmental problems.

Inference

Fossil fuels are the most widely used energy resources. Over the globe, all sections of community rely on these resources for their energy needs. However, extensive use of fossil fuels is causing extensive damage to the environment. Fossil fuels discharge more pollutants into the environment than any other energy resource. Yet, we cannot abrubtly stop using them. Nevertheless, we can make use of new and emerging technologies that obliterate the problem of pollution caused by fossil fuels.

One way to curtail the chemicalization of the environment is to make synthetic fuels from the existing fossil fuels. As has been described, coal may be converted into synthetic natural gas through the process of gasification. Contrary to coal, the synthetic natural gas burns without pollution.

The pollution problem may also be curtailed by developing alternative fossil fuel technologies. An example of such a technology is the *integrated coal gasification combined cycle* (IGCC). Here, coal is first subjected to gasification. The gasification requires high temperature and so the synthetic natural gas which emerges from the reactor carries with it a significant amount of heat. Normally this heat goes waste. However, the IGCC system taps this heat for making steam, which is then used to run a turbine of a power plant. Compared to thermal power plants, IGCC system emits negligible pollutants.

Yet another way to combat pollution is to develop alternatives to fossil fuels. Synthetic methanol and ethanol are being looked upon as the most attractive alternatives to fossil petrol. The alcohol fuels—called *gasohols*—cause less pollution than gasoline. Dimethyl ether is emerging as an alternative to diesel fuel. Dimethyl ether is comparable to conventional diesel fuel in terms of thermal efficiency—the efficiency of converting chemical energy to mechanical energy. However, in terms of auto-emissions, dimethyl ether gives better results. No soot is produced and the nitrogen oxides emissions are only 25% of those produced from conventional diesel engine.

Last, but far from least, it is also advisable to develop alternative energy resources—resources other than fossil fuels. Solar, geothermal and nuclear energy resources are less polluting than fossil fuels. Tapping these resources would be beneficial in yet another way—it shall put a brake on the fast rate of depletion of fossil fuels.

References

Atlas, R.M. (1996). Slick solutions, *Chem. Brit.*, **32,** 42–45.

Hall, C.A.S., Cleveland, C.J. and Kaufmann, R. (1986). *'Energy and Resource Quality'*, John Wiley & Sons, New York, p. 201–257.

Hammond, O.H. and Baron, R.B. (1976). Synthetic fuels: Prices, prospects and prior art, *Amer. Sci.,* **64,** 407–417.

Harker, J.H. and Allen, D.A. (1972), *'Fuel Science',* Oliver & Boyd, Edinburgh, p. 30–48.

Hirsch, R.L., Gallegher, J.E., Lessard, R.R. and Wesselhoft, R.D. (1982). Catalytic coal gasification: An emerging technology, *Science,* **215,** 121–127.

Keller, E.A. (1979), *'Environmental Geology',* 2nd Edn., Bell & Howell Co., Columbus, p. 421–426.

Laubier, L. (1992). Long term consequences of marine oil spills, *The World Scientist,* November, p. 36–47.

Metz, W.D. (1974). Oil shale: A huge resource of low grade fuel, *Science,* **184,** 1271–1275.

Pearce, F. (1995). Dirty rigs choke North Sea to death, *New Scientist,* May 6, p. 4.

Pearce, F. (1996). Black tide engulfs marine reserves, *New Scientist,* March 2, p. 5.

Perry, H. (1974). The gasification of coal, *Sci. Amer.,* **230**(3), 19–25.

Rouhi, M. (1995). Amoco, Haldor Topsoe develop dimethyl ether as alternative diesel fuel, *Chem. Eng. News,* May 29, p. 37–39.

Seewald, J.S. (1998). Sulphur greases the wheels, *Nature (London),* **391,** 128–129.

Spinks, A. (1982). Alternatives to fossil petrol, *Chem. Brit.,* **18,** 99–105.

Sundquist, E.T. and Miller, G.A. (1980). Oil shales and carbon dioxide, *Science,* **208,** 740–741.

Swannell, R.P.J. and Head, I.M. (1994). Bioremediation comes of age, *Nature (London),* **368,** 396–397.

Swinbanks, D. (1997). Oil spill threatens Japanese nuclear plants, *Nature (London),* **385,** 191.

CHAPTER 11

Terrestrial Energy

Two terrestrial energy resources, viz., geothermal energy and hydrogen-based economy are discussed in this chapter.

Geothermal Energy

Geothermal energy refers to the energy available from natural, underground reservoirs of steam and hot water. Natural heat of the earth represents a valuable and a vast resource. Tapping this resource to extract work from steam and hot water produces a minimal amount of air or water pollution. It is, therefore, not surprising that plans to utilize this clean source of energy are being made worldwide. The emphasis is on exploiting new geothermal reserves, so that the natural heat of the earth can be harnessed into commercially usable forms of energy.

Classification and Origin of Geothermal Reserves

The geothermal reserves on the earth may be classified into two broad types: *Hyperthermal* and *semi-thermal.* From hyperthermal areas, steam, as well as water, near its boiling point, may be extracted. Semi-thermal areas give out only water and that too much below its boiling point.

Hyperthermal Resources

The origin of hyperthermal fields is explained on the basis of the theory of *plate tectonics.* According to this theory there are several large mobile belts, called plates, below the earth crust. The plates, being submerged, are under a state of high stress. The temperature below the earth crust is not constant at all places. Due to the temperature difference the plates indulge in slow movements. When the moving plates collide with each other, their mechanical energy is converted into heat. The liberated heat is enough to melt the rock material and push it upwards, towards the crust. The molten rock, also called *magma,* heats up other rocks. The heat passes through a layer of porous rocks to semi-permeable rocks which are embedded immediately below the earth's surface (Fig. 11.1). The earth crust is made up of low thermal conductivity material. Therefore, it prevents the heat from escaping out.

Thus, geothermal energy is primarily stored in rocks. Since rocks are spread out over several square kilometers, the energy stored in them in quite diffused and must be concentrated and transported to the surface before it can be used.

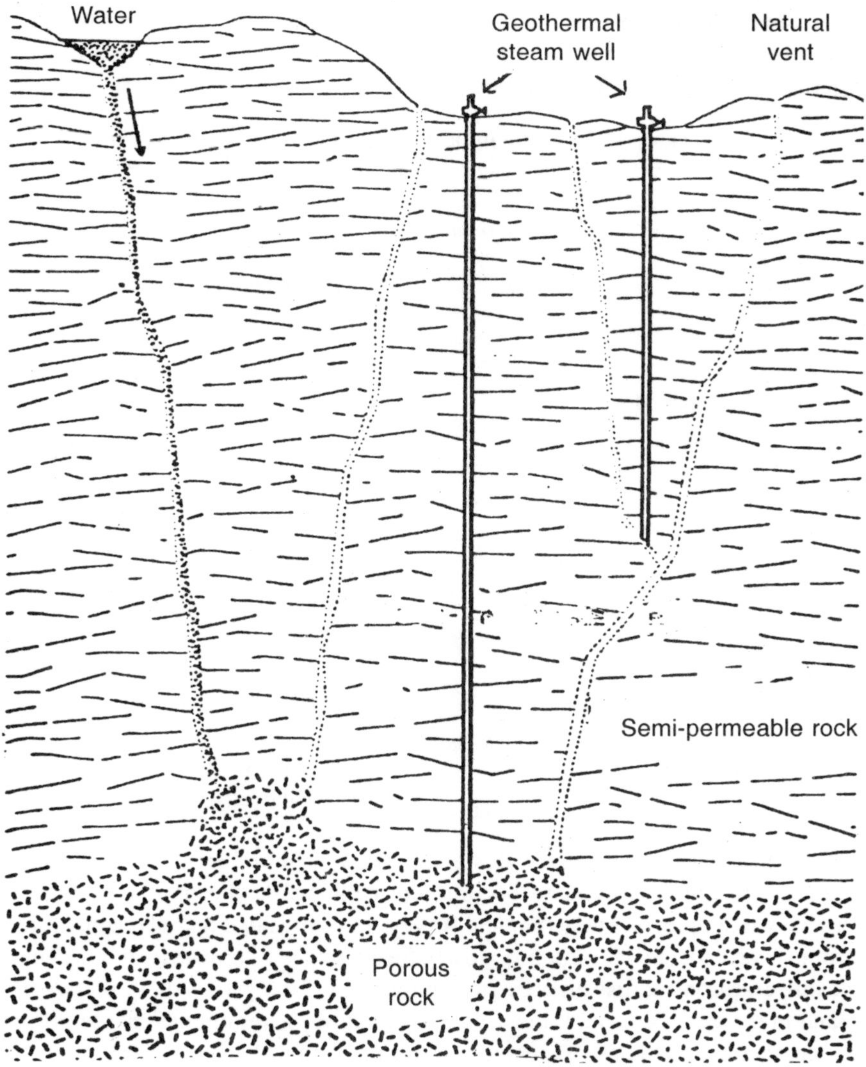

Fig. 11.1 A hyperthermal reserve

Mineral water stored in porous rocks and in cracks, fractures and faults provides the necessary medium for transfer of heat from the rocks.

The water in porous or fissured rocks is heated by geothermal energy to a temperature of about 250°C. Being at depth, the water is under high pressure and is, therefore, liquid. When the hot water can escape through natural vents it begins to boil, and part of it flashes off as steam. Such hot springs are called *geysers*. However, natural vents are scattered and scanty. In order to use geothermal energy, appropriate wells, 2–3 km deep, are drilled upto the porous rocks (Fig. 11.1). Both steam and hot water gush out of the wells. Steam is used for producing electricity and for space heating. Hot water is used for space heating.

The collision between highly stressed plates not only produces geothermal energy, but also causes volcanic activity. Hence, all the hyperthermal resources of geothermal energy are localized in active seismic zones. Some important hyperthermal sites are listed in Table 11.1

Table 11.1 Some important geothermal systems

Country	Field
Hyperthermal	
Iceland	Namafjall
	Reykjavik
Italy	Larderello
	Monte Amiatu
Mexico	Cerro Prieto
New Zealand	Kawerau
	Wairakei
USA	The Geysers
Semi-Thermal	
France	Melun
Hungary	Danubian
Iceland	Hengill
	Husavik
New Zealand	Rotorua
China	Cona
USA	Nevada
Hot Dry Rock Technology	
New Mexico	Fenton Hill
UK	Falmouth

Hyperthermal sites are not widespread. Yet, their number is not static. Over the eons, the old sites tend to become dormant, while the new sites became active. For example, recent geological surveys have revealed that beneath western Europe, a vast sheet of hot rock is slowly forcing its way upwards. It has caused volcanic eruptions across an area stretching from Canary Islands in the west to Poland in the east. No doubt, the volcanic threat in the region has increased, yet new geothermal fields are expected to crop up in the area.

Semi-Thermal Resources

Semi-thermal resources are not associated with volcanic activity. They lie outside the seismic zones, where geologic activity is less intense. Hence, not much heat can be tapped from these resources. Only hot water, generally in the temperature range 50–80°C, can be extracted from these fields. Nevertheless, semi-thermal sites are more widely distributed as compared to hyperthermal sites. Some important locations of semi-thermal fields are listed in Table 11.1

Semi-thermal fields originated because of one or more of the following mechanisms.

1. In fracture dominated geologic areas, deeply circulating water was heated by earth's natural heat and then transmitted to the surface through faults. Most of the hot springs originated in this manner.

2. In certain areas the concentrations of radioactive elements, such as uranium, thorium and potassium were higher than normal. These generated heat and increased the temperature locally.

3. The temperature of the earth increases with depth. However, the surface is not heated up by the embedded hot rocks because the clay material, which constitutes the bulk of the crust, has very low thermal conductivity. Yet, at some places the crustal material is not clay, but cystalline quartzine rocks. The latter have significantly high thermal conductivity. Thus, where the earth crust was made up of crystalline rocks, the surface was heated up and semi-thermal fields originated.

Whatever be the mechanism of the origin of semi-thermal areas, in order to exploit geothermal energy, wells have to be dug to a depth of 2–3 km. These wells give out only hot water, not steam. Hot water cannot run the turbine. Therefore, energy from semi-thermal areas can be tapped only for space heating and not for producing electricity. Yet, at a depth of 6–7 km below any semi-thermal area, there are vast dry and hot rock formations. By injecting water that deep, steam can be produced to generate electricity. However, to drill wells to a depth of 6–7 km is expensive and not always feasible. Nevertheless, where it is absolutely necessary to produce steam for power generation, the energy of the hot and dry rocks can be tapped by a process called hot dry rock (HDR) technology.

Hot Dry Rock Technology

Most of the semi-thermal fields contain vast, hot, dry rock formations at a depth of 6–7 km. These fields possess more stored heat per unit volume than any of the other geothermal system. However, they lack circulating water which can transfer the heat energy to the surface. In order to make a major impact on energy supply, hot dry rock technology attempts to tap the heat energy of the hot, dry rocks.

This technology engineers drilling of wells upto the hot, dry rocks and then fracturing them by using explosives. Then cold water, under pressure, is injected into the rocks at one location and steam flashes out at another location (Fig. 11.2). The induced circulation of water thus mines the heat of dry rock system. The steam so produced can be used for electricity generation.

Hot dry rock technology has several drawbacks. As has been stated, drilling to a depth of 6–7 km is not possible at all locations. A significant fraction of water injected into the rocks leaks away and cannot be recovered. The cost of engineering hot dry rock technology is enormously high. It is also important to ascertain the thermal lifetime of the rocks. If the rock formations cool too quickly, the economics of the scheme does not permit pursuit. The project is considered useful only if the rocks are able to supply steam for 15 years or more. Hot dry rock technology has been put to use only at a very few locations (Table 11.1).

Applications of Geothermal Energy

The applications of geothermal energy fall into two broad categories depending upon whether the energy is being harnessed from steam or hot water.

1. Energy extracted from steam

Steam from hyperthermal resources or those semi-thermal resources which are

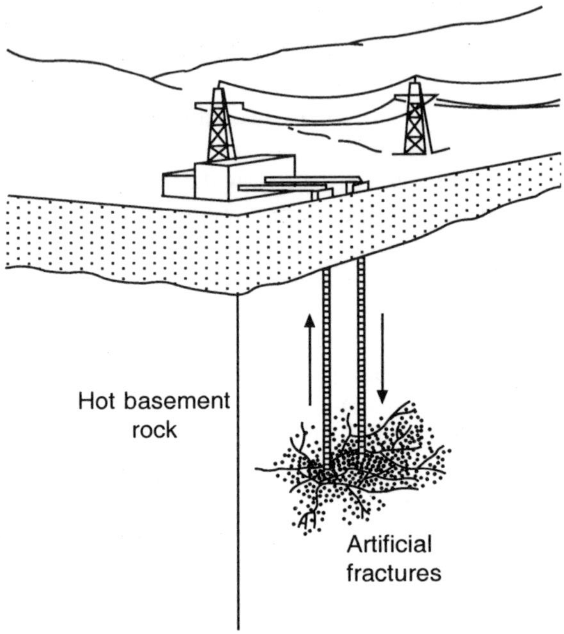

Fig. 11.2 Extraction of geothermal energy by hot dry rock technology

subject to hot dry rock technology can be exploited for power generation. The steam from these resources is usually at pressures ranging from 5 to 20 atmosphere. It is piped directly to the turbines. In order to minimize piping costs the plant must be located close to the steam well. Moreover, there is a pressure loss after long distance transportation of steam and this decreases the efficiency of power production. Hence, it is appropriate to construct the power generating units within the geothermal field itself.

Generation of electrical power is the primary objective of steam producing geothermal fields. However, steam from these fields can be put to other uses as well. In locations near the ocean or a salt water lake, the steam is employed as a heating medium in distillation plants for producing potable water by subatmospheric boiling. Steam is also used for drying of timber, drying of diatomaceous earth and evaporation of highly concentrated solutions.

2. Energy extracted from hot water

Hot water from semi-thermal resources can be used for space heating. Examples include the control heating systems for houses. As shown in Fig. 11.3, geothermal water is pumped to a storage tank from where it flows to a series of houses. Geothermal water may also be used in air conditioning. Here, water is used as a refrigerant and a solution of lithium bromide as a low temperature absorbent fluid. The refrigerant is vaporized by extracting heat from the surroundings. The surroundings, therefore, become cool. The refrigerant vapours are absorbed by lithium bromide solution.

External heat, supplied by geothermal water, drives the refrigerant off the

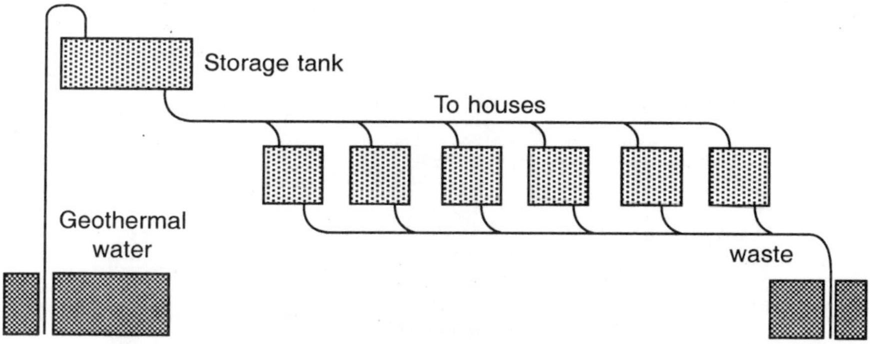

Fig. 11.3 Residential heating system by geothermal energy

absorbent as vapours. The vapours are condensed to liquid, which returns to the evaporator to begin the cycle again.

Hot water is also used for maintaining greenhouses and swimming pools at appropriate temperatures. In cold countries, fish farming and mining operations are carried out at optimum temperatures using geothermal water as a heating source. Spas and therapeutic bathing is also an important traditional application of space heating by hot water.

Geothermal Resources in Iceland and New Zealand

Iceland is richly endowed with geothermal resources. These resources have been developed commercially by the National Energy Authority of Iceland. They have now been exploited to such a degree that, on a *per capita* basis, Iceland is the largest user of geothermal energy. A flow diagram showing utilization of geothermal energy is Iceland is depicted in Fig. 11.4

The plant at Namafjall produces about 240 tonnes h^{-1} of steam from wells. The steam is used to dry diatomite from a nearby lake bed and to generate 2.5 MW of electricity, while the separated hot water is used for community space heating. The combination of electricity produced by steam and space heating by separated hot water is highly efficient. The Reykjavik plant supplies hot water to cater to most of the city's heating needs and recreational activities (Fig. 11.5). In the nearby Hengill field, water heats houses and greenhouses.

New Zealand too enjoys an enormous geothermal resource. The Kawerau plant produces about 200 tonnes h^{-1} of steam. A part of the steam is used for drying timber, while the remaining drives a turbine to produce 10 MW of electricity for the timber and paper industry. At Rotorua, more than a thousand hot water wells supply thermal energy to individual homes, as well as to schools, hospitals and industries. A major hotel with 100 rooms uses a geothermal well to produce air conditioning by means of a lithium bromide absorption unit.

Environmental Impacts of Geothermal Resources

Consumption of energy from any of its sources leaves scars on the environment. Geothermal energy is no exception. Yet, geothermal resources produce energy with minimum damage to the environment. The impacts of geothermal resources

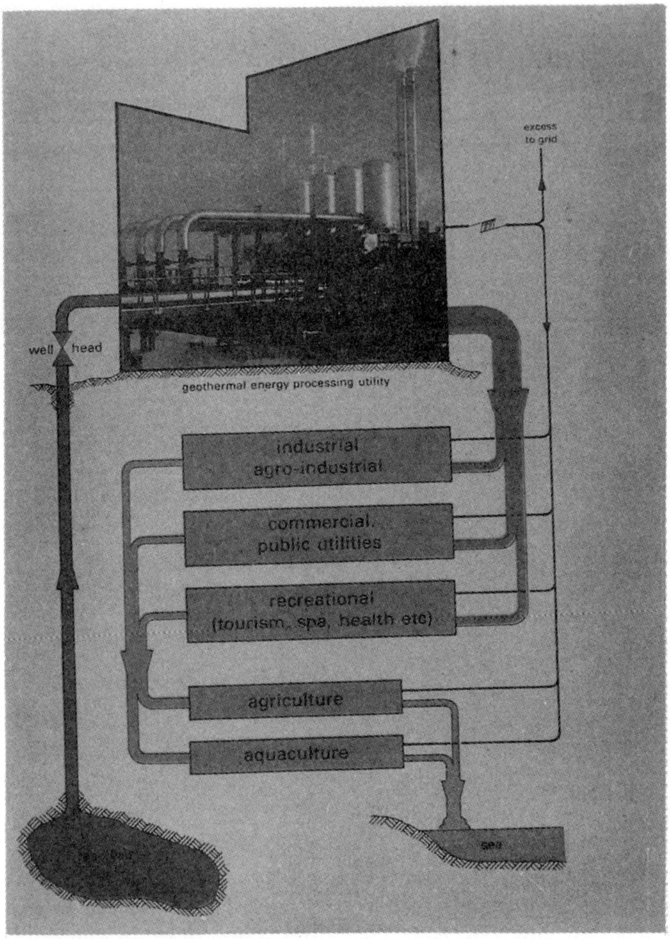

Fig. 11.4 A geothermal energy utilization plan (Reproduced from Orkustornum with the permission of National Energy Authority of Iceland, Reykjavik)

on the environment are listed below. An assessment of these deleterious effects is also examined.

1. Landscape scarring

Landscape scarring results from such diverse activities as laying roads, drilling wells, installing pipeline and erecting power plants—all factors in the commercial exploitation of geothermal resources. These effects have a great influence upon land utilization patterns in any region.

The visual pollution is a problem associated with all energy sources producing electricity. Yet, in geothermal power production, every phase of the fuel cycle is localized at one site. Other kinds of power plants need considerable industrial back-up, such as mines, transportation facilities and processing plants. All this enlarges the area of environmental damage.

Moreover, installations of geothermal units do not interfere with other activities.

Fig. 11.5 A revolving restaurant in Reykjavik built on top of the city's geothermal water tanks (Reproduced from Orkustornum with the permission of National Energy Authority of Iceland, Reykjavik)

For example, the Larderello plant in Italy exists amidst agricultural region. The unused land in the vicinity of the Geysers plant in the USA is employed as a deer reserve and for cattle grazing.

2. Land subsidence

One of the hazards of tapping geothermal energy is that the land may subside when large amounts of water are withdrawn from geothermal reservoirs. Land subsidence occurred at the Wairakei field in New Zealand from where water was drawn out at a rate of 70 million tonnes per year.

This problem can be controlled by limiting withdrawals from the fields to a safe rate. Moreover, after the heat energy from water has been used, the water may be reinjected into the geothermal system. This methodology has been adopted by most hyperthermal plants. First, steam is separated from water and is used to generate electric power. Then, water is injected back into the earth by a disposal well (Fig. 11.6). Continuous withdrawal of petroleum also causes land subsidence. This problem too is avoided by injecting water into the earth so as to compensate for the withdrawn oil.

3. Air pollution

Geothermal energy resources emit two gaseous pollutants, viz., carbon dioxide and hydrogen sulfide. The amount of carbon dioxide released per unit of power produced is, however, much less than that emitted by fossil fuel plants. Hydrogen sulfide, no doubt, has degradative effects on the environment. Nevertheless, hydrogen sulfide can be easily liquified and, therefore, easily removed by condensation. Moreover, most geothermal plants are located far away from

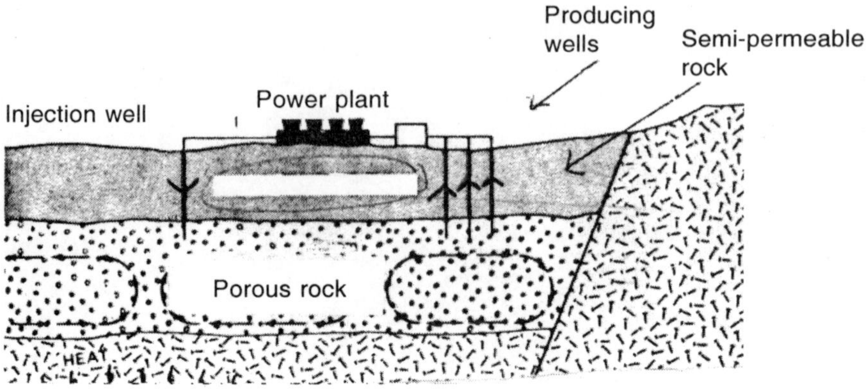

Fig. 11.6 Reinjection of geothermal water

habitable areas. So the hydrogen sulfide problem is not considered very acute. Further, unlike fossil fuel plants, geothermal units do not emit sulfur oxides, nitrogen oxides or particulates.

4. Water pollution

The hot water from many fields dissolves appreciable amounts of boron compounds, sodium chloride and other chemicals from rock materials. These compounds can elicit toxic effects if the geothermal water is discharged into streams. The stream water may become unfit for irrigation and drinking purposes. The problem may be avoided by reinjecting geothermal water into the earth, rather than discharging it into streams. This will not only put back toxic chemicals below the earth's surface but would also avert the chances of land subsidence.

Inference

The effects of geothermal units on the environment are ridiculously small in comparison with the impacts produced by fossil fuel plants. The latter pollute the biosphere with combustion products. Geothermal resources operate without combustion, using naturally occurring steam and hot water. Moreover, geothermal plants are compact, using comparatively small areas of land. The fossil fuel plants, on the other hand, need large land areas, taking into account the mining, transportation, handling and water disposal facilities.

When compared with nuclear power plants, geothermal resources do not produce any radiation hazards. Further, in nuclear plants high energy radioactive waste requires adequate and permanent insulation from accidental entry into the environment. This entails reserving large areas of surface land, as well as subterranean storage facilities. Geothermal plants are free from these complications.

Unlike hydropower systems, geothermal plants do not flood fertile lands or generate stress that may lead to earthquakes. Moreover, they do not have a dislocatory effect, as is common in the construction of dams.

It is thus obvious that geothermal energy can be used at little or no cost to the environment. Yet, at present, the technology for tapping geothermal energy is in the developing stage. Many geothermal fields have not been engineered to militate against the energy crisis. However, it is a matter of time before the geothermal resources make an increasing contribution towards alleviation of energy shortage and reduction of pollution.

Hydrogen-Based Economy

The dwindling reserves of fossil fuels and rising prices of oil demand alternative routes to overcome energy crisis. One possible route is a fuel economy based on hydrogen. Unlike fossil fuels, geothermal energy or solar energy, hydrogen is not a primary source of energy. It is an energy carrier. On combustion it produces only water. Hence, the advantage of hydrogen is its virtually pollution-free combustion. It is, therefore, destined to become the fuel of the future.

Production of Hydrogen

If hydrogen is to takeover as a fuel of the future, it must be produced economically and in large quantities. Hydrogen can be produced commercially by either of the following methods.

1. Cracking of hydrocarbons

During refining of petroleum, larger hydrocarbon molecules are fragmented into smaller molecules. This process is called *cracking*. One of the products of hydrocarbons' cracking is hydrogen.

$$C_nH_{2n+2} \rightarrow C_nH_{2n} + H_2 \tag{11.1}$$

The problem with this method of producing hydrogen is that the hydrocarbon fuels are already in short supply. Hence, there is a need to follow alternative synthetic routes to hydrogen production.

2. Oxidation of hydrocarbons

Hydrogen can also be produced by partial oxidation of hydrocarbons. Steam and hydrocarbons are passed over nickel catalyst, whereby the feed is converted into carbon monoxide and hydrogen. The mixture is treated with more steam and passed over an iron oxide-chromium oxide (Fe_2O_3–Cr_2O_3) catalyst.

$$CO + H_2 + H_2O \rightarrow CO_2 + 2H_2 \tag{11.2}$$

As shown in equation (11.2), carbon monoxide is converted to carbon dioxide. The yield of hydrogen increases. Carbon dioxide and hydrogen are easily separated. A stream of water is sprayed over the mixture. Carbon dioxide, being soluble in water, enters the solution phase. Hydrogen, being sparingly soluble, remains in the gaseous form. Like the previous method, this technique has a limited utility because it uses scarce hydrocarbon fuels.

3. Cracking of water

Water can be fragmented into hydrogen and oxygen under the influence of strong heat.

$$H_2O \xrightarrow{\text{Heat}} H_2 + \tfrac{1}{2}O_2 \tag{11.3}$$

The thermal energy required for reaction (11.3) may be provided by a *solar furnace*. A solar furnace incorporates a set of mirrors which focus solar radiations at a particular spot. Sufficiently high temperatures may be provided by this method. However, thermal cracking of water requires temperatures as high as 2500°C and the technology for producing such high temperatures with the help of a solar furnace is yet to be perfected.

Alternately, water may be decomposed by heat generated in a nuclear reactor. However, fission reactors cannot produce temperatures of the order of 2500°C. Fusion reactors, no doubt, can produce heat sufficient to fragment water, but this technology too is in the developing stage.

Thus, although thermal cracking of water seems to be a feasible method of producing large scale hydrogen, at present this method is of academic interest only. Nevertheless, recently the prospects of a cheap supply of hydrogen have moved closer to reality. Water has been split into its elements, hydrogen and oxygen, at room temperature, using cuprous oxide catalyst in presence of visible light of wavelength 460–650 nm. The catalyst works for 1900 hours.

4. Electrolysis of water

The most promising approach to large scale production of hydrogen is electrolysis of water containing a trace of alkali.

$$H_2O \xrightarrow{\text{Electrolysis}} H_2 + \tfrac{1}{2}O_2 \tag{11.4}$$

Two types of electrolyzers are used for production of hydrogen. The first type employs cells with monopolar electrodes. Alternate cathodes and anodes are separated by porous diaphragms to prevent mixing of gases. The cathodes and anodes are electrically connected in parallel. This arrangement requires a large floor area. The second type of electrolyzers employ cells with bipolar electrodes. Such electrodes are cathodic on one side and anodic on the other. The electrodes are connected in series and suitably insulated. This arrangement results in a more compact installation.

5. Thermochemical cycles

A thermochemical process which can decompose water at low temperature is also suited for producing hydrogen. The following multistep reaction sequence is an important thermochemical cycle which can cause low temperature decomposition of water.

$$CaBr_2 + 2H_2O \rightarrow Ca(OH)_2 + 2HBr \tag{11.5}$$

$$Hg + 2HBr \rightarrow HgBr_2 + H_2 \tag{11.6}$$

$$HgBr_2 + Ca(OH)_2 \rightarrow CaBr_2 + HgO + H_2 \tag{11.7}$$

$$HgO \rightarrow Hg + \tfrac{1}{2}O_2 \tag{11.8}$$

$$\text{Net reaction: } H_2O \xrightarrow[\text{Decomposition}]{\text{Thermochemical}} H_2 + \tfrac{1}{2}O_2 \tag{11.9}$$

All the steps (reactions 11.5–11.9) in the above sequence can be carried out below 750°C. The raw materials, calcium bromide and mercury, are regenerated in the course of thermochemical process. However, the process requires careful handling operations, since hydrogen bromide and mercuric bromide, if allowed to escape, would elicit characteristic toxic and corrosive effects.

Storage and Transportation of Hydrogen

In gaseous state, hydrogen has a relatively low density even at high pressures. Hence, very large containers are required to store it. Therefore, normally hydrogen is stored either in liquid state or in solid state.

Liquid hydrogen, called *cryogenic hydrogen,* can be stored in special containers in which extremely cold temperatures and high pressures can be maintained. In the solid form, hydrogen may be stored as metal hydrides.

$$Ti + H_2 \rightleftharpoons TiH_2 \tag{11.10}$$

$$4Pd + H_2 \rightleftharpoons 2Pd_2H \tag{11.11}$$

Hydrogen occupies interstitial sites in the metallic lattice of titanium or palladium. Reactions (11.10) and (11.11) are exothermic in nature. Hence, when the temperature is raised the hydrides release hydrogen through decomposition. This storage method would be ideal if hydrogen were to be used to operate automobiles. The waste heat from the engine could be used to generate hydrogen fuel, as and when desired.

Hydrogen can be transported safely either in the liquid state or in the solid form. Even gaseous hydrogen can be transported over long distances through the same pipeline which now transports natural gas.

Applications of Hydrogen-Based Economy

Hydrogen as an energy carrier has widespread applications. At present, however, all the applications cannot be put to practical usage. Yet, in near future these applications would negate, at least in part, the threat of energy crisis.

1. Fuel cells

In near future, fuel cells using hydrogen as the fuel, would be used for generation of electricity, as well as in electric vehicle propulsion. A fuel cell is an electrochemical device which converts chemical energy into electrical energy.

In hydrogen-based fuel cells, generally an aqueous solution of sodium hydroxide or potassium hydroxide is used as an electrolyte. Hydrogen and oxygen are bubbled through the carbon electrodes into the electrolyte. The following reactions take place.

$$\text{At anode:} \quad [H_{2(g)} + 2OH^-_{(ag)} \rightarrow 2H_2O_{(l)} + 2e^-] \times 2 \tag{11.12}$$

$$\text{At cathode:} \quad O_2 + 2H_2O_{(l)} + 4e^- \rightarrow 4OH^-_{(aq)} \tag{11.13}$$

$$\text{Net reaction:} \quad 2H_2(g) + O_2(g) \rightarrow 2H_2O_{(l)} \tag{11.14}$$

Equation (11.14) shows that water is the only product discharged by fuel cells. Thus, unlike conventional methods of generating electricity, fuel cells do not pollute the environment.

We have already stated that the most practical and feasible method of producing hydrogen is the electrolysis of water. No doubt, this method consumes electric power, yet during off-peak hours, when power is not being utilized to full extent, a part of the electrical energy may be directed to produce hydrogen. The hydrogen, in turn, can be used for producing electricity through the fuel cells when the demand exceeds the supply or in remote places where power generating resources are scarce.

Thus, electricity and hydrogen-based economy complement each other. Electricity can be produced efficiently and without damage to the environment from hydrogen through the fuel cells; hydrogen can be derived from electricity. The complementarity of hydrogen and electricity is further enhanced by the fact that hydrogen, unlike electricity, can be stored and easily transported over long distances.

Fuel cells can also be used in electric vehicle propulsion. In due course, vehicles using fuel cell systems will be quite common. These vehicles will not emit carbon dioxide, sulfur oxides or particulates. Moreover, if the cell reaction (11.14) is carried out at low temperatures, even nitrogen oxides will not be formed. In addition, this technology will reduce the burden on fossil fuels.

2. As indirect fuel

Hydrogen can be used for producing or refining other fuels. For example, it can react with carbon monoxide under a pressure of 300 atmosphere, in presence of zinc oxide-chromic oxide catalyst ($4ZnO–CrO_3$), heated to 300°C.

$$CO + 2H_2 \rightarrow CH_3OH \tag{11.15}$$

Methanol, so produced, can be used as a fuel.

Hydrogen can be used for coal gasification, a process which produces methane. The latter, being the principal constituent of natural gas, is an important fuel. Hydrogen is also used for refining oil shales. Hydrogenation of shale oil converts the undesirable unsaturated hydrocarbons into more combustible, saturated hydrocarbons, upgrading the fuel value of oil.

3. Space and process heating

In heating service, hydrogen offers advantage over natural gas because of its hotter flame. In conventional practice, gas used for heating requires a duct to carry away combustion products. With hydrogen this is unnecessary, since water is the only product of combustion. Thus, smaller appliances and simpler ventilating systems are required where hydrogen is being used for space or process heating.

Environmental Impacts of Hydrogen

Despite its potential to act as a clean energy carrier, the use of hydrogen leaves some impacts, on the environment. These impacts are, however, relatively minor and can be erased by adopting suitable technologies.

1. Production stage

The electrolysis of water produces a mixture of hydrogen and oxygen (reaction 11.4). The two gases form an explosive mixture. Hence there is a need to prevent the interaction of hydrogen and oxygen.

The threat that hydrogen can explosively react with oxygen is, no doubt, real, but even natural gas and oxygen form an explosive mixture. This fact has, however, not lessened natural gas's acceptability as a fuel. Moreover, it is quite easy to prevent the mixing of hydrogen and oxygen in an electrolyzer by partitioning the cathodic and the anodic compartments. In fact, it is being envisaged that oxygen, produced as a by-product in electrolysis of water, can be used for sewage treatment or for roasting of ores.

2. Storage and transportation stage

Hydrogen molecules, being small in size, are able to diffuse into the interstitial spaces in the metallic lattice, making the metal brittle and susceptible to cracking. Therefore, it is likely that in course of time, metallic containers in which hydrogen is stored or metallic pipeline through which hydrogen is transported will crack and the gas will leak out. The problem can be averted by applying a plastic coating on the inside of hydrogen containers or pipeline.

3. Application stage

It has been stated that the combustion of hydrogen produces only water. However, sometimes a side reaction also produces hydrogen peroxide.

$$H_2 + O_2 \rightarrow H_2O_2 \tag{11.16}$$

The hydrogen peroxide would discharge peroxy radicals to the atmosphere. These radicals have the potential to build-up photochemical smog.

Peroxy radicals, no doubt, are harmful, but the yield of hydrogen peroxide (reaction 11.16) is extremely low. Moreover, when hydrogen-based economy would be put to use on a large scale, catalytic converters can be installed wherever hydrogen is being burnt as a fuel. These converters would destroy hydrogen peroxide before it is able to escape into the atmosphere.

Inference

The only fear of extending the scope of hydrogen-based economy is that hydrogen, during its production, storage, transportation or application stage, can diffuse out of the equipment, mix with aerial oxygen and explode. This fear is often referred to as *Hindenburg Syndrome,* after an accident in New Jersey, USA, in 1937 when hydrogen leaked out of a container and caught fire. However, since then no major accident involving hydrogen has been reported. This may be attributed partly to better technological handling of hydrogen and partly to the fact that the gas has such a high rate of diffusion that it is rapidly removed from the site of spill, minimizing the time during which ignition can occur.

The high rate of diffusion of hydrogen also offers the advantage of an increased flow rate through the pipeline. That hydrogen can be carried over long distances in pipeline and that it is interchangeable with electricity, without polluting the atmosphere, offers an optimum environmental solution for future power requirements.

At present hydrogen-based economy, like geothermal energy sources, is in the developing stage. Yet, there are growing indications that hydrogen will be the principal fuel of the twenty first century.

References

Abraham, B.M. and Schreiner, F. (1973). A low temperature thermal process for the decomposition of water, *Science,* **180,** 959–960.

Barnea, J. (1972). Geothermal power, *Sci. Amer.,* **226**(1), 70–77.

Bauer, E. (1973), *'Wonders of the Earth',* Collins Publishers, London, p. 15–21.

Bowen, R. (1979), *'Geothermal Resources'*, Applied Science Publishers Ltd., London, p. 167–174.

Ellis, A.J. (1975). Geothermal systems and power development, *Amer. Sci.,* **63,** 510–521.

Fletcher, E.A. and Moen, R.L. (1977). Hydrogen and oxygen from water, *Science,* **197,** 1050–1056.

Gates, D.M. (1985), *'Energy and Ecology',* Sinauer Associates, Inc., Sunderland, p. 344–349.

Griffith, E.J. (1974). Hydrogen fuel, *Nature (London),* **248,** 458.

Hadfield, P. and Warden, R. (1998), Catalysts for change, *New Scientist,* February 28, p. 10.

Hammond, A.L. (1972). Geothermal energy: An emerging major resource, *Science,* **177,** 978–980.

Harrison, R. (1992). Applications of geothermal energy, *Endeavour,* **16,** 29–36.

Heiken, G., Murphy, H., Nunz, G., Potter, R. and Grigsby, C. (1981). Hot dry rock geothermal energy, *Amer. Sci.,* **69,** 400–407.

Jones, L.W. (1971). Liquid hydrogen as a fuel for the future, *Science,* **174,** 367–370.

Keller, E.A. (1979), *'Environmental Geology',* Bell & Howell Co., Columbus, p. 432–441.

Kordesch, K.V. and Simader, G.R. (1995). Environmental impact of fuel cell technology, *Chem. Rev.,* **95,** 191–207.

Marchetti, C. (1977). The hydrogen economy and the chemist, *Chem. Brit.,* **13,** 219–222.

Walker, G. (1995). Deep heat cracks up Europe's crust, *New Scientist,* March 11, p. 20.

Wentorf, R.H. and Hanneman, R.E. (1974). Thermochemical hydrogen generation, *Science,* **185,** 311–319.

CHAPTER 12

Solar Energy

The sun's energy is the most fundamental form of energy. It is also the largest pool of energy. The solar flux is enormous—about 1.8×10^{18} Kcal reach the earth each day. Moreover, it is a renewable resource. It is the only source that is completely and rapidly transformed into heat at ambient temperatures. This means that extraction of energy from solar radiation causes a minimal damage to the environment. The sun is also a reliable energy source in the sense that a given place will receive about the same amount of energy in form of sunshine every year. It is an unreliable source, however, in that sunshine is intermittant. It is not available at night or in unfavourable weather. Yet, at a time when the fossil fuels are fast depleting, nuclear energy resources have not gained acceptability and geothermal resources are not completely developed, solar energy is treated as a panacea for meeting the continuously growing human energy demands.

Utilization of Solar Energy

Solar energy may be harnessed in two ways: Firstly, it may be directly converted into useful work. This is referred to as the direct utilization of solar energy. Secondly, solar energy stored as a potential resource in water, wind and biomass may be tapped for useful purposes. This is referred to as the indirect utilization of solar energy.

Direct Utilization of Solar Energy

Direct utilization of solar energy offers a simple and quick way of converting sun's energy into thermal, thermochemical and electrical work.

Direct Utilization by Thermal Conversion

Direct conversion of solar energy into heat has the following applications.

1. Heating water. The solar flux on earth, though abundant, is dilute. It is unevenly and irregularly distributed and has a low energy density. Atmospheric particulates, cloud cover and fog further dilute the flux. Therefore, before the solar energy can be harnessed for performing useful work, it must be concentrated by solar collectors.

For heating water for domestic purposes, a *flat plate collector* is used. It consists of black radiation absorbing plates, covered with glass and attached to pipes that circulate water to remove useful heat. Hot water is emptied in storage tanks provided with built-in electric heating system, to be used on a cloudy day or during night time. Glazing the collector's surface improves its performance by greatly reducing convective and conductive heat loss to the surroundings. The glass covering allows the short wavelength solar radiation to strike the collector's plate, but retains the transmitted, longer wavelength radiation, thus trapping the heat. The abosrbing capacity of the plate can be increased with the application of a surface coating, such as black chrome.

Flat plate collectors can achieve working temperatures of 70–150°C. The collectors are mounted horizontally or are tilted to the south at an angle approximately equal to the latitude. Such collectors are quite economical and respond to even diffused sunlight.

2. Heating houses. Two distinct design strategies are usually adopted for heating houses: Active and passive. In *active heating,* first hot water is obtained using externally mounted flat plate collectors. Then blowers are used to transfer the heat from water to air. The solar heated air is circulated through a material, like pebble bed or gravel, which possesses heat storage characteristics. Using an electrical or mechanical device, the hot air is circulated as and when desired.

Passive heating uses natural mechanisms of radiation and convection to heat a residential building. In passive homes, south-facing windows, located adjacent to indoor spacing, provide a desired access to sunlight. Walls and floors are constructed from materials possessing heat-storing properties. Thus, sensible architectural design of passive buildings replaces expensive equipment of active houses.

Passive design has many advantages. It has no moving parts to wear out or fail, so is quite reliable and requires no maintenance. Since there are no pumping losses, a passive design is more efficient than an active one. In passive houses, a solar contribution of 80% of total household heating requirements are obtainable even in unfavourable climates. Payback time for the small added cost of design is very short. Finally, solar buildings can take many forms, so aesthetic or cultural acceptance is not a problem.

3. Drying. Solar drying of grains is a standard practice in agriculture. To ensure good quality harvest, the stored seeds must be dried. After the harvest, the food grains too must be dried before they are stored. Otherwise, the food crop tends to be infected by fungus or bacteria. Drying is carried out by blowing solar heated air through the grains. This also rids the crop of the adhering dust.

4. Desalination of sea water. In many parts of the world, an abundant supply of water is taken for granted. Not so in the semi-desert and desert areas. The supply of water is the limiting factor for growth in these areas. Yet, as the world's population grows and the pressure for living space increases, desert lands are being given increasing attention. Solar desalination of sea water is the only way to ensure an adequate water supply to desert lands. Solar distillation also holds a promise for those regions where a water supply that was formerly complacently considered adequate, but is now threatened by pollution.

Solar distillation of sea water involves a simple equipment. Sea water is placed in a black tray tightly covered with a transparent roof which allows the solar radiation to pass through and heat-up water. The water vaporizes and condenses on the air cooled roof of the tray. It is then collected in a trough which drains into a storage vessel.

5. Solar furnaces. Solar furnaces are used for producing very high temperatures for scientific investigations. As has been stated, flat plate collectors can achieve temperatures in the 70–150°C range. Solar furnaces, which operate upto 3000°C, use another type of collector called a *focusing collector.* The latter focuses the solar radiation on a small area. Usually this is done through the use of a set of parabolic mirrors. The mirrors are continuously turned in the direction of sun. Large parabolic reflectors and their heavy mountings are very expensive. Moreover, they cannot function in diffused sunlight and so can be used only on clear, sunny days. Nevertheless, they offer the advantage in that the heating is pollution free.

Some important applications of solar furnaces are:

(a) Studying physical and chemical properties of refractory materials at very high temperatures.
(b) Melting and sintering of ceramics, such as zirconia.
(c) Purifying materials by zone refining.
(d) Studying chemical kinetics using flash photolysis, followed by quick chilling, thus permitting the isolation of intermediate, unstable compounds, which are not isolable with conventional heating and cooling.

6. Cooling. In many isolated tropical countries, the cost of electricity is too high to afford air conditioning. In near future, solar air conditioning is speculated to substitute for conventional air conditioning. Solar cooling is based on the principle of absorption-desorption. Water is heated by flat plate collectors. The vapours so formed are absorbed by a concentrated solution of lithium bromide. The solution becomes dilute and hot. It causes liquid water in another compartment to vaporize and produce a cooling effect. The lithium bromide solution is concentrated again by heating the diluted solution with solar radiation and the system is operated on a continuous basis.

Once it becomes operational, solar cooling is likely to be used not only as a luxury but also for maintaining cold storages for fruits and vegetables.

7. Cooking. Solar cooking is a simple way of using solar heat to substitute for the scarce fossil fuels. A solar cooker consists of a plastic container having four utensils. The utensils are made of aluminium and are blackened from the outside, so as to effectively absorb solar radiation. The lid of the container holds a focusing mirror which directs the sunlight to the utensils. The lid has to be adjusted to follow the direction of the sun. Solar cookers have been seriously accepted in those regions where bright sunlight is reliable, wood is not easily available and conventional fuels are expensive.

Direct Utilization by Thermo-Mechanical Conversion

Conversion of solar energy into thermo-mechanical work has proved to be quite useful for irrigation in the tropical regions. It is based on the idea of a *solar pump*.

A solar pump incorporates flat plate collectors which heat-up water. The accumulated heat energy is then transferred to a low boiling liquid like butane in a heat exchanger. The pressure produced by butane vapours is sufficient to drive a pump. The gas is then cooled in a condenser and liquified. The liquified gas is reinjected into the evaporator and a new cycle starts.

Though a solar pump has a low thermodynamic efficiency, it has the advantage of great reliability and little need for maintenance. Unlike steam engines, there are no boiler hazards and no necessity of treating water for the boiler. Moreover, a solar pump can be operated even by unskilled labour.

Direct Utilization by Electrical Conversion

The energy of the sun can be directly converted into electricity by using solar or photovoltaic cells. These cells are already being used for producing electricity in space vehicles. However, their high manufacturing cost limits their use for producing electrical power for conventional applications. Nevertheless, it is speculated that in near future new kinds of fabrication may make them economically competitive. Once this is achieved, photovoltaics would serve as the most promising and practical alternatives for supplementing—or even replacing—fossil fuels for generating electricity.

Mechanism of photovoltaics. The solar cells that have been employed so far are made from silicon. In a crystal lattice of silicon, each atom is tetrahedrally coordinated. Thus, each of the four valence electrons of silicon is shared with four neighbouring atoms. Such a crystal is a perfect insulator because no electrons would be free to move if a voltage was applied to it.

In order to make it conducting, it is necessary to build into the silicon's lattice, atoms that supply either more or fewer electrons than the four that are required for the ideal structure. For example, if a few phosphorus atoms, having five valence electrons, are doped in a silicon crystal, the basic structure of the lattice will remain the same. However, it will have extra electrons as there are no bonding sites available for them. A phosphorus-doped silicon crystal is called an *n-type semiconductor* because it has extra negatively charged electrons.

Conversely, if silicon is doped with boron, which has three valence electrons, the fourth coordination site around boron remains vacant. A vacant bonding site is called a positive hole and a boron-doped silicon crystal is called a *p-type semiconductor.*

Doping a crystal makes it preferentially receptive to electrons or holes. An *n*-type crystal, for example, can readily absorb extra electrons; a *p*-type crystal has an affinity for holes.

A silicon solar cell is a wafer of *p*-type silicon with a thin layer—about half a micron—of *n*-type silicon at the top (Fig. 12.1). When a photon of light with appropriate amount of energy penetrates the cell near the junction of two types of crystals and encounters a silicon atom, it dislodges one of the electrons, leaving behind a hole. The electron migrates to the *n*-type layer; the hole moves into the *p*-type layer. The energy required to promote an electron from the n-p junction to *n*-type layer is called *band gap.*

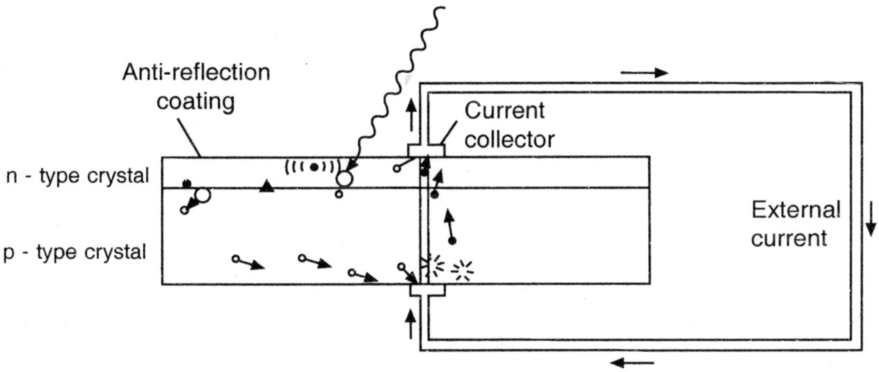

Fig. 12.1 A single crystal silicon solar cell

A current can be drawn from the cell through an external circuit that makes electrical contact with both the *n*-type and *p*-type layers. It is through the external circuit that the electrons trapped in the *n*-type layer find their way to the p-type layer and can then recombine with holes. If the external circuit has a very low resistance, the current that flows is a measure of the rate at which the electrons are separated from the holes.

Limitations of photovoltaics. There are many advantages to photovolatic electricity. Sunlight is converted directly into electrical energy without any intermediate processes. There are no moving parts to wear out or break. Solar cells are light and simple to operate. They require very little maintenance and can function at ambient temperatures, without auxilliary cooling. Moreover, photovoltaics represent the best power options available in remote regions, not yet committed to fossil fuel or nuclear technologies. In inaccessible locations, the photovoltaics are the only viable alternative for such applications as remote lighting, telecommunications, or security systems. However, solar cells suffer from two major drawbacks: Their manufacturing cost is high and their efficiency is low.

The cost factor. Photovoltaics are bound to be expensive. Though silicon is an abundant element in the earth's crust, it has to be chemically synthesized from its ore, silicon dioxide. Thereafter, it has to be refined to a highly purified state, grown as a single crystal and cut with a diamond saw. It is then doped with phosphorus and boron. All these steps add on to the cost of the solar cell.

Cost reduction may be achieved by using hydrogenated amorphous silicon (α-Si:H) in place of semiconductor grade, single crystal material. In amorphous form, silicon can be extracted relatively cheaply. It differs from crystalline silicon in that it has no lattice structure. The inclusion of hydrogen in amorphous silicon, however, compensates partly for the structural disorder. Hydrogen-occluded silicon can be doped with electron-rich or electron-deficient atoms and cut into thin layers, which can be deposited on a variety of inexpensive surfaces, such as steel, glass or plastic.

Amorphous silicon exhibits high optical absorption and photoconductivity. However, the solar cells made from amorphous silicon suffer from a fundamental problem. In the months immediately following their manufacture, some of the incorporated hydrogen escapes from the structure. This is called *Wronski effect.* As a result of this effect, the efficiency of amorphous silicon solar cell falls progressively. Once the diffusion of hydrogen stops, the efficiency attains a constant value, but not before it drops to about two-third of the original value.

The efficiency factor. The maximum theoretical efficiency of a silicon solar cell is about 21%. If solar cells are to replace the conventional energy sources, this figure must be improved. The efficiency of a photovoltaic is low because of the following three factors.

1. Silicon has a band gap of 1.12 eV. Thus, in a silicon photovoltaic device an electron needs an energy of 1.12 eV in order to move into the conduction band. This energy corresponds to a wavelength of 1.1 microns, which lies in the infrared region of the spectrum. Only about half the energy the earth receives from the sun arrives in form of radiations with wavelength near to 1.1 microns. The remaining radiations play no part in photoconductivity.

2. Not all the photons having wavelength of about 1.1 microns can produce photoconductivity. Some of them are reflected from the surface of the solar cell. The reflection may be minimized by using an anti-reflection coating. This coating is made up of a transparent material like glass of a thickness such that the light reflected from the surface of the crystal interferes destructively with the rays reflected from the surface of the coating. However, an anti-reflection coating can never completely eliminate reflection.

3. As has been stated in the section on mechanism of photovoltaics, sunlight must penetrate through the *n*-type layer of the solar cell and strike the n-p junction (Fig. 12.1). In order to achieve this step efficiently, the n-type layer must be made as thin as possible. However, as thickness of *n*-type layer is reduced, its resistance increases and the layer becomes easily heated-up by the current. Thus, most of the energy is lost as heat.

The efficiency of solar cells may be improved by using compound semiconductor materials instead of silicon. Typical compound semiconductors are gallium arsenide (GaAs), cadmium telluride (CdTe) and copper indium diselenide ($CuInSe_2$). The band gaps of these materials are a better match to the spectrum of sunlight. Thus, while photons having a wavelength of nearly 1.1 microns can be used only with the silicon based cells, radiations with a broader wavelength range are effective to produce photoconductivity with compound semiconductors.

Another advantage of compound semiconductors is that they may be converted into multijunction devices. In these devices, cells made from different compound semiconductor materials are used. Each cell has a different band gap. The cells are then stacked on top of each other in descending band gap order. With each cell tuned to a different portion of the solar spectrum, this configuration utilizes sunlight more effectively, thereby increasing the efficiency.

However, the manufacturing cost of compound semiconductor cells is even higher than single crystal silicon solar cell. By using compound semiconductors, the efficiency of the photovoltaics, no doubt, increases; but the cost also increases.

Indirect Utilization of Solar Energy

There are three important indirect sources of solar energy, viz., wind, water and biomass. These are more popular than the direct sources in the sense that they serve a larger section of humanity. While biomass energy is used essentially for heating purposes, the wind and water power find multipurpose utility.

Wind Power

Wind energy is one of the most ancient forms of power used by the people. Windmills have long been used for irrigation, threshing of corn and producing electricity. They were, however, abandoned when electrical power lines became available. The present energy crisis has once again focussed the attention on windmills.

Windmills are the cheapest means of harnessing energy. They produce no air pollution. They require no concentrators—the geography of the countryside provides natural concentrators for wind energy. Wind power systems routinely operate with 35% efficiency—much higher than that of photovoltaics.

The only handicap of the windmills is the catastrophic windless drought that may occur for extended periods. However, this problem may be solved by installing windmills in places rich in underground gas formation so that the naturally stored gases may drive the windmill if the need arises. Another alternative is that windmills may be put up near a hydroelectric dam so that in windless climate, hydro-based power may be switched on.

Water Power

Hydroelectric dams convert the potential energy of elevated water into electrical energy. Hydroelectric power is extremely clean—no fuel is required and no pollutants are produced. It can be turned on rapidly and, unlike any other electric generating technology, it can follow widely fluctuating loads rapidly. Therefore, intermittent technologies like windmills may be integrated with hydropower dams. The dams can also be used for flood control, irrigation, fisheries management and recreation. They have long life spans with respect to other forms of power generation and require little maintenance.

Biomass Energy

Biomass energy is the energy associated with carbon and hydrogen bonds present in vegetative material. These bonds are formed because of photosynthesis—a process which requires solar energy. The most obvious sources of biomass energy are trees, weeds and herbs. Biomass can be used as an energy resource in those places where other energy alternatives are either non-existent or too costly. Its main advantage is that it is more or less indefinitely renewable, for it can be grown on the same site, year after year. Moreover, burning of biomass emits only low level of pollutants. Its disadvantage is that it has a low energy density and therefore cannot serve as a substitute for fossil fuels.

Biomass energy resources can be divided into two broad categories: By-product sources and deliberately grown sources. An example of a by-product source is the logging residues. These are the remains of dead trees, tree branches and shrubs, which have been discarded by the logging industry. An example of deliberately grown sources includes short rotation energy plantation. Trees like eucalyptus and poplar grow very rapidly, but have little commerical value due to the low density of their fibres. Such species are sometimes harvested as biomass energy sources.

Environmental Impacts of Solar Energy

Solar energy utilization is considered to be essentially environmentally benign. Neither the precursors of acid rain nor any greenhouse gases are emitted when solar energy technologies are put to practical usage. No hazardous chemicals are discharged into the aquatic systems. Some negative impacts do result when solar energy resources are being used, but these are insignificant as compared to the hazards posed by fossil fuel or nuclear fuel utilization. The environmental impacts of solar energy utilization are associated with the following applications.

1. Solar Collectors

Solar collectors require a great deal of land and materials. To conserve land, the collectors may be installed on roof tops, but this normally does not appeal to the aesthetic taste of the users.

2. Photovoltaics

The extraction and refining of silicon for solar cells discharge significant quantity of particulates and fly ash to the atmosphere. These aerosols act as potential agents for direct occupational problems. In case of compound semiconductor-based solar cells, the occupational hazards mainly stem from extraction of cadmium or arsenic. Cadmium is an acutely toxic element. Chronic exposure to as low levels as 10^{-4} ppm may cause gastrointestinal disorders, kidney damage and bone mineralization. Increased incidence of cancer and hypertension have also been linked to low, but extended, exposure to cadmium. The long residence time of cadmium in the body (half life, 20 years) and the delay in appearance of health effects make the determination of dose-response relationship difficult to assess. Arsenic too is toxic and carcinogenic. Exposure at levels of 10^{-4} ppm can cause skin irritation, vomiting, diarrhoea, inflammation and ulceration of mucuous membrane and kidney damage.

3. Windmills

The major objections to windmill installations have been raised against their aesthetic impact and possible interference with television reception.

4. Hydroelectric Dams

Construction of dams obliterates vast areas of agricultural land. It also involves flooding of large areas, altering of river courses and rehabilitation of vast sections of people. The water stored in the dam's reservoir weighs a great deal. As a result, minor earthquakes occur occasionally. The great weight of impounded water may also cause the dam to break, producing disastrous results downstream.

Inference

Man has been using solar energy purposefully for millennia. In the pre-industrial era, it was the predominant source of power and the windmills, watermills and sailing ships of those times derived their energy ultimately from the sun. However, with the coming of industrial revolution, these devices were rendered obsolete. They were well suited for modest, local needs, but could not meet the demands of vast industrial infrastructure. Fossil fuels stepped in to serve the industry, and have continued to do so.

However, now the time has come to return to an economy directly powered by solar energy. The supplies of economically recoverable fossil fuels are approaching exhaustion. Even in places where the fossil fuels are not in short supply, their utilization imposes a tremendous burden on the environment. The fission reactor is not universally acceptable. The fusion reactor is proving elusive. Under these circumstances, solar energy is the only prepetual source capable of meeting the likely demands of the future.

There are several underdeveloped nations which have scanty deposits of fossil fuels and no opportunity of developing nuclear energy technologies. Many of these nations lie in the sun belt, near the equator. Solar radiation is the most important, potentially available, natural resource for these countries.

In recent years there has been an increasing emphasis on minimizing environmental degradation in all its diverse forms and protecting land, water and air resources, whilst sustaining economic productivity. In fact, access to uncontaminated land, to clean water and to fresh air must be considered a natural right of the people. Developing solar energy technologies shall help prevent contravention of this natural right.

References

Bakthavatsalam, V. (1995). Renewable energy technologies, *Asia Pacific Tech Monitor,* **12**(3), 9–14.

Brinkworth, B.J. (1974). Solar energy, *Nature (London),* **249,** 726–729.

Chalmers, B. (1976). The photovoltaic generation of electricity, *Sci. Amer.,* **235**(4), 34–43.

Charters, W.W.S. (1995). The use of solar energy for ecologically sustainable development, *Asia Pacific Tech Monitor,* **12**(3), 15–19.

Claverie, M. and Dessus, B. (1992). Solar electricity, *The World Scientist,* January, p. 28–38.

Daniels, F. (1967). Direct use of the sun's energy, *Amer. Sci.,* **55,** 15–47.

Duffie, J.A. and Beckman, W.A. (1976). Solar heating and cooling, *Science,* **191,** 143–149.

Duguay, M.A. (1977). Solar electricity: The hybrid system approach, *Amer. Sci.,* **65,** 422–427.

Hall, C.A.S., Cleveland, C.J. and Kaufmann, R. (1986), *'Energy and Resource Quality',* John Wiley & Sons, New York, p. 285–319.

Hoagland, W. (1995). Solar energy, *Sci. Amer.,* **273**(3), 136–139.

Hubbard, H.M. (1989). Photovoltaics today and tomorrow, *Science,* **244,** 297–304.

McEvoy, A.J. (1993). Outlook for photovoltaic electricity, *Endeavour,* **17,** 17–20.

Metz, W.D. (1977). Wind energy: Large and small systems competing, *Science,* **197,** 971–973.

Wolf, M. (1974). Solar energy utilization by physical methods, *Science* **184,** 382–386.

Zweibel, K. (1986). Photovoltaic cells, *Chem. Eng. News,* July 7, p. 34–48.

CHAPTER 13

Nuclear Energy

Nuclear energy is not a direct energy resource for civilian use. Indirectly, however, it is a means for producing electrical power. Its power to provide electricity is unmatched as compared to any other energy resource. Theoretically, one tonne of uranium can give as much electrical energy as 3 million tonnes of coal or 12 million barrels of oil. Hence, no technology shows more potential than nuclear energy for long-run supply of electricity. Yet, no source of energy today is more controversial than nuclear power. Some look upon it as a means for advancement of civilization. Others believe it to be the ultimate destroyer of civilization. Whether it is good or bad, we have no choice but to develop it further, for the conventional energy resources are fast depleting.

Nuclear Power Generation

There are two fundamentally different ways to produce electrical power from nuclear energy: Nuclear fission and nuclear fusion. The former involves the splitting of a heavy radioisotope, such as uranium-235 or plutonium-239, while the latter involves fusion of lighter atoms, such as hydrogen. Although fusion offers more advantages as compared to fission, it is technically complex and, at present, away from commercial reality.

Nuclear Fission

Most nuclear power generating plants use radioactive uranium-235 as the fission fuel. The isotope is fissioned when struck by a slow moving neutron.

$$^{235}_{92}\text{U} + \underset{\text{Slow}}{^{1}_{0}n} \rightarrow {}^{142}_{56}\text{Ba} + {}^{91}_{36}\text{Kr} + 3\,^{1}_{0}n + 200\ \text{MeV} \tag{13.1}$$

The energy released is used to drive a turbine. As indicated in reaction (13.1), average three neutrons are produced, whereas only one is required to initiate fission. This permits a chain reaction to occur in which the released neutrons initiate the fission of fresh uranium-235 nuclei, producing more energy. However, care must be taken to produce only that much energy which can be harnessed for electrical power generation. Unabated release of energy may cause serious nuclear accidents. Hence, reaction (13.1) must be carried out in a controlled manner. A device which permits a nuclear fission reaction to occur smoothly is called a *nuclear reactor.*

Components of a nuclear reactor

A nuclear fission reactor has the following components.

1. Fuel rods. The fuel rods of a nuclear reactor are made up of a *fissile isotope,* that is, an isotope which undergoes spontaneous fission by capturing neutrons. Uranium-235 is the most common fissile isotope to be used in the reactors. However, the natural abundance of uranium-235 is only 0.71%. Before it can be used as a nuclear fuel, its concentration must be increased to about 3%. This is done by extracting uranium from uranium oxide ore, purifying it and converting it into uranium hexafluoride, UF_6. The fluoride is concentrated by gaseous diffusion method and reconverted into uranium oxide, UO_2. The process of increasing the concentration of uranium-235 is called *enrichment.*

The uranium oxide is shaped in form of pellets, about 5 mm in diameter. The pellets are fed into zirconium alloy tubes, called *claddings.* A reactor normally contains about 40,000 claddings, each about 3.5 m long.

2. Control rods. In order to carry out reaction (13.1) in a controlled manner, the number of neutrons must be restricted. Otherwise, uranium-235 would be fissioned at an abnormally high rate, releasing excessive energy. This, in turn, will cause the reactor temperature to rise. Control rods are meant to reduce the number of neutrons, so that the nuclear reaction proceeds at an optimum pace. Generally, the control rods are made of cadmium since this element has the potential to absorb neutrons.

$$^{113}_{48}\mathrm{Cd} + {}^{1}_{0}n \rightarrow {}^{114}_{48}\mathrm{Cd} + \gamma \qquad (13.2)$$

The control rods are arranged alternately between the fuel rods. When the reaction initiates, the control rods are partially withdrawn so that the neutron flux is high. However, once the reaction begins, the control rods are reinserted so that the number of neutrons are reduced to such a level that they just maintain the chain reaction (13.1).

3. Moderator. Not only the number of neutrons in a fission chain reaction needs to be controlled, but their speed too must be regulated. Uranium-235 nuclei are split more efficiently by slow moving neutrons than by fast neutrons. Hence, in order to achieve high efficiency, the projectile neutrons must be slowed down to optimum velocities. Substances which reduce the velocity of neutrons are called moderators. Graphite or heavy water are the commonly used moderators in the nuclear reactors.

4. Coolant. The coolant serves to absorb the heat generated in the fission reaction. The heat is then utilized to make steam which, in turn, is used for performing useful work. The coolant also serves another purpose: In case the nuclear reaction begins to go faster than expected, it brings about emergency cooling. Most reactors use either water or liquid sodium as coolants because of their high thermal conductivities and large heat capacities. In some reactors helium gas is also employed as a coolant.

The fuel rods, the control rods, the moderator and the coolant together constitute the *core* of the nuclear reactor.

5. Containment vessel. Containment vessel is a screen which prevents the radiation and radioisotopes from escaping out of the reactor. It is made up of reinforced concrete to which salts of lead are added to absorb radiation. The containment is strong enough to withstand an external threat such as a tornado. Conventional explosives cannot fracture it.

Types of Reactors

In a typical fission reactor, neutrons, slowed down by a moderator, strike the fuel rods. A chain reaction ensues. The energy released is taken up by the coolant. If the coolant is water, it is directly converted into steam. If helium or liquid sodium are the coolants, they pass their energy to water, contained in a heat exchanger, and convert it into steam. The steam drives the turbine, which turns a generator to produce electric power. After passing through the turbine, the steam is cooled and the water is returned to the reactor to be heated again. Different types of reactors follow different methods for producing steam.

1. Light water reactors. Light water reactors use water as a coolant. There are two types of light water reactors: Boiling water reactors and pressurized water reactors.

A boiling water reactor is one in which steam is produced within the reactor and then transferred directly to the turbine. A boiling water reactor is shown in Fig. 13.1. Although such reactors are comparatively cheap, they suffer from a disadvantage in that the steam carries away some radioactive materials which may subsequently escape to the environment.

In a *pressurized water reactor,* water is kept under high pressure so that steam is not allowed to form in the reactor. A secondary loop transfers the pressurized water to a steam generator. The steam is used to run the turbine and generate electricity. Such an arrangement reduces the risk of radioactivity being carried by steam but adds to the cost of construction by requiring a secondary loop. A pressurized water reactor is shown in Fig. 13.2.

2. High temperature gas cooled reactors. These reactors use an inert gas, usually helium, in the primary coolant circuit instead of water. The helium gas circulates in pipes totally enclosed within the reactor vessel. It passes through a heat exchanger, where the energy produced by the reactor is transferred to water. The steam so produced is carried outside the reactor to the turbine. A high temperature gas cooled reactor is depicted in Fig. 13.3.

3. Liquid metal fast breeder reactors. A breeder reactor is one which generates more fissile material than it consumes, while at the same time generates electricity. The most common fuel in a breeder reactor is a mixture of 85% uranium-238 and 15% plutonium-239. Uranium-238 is a *fertile material,* that is, it does not undergo spontaneous fission like the fissile uranium-235. However, it may be converted into fissile material by absorbing fast moving neutrons.

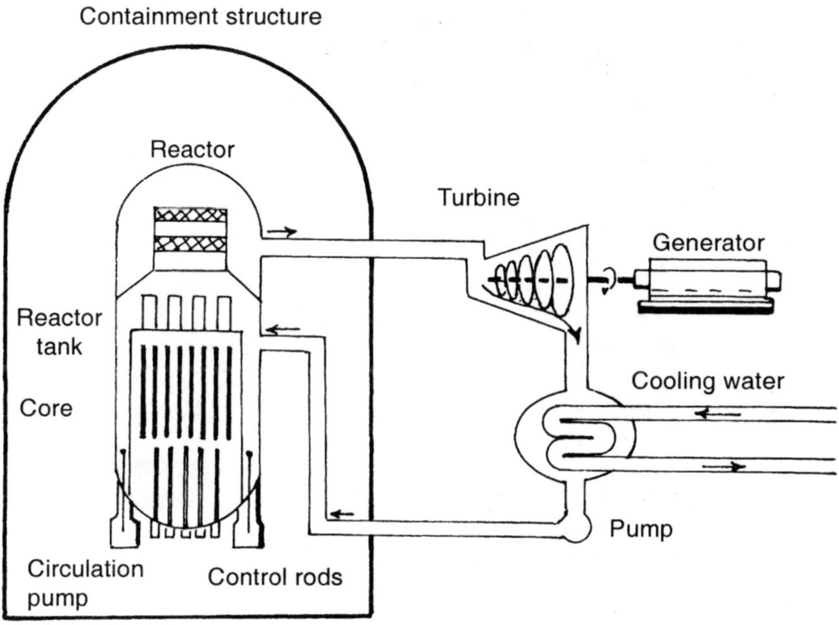

Fig. 13.1 A boiling water reactor

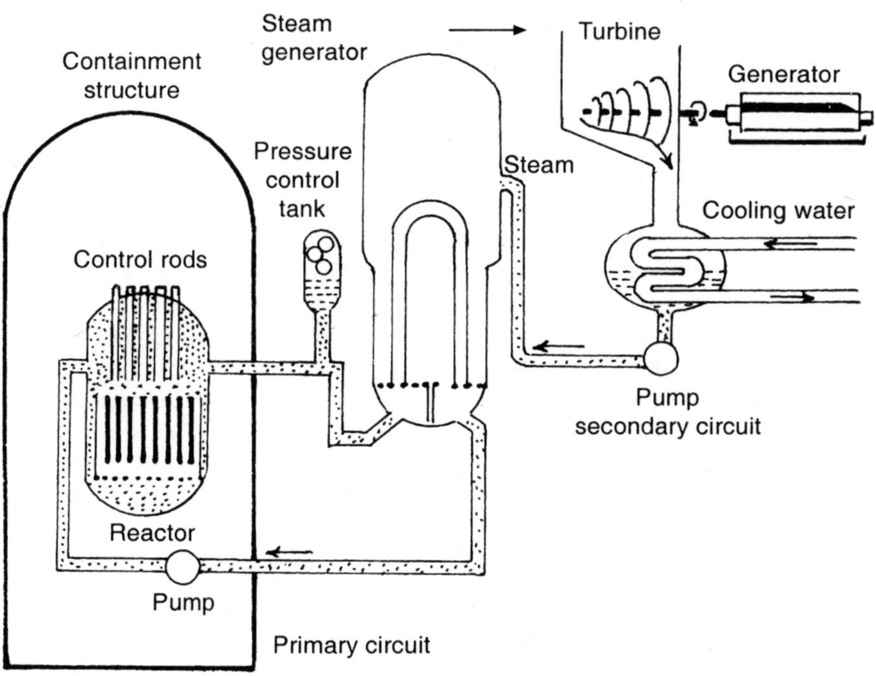

Fig. 13.2 A pressurized water reactor

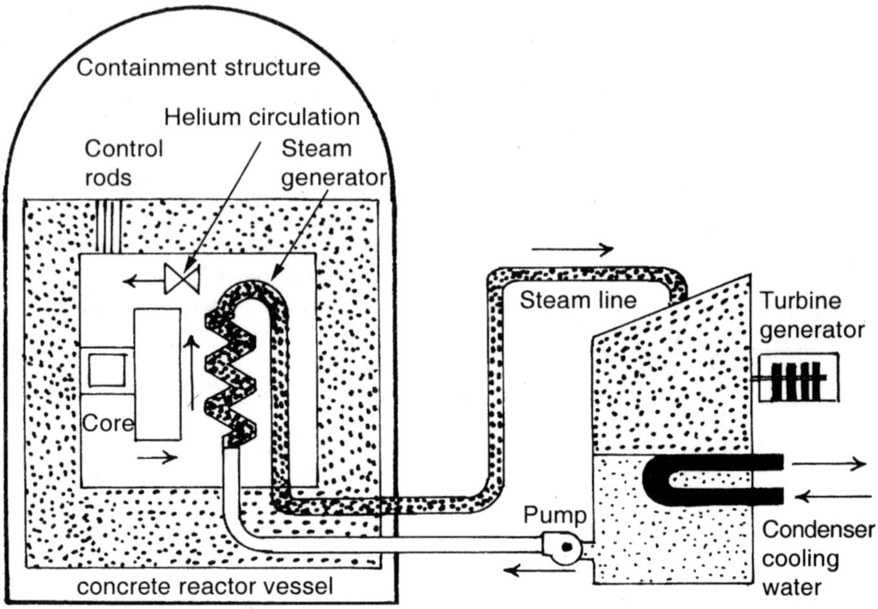

Fig. 13.3 A high temperature gas cooled reactor

$$^{238}_{92}\text{U} + \underset{\text{Fast}}{^{1}_{0}n} \rightarrow {}^{239}_{92}\text{U} \rightarrow {}^{239}_{93}\text{Np} \rightarrow {}^{239}_{94}\text{pu} \qquad (13.3)$$

$$\qquad\qquad {}^{\circ}_{-1}\beta \qquad {}^{\circ}_{-1}\beta$$

Plutonium-239 is fissile in nature. Like uranium-235, it can undergo spontaneous fission. The ratio of fissile plutonium-239 produced to the fertile uranium-238 consumed is called *conversion ratio.* For a breeder reactor the conversion ratio is greater than unity.

Most of the breeder reactors employ liquid sodium as a moderator, as well as a coolant. As a moderator it slows down only a few of the fast moving neutrons. This serves an important purpose, for both fast and slow neutrons are required in a breeder reactor; fast neutrons for converting uranium-238 to plutonium-239 and slow neutrons for fissioning plutonium-239.

As a coolant, liquid sodium circulates in the reactor core through the first loop. It absorbs the heat produced as a result of fission reaction. However, in this process it also becomes radioactive.

$$^{23}_{11}\text{Na} + {}^{1}_{0}n \rightarrow {}^{24}_{11}\text{Na} + \gamma \qquad (13.4)$$

To prevent radioactivity to leak out, the heat absorbed in the first sodium loop is transferred to the second sodium loop. The thermal energy of the second sodium loop is then transferred to a steam generator. The steam so produced drives the turbine (Fig. 13.4).

A breeder reactor has two important advantages. Firstly, it greatly expands the resource base for fuel production. Secondly, all the uranium recovered from

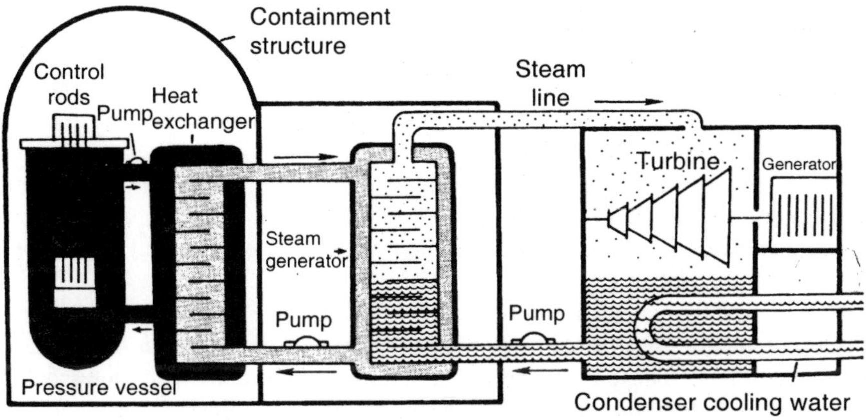

Fig. 13.4 A liquid metal fast breeder reactor

the ore can be directly used as a fuel. Most of the naturally occurring uranium is in form of uranium-238. Against this, uranium-235 occurs in extremely small amounts and hence has to be enriched before it can be used as a nuclear fuel. Since a breeder reactor employs uranium-238 as the fuel, the enrichment step is eliminated.

However, a breeder reactor suffers from several drawbacks. Firstly, sodium reacts explosively with air and water. Therefore, costly and highly specialized equipment is required to prevent this hazard. Secondly, the fission reaction occurring in a breeder reactor is extremely fast and, therefore, very difficult to regulate. It must be precisely and carefully monitored. Thirdly, a breeder reactor produces plutonium-239, a radionucleide which is very toxic and carcinogenic to humans. Moreover, plutonium-239 is used in production of nuclear weapons. There is always a fear that it may fall in the hands of unscrupulous elements, who may use it for non-peaceful purposes.

Nuclear Fusion

Another way to obtain nuclear energy is to fuse two light nuclei together to form a heavier element. The energy produced by sun and stars is the result of fusion reactions. For producing electrical power from nuclear energy, the most feasible fusion reaction is that of deuterium and tritium.

$$^{2}_{1}\text{D} + ^{3}_{1}\text{T} \rightarrow ^{4}_{2}\text{He} + ^{1}_{0}\text{n} + 17.6\ \text{MeV} \tag{13.5}$$

Deuterium is found in water to the extent of 1 part in 6500 parts. Tritium, a weakly radioactive isotope, is virtually non-existent in nature. It may, however, be produced by bombarding lithium-6 with neutrons.

$$^{6}_{3}\text{Li} + ^{1}_{0}n \rightarrow ^{3}_{1}\text{T} + ^{4}_{2}\text{He} \tag{13.6}$$

A fusion reactor is a breeder type of reactor. Its fuels are deuterium and lithium-6. The neutrons released by reaction (13.5) are absorbed by lithium-6 in reaction (13.6), breeding more tritium. The latter fuses with deuterium and the cycle goes on.

Limitations of fusion technology

In order to carry out nuclear fusion, the ingredients must be heated to such high temperatures that the fuel nuclei can overcome the electrostatic barriers and merge with each other. The temperature requirement for reaction (13.5) to occur is of the order of 10^7 K. At this temperature matter exists as plasma—atoms stripped of their electrons. The constituents of plasma, D^+ and T^+, have very high kinetic energy. They would strike the walls of the reactor and melt it. Simultaneously, the plasma will get cooled, its kinetic energy will be reduced and so will the probability of fusion.

Herein lies the scientific and engineering challenge to fusion technology. Reaction temperatures should be so high as to create a plasma state. The plasma must be confined. These two conditions, high temperature and confinement, are difficult to achieve simultaneously. There are two strategies to overcome this problem: Magnetic confinement and inertial confinement.

Magnetic confinement is a means to confine the plasma, containing ionized deuterium and tritium, not by material walls but by magnetic field. The most successful magnetic fusion device is the *Tokamak reactor* (Fig. 13.5). In this reactor, electric current is made to flow in coils that are arranged around a dough-nut shaped chamber. This current acts in concert with another one driven through the plasma to create a magnetic field. The charged deuterium and tritium nuclei are held within the magnetic field. The Tokamak device can confine the plasma at densities of about 10^4 fuel particles per cubic centimeter for roughly a second.

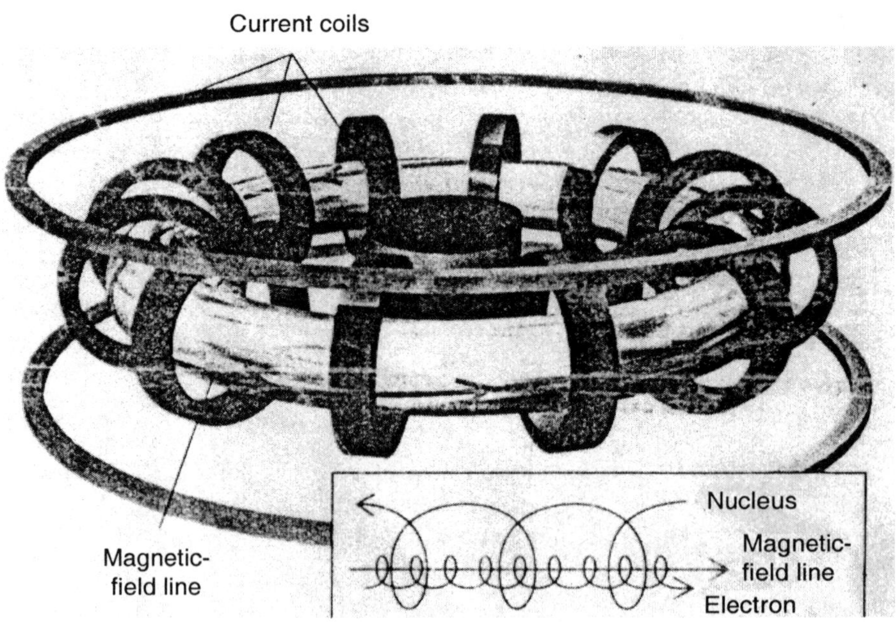

Fig. 13.5 A Tokamak fusion reactor

In *inertial confinement,* fusion fuel is heated so rapidly that the fuel nuclei fuse with each other before they can escape to the walls. In this technique, powerful laser beams are made to strike a spherical pellet containing deuterium-tritium mixture (Fig. 13.6). The radiation vaporizes the surface coating of the pellet, which explodes outwards. To conserve momentum the inner layers of the pellet shoot further inwards. The fuel is thus compressed and its density increases. Simultaneously, the laser beams cause the temperature to rise. The fuel at high temperature and high density is conducive to fusion. The whole process takes place in just 10^{-10} second.

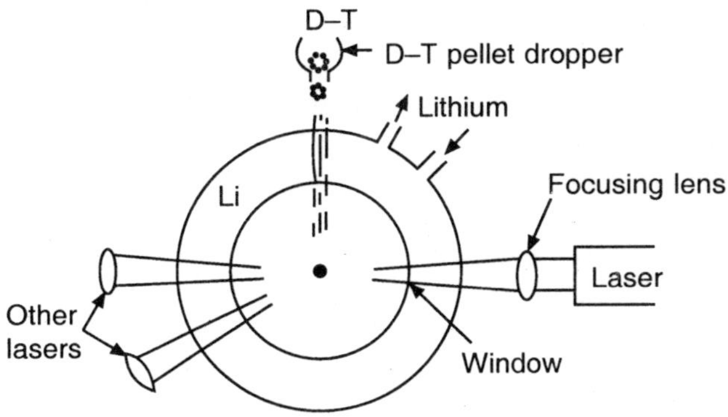

Fig. 13.6 A laser fusion device

Advantages of Fusion Technology

Once it comes into practice, the fusion technology will prove to be an excellent energy resource. Its fuel material is universally available and is obtainable with little environmental impact. No radioactive products are formed in a fusion reaction. Engineering safeguards and control devices are simple and easily manageable. The chances of a serious nuclear accident are negligible. Therefore, fusion reactors can be installed virtually in the heart of population centers, thereby cutting transmission and transportation costs. Moreover, there is no threat of proliferation of nuclear weapons.

However, at present all the advantages of fusion technology are only of academic interest. Technical difficulties appear to prevent its commercial use anywhere in the near future.

Nuclear Waste Disposal

Nuclear waste consists of radiologically hazardous isotopes, as well as the equipment which contains them. These isotopes are produced as by-products of fission reaction in a nuclear reactor. The nuclear waste must be separated from the biosphere to a sufficient degree, and for a long enough time, so that it presents no significant risk to life and human health. Depending on the level of radioactivity contained in them, the nuclear wastes are classified into two broad categories: Low level wastes and high level wastes. Different strategies have to be adopted to dispose off these two categories of nuclear wastes.

Low Level Waste

Low level waste contains trace amounts of short half-life isotopes such as strontium-

90, cesium-137, neptunium-236 and uranium-237. Materials containing low level wastes include plastic bottles, rubber gloves, glassware, floor sweepings and protective clothing. Although low level wastes are less hazardous than high level wastes, they are more bulky and plentiful.

The disposal of low level wastes is simple and straight forward. The material is packed in bags or drums and placed in deep trenches. A typical trench is 15 m wide and 10 m deep. Seabed also serves as a host medium for disposal of both the low level and high level wastes. The difference, however, is that while the low level waste is dumped *on* the seabed, the high level waste has to be buried *under* the seabed.

Between 1971 and 1982, Britain, Belgium, the Netherlands and Switzerland used a site 800 km south west of Land's End in the Atlantic Ocean to dump low level waste on the sea floor. In 1983 the dumping on the seabed was prohibited by an international moratorium. A 1994–95 survey of Atlantic Ocean revealed that plutonium-238, americium-241 and other radionucleides had started leaking from the drums. While it is quite early to assess the damage caused by the leakage, it now appears that the ban on seabed dumping has come not a day too soon.

High Level Waste

The high level waste includes appreciably high concentrations of a spectrum of radionucleides, ranging from short half-life ones, such as strontium-90, cesium-137 to long half-life ones, such as iodine-129 and plutonium-239. Mostly these radionucleides are a part of the *spent fuel*—the fuel which has run the reactor for a specific time. At frequent intervals the spent fuel is removed from the reactor and unused uranium is extracted from it. The extracted uranium is reconverted into fuel rods, while the remaining fission products are treated as high level waste. The materials containing high level waste include claddings and those parts of the reactor that need replacement.

High level waste poses a serious health risk if not controlled and managed properly. Additionally, its contents are very hot. Therefore, it is buried in some remote location. Before the burial, however, the waste is stored above-ground for about ten years so that its temperature is sufficiently reduced. Otherwise there is a danger that the burial site might fracture under intense heat. It has also been recommended that before burial, the long-lived isotopes in the high level waste should be transmuted into short-lived isotopes, so that the waste disintegrates at a faster rate.

For example, neptunium-237, one of the constituents of the spent fuel, takes a very long time to disintegrate.

$$^{237}_{93}\text{Np} \xrightarrow{t_{1/2} = 2.14 \times 10^6 \text{ years}} {}^{223}_{91}\text{pa} + {}^{4}_{2}\text{He} \tag{13.7}$$

However, a rapid disintegration may be brought about by bombarding neptunium-237 by neutrons and converting it into neptunium-236, which undergoes a fast decay to plutonium-236.

$$^{237}_{93}\text{Np} + {}^{1}_{0}n \rightarrow {}^{236}_{93}\text{Np} + 2\,{}^{1}_{0}n \tag{13.8}$$

$$^{236}_{93}\text{Np} \xrightarrow{t_{1/2} = 22.3 \text{ days}} {}^{236}_{94}\text{pu} + {}^{0}_{-1}\beta \tag{13.9}$$

Plutonium-236 too disintegrates quite fast. After transmutation, the waste is ready for burial. However, it must be ensured that once the waste is buried it should be totally immobilized. That is, it should under no circumstances re-enter the biosphere. The most likely mechanism which can bring the radioactive material back to the living environment is leaching by groundwater. The waste should, therefore, be protected from ground water. This is done by adopting a *multiple barrier approach.*

The liquid and solid wastes are first converted into a solid borosilicate glass like structure. In this form the waste becomes highly insoluble in water. This constitutes the first barrier. The borosilicate capsule is then enclosed in canisters made of highly corrosion resistant alloys, such as stainless steel. The canister is encased in a suitable absorbent material, capable of delaying the migration of any ion in the waste that may have dissolved inadvertently in the ground water. The canister and the encasement constitute the second barrier. The geologic environment of the burial site serves as the third barrier.

The treated high level waste is then buried at one of the following sites.

1. Geologic repository

The most common method of nuclear waste disposal is to bury the canisters containing high level waste in rock formations to a depth of 300–700 m. The rock formation should meet the following criteria.

(a) It should have high structural strength, good thermal conductivity and high heat capacity so that it may disperse the heat load from high level waste without undergoing any damage.
(b) It should have low permeability and low porosity, so that the ground water may not seep through it.
(c) It should have a reasonably high plasticity so as to allow healing in case the burial site is fractured.
(d) It should be located in a zone of low seismicity.
(e) It should be located in a zone of scarce subsurface water.
(f) It should not be used as a mineral resource in the future.

2. Salt deposits

Next to a geologic repository, the most common burial site for high level waste is a salt deposit. The following factors favour the salt deposits as disposal media.

(a) Salt possesses high thermal conductivity; it effectively disperses the heat load of radioactive waste without itself being damaged.
(b) It has extremely low permeability. Moreover, salt has no cracks in it to allow the flow of ground water.
(c) It possesses a high degree of plasticity. The degree of plasticity increases with depth. This allows it to self-heal any fractures that might develop.
(d) It can be mined at a low cost.

(e) The existence of several widespread deposits that have remained intact for millions of years reveal the long term stability of salt beds.

However, salt deposits also possess the following disadvantages.

(a) Unlike the rock formations, salt is highly soluble in water. Thus, ground water can reach the nuclear waste after dissolving the surrounding salt beds.
(b) All salt deposits contain some brine-fluids with very high concentrations of dissolved solids. Once the brine reaches the canisters, it is likely to corrode them, exposing the high level waste. Hence, whenever salt deposits are used as burial sites, the stainless steel canisters are replaced by containers made from more corrosion resistant titanium alloys.
(c) Salt possesses poor sorption qualities. It, therefore, does not have the potential to impede radionucleides' migration. Thus, before the waste-containing canisters are buried in salt mines, they are encased in a sorptive material like bentonite.

3. Beneath seabed

The high level waste remains safely embedded 30 m below the red clay deposits of the ocean basin. The following characteristics of red clay make it suitable for the long term storage of nuclear waste.

(a) It possesses a high degree of plasticity and so if any cracks develop, they are rapidly sealed.
(b) It has so low permeability that it would take 10^5 years for any waste material escaping from the canisters to reach the sea floor.
(c) It has high absorptivity and, therefore, in the eventuality of the canister being worn out, much of the radioisotopes would be absorbed and remain beneath the sea floor.

Japan and Great Britain have a relatively few terrestrial sites for burying the nuclear wastes. Hence, these countries favour burial of wastes under sea sediments.

4. Drill holes

We have stated that high level waste may be buried in a geologic repository to a depth of 300–700 m. If one goes still deeper, the rock formations become entirely devoid of ground water. Their permeability also decreases with depth. Hence, theoretically it would be safer to bury the waste at greater depths. Practically, however, it is not possible to do so since the temperature of the earth increases with depth. Beyond a depth of 1 km, the chances are that the geologic temperature of the earth plus the radiological heat of the canisters would be enough to melt the borosilicate glass barrier, paving the way for radionucleide leakage.

A new strategy for disposing nuclear wastes at greater depths seeks to replace the borosilicate glass with an artificial rock-type material called *synthetic rock*—SYNROC for short. SYNROC is essentially a titanate ceramic which is thermodynamically very stable and can withstand very high temperatures. It is a mixture of three main minerals: Hollandite, $BaAl_2Ti_6O_{16}$; zirconalite, $CaZrTi_2O_7$;

and pervorkite, $CaTiO_3$. These minerals have the capacity to take nearly all the radioisotopes present in the nuclear waste into their crystal structures in form of dilute solid solutions. Encasesd in SYNROC, the nuclear waste can be buried at a depth as low as 4 km since the geologic and radiological heat cannot melt the synthetic rock material. As there is no danger of SYNROC being melted, the burial does not require vast geologic trenches. Small drill holes about 1 m in diameter are suitable to bury the waste.

The drill hole method, in combination with SYNROC technology, offers the following advantages.

(a) The nuclear waste is buried at greater depths and in essentially non-permeable, dry geologic environment.
(b) SYNROC is an imitation of natural rocks. All natural rocks contain small amounts of radioactive elements, such as uranium-238, thorium-232, potassium-40 and rubidium-87. These elements remain locked-up in the crystal structure of parent rocks for billions of years. In the same manner, nuclear waste becomes practically immobilized in the SYNROC.
(c) Drill holes occupy less space than geologic repositories. Moreover, it is much easier to seal a drill hole than a vast repository.
(d) Drill holes cause a minimum damage to rock formations. On the other hand, digging large mines sometimes cause fractures in surrounding rocks, which serve as conduits for circulating ground water.

Transmittance of Radioactivity to Humans

Radioactivity may be transmitted to human beings in two forms: Ionizing radiation and radioisotopes.

Ionizing Radiation

The ionizing radiation is so named because when it hits neutral atoms it ionizes them, that is, strips them of their electrons. Examples of ionizing radiation include α-, β- and γ-rays. These radiations may be emitted from a nuclear power plant at any stage of electricity production. The radiations may then enter human body and transfer their energy to the tissues. The biomolecules absorb this energy and undergo fragmentation. The cellular medium loses its normal characteristics, causing a host of deleterious effects.

The amount of energy transmitted to the tissue by ionizing radiation is called *radiation absorbed dose* or *rad*. A rad is defined as 100 ergs of energy deposited in 1 g of tissue. The same dose in rads from different ionizing radiations produce different degrees of biological effects. Therefore, rad does not indicate the degree of biological damage that is caused by a specific dosage of any radiation. A better unit of radioactivity is *roentgen equivalent man* or *rem*. This unit compensates for the properties of a wide varieties of radiation by relating them to the effect caused by X-rays. A rem is defined as the amount of radiation of any type that causes the same biological injury as one rad of X-rays.

The maximum allowable dose of ionizing radiation for human beings is 0.17 rem $year^{-1}$. For workers in nuclear industry, it is 5 rem $years^{-1}$.

Radioisotopes

The amount of radioactivity contained in an isotope is measured in units of Curie (Ci). A curie is the quantity of radioactivity contained in 1 g of radium. A larger unit is million Curie (MCi).

Radioisotopes may enter the human body either through direct inhalation from air or through the food chain.

Airborne radioisotopes like radon-222 and xenon-133 are readily absorbed on suspended particles. When air is inhaled, these isotopes are deposited in the passageway of lungs where they release the energy produced by their decay.

Radioisotopes are also absorbed by soil and then taken up by plants. Humans ingest these radionucleides either by eating contaminated plants or by eating the meat of animals that had consumed such plants.

For example, iodine-131, an important by-product of fission reactors, is readily taken up by plants from contaminated soil. It is then appropriated by the grazing animals. Finally it appears in the milk of these animals. Humans consuming the contaminated dairy products take up iodine-131, which becomes concentrated in the thyroid gland. Likewise, strontium-90 finds its way into the food chain since it is readily accepted as a substitute for calcium, to which it closely resembles. Radioactive strontium becomes deposited in bones in place of calcium. Table 13.1 lists the radioisotopes which tend to be accumulated in various human organs.

Table 13.1 Human organs affected by radionucleides

Radionucleide	Organ
Sulfur-35	Skin
Potassium-42	Muscles
Cobalt-60	Liver
Krypton-85	Ovaries
Iodine-131	Thyroid
Polonium-210	Spleen
Radon-222	Lungs
Radium-226	Bones

Radioisotopes may also be ingested through the aquatic pathway, that is, by consuming contaminated water. For example, Severn Estuary in England has abnormally high concentration of radioisotopes. A survey carried out in 1995 revealed that babies born around the Estuary had unusually high levels of radioisotopes in their tissue. These children are feared to have high risk of developing cancers, particularly leukaemia.

Effects of Radioactivity on Human Health

The ionizing radiation, which directly enters the body or which emanates from the radioisotopes that are concentrated in specific organs, cause a multitude of health effects. Humans are subjected to low levels of ionizing radiation from the natural environment. This radiation is called *background radiation* and it emanates from rocks and soils containing radioisotopes or strikes the earth from outer space. Humans have evolved physiological mechanisms to repair the tissue damage caused by background radiation at about the same rate as the damage occurs. However, the dose of ionizing radiation from nuclear reactors is high enough to cause permanent damage.

It has also been observed that the biological effect of radiation is decreased when the total dose is divided into several fractions over a period of time. The reason being that a small dosage damages only a small number of cells. These repair by natural healing before the next dosage is received. However, if one large dose of ionizing radiation is impinged on the tissue, a large number of cells are damaged, physiological repair is no longer possible and permanent injury results.

Ionizing radiation causes two broad types of effects: Somatic and genetic. Somatic effects result from the direct cellular damage caused by ionizing radiation. They occur in organisms which themselves receive the radiation. Genetic effects result from molecular or physical abrasion of chromosomes by ionizing radiation. They appear in progeny, that is, in subsequent generations.

Somatic Effects

An intense dose of ionizing radiation causes a reduction of red and white blood cells, as well as blood platelets. Because of the drop in red cells, the blood loses its capacity to carry oxygen and the patient becomes anaemic. Because of the drop in white cells, the blood loses its capacity to combat invading microorganisms and the patient becomes prone to infections. Because of the drop in platelets, the blood loses its capacity to clot rapidly and the patient begins to bleed in various parts of the body. The victim declines in vitality until he dies from anaemia, infection and haemorrhage.

The radiation acts against the antibodies and partially destroys the body's immune response. The intestinal wall is very sensitive to the radiation. It partly breaks down and ceases to be a barrier to infection from the bacteria that inhabit the gut.

An increase in radiation above the natural background level increases the risk of cancer. The cancer most commonly associated with radiation is leukaemia. Concentration of iodine-131 in the thyroid gland induces thyroid cancer.

Ionizing radiation impairs the normal growth rate. It also causes a temporary impairment or a reduction in fertility. Heavy dosages cause permanent sterility.

Somatic effects may not appear immediately after the exposure to radiation. They may appear years later. Eye cataract, premature ageing and reduced longevity are some of the common delayed somatic effects.

Genetic Effects

Ionizing radiation has a greater effect on reproductive cells than on somatic cells. They cause mutations by damaging the nucleic acid molecules in the DNA of the germ cells. There is a linear relationship between the radiation dose and the mutation rate. That is, higher the radiation dose, higher is the mutation rate. Some of the mutations in the reproductive cells show up in the next generation.

Radiation also causes birth defects, as well as high mortality in neonatal period. It is, therefore, necessary that young people and pregnant women be guarded from exposure to ionizing radiation.

Risk Analysis of Nuclear Power Generation

Nuclear power is a cheap and an inexhaustible energy source. It causes a minimal of air and water pollution. It ensures an efficient use of fuel reserves. However, because of the ill effects associated with the ionizing radiation, its acceptability to the society has become a matter of dispute.

Yet, it must be emphasized that the ill effects would be caused only if the radioactivity leaks out to the biosphere. We would now identify various steps involved in nuclear power generation where radioactivity is likely to penetrate the environment. We would analyse the degree of risk likely to be incurred at each stage. We would also compare the hazards of nuclear power generation with the risk factors associated with coal since, at present, coal is the only major alternative to nuclear energy for power generation.

1. Mining

Workers engaged in the mining of uranium ore are likely to be exposed to ionizing radiation. However, the amount of radiation to which they might be exposed is generally below the allowable limits (5 rem year^{-1}). The reason being that the quantity of uranium required for power generation is quite small. Compared to this, coal has to be mined on a much larger scale. Workers engaged in coal mining are prone to be incapacitated by black lung disease. They are at a much greater risk as compared to their counterparts in nuclear industry.

The problem associated with uranium mining is significantly reduced with breeder reactor technology. The future fusion reactors would not employ uranium as fuel and the problem would be altogether eliminated.

2. Enrichment

The enrichment of uranium in the processing mill leaves stockpiles of radioactive thorium and radium compounds. Radon gas emitted from such piles poses a threat to the local populace. However, if the heap is covered by a half inch layer of asphalt and then buried in a secluded place, the chances of radioactive emissions are obviated.

3. Reactor Operation

During the operation of the reactor, most of the radioactive materials remain sealed within the core. However, sometimes gaseous radioisotopes, such as krypton, xenon and radon ooze out through tiny leakages in the claddings. This way the environment is exposed to the radiation by routine reactor operations.

However, the reactors incorporate many redundant safety systems. The containment vessel would not allow the fission products, including the gaseous isotopes, to leak out. New regulations require installation of charcoal traps within the reactors. These absorb the gaseous radioisotopes till the time their radioactivity is substantially decreased.

The reactor operation has a better safety record than burning of fossil fuels, especially coal. Coal contains small quantities of thorium-232, uranium-235 and uranium-238. During combustion, these radionucleides become a part of airborne effluents and emit ionizing radiation. In fact, radiation doses from coal-fired plants are greater than those from nuclear plants.

Moreover, stack emissions from coal-based power plants emit oxides of nitrogen and sulfur — the precursors of acid rain. The adverse effects of acid rain are described in Chapter 14, p. 224. Burning of coal also emits carbon dioxide to the atmosphere. This gas is responsible for the greenhouse effect—the potential problem of warming of the earth's surface over long periods of time (Chapter 18). No such problems are associated with nuclear power generation.

4. Spent Fuel Reprocessing

The residue left out after the reactor fuel has been consumed is designated as spent fuel. The spent fuel is stored for about 6 months in the power plant to allow short half-life radioisotopes to decay. It is then transported to a fuel reprocessing plant. Here the unused uranium is extracted for reuse. Whatever remains is classified as nuclear waste.

During reprocessing of spent fuel, the gaseous radioisotopes, mainly krypton, leak out to the atmosphere in extremely small amounts. Krypton is released from tall stacks so that insignificant amount of radiation enters the living environment.

Fears have been expressed that when the spent fuel is being transported from the power plant to the reprocessing unit, radioactivity might be released in case the transporting vehicle meets an accident. Experience has shown that such fears are unfounded. At a time only a small fraction of the reactor fuel is transported for reprocessing and that too after most of its radioactivity has decayed out. Moreover, the highest temperatures that would be encountered in the event of an accident are those of gasoline fire—about 800°C. The melting temperature of fissile material is nearly 2800°C. Hence most of the radioactivity shall remain trapped in the containers. The containers are designed to withstand a high speed crash or a gasoline fire or immersion in water. Therefore, the chances of a radioactivity leak during an accident are quite remote.

In order to increase its heat efficiency, coal too is sometimes subject to reprocessing and refining techniques. These techniques begin with washing and pre-treatment heating of coal. During these operations dust and sulfur oxides are released to the atmosphere. The water used for washing has to be treated before being discharged to aquatic systems.

5. Waste Dumping

The nuclear waste is disposed off by burial at a suitable site. Theoretically, there is a possibility that the waste might be contacted by ground water and enter into the food chain. However, the multiple barriers around the buried waste have proved to be very efficient in preventing the leaching by ground water.

We have stated that the nuclear waste dumped on the floor of Atlantic Ocean has started leaking out of the drums (p. 200). However, it must be remembered that the waste was placed *over* the seabed and not *under* it. The dumping of nuclear waste on the sea floor is now prohibited. Moreover, the waste in question was of low level type. Hence it is not expected to cause any serious damage.

Coal plants too produce solid wastes—and in much larger quantities. No doubt the toxicity of such wastes is much less as compared to the nuclear leftovers, yet the coal waste is not buried in geologic sites. It is dumped in landfills near the inhabited locations. Wind erosion often suspends the particulate waste from coal power plants into the atmosphere.

6. Reactor Mishaps

By far, the greatest amounts of ionizing radiation and radioisotopes are discharged to the environment is the event of a nuclear accident. Such accidents are almost always caused by human errors and hence are not inevitable. If all the regulations listed in the reactors' protocol are strictly adhered to, chances of a nuclear catastrophe are almost negligible. Nuclear catastrophes merit a separate discussion and are, therefore, discussed in the next section.

Nuclear Catastrophes

Under normal circumstances the radioactive materials are sealed within the reactor core. However, if the reaction temperatures rise uncontrollably, the core begins to melt-down and radioisotopes escape out to the environment. Nuclear accidents may be classified into two broad categories: Reactivity excursion accidents and loss of coolant accidents.

1. Reactivity Excursion Accidents

If an indifferent worker withdraws the control rods too far and/or too fast, the number of projectile neutrons increases manifold. As a result the uranium-235 nucleus begins to fission at abnormally fast rates. The temperature shoots up, causing the core to melt.

Most reactors employ water as a coolant. Water is also an excellent absorber of neutrons. In the absence of control rods, it absorbs some of the neutrons and keeps the reaction under control, but only for a while. As the fission reaction proceeds, the core temperatures rise and liquid water is converted into steam. Unlike liquid water, steam is a poor absorber of neutrons. In the absence of control rods and liquid water, the fission reaction proceeds at abnormally fast rates. This condition is called reactivity excursion. Under this condition the temperatures become very high and core melt-down begins. Reactivity excursion was the cause of the nuclear accident that occurred at Chernobyl.

Modern nuclear reactors are endowed with several safety systems to counter reactivity excursion. There is a provision for emergency insertion of control rods. Even if the computer or mechanical operations fail, the emergency rods can still be inserted by manual devices. If emergency insertion of control rods do not stop reactivity excursion, another emergency system, called *poison insertion system,* takes over. It injects a solution of boron compounds into the reactor. Boron absorbs the excess neutrons.

$$^{11}_{5}\mathrm{B} + ^{1}_{0}n \rightarrow ^{12}_{5}\mathrm{B} + \gamma \qquad (13.10)$$

The fission reaction slows down and the core temperature falls.

2. Loss of Coolant Accidents

Loss of coolant accidents occur more frequently than reactivity excursion mishaps. They result from leakages in the cooling water system. The water escapes from the ruptures, leaving the reactor with scarce amount of coolant. The temperature rises and if the coolant is not restored within a few seconds, the core melt-down begins. The Three Mile Island accident occurred because of loss of coolant from the reactor core.

Nuclear reactors are provided with an *emergency core cooling system* (ECCS). This is a protective device that is designed to feed large amounts of water into the coolant pipes. It is backed-up by a *multiple forced circulation circuit* (MFCC). This system consists of pumps and plumbing equipment to circulate the coolant through the reactor core. The ECCS and MFCC are normally controlled by a computer, although there is a provision of manual operation also.

The most important preventive against loss of coolant accidents, however, is the periodic inspection of reactor equipment and the surroundings. When water leaks out of the cooling system, there is an increase in the humidity in the space housing the reactor. Hence, a higher humidity indicates a possible leakage. The water which seeps out carries trace amounts of radioisotopes. As recently as 1994, a portable radiation detector called *Betascint sensor* has been fabricated to measure soilborne or airborne radioactivity in the vicinity of nuclear sites. The sensor is about 1.5 m long and is filled with plastic fibres laced by fluorescent compounds, such as anthracene or stilbene. When β-particles emitted by radioisotopes pass through the sensors, the fibres begin to glow. Betascint sensors take just a few minutes to detect radioactivity. Additionally, there is no dearth of scanning facilities like X-ray and ultrasound devices which can detect even hairline cracks in the coolant pipes. What is required is a vigilant observation.

We now describe two infamous nuclear accidents which occurred at Three Mile Island and Chernobyl.

Three Mile Island Episode

Three Mile Island, a nuclear reactor site in the USA, made news because of an accident that resulted from a complex operational failure compounded by a human error. The reactor was operating at 98% of its rated power when the mishap occurred.

Causes

At 4.00 AM on March 28, 1979, a pump in the secondary steam loop of unit 2 reactor of Three Mile Island stopped functioning. Under normal circumstances, the steam would have been used to turn the turbine and then condensed. Subsequently, the condensate would have been reinjected into the reactor as coolant water. Because of the pump failure, however, reinjection of coolant stopped. In such an event, auxilliary pumps are normally called into action. But a fortnight earlier, on March 14, 1979, the primary valve of the auxilliary pump had struck open. Therefore, when the coolant was circulated through the auxilliary systems it drained off. At this stage the emergency core cooling system (ECCS) was switched on. The coolant did start to flood the reactor, but a faulty gauge indicated that the reactor was flooded to an optimum level. A negligent worker shut the ECCS without verifying the coolant level. Without the coolant the reactor temperatures rose rapidly. A partial core melt-down initiated. The damage was arrested when the coolant level was restored.

Consequences

As the core melt-down proceeded, steam containing about 10 MCi of radioisotopes, mainly those of xenon and krypton, vented into the atmosphere. A part of the plant's ventilating system was also contaminated.

Enormous quantities of hydrogen gas were released at the reactor site. The reactor core is made up of a zirconium alloy. As the temperature of the reactor rose, the zirconium content of the alloy started reacting with steam, releasing hydrogen.

$$Zr + \underset{\text{Steam}}{2H_2O} \rightarrow ZrO_2 + 2H_2 \tag{13.11}$$

The steel walls of the reactor were indented from the pressure pulse of hydrogen gas. Other accessories showed flame damage due to combustion of hydrogen.

Aftermath

Immediately after the accident, pregnant women and children were evacuated to safer areas. Surprisingly, no major casualty was reported. The reason being that radioactive iodine, which normally escapes a failed reactor and causes toxic effects, was unexpectedly not released in the Three Mile Island episode. Subsequent analysis of the mishap revealed that radioactive iodine underwent dissolution in water and was passed into an auxilliary building through water pipes. This way its escape to the environment was prevented.

Chernobyl Episode

Chernobyl city (Fig. 13.7) in Russia is the site for the V.I. Lenin nuclear power plant. Only a few people outside the former Soviet Union had heard of this city before it shot into headlines because of what was the most serious nuclear accident of the last century.

Causes

The reactors installed at the Chernobyl nuclear power plant were designated RBMK 1000. These reactors employed graphite as a moderator, water as a coolant, and uranium(IV) oxide that is 2% enriched in uranium-235 as a fuel. They were designed to produce 1000 MW electrical energy.

On April 26, 1986 a major accident occurred at reactor 4 of the Chernobyl power plant. At the time of the accident, reactor 4 had 1659 fuel assemblies. The reactor had been in operation continuously for nearly two years. It was scheduled to be shutdown for intermediate maintenance at 1.00 AM on April 25, 1986. The schedule was, however, ill timed. It was to begin at an hour when the top brass of an organisation is normally not on duty and in the week during which the

Fig. 13.7 Location of Chernobyl and surrounding areas

Russian authorities and people were busy making preparations for their biggest national holiday—the May Day.

The shutdown operation began with a progressive reduction in the power supply produced by the reactor. This is done by inserting the control rods at regular intervals. However, for maintenance work the power supply is never completely cut-off. Some power is needed to run the coolant pumps because the reactor remains hot for a long time after the shutdown and, therefore, has to be fed with the coolant. The extent to which the power level should be reduced is controlled by a local automatic regulation system. This system was, however, switched off at the time when the maintenance work started.

The power dipped to an extremely low level. Yet, all the coolant pumps were kept operational. Since the power level was low, it was essential to inject only a small amount of water into the reactor. Smaller volume of water would have heated up rapidly and produced sufficient steam to run the turbine. However, because all the pumps were functional, the reactor received excessive water. The rate of steam production declined. The pressure of steam became too low to run the turbine. The power level dipped further.

At this stage two operations were carried out. Firstly, the speed of the pumps was reduced. The idea was to curtail the water supply to the reactor, so that steam could be rapidly produced. Secondly, nearly all the control rods were withdrawn. The idea was to immediately increase the power supply. Operational regulations for the RBMK 1000 reactors specify that at no time the reactor should have fewer than 30 control rods. This regulation was violated.

Reactor 4 reached a pernicious stage: There were neither control rods nor liquid water. As has been stated earlier, control rods are meant to keep the number of neutrons at a low level so that the fission reaction (13.1) remains under control. Like the control rods, liquid water also absorbs some of the neutrons and helps regulate the fission reaction. When both the control rods and liquid water were absent, reactivity excursion developed. The number of projectile neutrons increased and the power level registered a manifold increase. The core temperature shot up and the steam pressure became astronomically high. An operator attempted to reinsert the emergency control rods but it was too late. The poison insertion system, which tends to control the number of projectile neutrons by injecting a solution of boron salt (reaction 13.10), had been switched off at the time of shutdown.

At 01.23.40 hours on April 26, 1986 an explosion occurred in the reactor. The reactivity excursion had raised the power level 120 times the normal level. About 3 seconds later the power level had risen 440 times the normal value and a second explosion occurred.

Consequences

The explosions which rocked the Chernobyl power plant were so severe that the 1000 tonne concrete lid of the reactor 4 blew off. The coolant pipes were ripped apart and radionucleides were released to the atmosphere. About 30 fires started in the reactor complex due to the combusion of graphite rods. The first order of business was to extinguish the graphite fires and cover the remains of the reactor, thus stopping the release of radioactivity. About 5000 tonnes of boron, dolomite, sand, clay and lead were dropped by helicopter into the smouldering core. Initially, the core temperature fell, but after a week it rose again. Apparantly, heat was being generated from the decaying fission products. This caused the radioactive releases from the reactor to increase.

The total release of the fission products was initially estimated to be about 50 MCi. However, remodelling of the accident scenario revealed that the mishap released 170–185 MCi, 3–4 times above the initial Soviet estimates. This discrepancy arose because of two natural phenomena.

Firstly, the under-estimation was apparently due to the failure to register the size of the radioactive plume created by the explosion. This plume was missed by the monitoring station located 14 km away from Chernobyl. The station was not operating at night. The plume crossed Poland and Denmark. Ultimately, it was detected at the Forsmark Nuclear Power Plant in Stockholm, Sweden. By this time the plume was appreciably diluted.

Secondly, by an accidental luck the explosion plume was blown in a westerly direction. The remaining three directions around the power plant were thickly populated. However, on the western side the nearest village, Tolsty Les, which translates as 'thick forest', was 25 km away from the plant. As its name implies, the village was surrounded by dense pine grove. The trees played a buffer role. They absorbed significantly high doses of radionucleides. The plume lost some of its reactivity to the trees. The residents of the village had a miraculous escape. Trees indeed are the saviours of human beings!

On the first day of the accident 31 people died and 200 were injured. Since the plume was rich in iodine-131, cesium-134 and cesium-137 it was feared that the accident would cause long term health effects. Periodic surveys of areas around Chernobyl have proved that these fears were not misplaced. For example, thyroid cancer is emerging as one of the tragic consequences of Chernobyl disaster. This type of cancer is caused by iodine-131, which is ingested mainly through milk and milk products. Since children consume more milk than adults and since the thyroid glands in children are in growing stage, infants are more prone to protruding thyroid cancer. After the Chernobyl accident, the surrounding areas of Belarus, Ukraine and Russia (Fig. 13.7) registered a 200-, 100- and 8-fold increase, respectively, in the incidence of thyroid cancer among the children.

Aftermath

Immediately after the accident about one lakh people living around the plant were evacuated. In the absence of human beings, the population of wild boar, otters, wildfowl and voles boomed. The animals, however, could not escape the consequences of radioactive landscape. They developed a high degree of mutations. For example, voles from contaminated zone have been found to have a mutation rate about 40 times that of the background rate. The rate is high enough to pass the mutations from one generation to other.

The trees surrounding Tolsty Les would be cut down and burnt to produce electricity. A power plant fuelled by timber from around the crippled reactor is being constructed at Belarus. The power plant shall be equipped with modern filtering technologies and electrostatic precipitators to prevent radioactive fallout. This operation would decontaminate Belarus forests within 40 years, compared to centuries required for natural decay of radioactivity. It is feared that if nothing is done with the forest trees, they will eventually burn in forest fires with no clean-up at all. The area around Chernobyl has had already suffered several forest fires since 1986.

Inference

Nuclear energy utilization presents a paradoxy. On one hand it represents a cheap, inexhaustible and non-polluting energy resource. On the other, it has become a target of impassioned censure. This dilemma is not difficult to understand.

Technical advances have all but eliminated routine radioactive emissions from nuclear power plants. No dread factor is associated with normal operation of nuclear reactors. It is because of the power plant accidents that the nuclear energy has become synonymous with disaster. The Three Mile Island and Chernobyl types of disasters mould the public opinion against nuclear energy sources.

The nuclear power plants are provided with various safety systems which ensure that no harm befalls the populace. Yet, a human error can negate the protection afforded by safety systems. Therefore, the probability of a nuclear mishap may be negligible, but can never be zero. However, it must be remembered that the probability of a hydroelectric dam collapsing under the influence of seismic shocks is also not zero; it is much more than that of a nuclear accident. The casualty figure too would be much higher.

The critics of nuclear energy are those who believe in the maxim that "anything that can go wrong will go wrong". Such thinking impedes the progress of the society. About a century ago, when oil was being mooted to replace coal in the naval ships, the critics were quick to point out that oil was too inflammable and, therefore, too dangerous a fuel to be used in a vessel.

Today the oil-based vessels have given way to nuclear ships and submarines—an endeavour which would have been impossible had the critics had their way. This is not to imply that we support the construction of nuclear naval vessels. Yet, we do admire them as marvels of scientific progress. And if the energy which fuels an instrument of doom is held in esteem, how much more benevolent would it become if it is used for the welfare of humanity. In return, the only price this magical energy resource demands in a circumspect handling of nuclear reactors.

References

Angino, E.E. (1977). High level and long lived radioactive weste disposal, *Science,* **198,** 885–890.

Atwood, C.H. (1988). Chernobyl—what happened?, *J. Chem. Ed.,* **65,** 1037–1041.

Beard, J. (1995). Glowing fibres signal radiation, *New Scientist,* April 22, p. 22.

Cohen, B.L. (1976). Impacts of the nuclear energy industry on human health and safety, *Amer. Sci.,* **64,** 550–559.

Crawford, D. (1995). Leukaemia linked to radioactive rivers, *New Scientist,* September 23, p. 12.

Edwards, R. (1995). Leaky drums spill plutonium on ocean floor, *New Scientist,* July 22, p. 5.

Edwards, R. (1996). Radioactive forests will bring power to the people, *New Scientist,* October 26, p. 9.

Enger, E.D. and Smith, B.F. (1992), *Environmental Science,* 4th Edn., Wm. C. Brown Publishers, Dubuque, p. 198–220.

Furth, H.P. (1995). Fusion, *Sci. Amer.,* **273**(3), 140–142.

Gonzales, S. (1982). Host rocks for radioactive waste disposal, *Amer. Sci.,* **70,** 191–200.

Likhtarev, I.A., Sobolev, B.G., Kairo, I.A., Tronko, N.D., Bogdanova, T.I., Oleinic, V.A., Epshtein, E.V. and Beral, V. (1995). Thyroid cancer in the Ukraine, *Nature (London),* **375,** 365.

Medvedev, Z.A. (1994). Chernobyl: Eight years after, *Trends Ecol. Evol.,* **9,** 369–371.

Miller, W.F. (1993). Present and future nuclear reactor designs, *J. Chem. Ed.,* **70,** 109–114.

Mynatt, F.R. (1982). Nuclear reactor safety research since Three Mile Island, *Science,* **216,** 131–135.

Ringwood, T. (1982). Immobilization of radioactive wastes in SYNROC, *Amer. Sci.,* **70,** 201–207.

Salvatores, M. and Prunier, C. (1993). Transmutation and nuclear radioactive waste management: A perspective, *Endeavour,* **17,** 116–120.

Shcherbak, Y.M. (1996). Ten years of the Chernobyl era, *Sci. Amer.,* **274**(4), 32–37.

Webb, J. (1995). Thyroid cancer takes its toll on Chernobyl's children, *New Scientist,* April 1, p. 7.

Weinberg, A.M. (1972). Social institutions and nuclear energy, *Science,* **177,** 27–34.

Williams, N. (1995). Chernobyl: Life abounds without people, *Science,* **269,** 304.

Part IV

Air Pollutants

CHAPTER 14

Sulfur Oxides

In environmental chemistry, the term sulfur oxides includes sulfur dioxide and sulfur trioxide. These gases are emitted to the atmosphere as a result of large scale applications of sulfur compounds in industry and agriculture and their extensive involvement in biological processes. In the atmosphere, a part of sulfur oxides is converted into sulfate aerosols and sulfuric acid. The latter is washed down to the earth as acid rain. Hence the term sulfur oxides, in a broad sense, also includes the environmental implications of sulfate aerosols and acid rain.

Sources of Sulfur Oxides

The oxides of sulfur are released to the environment by natural sources, as well as by human activity. The natural contribution is much more than the amount released by human activity. However, the man-made sources of sulfur oxides have received more attention because they are highly localized and, therefore, seen as immediate threat to the regional environment. This is reflected in the striking difference between the amount of sulfur oxides emitted by man in the two hemispheres: In the highly populated and industrialized Northern Hemisphere it is 93%; in the sparsely populated Southern Hemisphere it is just 7%.

The natural and man-made sources of sulfur compounds are listed in Table 14.1, along with the estimated annual contribution.

Table 14.1 Principal sources of sulfur compounds

Source	Main Component	Annual Contribution ($\times 10^6$ tonnes)
Natural	Sulfur dioxide	1–2
Eruption of volcanoes		
Sea sprays	Sulfate aerosols	130
Biogenic decomposition		
on land	Hydrogen sulfide	100
in sea	Dimethyl sulfate	60–400
Man-made	Sulfur dioxide	150

Natural Sources

There are three main natural sources of sulfur oxides.

1. Eruption of volcanoes

During a volcanic eruption, the sulfur compounds commonly emitted to the atmosphere are sulfur dioxide, sulfur trioxide and hydrogen sulfide. The latter is oxidised to sulfur dioxide. Elemented sulfur is also discharged to the environment and subsequently it also is oxidised to sulfur oxides. The relative proportion of sulfur oxides, however, varies from one volcano to another and even from one eruption to another of the same volcano.

2. Sea sprays

The sulfur oxides present is sea water are mainly in the form of sulfate ions. Oceans contain approximately 2.5 mg of sulfate ions per gram of water. The sulfate ions are 'sprayed out' as aerosols from the sea during rough weather. These sulfate aerosols then accumulate in the lower atmosphere.

3. Biogenic decomposition

The biogenic decomposition of sulfur-containing organic matter occurs both on land and in sea. On land, the major sulfur compound generated is hydrogen sulfide. It is formed through the reduction of organic matter by anaerobic bacteria in swamps, muddy areas and coastal zones. In these regions the supply of oxygen is limited and hence these are ideal for the anaerobic reduction processes. Once in the atmosphere, hydrogen sulfide is easily oxidised to sulfur oxides.

The principal sulfur-containing compounds released from sea due to biogenic decomposition are demethyl sulfide, $(CH_3)_2S$ and hydrogen sulfide, the former being predominant. Dimethyl sulfide is formed through reduction of organic matter by marine algae. Unlike hydrogen sulfide, dimethyl sulfide is resistant to atmospheric oxidation.

Man-Made Sources

The major constituent of sulfur oxides released by man-made sources is sulfur dioxide. The important man-made sources of sulfur dioxide are:

1. Burning of fossil fuels, especially coal, petroleum and petroleum products.
2. Petroleum refinery operations.
3. Smelting of sulfide ores, especially those of copper and zinc.
4. Manufacture of sulfuric acid.
5. Paper making
6. Burning of domestic refuse.

Table 14.2 Man-made sources of sulfur oxides

Source	Contribution to Sulfur Oxides (%)
Coal combustion	60
Petroleum combustion	21
Petroleum refining	6.8
Ore smelting	7.4
Other industrial applications	4.8

As seen in Table 14.2, coal combustion is the predominant man-made source of sulfur oxides. Coal contains 0.1–6% of sulfur. During combustion, its sulfur content is oxidised to sulfur dioxide, which subsequently disperses in the atmosphere. Coal is used in vast quantities for electricity generation. Hence, power plants are largely to be blamed for sulfur oxides pollution problem.

Fate of Sulfur Oxides in the Environment

The sources which emit oxides of sulfur also release hydrogen sulfide to the atmosphere. Hydrogen sulfide is oxidised to sulfur dioxide by atomic oxygen,

$$H_2O + O \rightarrow HS + OH \tag{14.1}$$

$$HS + OH + O_2 \rightarrow SO_2 + H_2O \tag{14.2}$$

or by molecular oxygen, $$H_2S + \tfrac{3}{2}O_2 \rightarrow SO_2 + H_2O \tag{14.3}$$

or by ozone, $$H_2S + O_3 \rightarrow SO_2 + H_2O \tag{14.4}$$

The sulfur dioxide so formed and the sulfur dioxide directly emitted to the atmosphere persists for 2–4 days. During this time it is not only dispersed to far-away locations, but is also oxidised to sulfur trioxide. The oxidation of sulfur dioxide occurs both photochemically and non-photochemically.

Sulfur dioxide in its electronic ground state does not react with ozone in the gas phase. However, it absorbs strongly in the near ultraviolet region to form electronically excited molecules which react photochemically with ozone.

$$SO_2 + O_3 \xrightarrow{hv} SO_3 + O_2 \tag{14.5}$$

Similarly, ground state of sulfur dioxide does not react with oxygen in the gas phase. However, excited sulfur dioxide reacts.

$$SO_2 + \tfrac{1}{2}O_2 \xrightarrow{hv} SO_3 \tag{14.6}$$

Non-photochemically, sulfur dioxide may be oxidised by molecular oxygen in presence of powdered oxides of aluminium, calcium, chromium, iron, lead and vanadium. Dust and soot particles have a similar effect. The soot particles are particularly important because near the combustion source, the concentration of both sulfur dioxide and soot particles are significantly high.

$$SO_2 + \tfrac{1}{2}O_2 \xrightarrow{\text{Particulates}} SO_3 \tag{14.7}$$

Sulfur dioxide may also undergo a three-body non-photochemical reaction with atomic oxygen,

$$SO_2 + O + M \rightarrow SO_3 + M \tag{14.8}$$

where M is a molecule of oxygen or nitrogen.

The sulfur trioxide formed by both photochemical and non-photochemical reactions immediately reacts with water to form sulfuric acid.

$$SO_3 + H_2O \rightarrow H_2SO_4 \tag{14.9}$$

Thus, sulfur trioxide never exists to an appreciable extent in the atmosphere.

It is also possible that sulfur dioxide gets dissolved in fog or cloud droplets and is converted into sulfurous acid,

$$SO_2 + H_2O \rightarrow H_2SO_3 \tag{14.10}$$

which is then oxidised to sulfuric acid by oxygen, in presence of metal salts.

$$H_2SO_3 + \tfrac{1}{2}O_2 \xrightarrow{\text{Metal salts}} H_2SO_4 \tag{14.11}$$

Some of the sulfuric acid precipitates as acid rain. The remaining reacts with ammonia or sodium chloride from sea sprays to produce sulfates.

$$H_2SO_4 + 2NH_3 \rightarrow (NH_4)_2SO_4 \tag{14.12}$$

$$H_2SO_4 + 2NaCl \rightarrow Na_2SO_4 + 2HCl \tag{14.13}$$

Analysis of Sulfur Oxides

We have seen that hydrogen sulfide and sulfur trioxide do not persist in the environment for long. Sulfur dioxide, however, is stable in the atmosphere for 2–4 days. The concentration of sulfur dioxide can, therefore, be determined. The test sample of air is bubbled through a dilute aqueous solution of sodium tetrachloromercurate(II), $Na_2[HgCl_4]$, containing sulfamic acid. The latter destroys any nitrogen oxides that may be present in the air. Sulfur dioxide, in the air sample, reacts with sodium tetrachloromercurate(II) and is quantitatively converted into a stable, non-volatile complex, sodium dichlorosulfitomercurate(II), $Na_2[HgCl_2(SO_3)]$.

$$Na_2[HgCl_4] + SO_2 + H_2O \rightarrow Na_2[HgCl_2(SO_3)] + 2HCl \tag{14.14}$$

A mixture of p-rosaniline dye, formaldehyde and phosphoric acid is then added to the sodium dichlorosulfitomercurate(II) complex. The p-rosaniline dye is converted to p-rosaniline methylsulfonic acid.

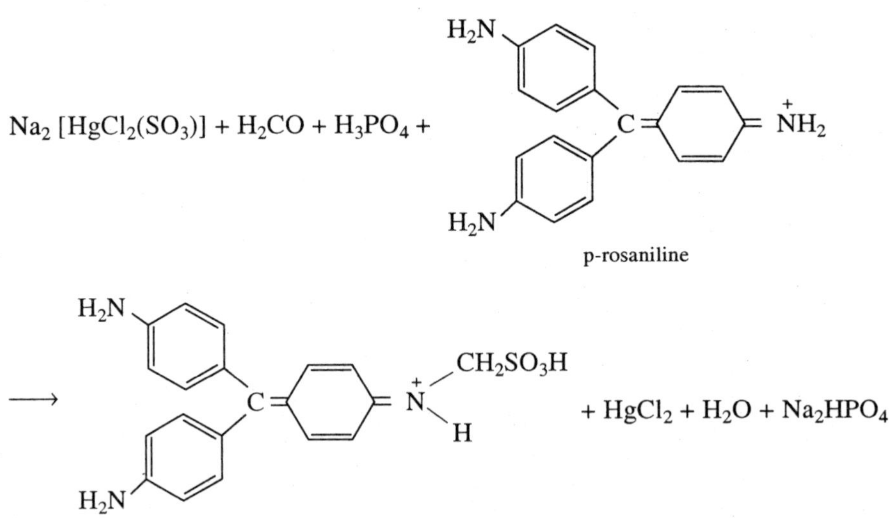

(14.15)

The functions of phosphoric acid are, to liberate sulfur dioxide from sodium dichlorosulfitomercurate(II) complex, to maintain the pH of the reaction at 1.0 and to tie-up any heavy metal ions that may be present, along with sulfur dioxide, in the environment. For example,

$$Fe^{3+} + 3PO_4^{3-} \rightarrow [Fe(PO_4)_3]^{6-} \tag{14.16}$$

The p-rosaniline methylsulfonic acid, formed in reaction (14.15), has a red-violet colour. At pH 1.0, it absorbs at 575 nm. Its concentration can, therefore, be determined colorimetrically, following Lambert-Beer's law. The concentration of p-rosaniline methylsulfonic acid is proportional to the concentration of sulfur dioxide in the the atmosphere.

Effects of Sulfur Oxides

The sulfur oxides have the following deleterious effects in the environment.

1. Human Health

The primary standard of sulfur oxides, promulgated to protect human health is 0.03 ppm. However, sulfur oxides alone produce no acute response in humans upto 1.0 ppm. Nevertheless, adverse effects on human health result in circumstances where sulfur oxides condense on smoke or particulates. Thus, there exists a synergistic effect between sulfur oxides and particulates. That is, the total health hazards of both combined is greater than the sum of their individual effects.

The sulfur oxides adsorbed on particulates can be inhaled quite deeply into the respiratory tract. The adsorbed sulfur oxides then become solubilized in body fluids and enter the blood stream, the lymphatic system or the connective lung tissue. As a result, the bronchioles contract spasmodically in order to prevent admission of irritants. This makes breathing difficult and strains the victim's heart. Higher concentrations of sulfur oxides and smoke lead to bronchitis and lung cancer.

2. Vegetation

Sulfur oxides are quite harmful to plants. They destroy plant cells and interfere with chlorophyll synthesis. Moreover, leaf blotching and reduction in crop yield occur even when the atmospheric concentration of sulfur oxides is less than 1.0 ppm. The sulfur oxides also kill the tips of evergreen trees, cause their needles to drop and deform their growth. In presence of nitrogen oxides or ozone, the adverse effects of sulfur oxides on vegetation tend to increase.

3. Non-living Materials

Following exposure to sulfur oxides,

(a) Building stones, particularly carbonate-based stones, suffer severe erosion. This phenomenon is called *stone leprosy.*

(b) The rate of corrosion of metallic structures increases. For example, at a concentration of 0.10 ppm sulfur oxides and in presence of particulates, steel panels corrode 50% more rapidly.

(c) Drying of paints is retarted.

(d) Discolouration and embrittlement of paper results. This leads to deterioration of mechanical strength of books.
(e) Fabric and leather become discoloured and weathered.
(f) Deterioration of electrical equipment results.

4. Visibility

The sulfate aerosols can cause loss of visibility. When the relative humidity is 50% and the individual concentrations of sulfur oxides and particulates are about 0.10 ppm, visibility is reduced to about 5 miles. Thus, sulfur oxides pollution curtails such activities as scheduled airline flights which depend on clear lines of sight.

5. Acid Rain

Acid rain is the most insidious and visible effect of sulfur oxides pollution and is discussed in a separate section of this chapter (p. 224).

Control Measures for Sulfur Oxides

Since sulfur oxides have several undesirable effects, considerable attention has been paid to develop efficient methods for controlling the emission of these compounds to the environment. There are two broad methods to reduce the amount of sulfur oxides released to the atmosphere. These are, the fuel desulfurization techniques and the flue-gas desulfurization techniques.

1. Fuel Desulfurization Techniques

In these techniques, sulfur is removed from coal before combustion takes place. The following processes are based on fuel desulfurization techniques.

(a) Physico-chemical method

About half the sulfur present in coal is in inorganic form. The remaining half is organically bound. The inorganic sulfur in coal is usually present as iron pyrites, FeS_2. To remove it, the coal is grinded and subjected to hydraulic washing. Iron pyrites, being heavier, tends to settle down under the influence of gravity, while the inorganic sulfur-free coal floats on the surface form where it is collected.

The organically bound sulfur in coal is removed by passing hydrogen gas over the fuel in presence of a mixed cobalt-molybdenum oxide catalyst. The organic sulfur is converted into hydrogen sulfide. The latter is absorbed in a solution of diethylanolamine. It can subsequently be regenerated and oxidised to sulfur.

$$H_2S + (O) \rightarrow H_2O + S \tag{14.17}$$

Sulfur is a useful raw material. However, it is a relatively cheap chemical and its commercial value does not support the high cost expended on fuel treatment.

(b) Microbial method

The organic sulfur from coal can be removed by a relatively recent method that makes use of a bacterial strain. The strain, designated as IGT-57, removes sulfur from coal without degrading coal itself. IGT-57 is cultured in a sulfur-limited chemostat, that is, a bacterial growth culture which supplies all nutrients except sulfur. The needed sulfur is extracted from coal. Thus, IGT-57 metabolizes sulfur selectively from coal, without simultaneously degrading carbon.

The disadvantage of this method is that it is ineffective in removing inorganic sulfur from coal. Moreover, the rate of organic sulfur removal is too low to be of commercial utility.

2. Flue-Gas Desulfurization Techniques

In these techniques, the fuel is burnt in such a way that the sulfur oxides are not emitted to the atmosphere. The following processes are based on flue-gas desulfurization techniques.

(a) Wet lime-limestone scrubbing

In this method, powdered limestone and powdered coal are injected into the boiler. At high temperatures, limestone decomposes into lime.

$$CaCO_3 \xrightarrow{\Delta} CaO + CO_2 \tag{14.18}$$

The lime then ties-up the sulfur dioxide generated from coal combustion.

$$CaO + SO_2 \rightarrow CaSO_3 \tag{14.19}$$

$$CaO + SO_2 + \tfrac{1}{2}O_2 \rightarrow CaSO_4 \tag{14.20}$$

Calcium sulfite and calcium sulfate are removed by scrubbing with a slurry of limewater. The latter is obtained by the hydrolysis of lime.

$$CaO + H_2O \rightarrow Ca(OH)_2 \tag{14.21}$$

The wet lime-limestone scrubbing method suffers from several demerits. The scrubbers create huge pile of sludge waste, thus constituting another environmental problem. According to one estimate, a 1000 MW power plant that burns coal with a sulfur content of 3% produces enough sludge in one year to cover a square mile to a depth of one foot! Wet scrubber systems also require a large amount of water and, therefore, problems arising from fouling and corrosion of equipment are frequently encountered. Lastly, about 3–6% of energy output is expended to run pumps and fans to reheat flue gas in order to prevent corrosive condensation in the chimney.

(b) Catalytic reduction of sulfur dioxide

Direct reduction of sulfur dioxide to elemental sulfur is also feasible. Such a reaction can be carried out by passing a mixture of carbon monoxide (from another polluting unit) and sulfur dioxide over a catalyst.

$$2CO + SO_2 \xrightarrow{\text{Catalyst}} 2CO_2 + S \tag{14.22}$$

Formerly, palladium, silver, cobalt or nickel, supported on alumina were used as catalysts. However, in presence of these catalysts, carbon monoxide reacted with elemental sulfur to give highly toxic carbonyl sulfide, COS.

$$CO + S \xrightarrow{\text{Catalyst}} COS \tag{14.23}$$

Alternately, lanthanum titanyl, $LaTiO_3$, can be used as a catalyst. Though it suppresses the formation of carbonyl sulfide to a significant extent, yet the evolution of this toxic compound is not completely avoided. A copper/alumina catalyst is very efficient. However, it suffers from the limitation that the reacting gases (carbon monoxide and sulfur dioxide) must be completely dry, since traces of water result in catalyst's deactivation.

Assessment of Control Methods

We have seen that each remedial method to control the evolution of sulfur oxides suffers from one or more drawbacks. Apart from the drawbacks that have been listed, there are two common demerits in all the processes. Firstly, they are not economicaly attrahent and hence there is a great reluctance to put them into practice. Secondly, and this is more important from the point of view of environmental conservation, these techniques do nothing to reduce the emission of oxides of nitrogen to the atmosphere. Since nitrogen oxides are as hazardous a contaminant as are sulfur oxides, these techniques need to be modified so as to curtail the emission of undesirable nitrogen compounds. Two modern methods *coal conversion technology* and *fluidized bed combustion* reduce the emission of both sulfur oxides and nitrogen oxides during coal combustion. These are detailed in Chapter 15, p. 234.

Acid Rain

Rain water normally has a pH of 5.6. If the pH drops below 5.6, the rain is described as acidic. Acid rain is a complex environmental problem, causing deleterious ecological changes due to depositions traceable to pollutant acid precursors that result from burning of fossil fuels. The main pollutants resulting from combustion are sulfur oxides and nitrogen oxides. These may be absorbed by droplets of water that make up clouds and get chemically converted to sulfuric acid and nitric acid respectively. When precipitation coalesces from cloud water, it washes the acid contaminants out of the atmosphere. This is called *wet deposition.* Alternately, the sulfur oxides and nitrogen oxides may be absorbed by a wet surface of land, vegetation or a structure and get chemically transformed to sulfuric acid and nitric acid. This is called *dry deposition.* Trees and vegetation in the upland areas may also capture acidic droplets in mists and clouds. This is called *occult deposition.*

Causes

Acid rain is caused by the emission of oxides of sulfur and nitrogen into the atmosphere. We have seen that sulfur oxides are released to the environment largely because of fossil fuel combustion. Ore smelting, petroleum refining and sulfur-based industries are the other sources of sulfur oxides. Nitrogen oxides are emitted to the atmosphere mainly as a result of automobile exhaust and fossil fuel combustion.

The major problem with acid rain is that it does not remain confined to the region where oxides of sulfur and nitrogen are emitted. In fact, it may fall hundreds of miles from the pollution source. There are two reasons for the large scale migration of pollutants before culminating in wet or dry deposition. Firstly, in recent years there has been an increasing tendency to install taller chimneys. This causes a wider dispersion of sulfur oxides and nitrogen oxides and thus protects the local environment. However, the use of taller chimneys also causes the pollutants to travel great distances. Secondly, thunderstorms too cause extensive dispersion of sulfur oxides and nitrogen oxides. It has been estimated that a thunderstorm transports the atmospheric pollutants more than 1000 km away from the point of their origin within 2–4 days.

Thus, tall chimneys and thunderstorms transform a local environmental hazard into a global problem. If the pollutants remain localized, it is easier to institute control measures. If, on the other hand, the pollutants manifest their adverse consequences hundreds of miles from their origin, it is very difficult to reverse the ecological damage caused by them.

Consequences

Acid rain has caused visible damage to the environment downwind of a large emission source of sulfur oxides and/or nitrogen oxides. What has taken mankind centuries to build and nature millennia to evolve, is being degraded and destroyed in a matter of a few years. The effects of acid rain on forests, fisheries, drinking water supply and monuments are particularly pronounced and these are analysed in details.

1. Forests

Forest death and decline in Germany, North America, Canada and Czechoslovakia has been attributed to acid deposition. It is a matter of time before the forests in other parts of the globe are affected. In fact, the tree losses take place over a long period of time and, therefore, it is believed that many forested areas could already be affected by acid rain, with the results becoming manifested in the future. Silver fir, spruce and coniferous forests are particularly sensitive to acid rain. Forest decline is caused by the following two mechanisms.

(a) Leaching of nutrients from the soil. To grow adequately, trees require optimum supply of sixteen elements, calcium, magnesium, potassium, sodium and manganese being chief among them. These are taken up in cationic forms by tree roots from moist soil. When acid rain occurs, H^+, SO_4^{2-} and NO_3^- ions are added to the soil. These cause the leaching of the nutrient cations from the soil. Leaching is a two step process: In the first step, H^+ ions displace the nutrient cations from soil exchange sites. In the second step, the displaced cations combine with free anions and are mobilized away from the rooting zone. Theoretically, both the sulfate and nitrate anions can mobilize the nutrient cations. However, the nitrate anions are immediately taken up by soil as fertilizers and, therefore, these do not leach nutrient cations. In soils which are rich in iron or aluminium oxide, the sulfate anions are adsorbed on these oxides and are ineffective in leaching cations. However, in soils which are low in iron or aluminium oxide content, the sulfate anions combine with nutrient cations and carry them away from the roots. Thus, the soil becomes depleted in nutrient elements and this adversely affects the forest health. The leaching phenomena is shown in Fig. 14.1.

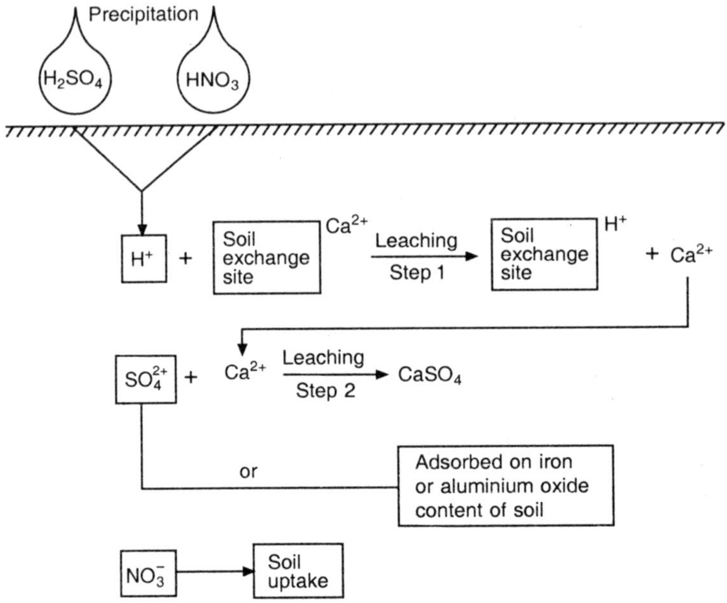

Fig. 14.1 Leaching of nutrients from soil

(b) Biological uptake of aluminium from soil. Aluminium, the third most abundant element in the earth's crust, occurs primarily in soil as alumino silicates. These minerals are insoluble and contain aluminium bound covalently to silicon and oxygen atoms. Therefore, aluminium content of soil does not interfere with nutrient uptake or cause biological toxicity. However, when acid rain falls on forest soil, the alumino silicates release aluminium either as free ions, Al^{3+}, or as monomeric hydroxides, $[Al(OH)]^{2+}$, $[Al(OH)_2]^+$, $[Al(OH)_4]^-$. In these forms, aluminium is soluble in water. It is, therefore, mobilized and taken up by trees through the roots. Aluminium is toxic to plants. It kills the roots and eliminates the vegetation-favourable microbes. It also interferes with nutrient cations uptake. The net result is that the forest areas start declining.

2. Fisheries

The number of lakes with no or poor fisheries has increased progressively over the past hundred years. This has been attributed to acid rain. We have seen that acid deposition releases aluminium in soluble forms from the soil. A part of the dissolved aluminium passes into lakes. Subsequently, it accumulates in the body of the fish, where it elicits characteristic toxic effects. It leads to necrosis of gill filaments, loss of plasma chloride, decrease in blood oxygen levels and increase in blood carbon dioxide levels. Further, at pH less than 5.5, the reproduction success of fish is impaired. At still lower pH (4.5–5.5), the hatching success and survival of embryos decreases. Species like salmon and trout are particularly sensitive to acid rain.

3. Drinking water supply

The potential effects of acid deposition on drinking water supply fall into the following two categories.

(a) Increased residue of heavy metals in well waters. Heavy metal ions, such as copper, mercury, lead and aluminium are leached from soil on which acid deposition accumulates. These solubilized metal ions then enter well waters and produce varied toxic effects. For example, homeodialysis patients are at risk in areas where drinking water contains even a trace of aluminium ions.

(b) Corrosion of water pipes. Acid rain causes corrosion of water conduit pipes. This results in physical deterioration of pipes and leaching of heavy metals, such as iron, lead and copper in the drinking water.

4. Monuments

Monuments constructed from marble are readily attacked by acid rain, with the formation of soluble calcium sulfate.

$$CaCO_{3(s)} + H_2SO_{4(aq)} \rightarrow Ca^{2+}_{(aq)} + SO^{2-}_{4(aq)} + H_2O_{(l)} + CO_{2(g)} \qquad (14.24)$$

Monuments constructed from sandstone generally have a coating of iron oxide which holds the quartz grains together. The iron oxide too dissolves in acid rain and the building becomes disfigured.

$$Fe_2O_{3(s)} + 3H_2SO_{4(aq)} \rightarrow 2Fe^{3+}_{(aq)} + 3SO^{2-}_{4(aq)} + 3H_2O_{(l)} \qquad (14.25)$$

Sulfur dioxide also attacks the monuments in a similar way, causing stone leprosy. But the reactions occur at a very slow pace compared to the attack by acid rain.

The disfigurement of monuments runs parallel to the industrial development. In fact, there is a definite correlation between occurrence of acid rain and deterioration of marble and sandstone buildings. Greek and Italian monuments, that have withstood centuries of weathering, have degraded rapidly over the past hundred years. Acid rain is envisaged to accelerate the deterioration of Taj Mahal as a result of the emission of sulfur and nitrogen oxides from the nearby Narora Power Project (see box).

The Taj Trapezium

The air in the city of Agra, where Taj Mahal is located, contains very high levels of sulfur and nitrogen oxides. The resulting acid rain is slowly eating away the monument, leaving the marble discoloured and lustreless. The Government of India announced an action plan in early 1995 to clean-up the pollution which is destroying the country's best loved landmark.

The new plan aims to clear the air in the 'Taj Trapezium'—an area that includes the towns of Agra, Firozabad, Mathura and Bharatpur. More than 2000 polluting industries lie inside the trapezium including cast iron foundries, rubber manufacture, brick kilns, glass works and the Government owned oil refinery at Mathura. These industries would switch over from using coal or oil to natural gas or liquefied petroleum gas. A new natural gas pipeline would bring more than half a million cubic meters of natural gas a day to the area covered under the Taj Trapezium. People living in the city will also be encouraged to use liquefied petroleum gas in place of coal, kerosene or firewood. Vehicles plying on highways in the vicinity of Taj would be encouraged to use low-sulfur content diesel.

Mitigation

Liming of lakes and forest soil has been successful in reversing the adverse effects of acid rain. Lime, being alkaline, neutralizes the acid content of rain water.

$$Ca(OH)_{2(s)} + H_2SO_{4(aq)} \rightarrow CaSO_{4(aq)} + 2H_2O_{(l)} \tag{14.26}$$

This restores the pH of water or soil to an optimum level. Moreover, addition of lime to lakes or soil lowers the dissolved aluminium content (rendered free by the action of acid rain on alumino silicate minerals) by precipitating it out as aluminium hydroxide.

$$2Al^{3+}_{(aq)} + 3Ca(OH)_{2(s)} \rightarrow 2Al(OH)_{3(s)} + 3Ca^{2+}_{(aq)} \tag{14.27}$$

The biological uptake of aluminium by forest trees and fish is thus averted and so are the hazardous consequences of acid rain.

However, in the long run, liming strategies are behest with several problems. In case of lakes, its application needs to be repeated with a frequency depending on the turnover of water. This can be as frequent as once a year. Further, liming of lakes must be strictly monitored. Extensive liming results in fish mortality. Similarly, in case of forest soil, addition of lime immobilizes the nitrogen content of soil. This means that in order to restore forest growth, fertilizers have to be added whenever the soil is limed.

Corrosion of drinking water pipeline is protected against acid rain by coating a protective film of an inert polymer along the interior of pipes. However, no such preventive technique is available to protect monuments from acid rain. No doubt, the monuments can be repaired, but that is only after the acid deposition has disfigured them. This entails high expenditure. Moreover, it may not restore the monument to its original glory.

Since liming strategies suffer from several drawbacks and since the protection of monuments from acid rain is really not possible, the best approach is to prevent acid rain from occurring, rather than mitigating it after it has occurred. For this we have to retrieve back to the control measures that reduce the emission of sulfur and nitrogen oxides to the environment.

References

Baedecker, P.A. and Reddy, M.M. (1993). The erosion of carbonate stone by acid rain, *J. Chem. Ed.*, **70,** 104–108.

Bonga, S.E.W. and Dederen, L.H.T. (1986). Effect of acidified water on fish, *Endeavour,* **10,** 198–202.

Cullis, C.F. (1978). Sulfur compounds in the atmosphere, *Chem. Brit.*, **14,** 384–389.

Del Monte, M. and Vittori, O. (1985). Air pollution and stone decay: The case of Venice, *Endeavour,* **9,** 117–122.

Ember, L.R. (1988). New acid rain study stirs interest, concern, *Chem. Eng. News,* May 16, p. 19–21.

Freemantle, M. (1995). The acid test for Europe, *Chem. Eng. News,* May 1, p. 10–17.

Friend, J.P. (1972). Sulfur mobilization as a result of fossil fuel combustion, *Science,* **175,** 1278–1279.

Gorham, E. (1994). Neutralizing acid rain, *Nature (London),* **367,** 321.

Hoggin, J. (1988). Newly developed bacterial strain removes organic sulfur from coal, *Chem. Eng. News,* August 29, p. 36–37.

Howells, G.D. and Kallend, A.S. (1984). Acid rain—the CEGB view, *Chem. Brit.,* **20,** 407–412.

Kumar, S. (1995). Save the Taj campaigners win first round, *New Scientist,* March 18, p. 10.

Leh, F. and Chan, K.M. (1973). Sulfur compounds, *J. Chem. Ed.,* **50,** 246–247.

Mohnen, V.A. (1988). The challenge of acid rain, *Sci. Amer.,* **259**(2), 14–22.

Moore, J.W. (1986). *'The Changing Environment',* Springer-Verlag, New York, p. 203–233.

Novakov, T., Chang, S.G. and Harker, A.B. (1974). Sulphates as pollution particulates: Catalytic formation on carbon (soot) particles, *Science,* **186,** 259–261.

O'Neill, p. (1985), *'Environmental Chemistry',* George Allen & Unwin, London, p. 108–124.

Scaringelli, F.P., Saltzman, B.E. and Frey, S.A. (1967). Spectrophotometric determination of atmospheric sulfur dioxide, *Anal. Chem.,* **39,** 1707–1719.

Schindler, D.W. (1988). Effects of acid rain on fresh water ecosystems, *Science,* **239,** 149–156.

Smith, W.H. (1991). Air pollution and forest damage, *Chem. Eng. News,* November 11, p. 30–42.

Stoiber, R.E. and Jepsen, A. (1973). Sulfur dioxide contributions in the atmosphere by volcanoes, *Science,* **182,** 577–578.

CHAPTER 15

Nitrogen Oxides

Nitrogen combines with oxygen to form a family of oxides: Nitrous oxide (N_2O), nitric oxide (NO), nitrogen dioxide (NO_2), dinitrogen trioxide (N_2O_3), dinitrogen tetroxide (N_2O_4) and dinitrogen pentoxide (N_2O_5). From the point of view of environmental pollution, however, only nitrous oxide, nitric oxide and nitrogen dioxide are important. They are collectively symbolized as NO_x. These oxides play a multiple role in air pollution. At a primary level, they elicit several toxic effects. At a secondary level, they enter into chemical reactions with other constituents of the atmosphere, producing acid rain, photochemical smog and aerosols.

Sources of Nitrogen Oxides

Nitrogen oxides are emitted to the atmosphere both by natural sources and man-made sources. In global terms, anthropogenic emissions account for only 8% of naturally produced nitrogen oxides. However, in urban areas the local concentration from anthropogenic sources are elevated several hundred times, depending on the regional conditions. Thus, whereas the global concentration of nitrogen oxides is 3×10^{-3} ppm, the concentration in some cities is as high as 1–2 ppm. Under natural conditions, oxides of nitrogen are produced during thunderstorms, when the temperature is quite high. They are also produced by soil microorganisms.

There are two major anthropogenic sources of nitrogen oxides: Fossil fuel based power plants and automobiles. In fossil fuel based power plants, the combustion temperatures are sufficiently high to allow nitrogen and oxygen to combine. Higher the combustion temperature, greater are the quantities of nitrogen oxides produced. Nitrogen oxides are also emitted from the gasoline and diesel engines of automobiles. The high temperature and pressure in the combustion chamber provide suitable conditions for the production of oxides of nitrogen.

Fate of Nitrogen Oxides in the Environment

The combination of nitrogen and oxygen to give nitric oxide occurs only at high temperatures.

$$N_2 + O_2 \rightarrow 2NO \qquad (15.1)$$

The nitric oxide in the atmosphere is converted to nitrogen dioxide by aerial oxidation.

$$2NO + O_2 \rightleftharpoons 2NO_2 \tag{15.2}$$

Ozone too can oxidise nitric oxide to nitrogen dioxide.

$$NO + O_3 \rightarrow NO_2 + O_2 \tag{15.3}$$

These gases are also produced by the photolysis of nitrous oxide in the stratosphere. Nitrous oxide is produced by the action of aerobic and anaerobic bacteria in soil on nitrate-based fertilizers. It is relatively inert in the atmosphere. However, when it reaches the stratosphere, it undergoes photolysis, producing either nitric oxide,

$$N_2O + h\nu \rightarrow NO + N \tag{15.4}$$

or free oxygen atoms.

$$N_2O + h\nu \rightarrow N_2 + O \tag{15.5}$$

The free oxygen atoms interact with nitrous oxide, producing nitric oxide,

$$N_2O + O \rightarrow 2NO \tag{15.6}$$

which is subsequently oxidised to nitrogen dioxide (reactions 15.2–15.3).

Nitrogen dioxide reacts with water vapours, producing nitric acid.

$$3NO_2 + H_2O \rightarrow 2HNO_3 + NO \tag{15.7}$$

Nitric acid is either washed down as acid rain or combines with atmospheric ammonia, producing ammonium nitrate aerosols.

$$NH_3 + HNO_3 \rightarrow NH_4NO_3 \tag{15.8}$$

In addition, oxides of nitrogen may interact with hydrocarbons in presence of sunlight to produce photochemical smog.

Analysis of Nitrogen Oxides

Since in the atmosphere both nitrous oxide and nitric oxide are ultimately converted to nitrogen dioxide, the analysis of nitrogen oxides is basically the analysis of nitrogen dioxide. Contaminated air, containing nitrogen dioxide, is bubbled through a solution of sodium hydroxide over a period of 24 h. Nitrogen dioxide content of the air is converted into sodium nitrite and sodium nitrate.

$$2NO_2 + 2NaOH \rightarrow NaNO_2 + NaNO_3 + H_2O \tag{15.9}$$

The contents are added to a mixture of hydrogen peroxide, sulfanilic acid, hydrochloric acid and N (1-napthyl)-ethylenediamine dihydrochloride. Sodium nitrite formed in reaction (15.9) diazotizes with sulfanilic acid.

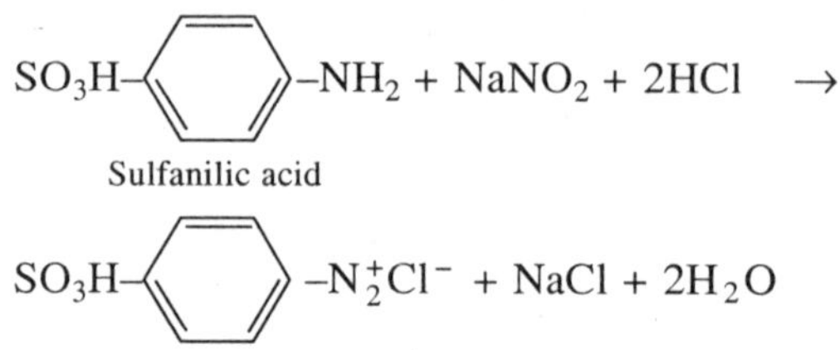

(15.10)

The diazonium salt couples with N(1-napthyl)-ethylenediamine dihydrochloride.

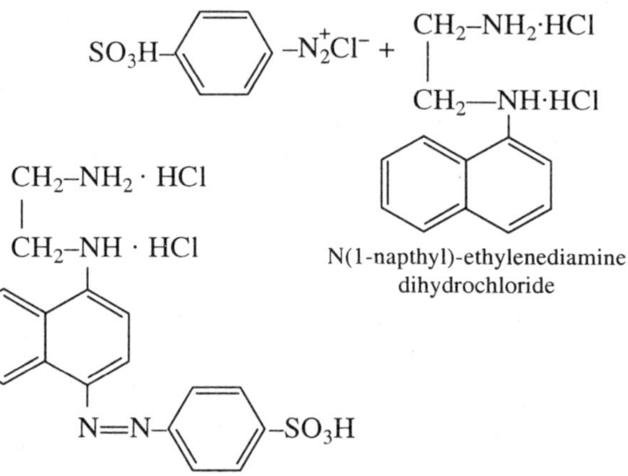

(15.11)

The coupled compound absorbs at 540 nm. Its concentration is determined by Lambert-Beer's law. One mole of coupled salt corresponds to one mole of sodium nitrite which, in turn, corresponds to one mole of nitrogen dioxide.

Effects of Nitrogen Oxides

We have seen that in the atmosphere, nitrous oxide and nitric oxide are ultimately converted to nitrogen dioxide. The major environmental hazards of nitrogen oxides are, therefore, attributed to nitrogen dioxide. The latter may elicit direct toxic effects as a primary pollutant or may interact with other pollutants producing secondary effects.

1. Primary Effects

Nitrogen dioxide is a deadly poison and one of the most obnoxious gases known. As a primary pollutant, it affects the humans, as well as vegetation.

(a) Humans

Humans exposed to nitrogen dioxide are likely to develop inflammation of lungs. However, the infection does not cause immediate pain. Therefore, exposure may go unnoticed until it is too late for the victim to recover from the delusion of well being. Edema (accumulation of fluid in the lungs) may occur several days later, resulting in death. In some cases the effects may be prolonged for weeks or even months.

The maximum allowable concentration for occupational exposure to nitrogen dioxide has been set at 8 ppm for an 8 h period. However, even at much lower concentrations, nitrogen dioxide causes increased incidence of acute bronchitis in children and increased airway resistance in adults. Cigarette and cigar smoke contains upto 1000 ppm nitrogen oxides which, when inhaled, are absorbed by lungs. Thus, smokers are more prone to be affected by the oxides of nitrogen than non-smokers.

(b) Vegetation

Nitrogen dioxide is injurious to plants. At as low a concentration as 0.3 ppm, it suppresses plant growth. At higher concentrations it causes visible leaf injury to sensitive plants. In case of food crops, exposure to nitrogen dioxide results in a decreased yield, as well as chlorosis. Nitrogen dioxide causes more grievous injury to those plants which are exposed to low light intensity.

2. Secondary Effects

As a secondary pollutant, nitrogen dioxide has the following environmental implications.

(a) Acid rain

In the atmosphere, a part of nitrogen oxides is converted into nitric acid (reaction 15.7), which may be washed down as acid rain. The ill effects of acid rain are listed in Chapter 14, p. 224.

(b) Photochemical smog

When associated with hydrocarbons and sunlight, oxides of nitrogen lead to the formation of photochemical smog. The smog contains a wide variety of organic compounds—such as peroxides, aldehydes, ketones, acylnitrates etc.—that are harmful in varying degrees. The obnoxious effects of photochemical smog are detailed in Chapter 17, p. 252.

(c) Aerosols

A part of nitric acid, formed in the atmosphere from the oxides of nitrogen, interacts with ammonia, producing ammonium nitrate aerosols (reaction 15.8). The aerosols have several hazardous effects (listed in Chapter 20, p. 297).

Control Measures for Nitrogen Oxides

As has been stated, there are two important anthropogenic sources of nitrogen oxides: Fossil fuel based power plants and automobile exhaust. In order to curtail pollution by nitrogen oxides, control measures have to be instituted at both the sources.

Abatement of Nitrogen Oxides Pollution from Power Plants

There are four important technologies to suppress the release of nitrogen oxides from power plants.

1. Coal conversion technology

This technology reduces the emission of both sulfur oxides and nitrogen oxides from the power plants. It incorporates an adjunct sulfur removal system which extracts essentially all the sulfur contained in the coal, to yield pure elemental sulfur. This is affected as follows: The coal is crushed and heated with sodium hydroxide solution under moderate pressure. This removes nearly all the sulfur and many trace metal components from the coal. Once the coal is devoid of sulfur, it will not produce sulfur oxides on combustion.

The advantage of this method is that sulfur and trace metal salts can be recovered. Moreover, sodium hydroxide solution can be regenerated. The sulfur-free coal is then reacted with steam to produce water gas, a mixture of mainly carbon monoxide and hydrogen. The gas is then sprayed with water and burned. Due to the presence of moisture, the combustion temperature is lowered. At low temperatures, nitrogen and oxygen do not combine and hence nitrogen oxides are not formed. Obviously, coal cannot be sprayed with water and then burned. But water gas may be subjected to water treatment so as to avert the formation of nitrogen oxides.

2. Fluidized bed combustion

In fluidized bed combustion, a bed of ash or sand rests on a grate at the bottom of a boiler. A fan is used to force the air upwards through the grate. The air causes the mixture of ash or sand to churn and flow like a fluid. At the same time, crushed solid fuel is fed into the boiler. The fuel heats up the particles of ash or sand to an even temperature. The pipes within the boiler pick up some of the heat directly from the fluidized particles by means of conduction, as well as by the usual methods of radiation and convection. A fluidized bed combustion system is shown in Fig. 15.1.

Limestone is added to the fluidized mixture to reduce the amount of sulfur dioxide emitted in the flue gases. Limestone captures sulfur dioxide by means of the following reactions:

$$CaCO_3 \xrightarrow{\Delta} CaO + CO_2 \tag{15.12}$$

$$2CaO + 2SO_2 + O_2 \rightarrow 2CaSO_4 \tag{15.13}$$

Conventional scrubbing methods remove sulfur dioxide from the flue gases after the combustion has taken place. Fluidized bed combustion system suppresses the level of sulfur dioxide at the time of combustion. Moreover, the system operates efficiently at 800–1000°C. At such low temperatures the oxides of nitrogen are not formed. Thus, the flue gases are free from both the sulfur oxides and nitrogen oxides.

3. Noxout technology

While the coal gasification technology removes both sulfur oxides and nitrogen oxides from the flue gases, Noxout technology eliminates only the oxides of nitrogen from the power plant emissions. It is a relatively modern technology which uses urea to rid flue gases of nitrogen oxides. The technique involves

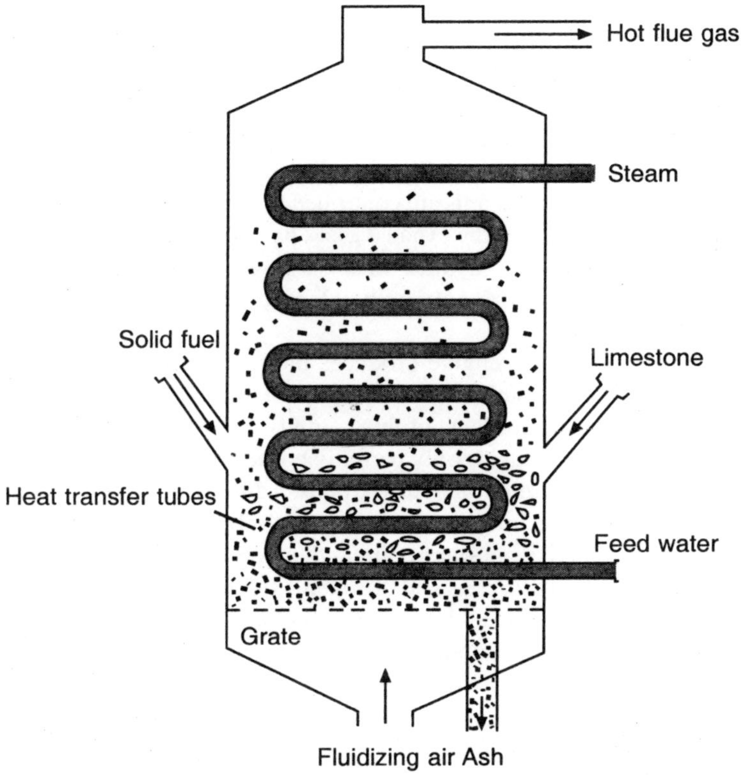

Fig. 15.1 A fluidized bed combustion

injecting an aqueous solution of urea at a number of points in the combustion system. Urea interacts with nitrogen oxides, forming carbon dioxide, nitrogen and water.

$$NH_2CONH_2 + NO + NO_2 \rightarrow 2N_2 + CO_2 + 2H_2O \quad (15.14)$$

Trace amounts of ammonia are also produced, but are removed by scrubbing with water. Almost 80% of nitrogen oxides are removed by Noxout technology.

4. Selective catalytic reduction technology

This technology too removes only nitrogen oxides from flue gases. The technique involves addition of ammonia to the flue gases and passing the mixture over the selective reduction catalyst. Oxides of nitrogen are reduced to nitrogen.

$$6NO + 4NH_3 \rightarrow 5N_2 + 6H_2O \quad (15.15)$$

$$6NO_2 + 8NH_3 \rightarrow 7N_2 + 12H_2O \quad (15.16)$$

Most commonly used catalysts in this technology are metallic platinum, which functions between 230 and 290°C and vanadium pentoxide on a titanium dioxide support, which functions between 300 and 450°C. The disadvantage of this method is that over long periods ammonia tends to poison the catalysts.

Abatement of Nitrogen Oxides Pollution from Vehicular Exhaust: The Chemistry of Catalytic Converters

With the enormous increase in the number of motor-vehicles, their exhaust has become a major source of atmospheric pollution, especially in urban areas where traffic density is high. Increasingly stringent regulations are being introduced worldwide to limit the emission of such pollutants.

Automotive exhaust pollutants are controlled by installing *catalytic converters* in the exhaust system, so that the exhaust gases pass through them. These harmful gases are converted catalytically into harmless gases. The limitation, however, is that catalytic converters work only with unleaded petrol since lead poisons the catalysts.

An internal combustion engine produces motive power as the hydrocarbon fuel (petrol or diesel) is combusted in presence of air. Carbon dioxide and water vapours are produced as products of this combustion reaction. However, during combustion, a part of the fuel is only partially burned, resulting in the production of carbon monoxide and unburnt hydrocarbons. Additionally, at the high reaction temperature of the engine, nitrogen and oxygen combine to form nitrogen oxides.

Thus, there are three important pollutants emerging from the auto-exhaust: Carbon monoxide, hydrocarbons and nitrogen oxides. These emissions from the engine are mainly influenced by the air: fuel ratio. Figure 15.2 shows the concentration of various pollutants and the variation of engine power *vis-a-vis* the air: fuel ratio. At the extreme left side of Fig. 15.2, the air: fuel ratio is very low, that is, the quantity of air is less. In presence of low amounts of air (or oxygen) the combustion of fuel is not complete. Therefore, large amounts of hydrocarbons and carbon monoxide are evolved. However, since oxygen is not present in sufficient quantity, nitrogen oxides are not formed to an appreciable extent. At the extreme right side of Fig. 15.2, the air: fuel ratio is large, that is, the amount of air is high. Therefore, it is expected that the fuel should burn very effectively. However, it is not so. Since the air: fuel ratio is high, it means that

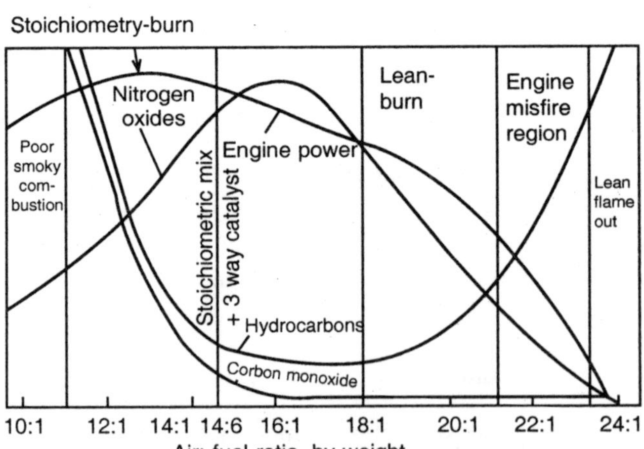

Fig. 15.2 Variation of the pollutants' concentration and engine power with air: fuel ratio

the quantity of fuel is low. There is thus not enough fuel for the mixture to burn properly. The flame may miss some droplets of fuel and the engine may misfire. Under such conditions high concentrations of hydrocarbons are evolved. Since the burning of fuel is not proper, the combustion temperature is not very high and, therefore, the emission of nitrogen oxides is maintained at a low level. On the other hand, in presence of a large dose of air, carbon monoxide is oxidised to carbon dioxide. Thus, the emission of carbon monoxide at a high air: fuel ratio is minimum.

Auto engines are tuned to work under conditions where the air: fuel ratio is neither too low nor too high. They operate effectively under two conditions: Lean-burn conditions, when the air: fuel ratio is about 20:1; and stoichiometric conditions, when the air: fuel ratio is about 14.6:1.

Under lean-burn conditions, the engine power, as shown in Fig. 15.2, is not maximum; yet fuel economy is better (higher air: fuel ratio means low quantity of fuel). The emission of nitrogen oxides is moderate but within safe limits and, therefore, may be ignored; the emission of hydrocarbons is quite high and, therefore, must be controlled; the emission of carbon monoxide is minimum and much less than that of nitrogen oxides. Even then control measures have to be instituted to prevent the emission of carbon monoxide since it is much more toxic than the oxides of nitrogen. Thus, the emission of two pollutants, hydrocarbons and carbon monoxide, need to be controlled when the engine operates under lean-burn conditions. This requires a *two way catalytic converter.*

Under stoichiometric conditions, priority is given to engine peak performance and not to fuel economy (lower air: fuel ratio means larger quantity of fuel). The emission of hydrocarbons, under these conditions is moderately high (Fig. 15.2) and must be suppressed; the emission of carbon monoxide is minimum, yet cannot be ignored; the emission of nitrogen oxides is extremely high and has to be prevented. Thus, the emission of all three pollutants, hydrocarbons, carbon monoxide and nitrogen oxides, need be controlled when the engine operates under stoichiometric conditions. This is done by a *three way catalytic converter.*

A two way catalytic converter removes hydrocarbons and carbon monoxide from the exhaust. It consists of a mixture of platinum and palladium, supported on a ceramic or metal honeycomb bed. It operates at an exhaust temperature of about 450°C and removes about 50% of the pollutants. A two way catalytic converter is an oxidation catalyst which converts carbon monoxide to carbon dioxide and hydrocarbons to carbon dioxide and water when exhaust gases, including oxygen, are forced through it.

$$2CO + O_2 \xrightarrow{\text{Catalyst}} 2CO_2 \qquad (15.17)$$

$$\underset{\text{Hydrocarbons}}{HC} + O_2 \xrightarrow{\text{Catalyst}} CO_2 + H_2O \qquad (15.18)$$

A three way catalyst removes hydrocarbons, carbon monoxide and nitrogen oxides from the auto-exhaust. It is an example of oxidation-reduction catalysis: Hydrocarbons and carbon monoxide are oxidised, while the nitrogen oxides are reduced. However, the oxidation and reduction reactions do not occur in isolation of each other. Nitric oxide is made to react with carbon monoxide in the exhaust system. In this reaction carbon monoxide is oxidised to carbon dioxide, while the nitric oxide is reduced to nitrogen.

$$2NO + 2CO \xrightarrow{\text{Catalyst}} N_2 + 2CO_2 \tag{15.19}$$

Partial reduction to nitrous oxide would also suffice,

$$2NO + CO \xrightarrow{\text{Catalyst}} N_2O + CO_2 \tag{15.20}$$

since nitrous oxide is comparatively easily decomposed.

$$2N_2O \xrightarrow{\text{Catalyst}} 2N_2 + O_2 \tag{15.21}$$

Similarly, nitric oxide and nitrogen dioxide interact with hydrocarbons.

$$NO_x + \underset{\text{Hydrocarbons}}{HC} \xrightarrow{\text{Catalyst}} N_2 + CO_2 + H_2O \tag{15.22}$$

A three way catalyst consists of a mixture of platinum, palladium and rhodium, deposited on a high surface area ceramic and metal honeycomb or alumina pelleted supports. On passing the exhaust gases over the bed of a three way catalyst, reactions (15.19) to (15.22) occur readily. A flow diagram of a three way catalyst is shown in Fig. 15.3. A three way catalytic converter operating

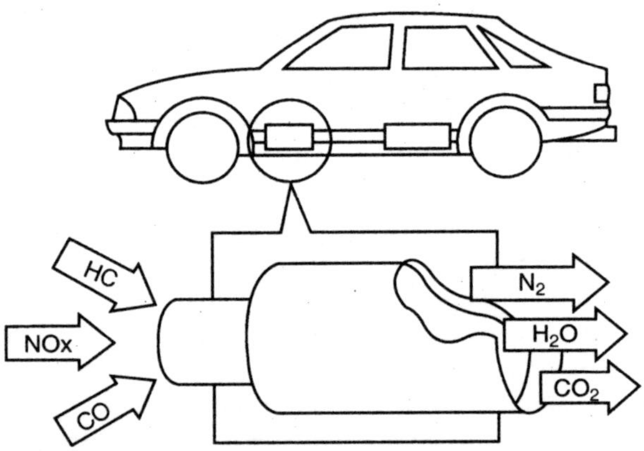

Fig. 15.3 Flow diagram of a three way catalyst

under the stoichiometric conditions of air: fuel ratio 14.6:1 and at 400–600°C can achieve about 90% efficiency in removing all three obnoxious emissions from auto exhaust (Fig. 15.4).

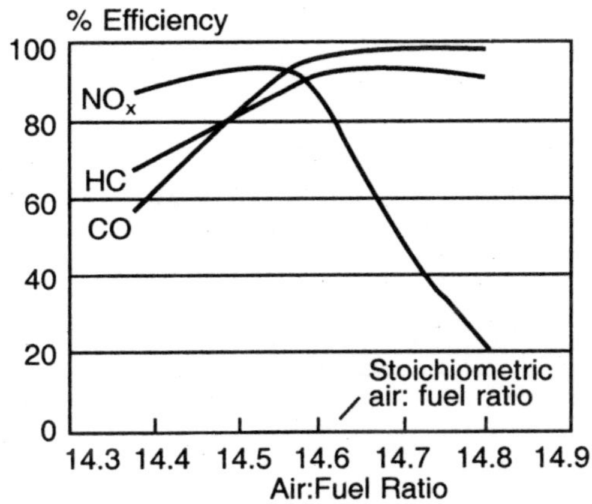

Fig. 15.4 Variation of the efficiency of a three way catalyst with air: fuel ratio

References

Balzhiser, R.E. and Yeager, K.E. (1987). Coal-fired power plants for the future, *Sci. Amer.,* **257**(3), 100–107.

Beaton, S.P., Bishop, G.A., Zhang, Y., Ashbaugh, L.L., Lawson, D.R. and Stedman, D.H. (1995). On-road vehicle emissions: Regulations, costs and benefits, *Science,* **268,** 991–993.

Butler, J.D. (1972). Controlling emissions from motor vehicle exhausts, *Chem. Brit.,* **8,** 258–260.

Dickerson, R.R., Huffman, G.J., Luke, W.T., Nunnermacker, L.J., Pickering, K.E., Leslie, A.C.D., Lindsey, C.G., Slinn, W.G.N., Kelly, T.J., Daum, P.H., Delany, A.C., Greenberg, J.P., Zimmerman, P.R., Boatman,. J.F., Ray, J.D. and Stedman, D.H. (1987). Thunderstorms: An important mechanism in the transport of air pollution, *Science,* **235,** 460–465.

Farrauto, R.J., Heck, R.M. and Speronello, B.K. (1992). Environmental catalysts, *Chem. Eng. News,* September 7, p. 34–44.

Fennelly, P.F. (1984). Fluidized bed combustion, *Amer. Sci.,* **72,** 254–261.

Katz, M. (1976) in, *'Air Pollution',* Volume III, Stern, A.C. (Ed.), Academic Press, New York, p. 259–305.

Lave, L.B. and Seskin, E.P. (1970). Air pollution and human health, *Science,* **169,** 723–733.

McFarland, J.H. and Benton, C.S. (1972). The oxides of nitrogen and their detection in automotive exhaust, *J. Chem. Ed.,* **49,** 21–24.

Mohnen, V.A. (1988). The challenge of acid rain, *Sci. Amer.,* **259**(2), 14–22.

O'Neill, P. (1985), *'Environmental Chemistry',* George Allen & Unwin, London, p. 92–107.

O'Sullivan, D. (1988). Method uses urea to rid flue gas of NO_x, *Chem. Eng. News,* April 18, p. 22.

Searles, R.A. (1989). Car exhaust pollution: The role of precious metal catalysts in its control, *Endeavour,* **13,** 2–7.

Sodhi, G.S. (1991). Coal treatment methods to counter acid rain, *Chemistry Education,* **8**(2), 25–26.

Wildeman, T.R. (1974). The automobile and air pollution, *J. Chem. Ed.,* **51,** 290–294.

CHAPTER 16

Carbon Monoxide

Carbon monoxide is a colourless, odourless, tasteless and non-irritating gas. It is generated by incomplete combustion processes. Although carbon monoxide is highly toxic, yet its presence is an index of affluence since it occurs in industry, in tobacco smoke, in household heating and in motor-vehicle exhaust. Since its emission to the atmosphere from motor-vehicle exhaust is significantly high, carbon monoxide is considered to be one of the most important urban atmospheric pollutants.

Sources of Carbon Monoxide

Carbon monoxide is generated both by natural sources and man-made sources.

Natural Sources

On a worldwide basis, natural processes produce a larger amount of carbon monoxide than man-made sources. The following natural sources contribute carbon monoxide to the atmosphere.

1. Vegetation

Carbon monoxide is produced by the oxidation of methane, a gas released in copious amounts in swamps, bogs and rice paddies. It is also produced as a result of the decomposition of chlorophyll, the green respiratory pigment of plants which contains a cyclic tetrapyrrole structure, similar to heme. Mature leaves have been shown to produce large quantities of carbon monoxide due to degradation of chlorophyll.

2. Ocean

The surface waters of the western Atlantic Ocean have been found to be supersaturated with carbon monoxide. The surface waters of other oceans too have high concentration of carbon monoxide. As a result, transportation of carbon monoxide across the sea-air interface is from the ocean to the atmosphere.

$$CO_{(sea)} \rightarrow CO_{(air)} \tag{16.1}$$

Thus, the ocean acts as a natural source of carbon monoxide. In fact, it is the largest natural source of this gas, contributing as much as 5% of its amount generated by burning of fossil fuels by man.

It has been observed that in sea water with high level of nutrients and/or abundant sunshine, greater amount of carbon monoxide is produced. This suggests that the mechanism of production of carbon monoxide in the ocean is biological in nature. Marine algae and marine siphonophore produce carbon monoxide. The portuguese man-of-war, a poisonous, stinging jellyfish, is also a releaser of carbon monoxide.

3. Endogenous production

Higher animals produce carbon monoxide from the breakdown of haemoglobin. This is termed as endogenous production of carbon monoxide. In human beings, an average person produces about 0.4 ml h^{-1} of carbon monoxide. The evidence for the endogenous production of carbon monoxide comes from the fact that the concentration of this gas is higher in expired air than in inspired air. This shows that carbon monoxide is actually produced within the body. It has also been found that endogenous production of carbon monoxide is greater in patients with haemolytic anaemia, extensive trauma and transfusion reactions.

Man-Made Sources

Although the natural sources contribute many times more carbon monoxide to the atmosphere than the man-made sources, yet the pollution caused by anthropogenic sources cannot be ignored. The reason is that the background concentration of carbon monoxide is increasing every year due to human activity. The anthropogenic sources of carbon monoxide are linked to the combustion processes, vehicular exhaust and cigarette smoking. These sources tend to increase the atmospheric concentration of carbon monoxide by 2–4% each year.

1. Combustion processes

Various forms of indoor combustion devices, such as stoves, furnances, gas-fired heaters and charcoal braziers emit carbon monoxide. In addition, carbon monoxide is ejected from waste disposal combustion, coal mines and power plants.

In any combustion process, carbon is oxidised to carbon monoxide.

$$2C + O_2 \xrightarrow{\text{Fast}} 2CO \qquad (16.2)$$

Subsequently, carbon monoxide is oxidised to carbon dioxide.

$$2CO + O_2 \xrightarrow{\text{Slow}} 2CO_2 \qquad (16.3)$$

The structure of carbon monoxide involves a triple bond (C≡O). Therefore, its oxidation to carbon dioxide requires high activation energy. Hence, reaction (16.3) occurs slowly. Unless the temperature is very high and/or oxygen is present in excess, reaction (16.3) will not proceed to completion and large amounts of carbon monoxide will persist.

Carbon monoxide is also a constituent of several industrial gases. For example, coal gas, obtained from distillation of coal, contains upto 10% carbon monoxide. Water gas, obtained by passing steam over incandescent coke, contains 30–40% carbon monoxide.

2. Vehicular emission

Carbon monoxide is the major pollutant from automobiles. In urban areas, it is observed that there is a relatively high concentration of carbon monoxide during peak hours, both in the morning, as well as in the evening. Moreover, basement garages have been found to contain 100–200 ppm carbon monoxide.

It has been estimated that for every liter of petrol consumed by automobiles, about 370 g carbon monoxide are emitted as vehicular exhaust. The consumption of gasoline and the number of motor-vehicles are increasing at an alarming pace. Therefore, carbon monoxide pollution is to be considered a problem affecting the entire community, rather than a problem in the immediate vicinity of traffic.

3. Cigarette smoking

Cigarette smoke contains upto 2% carbon monoxide due to incomplete combustion of tobacco. About 0.04% of the gas from the smoke is inhaled into the lungs of the smoker. As a result, the concentration of carbon monoxide in expired air of smokers is higher than with non-smokers. Heavy smokers expire more carbon monoxide than light smokers. Even ex-smokers expire more carbon monoxide than non-smokers, which suggests that a long period of recovery from the ill effects of carbon monoxide may be needed. Pipe and cigar smokers exhale less carbon monoxide, since less smoke is inhaled when tobacco is used in these forms.

Fate of Carbon Monoxide in the Environment

One of the most important natural sources of carbon monoxide is the oxidation of methane. Methane, in turn, is produced by various processes, chiefly biological in nature. Rice paddy fields are a major contributor. Methane is converted to carbon monoxide by hydroxyl radicals present in the atmosphere by the following mechanism.

Hydroxyl radicals interact with methane and convert it into methyl radicals, $CH_3^{\cdot}$.

$$CH_4 + \dot{O}H \rightarrow H_2O + CH_3^{\cdot} \tag{16.4}$$

The methyl radical is oxidised by atmospheric oxygen,

$$CH_3^{\cdot} + O_2 \rightarrow CH_3O_2^{\cdot} \tag{16.5}$$

followed by dissociation into formaldehyde and formyl radical.

$$CH_3O_2^{\cdot} \rightarrow HCHO + {}^{\cdot}OH \tag{16.6}$$

$$CH_3O_2^{\cdot} \rightarrow HCO^{\cdot} + H_2O \tag{16.7}$$

This is followed by photochemical dissociation of formaldehyde,

$$HCHO + h\nu \rightarrow H_2 + CO \tag{16.8}$$

and formyl radical.

$$HCO^{\cdot} + h\nu \rightarrow H^{\cdot} + CO \tag{16.9}$$

Thus, ultimately methane is oxidised to carbon monoxide. Add to this the carbon monoxide produced from other natural sources, as well as man-made sources. The average residence time of carbon monoxide in the atmosphere is about 0.1 year. Thus, it is expected that over the years, carbon monoxide would continue to accumulate in the atmosphere and produce long term hazards. Mercifully, this is not so. Nature has its own processes which remove carbon monoxide from the environment. There are three main removal processes.

1. We have seen that hydroxyl radicals act on methane and produce carbon monoxide (reactions 16.4–16.9). However, these very hydroxyl radicals are instrumental in removing carbon monoxide from the atmosphere. They oxidise it to carbon dioxide.

$$CO + {}^{\cdot}OH \rightarrow CO_2 + H^{\cdot} \tag{16.10}$$

It is estimated that, on an average, the background concentration of carbon monoxide increases at a rate of 1.2 ± 0.6% per year. However, from 1988 onwards the carbon monoxide concentration has declined at a rate of 2.6 ± 0.8% per year. This has been attributed, in part, to a reduction in tropical biomass burning. It is also believed that large depletion of stratospheric ozone (Chapter 19) would have exposed the troposphere to more ultraviolet radiation which, in turn, would have increased the concentration of hydroxyl radicals (by photodissociation of water vapours). The hydroxyl radicals would have destroyed carbon monoxide as shown in reaction (16.10).

2. The carbon monoxide may be transported from troposphere to stratosphere, where it is rapidly oxidised to carbon dioxide by oxygen free radicals.

$$CO + O \rightarrow CO_2 \tag{16.11}$$

3. Soil acts as an important sink for carbon monoxide. A number of soil bacteria, such as *Methanosarcina barkerii* and *Methanobacterium formicum* are known to remove carbon monoxide. In presence of these microorganisms, carbon monoxide is either oxidised to carbon dioxide,

$$CO + \tfrac{1}{2}O_2 \xrightarrow[\text{bacteria}]{\text{Soil}} CO_2 \tag{16.12}$$

or is reduced to methane.

$$CO + 3H_2 \xrightarrow[\text{bacteria}]{\text{Soil}} CH_4 + H_2O \tag{16.13}$$

Thus, the urban areas face a greater threat of carbon monoxide poisoning, not only because of greater vehicular traffic, but also because in the cities the soil is covered by concrete and buildings and, therefore, the soil bacteria are ineffective in removing carbon monoxide.

The natural processes for the removal of carbon monoxide proceed at a rate which is nearly equal to the rate of production of carbon monoxide. However, this does not mean that we can ignore the issue of carbon monoxide pollution, leaving it for the nature to tackle the problem. If the background concentration of carbon monoxide increases, larger number of hydroxyl radicals would be required to oxidise carbon monoxide to carbon dioxide (reaction 16.10). Thus, lesser concentration of hydroxyl radicals would be available for removing methane from the environment (reaction 16.4). This would lead to an increase in the abundance of methane. Other trace gases, such as hydrogen sulfide and halogenated hydrocarbons, which are destroyed by hydroxyl radicals, would also register an increase in concentration. Thus, the hazardous consequences of greenhouse effect (Chapter 18) and stratospheric ozone depletion (Chapter 19) would increase manifold.

Analysis of Carbon Monoxide

The concentration of atmospheric carbon monoxide is determined by a method called *nondispersive infrared analysis.* This method is based on the fact that molecules like carbon monoxide, which possess a dipole moment, absorb infrared radiation (infrared active), while those like nitrogen or oxygen, which have no dipole moment, do not absorb infrared radiation (infrared inactive). The main constituents of air are oxygen and nitrogen, which are infrared inactive and water vapours and carbon dioxide, which are infrared active (although net dipole moment of carbon dioxide is zero, each of the two C=O bonds possesses a positive dipole and therefore carbon dioxide molecules absorb infrared radiation).

The nondispersive infrared analyser (Fig. 16.1) consists of a sample cell containing air contaminated with carbon monoxide, and a reference cell containing pure air. Infrared beam from a suitable source is reflected by choppers alternately through the sample cell and the reference cell. Before the beam enters the cells, it is made to pass through two filters which remove the wavelengths that are absorbed by water vapours and carbon dioxide. This minimizes the interference due to water vapours and carbon dioxide.

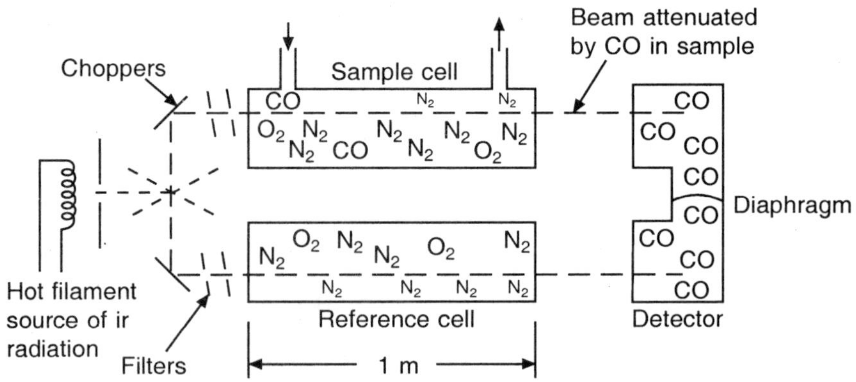

Fig. 16.1 A Nondispersive infrared analyser

The intensity of the beam emerging from sample cell is decreased since carbon monoxide absorbs infrared radiation. The intensity of the beam emerging from reference cell, however, remains unaffected since no infrared active molecule is present in this cell. Both the beams strike a detector containing carbon monoxide and a movable diaphragm. The carbon monoxide in the detector absorbs infrared radiation and becomes hot. However, carbon monoxide in the portion of the detector facing sample cell receives infrared radiation of lower intensity and is, therefore, heated to a lesser extent as compared to the gas present in the portion of the detector facing reference cell. As a result, a pressure gradient is set up. The diaphragm moves from high pressure-low temperature portion of the detector (facing the sample cell) to the low pressure-high temperature portion (facing the reference cell). The distension of diaphragm is proportional to the difference in intensity of the infrared radiation passing through sample cell and reference cell and this, in turn, is proportional to the amount of carbon monoxide present in contaminated air.

Effects of Carbon Monoxide

Carbon monoxide, when inhaled, reduces the efficiency of the blood in carrying oxygen from the lungs to all parts of the body. The oxygen carrying capacity of the blood is due to an affinity of the haemoglobin for oxygen. The haemoglobin, Hb, and oxygen combine, forming oxyhaemoglobin, HbO_2, which releases oxygen to various organs.

$$Hb + O_2 \rightleftharpoons HbO_2 \qquad (16.14)$$

The impaired transportation of oxygen in presence of carbon monoxide results through two distinct mechanisms. Firstly, carbon monoxide competes with oxygen in combining with haemoglobin. The product formed by the combination of carbon monoxide and haemoglobin is called *carboxyhaemoglobin,* COHb.

$$Hb + CO \rightarrow COHb \qquad (16.15)$$

Haemoglobin has about 210 times more affinity for carbon monoxide than for oxygen. Hence, less oxygen is taken up by haemoglobin at any moment when carbon monoxide is present. Secondly, when some carbon monoxide is present in the blood, it becomes harder for the oxygen to be released from haemoglobin to various parts of the body.

The net effect of the two mechanisms is that the efficiency of oxygen transportation through the body is hindered. Greater the concentration of carboxyhaemoglobin in the blood, lower is the efficiency of oxygen transportation. That is, the signs and symptoms of carbon monoxide toxicity depend on the concentration of carboxyhaemoglobin in the blood. The normal level of carboxyhaemoglobin in blood cells is about 0.5%, which results from endogenous production of carbon monoxide. The average smoker of a pack-a-day has about 5% carboxyhaemoglobin level.

Increased levels of carboxyhaemoglobin result in the impairment of physical abilities. For example, high levels of carbon monoxide in blood reduce the ability of a person to see small objects at a distance. The night vision also reduces. Time discrimination ability, as well as mental performance are impaired. Hearing ability is adversely affected. These factors increase the incidence of accidents. In fact, there exists a correlation between carbon monoxide pollution and the number of automobile and aeroplane accidents.

Prolonged exposure to low levels of carbon monoxide causes extreme discomfort, headache, dizziness and lassitude. Carbon monoxide is known to cross the placenta in pregnant women. As a result, less oxygen becomes available to fetus. Less oxygen means lower birth weight. Thus, the birth weight of the babies whose mothers smoke cigarettes have been found to be lower than those of non-smoking mothers.

Higher levels of carbon monoxide result in an increased *haematocrit*, that is percent volume of red blood cells. Thus, cigarette smokers have been found to have an increase in haematocrit within minutes of smoking. This increases the possibility of clot growth in blood vessels. Smokers having 10–20% carboxyhaemoglobin level begin to have artery blockage at a very young age. This explains the high cardiovascular mortality of cigarette smokers.

We have seen that carbon monoxide poisoning is due to the interference with normal functioning of haemoglobin as a transporting agent for oxygen. Under conditions where oxygen supply is already deficient, the effects of carbon monoxide poisoning increase. Therefore, the population group most susceptible to the adverse effects of atmospheric carbon monoxide pollution includes persons with severe anaemia, chronic lung disease or impaired circulatory system. Further, those living in high altitudes, where the atmospheric oxygen concentration is low, suffer more because of carbon monoxide poisoning. Thus, while the maximum allowable concentration of carbon monoxide is 50 ppm for 8 h exposure, its threshold limit for those living at 5000–8000 feet is just 25 ppm.

Control Measures for Carbon Monoxide

In order to prevent adverse effects of carbon monoxide on human health, adequate control measures must be instituted. Any confined space in which carbon monoxide is likely to accumulate presents danger. The toxic gas is, therefore, a potential hazard in space flights and submarines. For this reason catalytic burners are provided in such systems to oxidise carbon monoxide to harmless carbon dioxide.

However, on a large scale, abatement of carbon monoxide pollution depends on controls which must be imposed primarily on the automobiles. In order to reduce the emission of carbon monoxide from vehicles, the automobiles have to be provided with catalytic converters. These converters reduce not only the emission of carbon monoxide, but also of nitrogen oxides. Therefore, the chemistry of catalytic converters is discussed in the chapter on nitrogen oxides (Chapter 15, p. 236).

References

Bekki, S., Law, K.S. and Pyle, J.A. (1994). Effect of ozone depletion on atmospheric CH_4 and CO concentrations, *Nature (London),* **371,** 595–597.

Goldsmith, J.R. and Landaw, S.A. (1968). Carbon monoxide and human health, *Science,* **162,** 1352–1359.

Hexter, A.C. and Goldsmith, J.R. (1971). Carbon monoxide: Association of community air pollution with mortality, *Science,* **172,** 265–267.

Holum, J.R. (1977), *'Topics and Terms in Environmental Problems',* John Wiley & Sons, New York, p. 83–88.

Inman, R.E., Ingersoll, R.B. and Levy, E.A. (1971). Soil: A natural sink for carbon monoxide, *Science,* **172,** 1229–1231.

Khalil, M.A.K. and Rasmussen, R.A. (1984). Carbon monoxide in the earth's atmosphere: Increasing trend, *Science,* **224,** 54–56.

Khalil, M.A.K. and Rasmussen, R.A. (1994). Global decrease in atmospheric carbon monoxide concentration, *Nature (London),* **370,** 639–641.

Maugh, T.H. (1972). Carbon monoxide: Natural sources dwarf man's output, *Science,* **177,** 338–339.

Maxwell, K.E. (1973), *'Environment of Life',* Dickenson Publishing Co., Inc., California, p. 136–142.

McConnell, J.C., McElroy, M.B. and Wofsy, S.C. (1971). Natural sources of atmospheric CO, *Nature (London),* **233,** 187–188.

Novelli, P.C., Masarie, K.A., Tans, P.P, and Land, P.M. (1994). Recent changes in atmospheric carbon monoxide, *Science,* **263,** 1587–1590.

Swinnerton, J.W., Linnenbom, V.J. and Lamontange, R.A. (1970). The ocean: A natural source of carbon monoxide, *Science,* **167,** 984–986.

Sze, N.D. (1977). Anthropogenic CO emissions: Implications for the atmospheric CO-OH-CH_4 cycle, *Science,* **195,** 673–675.

Weinstock, B. and Niki, H. (1972). Carbon monoxide balance in nature, *Science,* **176,** 290–292.

Wilson, D.F., Swinnerton, J.W. and Lamontage, R.A. (1970). Production of carbon monoxide and gaseous hydrocarbons in sea water: Relation to dissolved organic carbon, *Science,* **168,** 1577–1579.

CHAPTER 17

Photochemical Smog

In the late 1940s, an unpleasant environmental phenomenon was observed in Los Angeles. On clear, sunny days, when the automobile traffic level was high, a mist of aerosols descended over the area. The visibility dropped drastically and the residents complained of eye irritation. Originally, the mist of aerosols was named *Los Angeles smog*. Later it was found that such smoggy weather was not typical of Los Angeles alone. Since the chemical constituents of the smog—mainly ozone, nitrogen dioxide and peroxyacetylnitrate—are produced by photochemical reactions and since such reactions can occur at places other than Los Angeles, the phenomena came to be named simply as *Photochemical smog*. Over the years, the frequency and global distribution of photochemical smog has increased dramatically. Under appropriate conditions of emission, meteorology and topography, it constitutes a serious environmental problem in the major urban areas of the world. Moreover, its impact can extend to hundreds of kilometers downwind.

Formation of Photochemical Smog

The main constituents of photochemical smog originate from automobile exhaust, although there is a definite contribution from industrial activity. Non-methane hydrocarbons, especially isoprene and α-pinene, are emitted biogenically from trees. The constituents of photochemical smog, along with their average concentrations are listed in Table 17.1. Of these, the oxides of nitrogen, nitric oxide and nitrogen dioxide, play a key role in smog formation.

Formation of smog initiates with the photolysis of nitrogen dioxide.

$$NO_2 + h\nu \rightarrow NO + O \tag{17.1}$$

Most of the oxygen atoms formed in reaction (17.1) (~ 99%) react with molecular oxygen to produce ozone.

$$O + O_2 + M \rightarrow O_3 + M \tag{17.2}$$

where M is atmospheric nitrogen or oxygen (the species M carries away excess energy, whilst it itself remains chemically unchanged). The ozone formed in reaction (17.2) reacts rapidly with nitric oxide formed in reaction (17.1) to regenerate nitrogen dioxide.

Table 17.1 Constituents of photochemical smog

Constituent	Concentration (ppm)
Water vapours	20,000
Carbon dioxide	400
Carbon monoxide	40
Methane	2.5
Aldehydes and ketones	0.60
Ozone	0.50
Ethylene	0.50
Higher olefins	0.25
Nitrogen oxides	0.20
Sulfur dioxide	0.20

$$NO + O_3 \rightarrow NO_2 + O_2 \tag{17.3}$$

Reaction (17.2) is the only known source of ozone in urban air. Reaction (17.3), called *titration reaction,* represents a photostationary state between ozone, nitric oxide and nitrogen dioxide. This reaction explains why the concentration of ozone does not rise significantly unless that of nitric oxide has fallen to low values. This is also depicted in the urban diurnal cycle of ozone, nitric oxide and nitrogen dioxide, as shown in Fig. 17.1. The mid-morning rush hours add nitrogen dioxide from automobile exhaust to the atmosphere. During early afternoon, the intensity of sunlight is strong enough to cause the photodissociation of nitrogen dioxide (reaction 17.1), producing free oxygen atoms. The peak value of ozone,

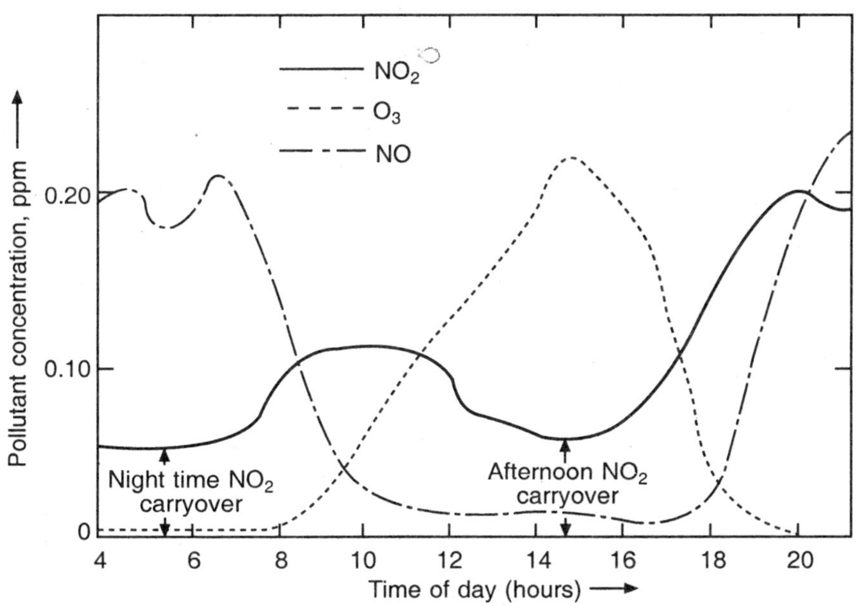

Fig. 17.1 Urban diurnal cycle of nitrogen oxides and ozone

formed by free oxygen atoms through reaction (17.2), therefore, also occurs during early afternoon hours.

Ozone is formed in reaction (17.2) and consumed in reaction (17.3). If these two were the only reactions occurring, a steady state concentration of ozone would have resulted. However, a part of ozone produced in reaction (17.2) reacts with hydrocarbons (RH)—especially olefins—released from automobile exhaust. This results in the formation of carboxyl free radicals ($RCO_2^{\cdot}$), as well as aldehydes and ketones.

$$O_3 + RH \rightarrow RCO_2^{\cdot} + RCHO + R_2CO \tag{17.4}$$

Likewise, a part of the free oxygen atoms produced in reaction (17.1) react with hydrocarbons to produce alkyl/aryl free radicals ($R^{\cdot}$), as well as acyl free radicals ($RC\dot{O}$)·

$$O + RH \rightarrow R^{\cdot} + RCO^{\cdot} \tag{17.5}$$

The acyl free radicals react with atmospheric oxygen, producing acylperoxy radicals ($RCO_3^{\cdot}$).

$$RCO^{\cdot} + O_2 \rightarrow RCO_3^{\cdot} \tag{17.6}$$

The acylperoxy radicals further react with nitrogen dioxide, giving peroxyacetylnitrate (PAN).

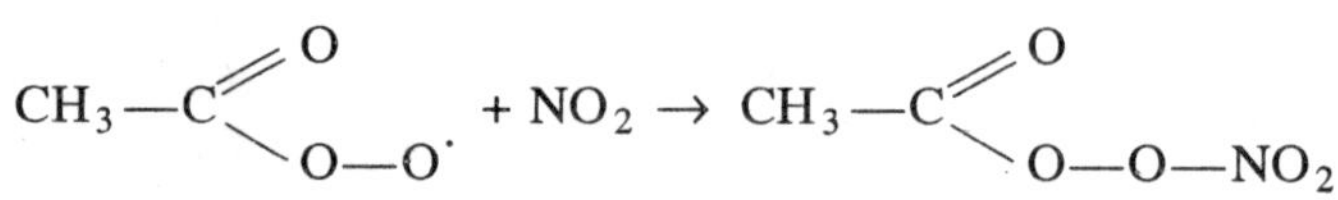

(17.7)

PAN

When the peroxy radical has an aromatic substituent, peroxybenzoylnitrate (PBzN) is formed.

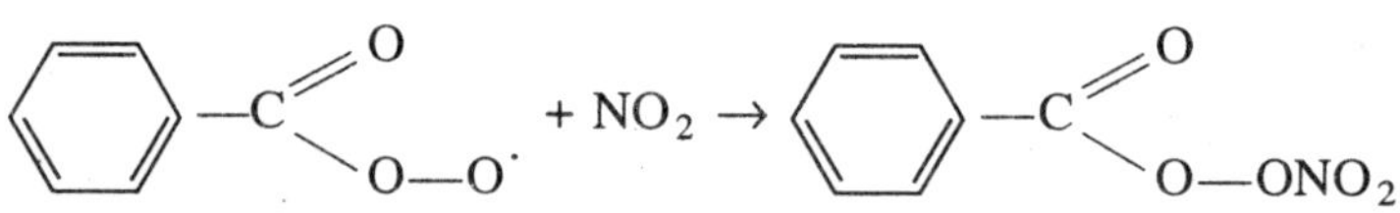

(17.8)

(PBzN)

Another set of compounds, called acylnitrates (RCO_2NO_2), are formed by the reaction of nitrogen dioxide with carboxyl free radicals.

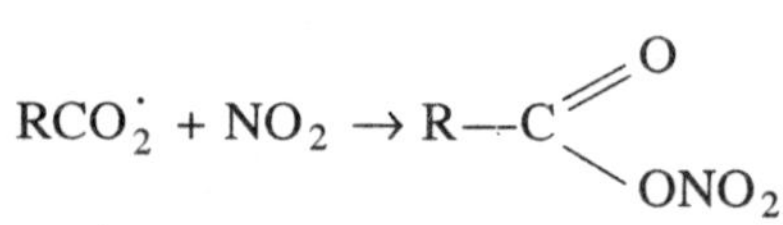

(17.9)

We have seen that nitrogen dioxide is consumed in reactions (17.7–17.9). It is, therefore, expected, that these chain terminating reactions would take up all the nitrogen dioxide and put an end to the smog formation. However, this does not happen because nitrogen dioxide is regenerated by several other reactions.

For example, the carboxyl free radicals may oxidise nitric oxide to nitrogen dioxide.

$$RCO_2^{\cdot} + NO \rightarrow RCO^{\cdot} + NO_2 \tag{17.10}$$

Further, the hydroxyl radicals ($^{\cdot}OH$) and hydroperoxy radicals ($HO_2^{\cdot}$) also produce nitrogen dioxide. The hydroxyl radicals first react with alkyl hydrocarbons to give alkyl free radicals,

$$^{\cdot}OH + RH \rightarrow H_2O + R^{\cdot} \tag{17.11}$$

which are then oxidised by atmospheric oxygen to alkylperoxy radicals, $RO_2^{\cdot}$.

$$R^{\cdot} + O_2 \rightarrow RO_2^{\cdot} \tag{17.12}$$

The alkylperoxy radicals react with nitric oxide, producing nitrogen dioxide and alkoxy radicals ($RO^{\cdot}$).

$$RO_2^{\cdot} + NO \rightarrow RO^{\cdot} + NO_2 \tag{17.13}$$

The hydroperoxy radicals can directly oxidise nitric oxide to nitrogen dioxide.

$$HO_2^{\cdot} + NO \rightarrow NO_2 + {}^{\cdot}OH \tag{17.14}$$

Wherefrom do the hydroxyl and hydroperoxy radicals enter the atmosphere? There are three sources of hydroxyl radicals.

1. Photodissociation of nitrous acid.

$$HONO + h\nu \rightarrow {}^{\cdot}OH + NO \tag{17.15}$$

2. Photodissociation of hydrogen peroxide.

$$H_2O_2 + h\nu \rightarrow 2^{\cdot}OH \tag{17.16}$$

3. Photodissociation of ozone,

$$O_3 + h\nu \rightarrow O_2 + O \tag{17.17}$$

followed by reaction of free oxygen atoms with water vapours.

$$O^{\cdot} + H_2O \rightarrow 2\dot{O}H \tag{17.18}$$

Hydroperoxy radicals are formed mainly from photodissociation of formaldehyde. The latter is not only emitted directly to the atmosphere from automobile exhaust but is also formed in the photochemical smog. Its photodissociation gives hydrogen atoms ($H^{\cdot}$) and formyl radicals ($HCO^{\cdot}$).

$$HCHO + h\nu \rightarrow H^{\cdot} + HCO^{\cdot} \tag{17.19}$$

Both, the hydrogen atoms and the formyl radicals, interact with oxygen, producing hydroperoxy radicals.

$$H^{\cdot} + O_2 \rightarrow HO_2^{\cdot} \tag{17.20}$$

$$HCO^{\cdot} + O_2 \rightarrow HO_2^{\cdot} + CO \tag{17.21}$$

Analysis of the Constituents of Photochemical Smog

The intensity of photochemical smog is gauged by measuring the concentrations of ozone and peroxyacetylnitrate. For determining the concentration of ozone, polluted air is bubbled through an acidified solution of potassium iodide. Ozone in the air reacts with iodide ions to give triiodide ions, I_3^-.

$$O_3 + 3I^- + 2H^+ \rightarrow I_3^- + O_2 + H_2O \tag{17.22}$$

The triiodide ions may be determined spectrophotometrically at 352 nm.

However, this method suffers from the limitation that reducing agents, like sulfur dioxide, which are present in polluted air, react with triiodide ions. Hence, the reducing species must first be removed from air. Since this is not easy to be accomplished, it is better to use an alternative method, which is based on the chemiluminescence technique. In this method the ozone present in the air is made to react with rhodamine B (Fig. 17.2). The reaction products are formed in excited electronic states. These decay to the ground state by emitting characteristic visible radiations. The intensity of these radiations, measured by means of a photomultiplier, determines the extent of reaction and hence the concentration of ozone.

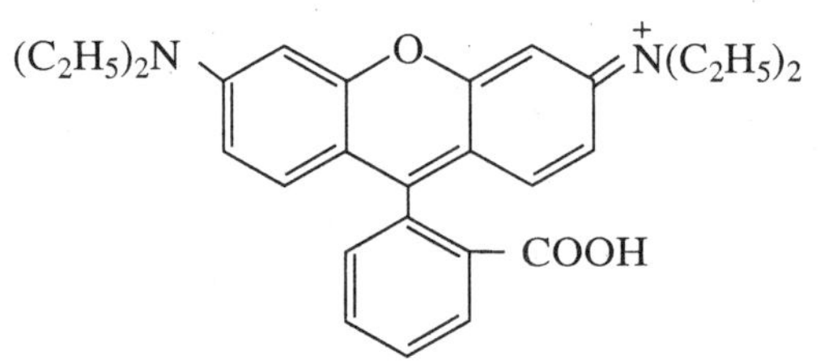

Fig. 17.2 Rhodamine B

The concentration of peroxyacetylnitrate (PAN) may be determined from the amount of ozone present in the smog. A linear correlation has been observed between these two components of the smog. The average ratio between the concentrations of ozone and PAN during daytime is 45. That is,

$$\frac{[\text{Ozone}]}{[\text{PAN}]} = 45 \tag{17.23}$$

Knowing the concentration of ozone, that of PAN may be calculated.

Effects of Photochemical Smog

The adverse effects of photochemical smog pose a potential threat not only to the local environment, but to the faraway regions as well. Wind transports the precursors, as well as the reaction products of photochemical smog to distant places. Thus, although photooxidants are formed mainly in highly industrialized and urban areas, photochemical air pollution has been observed even at a distance of about 1000 km from the source region.

The three main components of photochemical smog are, nitrogen oxides, ozone and organic derivatives (such as aldehydes, ketones, PAN, etc.). Each contributes to the hazardous effects of the smog. The harmful effects of nitrogen oxides have already been discussed (Chapter 15, p. 232). The deleterious effects of ozone and organic derivatives are discussed below.

Effects of Ozone

Ozone in the atmosphere has the following environmental implications.

1. Human health

Ozone is known to cause eye irritation, cough, chest discomfort and fatigue in humans. It damages the membranes surrounding pulmonary capillaries. As a result, water passes from blood vessels to lung alveoli, causing pulmonary edema. Ozone also tends to increase the incidence and severity of asthama attacks.

2. Vegetation

Compared to human beings, vegetation is more prone to the harmful consequences of ozone. The effects of ozone on plants are often proportional to dosage, that is, higher the concentration of ozone, greater is the damage caused to vegetation. Ozone causes a reduction in plant growth. It has been suggested that increasing concentration of ozone in the atmosphere is one of the causes of the forest deaths in Europe. Ozone also reduces crop productivity. The yield of some vegetables, such as head lettuce and spinach, are lowered in presence of ozone. Further, ozone produces visible foliar symptoms in plants.

3. Materials

Besides its damaging effects on humans and vegetation, ozone may cause harm to fibres, dyes, paints and polymers. Natural rubber and synthetic polymers, for example, suffer from deterioration of mechanical strength in presence of ozone. Atmospheric ozone adds on to the double bonds in materials containing unsaturated organic compounds. Such materials become brittle and crack. Ozone also damages exterior household paints and pigments.

Effects of Organic Derivatives

The aldehydes and ketones, which are formed as a result of secondary atmospheric reactions, are the main cause of eye irritation in the smoggy weather. Peroxyacetylnitrate too is an eye irritant. It is also a possible etiological agent in the growing incidence of skin cancer. It is known to react with the sulfhydryl groups of enzymes and amino acids, causing toxic biochemical effects in humans. It is toxic to plants as well.

Control of Photochemical Smog

Photochemical smog formation may be suppressed by the following technologies.

1. Catalytic Converters

The most straightforward way of reducing smog formation is to prevent the release of nitrogen oxides and hydrocarbons to the atmosphere. This involves installation of efficient catalytic converters in the automobiles (Chapter 15, p. 236).

2. Chemical Method

Photochemical smog can be suppressed by certain compounds which act as free radical traps. That is, when these compounds are sprayed to the atmosphere, they generate free radicals which readily combine with free radical precursors of photochemical smog (such as $O^{\cdot}$, $R^{\cdot}$, $RO^{\cdot}$, $H^{\cdot}$ etc.). The introduction of such compounds slows down the conversion of nitric oxide to nitrogen dioxide from the usual 2–6 h to 6–12 h. Therefore, the conversion will occur near the time of sunset and photochemical decomposition of nitrogen dioxide (reaction 17.1) would be avoided. As a result, the smog initiating step would be inhibited.

The compound, diethylhydroxylamine (Fig. 17.3), has been found to possess significant smog inhibiting characteristics. When sprayed to the atmosphere on a smoggy day, it freshens up the polluted urban air.

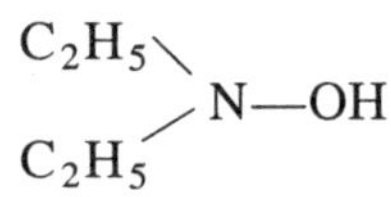

Fig. 17.3 Diethylhydroxylamine

The main advantage of diethylhydroxylamine is that it is cheaper than catalytic converters. Moreover, it can be used only when and where it is needed. Its main disadvantage is that it produces an unpleasant odour at concentrations greater than 0.5 ppm. Further, it is suspected that a delay in photochemical smog formation by using inhibitors like diethylhydroxylamine may merely transfer the problem downwind.

Till now diethylhydroxylamine has been used only in laboratory scale experiments. Extrapolation of the results of these experiments to real, urban atmosphere can be carried out only after the health effects of diethylhydroxylamine on general public are investigated in details—and in all seriousness.

3. Petrol Reformulation

Reformulated petrol has been used in Vienna in a bid to clean up the city's summer smog. Since September 1995, the leaded petrol has been replaced by *summer petrol,* a reformulated gasoline in which one-tenth of the aromatic hydrocarbons have been replaced by fuel oxygenates, such as methyl tertiary butyl ether. These additives allow the fuel to burn more completely and efficiently. As a result, the emission of volatile organic hydrocarbons is suppressed. Moreover, the emission of nitrogen oxides is reduced since petrol containing oxygenates burns at a lower temperature. A curtailed emission of hydrocarbons and nitrogen oxides suppresses smog formation.

References

Cadle, R.D. and Allen, E.R. (1970). Atmospheric photochemistry, *Science,* **167,** 243–249.

Chamaides, W.L., Lindsay, R.W., Richardson, J. and Kiang, C.S. (1988). The role of biogenic hydrocarbons in urban photochemical smog: Atlanta as a case study, *Science,* **241,** 1473–1475.

Demerjian, K.L. (1972). The mechanism of photochemical smog formation, *Chem. Brit.*, **8,** 252–257.

Finlayson, B.J. and Pitts, J.N. (1976). Photochemistry of the polluted troposphre, *Science,* **192,** 111–119.

Güsten, H. (1986) in, *'The Handbook of Environmental Chemistry',* Hutzinger, O. (Ed.), Springer-Verlag, Berlin, p. 53–105.

Haagen-Smit, A.J. (1964). The control of air pollution, *Sci. Amer.,* **210**(1), 25–31.

Lents, J.M. and Kelly, W.J. (1993). Clearing the air in Los Angeles, *Sci. Amer.,* **269**(4), 32–39.

Mackenzie, D. (1995). Summer petrol cleans up Viennese smog, *New Scientist,* August 19, p. 10.

Neiburger, M. (1957). Weather modification and Smog, *Science,* **126,** 638–645.

Russell, A., Milford, J., Bergin, M.S., McBride, S., McNair, L. and Yang, Y. (1995). Urban ozone control and reactivity of organic gases, *Science,* **269,** 491–495.

Seinfeld, J.H. (1989). Urban air pollution: State of the science, *Science,* **243,** 745–752.

CHAPTER 18

Greenhouse Gases

The glass roof of a greenhouse is permeable to radiations from the sun. Once inside the greenhouse, the radiations strike a surface and are transformed into heat. The glass roof of the greenhouse impedes the heat to pass out. Since much heat is trapped, the inside of a greenhouse is warmer than the outside. Like glass, a few gases—chiefly carbon dioxide, but also methane, nitrous oxide, ozone and chlorofluorocarbons—allow the sun's radiations to pass through them and strike the surface of the earth. But like the glass roof of a greenhouse, these gases prevent the resulting heat from escaping. The net result is that the earth is warming up.

The Greenhouse Effect

Atmospheric carbon dioxide, and a host of other gases, such as methane, nitrous oxide, tropospheric ozone and chlorofluorocarbons (CFCs), transmit short wavelength radiations from the sun, whilst at the same time absorb long wavelength infrared radiations from the earth. As the concentration of these gases increases, less of the earth's long wavelength radiations escape into space. The result is that the earth's atmosphere is warming up. This global warming—called *greenhouse effect*—is said to be one of the most important environmental problems, the pernicious implications of which must be resolved on an urgent basis.

Causes of Greenhouse Effect

It has been predicted that by the second decade of this century, the climate would be warmer than it has been over the past few centuries. The amount of warming which a greenhouse gas imparts to a square meter of earth is called *radiative forcing*. The radiative forcing exerted by each greenhouse gas depends on its concentration and its ability to absorb long wavelength infrared radiations.

As shown in Fig. 18.1, the greenhouse contribution of carbon dioxide is 50%, while the remaining 50% is augmented by other radiatively active trace gases. The greenhouse role of carbon dioxide and trace gases must be considered in all seriousness, for these atmospheric constituents may induce environmental perturbations that could undermine the integrity of the biological systems on which all *Homo sapiens* depend.

Greenhouse Role of Carbon Dioxide

The atmospheric concentration of carbon dioxide at present is about 356 ppm

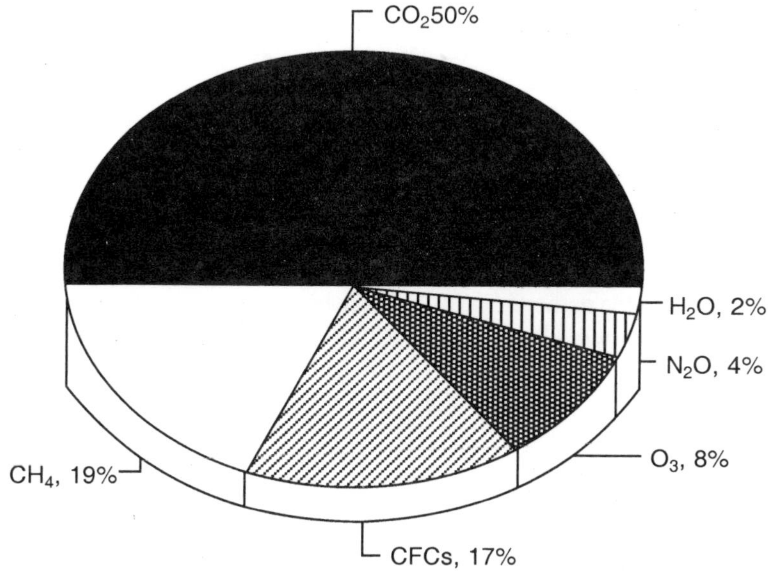

Fig. 18.1 Relative contribution of radiatively active gases to greenhouse effect

(Fig. 18.2). Around the year 1880, its concentration was approximately 280 ppm. Thus, there has been an increase of almost 25% as compared to the pre-industrial times. In fact, the atmospheric concentration of carbon dioxide has

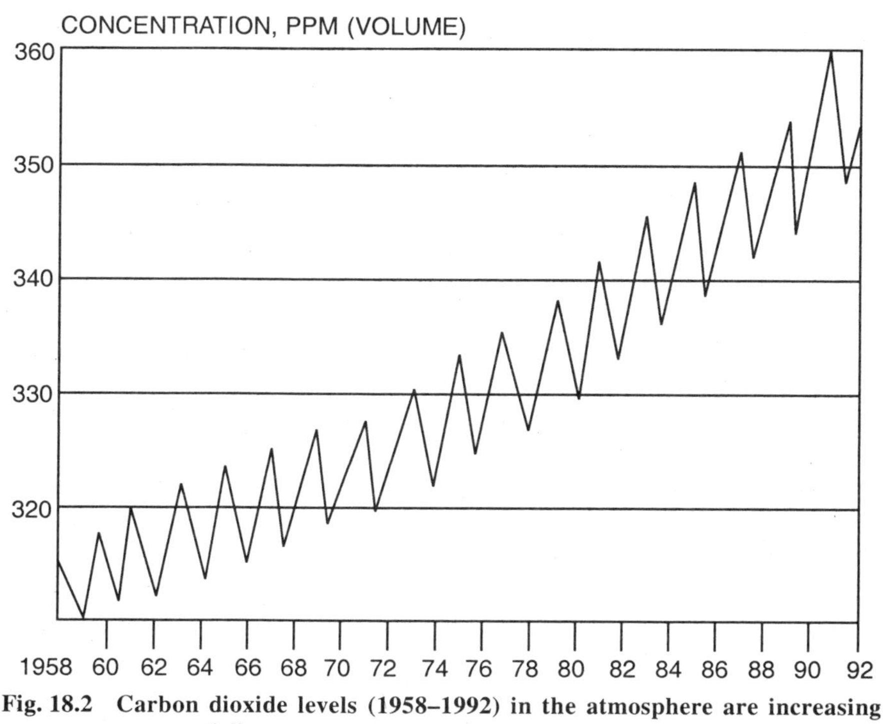

Fig. 18.2 Carbon dioxide levels (1958–1992) in the atmosphere are increasing exponentially

increased by 45 ppm since the beginning of last century. The rate of increase of carbon dioxide in the atmosphere has also escalated sharply. It is currently increasing at a rate of about 0.4% per year, in comparison to the 1960s rate of about 0.2% per year. The following factors are largely responsible for the increase in atmospheric carbon dioxide concentrations.

1. Burning of fossil fuels

By far, the most important and significant source of atmospheric carbon dioxide is the burning of fossil fuels. With the escalation of population and increase in industrial growth, the demand for fossil fuel has greatly increased. It is estimated that for each tonne of coal, mined and used, 0.693 tonne of carbon is converted into carbon dioxide. For each tonne of crude petroleum produced, 0.769 tonne of carbon is pumped into the atmosphere as carbon dioxide. Likewise, for each million cubic meters of natural gas, 524 tonnes of carbon are transformed into carbon dioxide.

Future trends of atmospheric carbon dioxide will depend on the continued use of fossil fuels. While the consumption in the developed countries is expected to remain essentially constant, the underdeveloped and developing nations are likely to use increasingly more fossil fuels as energy sources.

2. Deforestation

Plants use large quantities of atmospheric carbon dioxide in photosynthesis. As a result of deforestation, an important sink of carbon dioxide is being eliminated. Deforestation is a worldwide phenomenon. Every year about 11 million hectares of forests, an area larger than Australia, are destroyed. In India, the pace of deforestation is estimated to be 1.5 million hectares annually. That is, every five years, forest area equivalent to the size of Haryana vanishes.

Deforestation adds carbon dioxide to the atmosphere in two ways: Firstly, most of the trees are either burned or decomposed by bacteria, emitting carbon dioxide directly to the air. Secondly, the deforested land is unable to sequester carbon dioxide through photosynthesis. As a result of these two phenomena, deforestation contributes 10 to 30% as much carbon dioxide to the atmosphere as fossil fuel emissions do.

3. Volcanoes

Some volcanoes emit large quantities of carbon dioxide to the atmosphere. Mount Etna, an active but at present a relatively subdued volcano in Sicily, is a case in point. It is one of the most potent natural sources of carbon dioxide. Every year it adds about 25 million tonnes of carbon dioxide to the atmosphere. The entire region around the volcano is, therefore, enriched in carbon dioxide.

Greenhouse Role of Trace Gases

The important trace gases which contribute to the greenhouse effect are methane, nitrous oxide, ozone and chlorofluorocarbons. The greenhouse role of these gases was not realised unitl recently, when it was found that the addition of one molecule of chlorofluorocarbon can have the same greenhouse effect as the addition of 10^4 molecules of carbon dioxide to the atmosphere. Moreover, although the trace gases are minor constituents of the atmosphere, in recent years their concentration has increased due to various industrial and agricultural processes. Since their atmospheric levels are rising rapidly and since each molecule of these gases absorbs more infrared radiations than a carbon dioxide molecule, their combined greenhouse effect is almost equal to that of carbon dioxide (Fig. 18.1).

Methane

At present, the atmospheric concentration of methane is about 1.7 ppm. The principal sources of methane are:

1. Action of anaerobic bacteria on rice paddies and wetlands.
2. Leakage from coal mines and natural gas pipeline.
3. Decomposition of organic matter in landfills.
4. Incomplete combustion of forest or range fires.

The concentration of methane has been increasing at a rate of about 1.5% per year. Methane contributes to the greenhouse effect to an extent of 19%. In recent years (1990 onwards), the concentration of methane has tended to decline. This has been attributed, in part, to the reduced tropical biomass burning. Moreover, it is believed that large depletion of stratospheric ozone has exposed the troposphere to a large dose of ultraviolet radiation. These have increased the concentration of hydroxyl radicals (by photodissociation of water) which act as a major sink for methane (Reaction 16.4, p. 242).

Nitrous oxide

At present, the atmospheric concentration of nitrous oxide is about 0.31 ppm. The major sources of nitrous oxide are:

1. Microbial action on nitrogenous fertilizers in soil.
2. Burning of biomass, fossil fuels and forests.

The concentration of nitrous oxide has been increasing at a rate of about 0.2% per year. Its contribution to greenhouse effect is about 4%.

Ozone

The concentration of ozone in the troposphere is about 0.02 ppm. Its sources are:

1. Reaction between hydrocarbons and nitrogen oxides emitted from the burning of fossil fuels.
2. Reaction between nitric oxide, made by soil microbes, and terpenes or other hydrocarbons released by trees.
3. Deforestation. Tropical trees absorb ozone from the atmosphere. Thus the removal of forests reduces the uptake of ozone, which then is added to the troposphere.
4. Biomass burning. Although ozone is not directly produced by burning of fossil fuels, its precursors are emitted as combustion products.

The Concentration of tropospheric ozone is increasing at a rate of 1% in

Northern Hemisphere. The rate of its increase in the Southern Hemisphere is, however, insignificant. Tropospheric ozone contributes to an extent of about 8% to the greenhouse effect.

Chlorofluorocarbons

The chlorofluorocarbons which mainly contribute to the greenhouse effect are $CFCl_3$ (CFC-11) and CF_2Cl_2 (CFC-12). Their atmospheric concentrations are 2.3×10^{-4} and 3.8×10^{-4} ppm, respectively. Chlorofluorocarbons are used as:

1. Solvents
2. Refrigerants
3. Spray-can propellants

The rate of increase of chlorofluorocarbons is the fastest, 6% per year. Their contribution to the greenhouse effect is of the order of 17%.

Consequences of Greenhouse Effect

Before the greenhouse gases—mainly carbon dioxide and water vapours—started trapping the solar energy, the average temperature of the earth was –15°C. The greenhouse gases raised the earth's mean temperature to the present day 15°C value. Thus, the most trivial consequence of the greenhouse effect is that without it our planet would have been uninhabitable. However, any further increase in the atmospheric levels of greenhouse gases must be stopped, for otherwise the greenhouse effect would have adverse environmental implications.

1. Effects on Global Climate

If the present emission trends of greenhouse gases continue, a global warming of 3.5 to 4.5°C is likely to occur. This may seem to be a small and insignificant temperature change. However, records from the geologic past indicate that average global temperatures have varied over a range of no more than 2°C since the end of last glaciation, about 10,000 years ago. In fact, the period 1550–1850, when the average temperature was just 0.4°C less than what it was in 1900, is termed as the *Little Ice Age*. This period saw extensive glacial advances in almost all Alpine regions of the world. Global warming would have the following environmental fallouts on global climate.

(a) Rise in sea level

The most severe outcome of the global warming would be a rise in sea level. It has been estimated that the sea level may rise 0.5 to 1.5 m in the next 50 to 100 years. The rise will come from thermal expansion of ocean water and from melting of mountain glaciers. The ice cap in east Antarctica is not expected to melt in foreseeable future, but the west Antarctica ice sheet, which is partially in contact with ocean water, may break up and melt, causing sea level to rise even more than 1.5 m.

Higher sea levels would

(i) Increase the frequency and severity of floods
(ii) Damage coastal areas

(iii) Cause loss of soil replenishment

(iv) Cause sea water intrusion into river and other aquatic systems near the ocean

One third of the world's population lives in low-lying coastal areas. A large part of the population of some countries is at risk if sea level rises substantially. For example, half the population of Bangladesh lives at elevation less than 5 m. Bangladesh is already experiencing major flooding from sea level rise. If present trends continue, Bangladesh will lose upto 18% of its habitable land by 2050. Likewise, a combination of sea level rise and loss of soil replenishment on the Nile delta is expected to destroy as much as 19% of Egypt's arable land by the middle of this century. Almost all the 1, 196 islands of Maldives are less than 3 meters high, and most people there live less than 2 m above the waves. If the west Antaractica ice sheet begins to melt, the Pacific Ocean would rise to such dangerous levels that all the densely populated coastal cities from Shangai to San Francisco would be threatened. If the Arctic ice cap begins to melt, then Greenland, Iceland, Norway, Sweden, Finland, Siberia and Alaska would be adversely affected.

(b) Evaporation of water from aquatic systems

Global warming will cause more water to evaporate from the aquatic systems and create water vapours. The extra water vapour will amplify the warming because it itself is a greenhouse gas (Fig. 18.1).

(c) Changing patterns of rainfall

Patterns of rainfall would change worldwide causing large shifts in agriculturally productive areas.

2. Effects on Plants

The increasing concentrations of carbon dioxide and the consequential increase in temperature would have the following effects on plants.

(a) Carbon dioxide fertilization

At first glance, elevated carbon dioxide levels might seem an agricultural blessing. A carbon dioxide rich environment is expected to enhance plant growth by accelerating the pace of photosynthesis. During photosynthesis, plants capture carbon dioxide molecules and solar energy. The former diffuse into the plant through the pores in the outer layer of leaf cells (these pores are called *stomata*). The gas ultimately reaches chloroplasts—organelles in which photosynthesis takes place. The increase in rate of photosynthesis as a result of enhanced carbon dioxide levels is called carbon dioxide fertilization.

Studies on wheat, mungbean and mustard indicate that plant biomass may increase by 25 to 30% by carbon dioxide fertilization. Similar studies on citrus trees and cotton plants also show faster growth rates. However, these are laboratory studies. The carbon dioxide concentration is artificially increased, but care is also taken that other ingredients of plant growth—nutrients, light and water—are supplied in abundance.

Positive response to increased carbon dioxide levels in isolated laboratory experiments do not necessarily translate into increased growth for the plants in natural environment. Competition for scarce resources constrains the plant's response to higher levels of carbon dioxide. The requirements for fertilizers and biocides put the underdeveloped and developing nations at a disadvantage. For many countries, even water is a limited and expensive resource.

Even if nutrients, light and water are present in abundance, it does not mean that the rate of photosynthesis would always increase with increasing carbon dioxide concentrations. Most plants record an increase in photosynthesis only initially. Later, this rate tends to fall. This may be due to one or both of the following reasons.

(i) Increased photosynthesis results in an excess accumulation of starch in chloroplasts, thereby hindering the organelles' ability to function.

(ii) In the presence of greater amounts of carbon dioxide, a plant's ability to produce carbohydrates exceeds its ability to move the starch by-product to actively growing parts of the plants, thus slowing down photosynthesis.

Of course, such mechanisms do not operate in all plants. Where these are not applicable, enhanced carbon dioxide levels do increase the pace of photosynthesis. Such is the case with most herbs and weeds, the higher yields of which are in no way beneficial to the humanity.

(b) Lower nitrogen content

The dead plant material, such as fallen leaves and twigs, are rich in nitrogen. These act as natural fertilizers, providing nitrogen-based nutrients to the soil and thereby increasing soil productivity. However, plants grown in higher concentrations of carbon dioxide have less nitrogen and more carbon content. The litter of such plants, therefore, fails to increase soil fertility and nutrient cycling.

Moreover, less nitrogen in plants means less protein content. Insect pests that feed on carbon dioxide fertilized plants, therefore, would eat more leaf to obtain enough nitrogen. This negates any benefit of crop yield boon in a carbon dioxide rich environment, if at all such a boon occurs.

(c) Increased rate of decomposition

As a result of increased global temperature due to greenhouse effect, the dead plant matter and soil organic matter would decompose at a rate higher than normal. The decomposition shall yield more carbon dioxide, which would supplement the greenhouse phenomena.

(d) Increased threat of pests

As a result of global warming, farmers in temperate countries are likely to be plagued by a growing army of insecticide-resistant pests. For example, peach potato aphid, *Myzus persicae,* a major agricultural pest, feeding on the sap of potatoes and sugar-beet, has become more active in recent years. It is resistant to virtually all the insecticides. Earlier, the biggest challenge facing the aphids was to survive the winters. Aphids cannot move quickly when the temperature is low and so if they fall to the ground they die of starvation before they get back to their host plant. Now, however, global warming has removed the brake that winter had so far provided.

(e) Evaporation of water from soil

Due to increase in temperature, the moisture content of soil would decrease, and so would its fertility towards many crops.

3. Effects on Human Health

An increase in average global temperature is likely to increase the incidence of infectious diseases, such as malaria, schistosomiasis, sleeping sickness, dengue and yellow fever. The developing countries, where these diseases are already prevalent would receive the most deadly punch from the greenhouse effect. An increase in the global temperature is suspected to extend the range of vectors—the mosquitoes, flies and snails—that transmit infectious discases. A one degree increase in the average temperature in Rwanda (Central Africa) in 1987 led to a 337% rise in the incidence of malaria. The disease spread to the mountainous regions of Rwanda where previously the cases of malaria were negligible.

Due to the greenhouse effect, one of the prime carriers of dengue and yellow fever—the *Aedes aegypti*—has extended its range in such diverse regions as Coasta Rica, Colombia, Kenya and India. Likewise, the outbreak of cholera in Latin America in 1991, the outbreak of pheumonic plague in India in 1994, and the huntavirus epidemic in South America in 1994 can be directly linked to global warming.

4. Effects on Wildlife

Greenhouse effect shall seriously affect wildlife. With every rise of 1°C, plant and tree species will have to move about 90 kilometers polewards to survive. Many will simply not be able to spread fast enough. The strain will be greatest in the higher latitudes because they will heat up fastest; winter temperatures in latitudes between 60° and 90° are expected to warm up more than twice as fast as the global average; and Arctic Tundra may disappear altogether. Changing rainfall patterns will compound the ecological disaster, while rise in sea levels will swamp coastal habitats. As trees and plants die out and habitats disappear, so will the animals that depend on them. And as the world continues to get warmer there will be nowhere for habitats to re-establish themselves.

Abatement of Greenhouse Effect

Certain natural phenomena are known to offset the adverse implications of greenhouse effect. Strategies are also being worked out to restrict the emission of greenhouse gases.

1. Abatement by Natural Phenomena

The following natural factors prevent the concentration of greenhouse gases from rising abnormally high.

(a) Sink for carbon dioxide

Burning of fossil fuel contributes about 5 billion metric tonnes of carbon dioxide to the atmosphere annually. Deforestation contributes an additional one billion metric tonne. The net annual increase of carbon dioxide in the atmosphere is, however, only about 3 billion metric tonnes. The remaining 3 billion metric tonnes are absorbed by the ocean. Thus, the ocean acts as the most important sink for carbon dioxide. Oceans absorb carbon dioxide by way of direct solubility, algal photosynthesis or by means of chemical reactions that convert carbon dioxide into limestone, which acts as a shell material for many marine animals.

The average lifetime of a carbon dioxide molecule in the atmosphere is about 500 years. With such a long residence time, carbon dioxide could have played havoc with the environment, had the oceans not acted as its natural sink. However, there is a limit to which oceans can absorb carbon dioxide. Even at present, oceans contain 60 times as much carbon dioxide as the atmosphere. Like other gases, the solubility of carbon dioxide in water is inversely proportional to temperature. As the global temperature increases due to the greenhouse effect, not only lesser amounts of carbon dioxide would be absorbed in the oceans, but the already absorbed carbon dioxide would be released to the atmosphere, thereby reinforcing the greenhouse effect.

This is particularly true of the carbon dioxide which is present in the warm, top 100 m layer of the ocean. This layer absorbs carbon dioxide directly from the atmosphere. If the global temperature increases, the carbon dioxide is easily ejected from the top ocean layer into the atmosphere. However, if carbon dioxide enters the cold depths of the ocean, it would not be easily exchanged with the atmosphere. Carbon dioxide passes from the upper to lower layer of the ocean largely by mixing of marine waters. The mixing is a relatively slow process, taking anything from 500 to 1500 years. Thus, most of carbon dioxide absorbed in the last few decades would still be present in the upper 100 m of the ocean, from where the chances of its ejection into the atmosphere are quite high.

(b) Sink for trace gases

Methane molecules have an average residence time of 9 years. These molecules are destroyed by reaction with hydroxyl radicals in the atmosphere. The molecules of nitrous oxide have an average life time of 150 years. Nitrous oxide is not destroyed until it reaches stratosphere. There, it is acted upon by ultraviolet radiation and decomposed to nitric oxide. Ozone is stable in the troposphere only for a few hours, during which time it is broken down into oxygen by solar radiation. The chlorofluorocarbons, $CFCl_3$ (CFC-11) and CF_2Cl_2 (CFC-12), have residence times of 75 and 110 years respectively. There are no important sinks for CFCs in the troposphere. However, they are ultimately broken down in stratosphere by photolysis, yielding reactive chlorine radicals.

(c) Volcanic debris and aerosols

Volcanic eruptions inject tremendous amounts of debris particles in the atmosphere. These particulates act as anti-greenhouse material by blocking the sunlight and causing the temperature to drop.

Volcanoes also inject significant amounts of sulfur oxides into the stratosphere. These are eventually oxidised to sulfate aerosols. The sulfate aerosols play an important climatic role in that they reflect the solar radiation back to space. Thus, these aerosols prevent the warming effect of the greenhouse gases.

For example, in 1992, the opposing effect of volcanic eruptions interrupted the strong global warming trend that had been observed since 1980. The Mount Pinatubo (Philippines) and Mount Hudson (Chile) eruptions together injected about 30 million tonnes of sulfur as sulfur dioxide into the stratosphere. The sulfur dioxide reacted with water vapours to produce sulfate aerosols, which spread around the globe, dissipated by stratospheric winds. The sulfate aerosols reflected the incoming sunlight and thus cooled the earth.

The sulfate aerosols which are injected into the troposphere as a result of volcanic eruptions, along with the debris particles, can cool the earth in yet another way. They act as *cloud condensation nuclei.* That is, their presence in the atmosphere triggers cloud formation. Clouds reflect the incoming solar radiation, thereby cooling the earth.

(d) Phytoplanktons

During greenhouse warming, phytoplanktons can increase the cloud cover and thus create a negative feedback to greenhouse effect. Some species of phytoplankton excrete dimethylsulfide into the sea water. A part of dimethylsulfide then enters the atmosphere and oxidises to form sulfate aerosols, which serve as condensation nuclei for developing cloud droplets. Clouds cool the atmosphere immediately below them by reflecting back about 30% of the sunlight. A direct correlation between dimethylsulfide levels and the number of cloud nuclei over portions of the Pacific Ocean has been observed.

2. Restricting the Emission of Greenhouse Gases

In order to cut emission of greenhouse gases, the following steps have been suggested.

(a) Worldwide, about one-third of all carbon dioxide emissions come from thermal electric generating plants. A typical power plant flue gas contains about 15% by volume of carbon dioxide. The latter may be separated from the flue gas. The captured carbon dioxide can be used as a repressurising fluid for oil reservoirs. It may also be injected into the ocean at depths below 500 m so that its chances of being released to the atmosphere are almost negligible.

(b) It has been hypothesized that the presence of iron in the ocean would enhance the carbon dioxide uptake by phytoplankton and macroalgae (sea weeds), thereby reducing global warming. In presence of iron, the chlorophyll content of the phytoplankton and macroalgae increases which, in turn, results in an increase in the photosynthetic uptake of carbon dioxide. Thus, fertilizing the ocean with iron might be able to control, in part, the level of carbon dioxide in the atmosphere.

This idea originated from the observation that the coastal water of Antarctic Ocean, which has a high iron content, possesses a rich ecosystem of phytoplankton and macroalgae. As a result, the coastal water supports abundant species of krill, fish, seal and penguin. However, the offshore water of Antarctic Ocean, which is devoid of iron content, does not support population of phytoplankton, macroalgae and marine animals.

It has also been observed that adding iron to samples of Antarctica water, greatly enhances their carbon dioxide content. However, these are the results of laboratory experiments. At present, it is difficult to predict whether these experiments can be extended to as large an ecosystem as an ocean.

(c) Recently, a process has been devised to convert the main greenhouse gas, carbon dioxide, into fuel. It has been found that carbon dioxide reacts with water at 50°C, in presence of titanium dioxide catalyst and ultraviolet light to give methane, hydrogen and carbon monoxide. All these products can be used as fuels.

(d) Restrict the use of fossil fuels, such as coal, oil and natural gas.

(e) Increase auto-fuel efficiency.

(f) Phase out the production of chlorofluorocarbons.

(g) Encourage research into agricultural techniques, such as feed additives for cattle and water management in rice production, that could reduce emissions of methane.

Tie-in-Strategies

Scientists and policy makers respond to the greenhouse effect in basically two ways. Some say that the global warming would be so severe that fast action is necessary to counteract it. Others hold that so many uncertainties are involved in the issue that it is not possible to predict what the consequences will be. It is, therefore, useless to invest time, money and energy to reduce the emission of greenhouse gases until the problem is better understood.

Between these two extremes, a third lobby has emerged with an action plan called tie-in-strategies. These strategies aim at taking such steps which would not only counter greenhouse effect, but would produce other beneficial environmental effects as well. Thus, if global warming occurs, these strategies would act as an antidote. If global warming does not occur, we do not lose anything. Rather, the beneficial aspects of the plan would augment environmental conservation. Some of the components of the tie-in-strategies are discussed below.

1. Develop renewable alternatives to fossil fuels

In the section, 'Restricting Emissions of Greenhouse Gases', we had suggested that restrictions should be imposed on the use of fossil fuels. This is easier said than done. The underdeveloped and developing countries rely almost exclusively on fossil fuels as energy resources. Discouraging the use of fossil fuels shall be acceptable to these countries only if it is accompanied by simultaneous transfer of renewable energy technologies, such as solar, wind and hydro energy resources. Imposing restrictions alone would lead to social catastrophes as unbearable as the climatic catastrophes that would otherwise occur.

Once these clean and renewable energy resources are made available to the underdeveloped and developing countries, they would benefit even if no greenhouse warming occurs. These energy resources would reduce the emission of oxides of sulfur and nitrogen that cause acid rain, as well as help the economy at a time when oil and gas are in a short supply globally.

2. Reducing deforestation

Reducing deforestation by growing more trees in the tropics is valuable even if global warming does not occur because it would help preserve the rich abundance of species in the tropics. Often deforestation causes drought and reduces the fertility of land.

3. Halting chlorofluorocarbons production

The chlorofluorocarbons are known to cause depletion of stratospheric ozone which, in turn, would have severe environmental implications. Halting chlorofluorocarbons production would avert the grave situation arising out of stratospheric ozone depletion even if no global warming occurs.

4. Stabilizing world's population

Stabilizing world's population, which at present is increasing at a rate of 8.7 million people per year, is essential to stabilize the atmospheric burden of greenhouse gases. Even otherwise, it could help in creating better living conditions in many parts of the world and in allowing production to catch up with population.

5. Surveillance for public health

In view of the increased incidence of infectious diseases with global warming, it has become necessary to establish a formal global surveillance network for epidemics. It is equally necessary to integrate global environmental medicine into medical and public health education. Even if global warming does not occur, these initiatives would prove to be a boon for public health.

The Kyoto Protocol

The kyoto protocol refers to the covenant that emerged at the culmination of the *United Nations Climate Conference* held at Kyoto, Japan in December 1997. The protocol demands that the developed countries should reduce the emission of three greenhouse gases, carbon dioxide, methane and nitrous oxide, by an average 5% till 2010. There are three main components of the Kyoto Protocol.

1. Clean Development Mechanism

The clean development mechanism allows the industries in developed countries to gain credits towards their national targets by assisting the developing countries in cutting their greenhouse emissions. Thus, if a corporate is struggling to meet its domestic emission limits, it can provide funds and know-how to build a clean energy resource in an underdeveloped nation. Everyone benefits from this endeavour. The industry meets its target without restricting its operations, the developing country gets the much needed investment and access to modern technology and overall emissions are kept within limits.

2. Emission Trading Mechanism

This mechanism treats emission reduction as a commodity. Suppose, for example, a country is bound to reduce the emission of greenhouse gases by 5%, but it reaches a voluntary target of 10%, then it can sell permits equivalent to the excess to a country that is finding it impossible to meet its own target. Emission trading assumes importance in the present day scenario because the newly recognized fragments of the former Soviet Union have gained immense surplus value due to the collapse of their economics and the closure of their energy-intensive industries. The industrialized states, like the U.S., can meet their obligations by buying permits from these new nations.

3. Joint Implementation Mechanism

Joint implementation mechanism allows industrialized countries to gain credits by collaborating in large projects. For example, if two nations individually cannot meet the targets laid by the Kyoto Protocol, they can close down their conventional enterprise and instead set up joint ventures is mutually beneficial areas. Obviously, the new projects would incorporate modern technologies which would be more cost-effective and less polluting. This way the peoples of collaborating nations would better understand the global nature of the problem. Moreover, the concepts of restrictive trade practices and economic barriers would fade out, at least partially, paving the way for socio-economic equality.

References

Aldhous, P. (1995). Global heat breeds superaphids, *New Scientist,* December 23/30, p. 17.

Baum, R. (1990). Adding iron to ocean makes waves as way to cut greenhouse CO_2, *Chem. Eng. News,* July 2, p. 21–24.

Bazzaz, F.A. and Fajer, E.D. (1992). Plant life in a CO_2–rich world, *Sci. Amer.,* **266**(1), 18–24.

Bekki, S., Law, K.S. and Pyle, J.A. (1994). Effect of ozone depletion on atmospheric CH_4 and CO concentrations, *Nature (London),* **371,** 595–597.

Coghlan, A. (1995). Sunlight could recycle greenhouse gas, *New Scientist,* April 1, p. 22.

Frost, B.W. (1996). Phytoplankton bloom on iron rations, *Nature (London),* **383,** 475–476.

Gribben, J. (1986). Temperature rise in the global greenhouse, *New Scientist,* May 15, p. 31–32.

Gribben, J. (1995). Absorbent clouds cast shadow on climate models, *New Scientist,* March 25, p. 20.

Hileman, B. (1992). Web of interactions makes it difficult to untangle global warming data, *Chem. Eng. News,* April 27, p. 7–19.

Hileman, B. (1995). Scientists warn that disease threats increase as earth warms up, *Chem. Eng. News,* October 2, p. 19–20.

Kerr, R.A. (1983). Trace gases could double climate warming, *Science,* **230,** 1364–1365.

Kiehl, J.T. and Briegleb, B.P. (1993). The relative roles of sulphate aerosols and greenhouse gases in climate forcing, *Science,* **260,** 311–314.

Lelieveld, J. and Heintzenberg, J. (1992). Sulfate cooling effect on climate through in-cloud oxidation of anthropogenic SO_2, *Science,* **258,** 117–120.

Siegenthaler, U. and Sarmiento, J.L. (1993). Atmospheric carbon dioxide and the ocean, *Nature (London),* **365,** 119–125.

Stone, R. (1995). If mercury soars, so may health hazards, *Science,* **267,** 957–958.

Woodwell, G.M., Hobbie, J.E., Houghton, R.A., Melillo, J.M., Moore, B., Peterson, B.J. and Shaver, G.R. (1983). Global deforestation: Contribution to atmospheric carbon dioxide, *Science,* **222,** 1081–1086.

Wang, W.C., Yung, Y.L., Lacis, A.A., Mo, T. and Hansen, J.E. (1976). Greenhouse effects due to man made perturbations of trace gases, *Science,* **194,** 685–690.

CHAPTER 19

Depletion of Stratospheric Ozone

The ozone layer in the stratosphere is a natural feature of the earth's environment. This layer shields the earth from the harmful ultraviolet radiations of the sun (Fig. 19.1). It thus acts as the earth's most important life support system. In recent years, however, human activities have been projected to deplete stratospheric ozone through anthropogenic increase in the global concentrations of some atmospheric chemicals. A depletion of the ozone shadow is considered as a threat to all forms of life. The possible outcome of depletion is, therefore, sufficiently serious to justify intensive study of potential hazards, and to counsel caution regarding emission of chemicals which might be contributing to depletion.

Introduction

When ultraviolet light from the sun strikes the earth's atmosphere, it causes photolysis of oxygen molecules into reactive oxygen atoms.

$$O_2 + h\nu \rightarrow 2O \quad (19.1)$$

This reaction is followed by another in which a reactive oxygen atom recombines with an oxygen molecule.

$$O_2 + O \rightarrow O_3 \quad (19.2)$$

The ozone formed in reaction (19.2) distributes itself in the stratosphere, forming a layer which is thin in some places, thick in others. Stratospheric ozone then absorbs harmful ultraviolet radiations.

$$O_3 + h\nu \rightarrow O_2 + O \quad (19.3)$$

In absence of ozone these damaging solar radiations would have reached the earth, causing several ill effects. Thus, the ozone layer filters the incoming ultraviolet radiations and affords our natural protection. No doubt, reaction (19.3) destroys ozone, but it also produces reactive oxygen atoms, which combine with oxygen (reaction 19.2) producing more ozone to compensate the loss.

Thus, in the stratosphere, ozone is continuosly created (reaction 19.2) and destroyed (reaction 19.3) by the sun's radiations. This results in an equilibrium concentration of ozone. The equilibrium is, however, disturbed when reactive chlorine atoms, released from photolysis of chlorofluorocarbons, waft into the

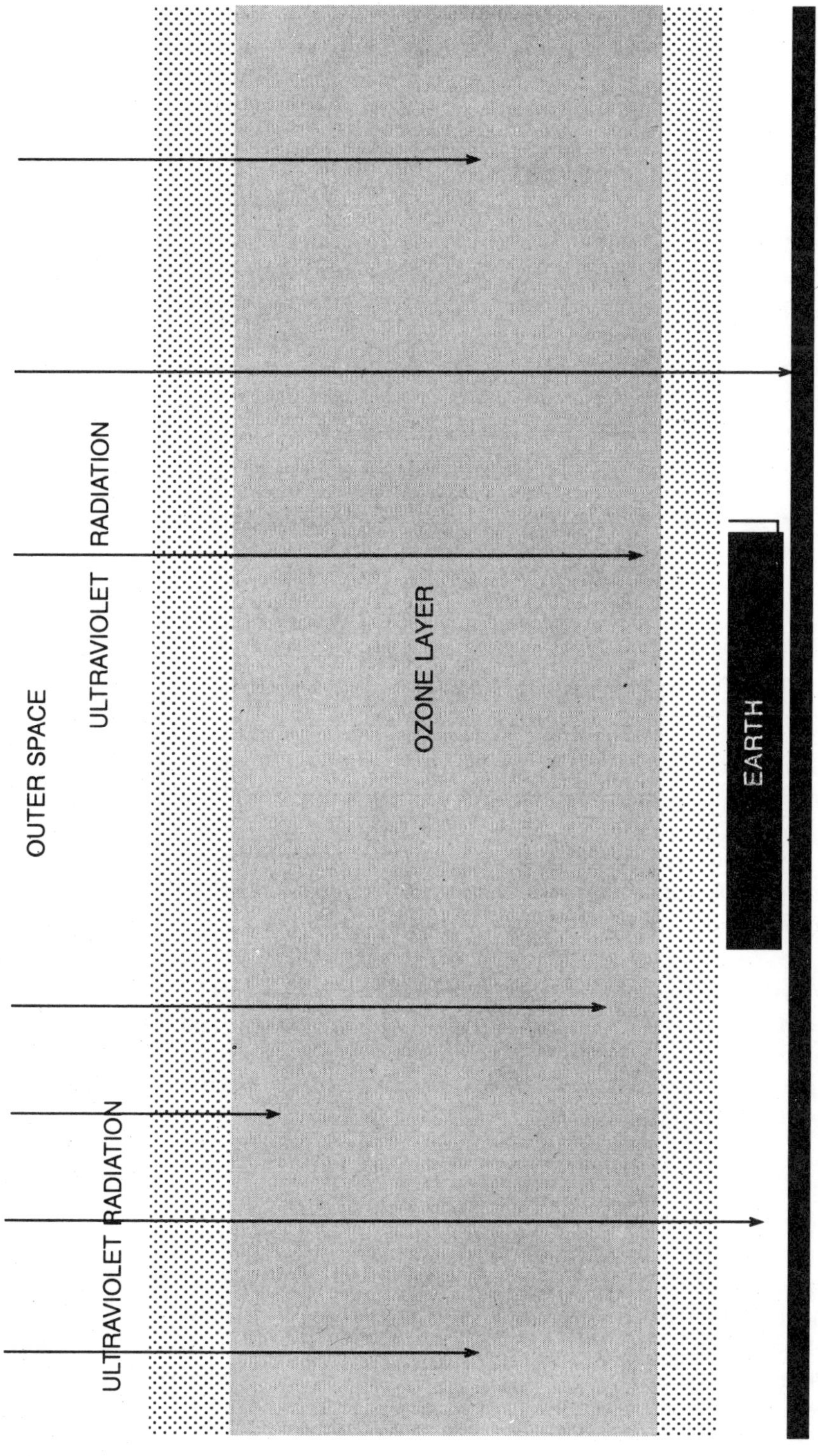

Fig. 19.1 Significance of stratospheric ozone layer

atmosphere. These atoms create an imbalance by destroying ozone molecules. The loss of ozone molecules in the upper atmosphere is termed *depletion of stratospheric ozone.* When this happens, the ozone layer's capacity to filter out the harmful ultraviolet rays from the sun decreases.

Mechanism of Ozone Depletion

Reactive chlorine atoms, $Cl^{\cdot}$, initiate the mechanism of ozone depletion in the stratosphere. The reactive chlorine atoms are generated from chlorofluorocarbons, a group of chemicals having wide industrial applications. Chlorofluorocarbons are stable in the lower atmosphere. However, when they reach stratosphere, the solar radiation splits them into reactive chlorine atoms. This is followed by the following chain reactions.

$$Cl^{\cdot} + O_3 \rightarrow ClO^{\cdot} + O_2 \qquad (19.4)$$

$$ClO^{\cdot} + ClO^{\cdot} + M \rightarrow (ClO)_2 + M \qquad (19.5)$$

$$(ClO)_2 + h\nu \rightarrow Cl^{\cdot} + ClOO^{\cdot} \qquad (19.6)$$

$$ClOO^{\cdot} + M \rightarrow Cl^{\cdot} + O_2 + M \qquad (19.7)$$

$$Cl^{\cdot} + O_3 \rightarrow ClO^{\cdot} + O_2 \qquad (19.8)$$

$$\text{Net reaction: } 2O_3 + h\nu \rightarrow 3O_2 \qquad (19.9)$$

The species 'M' represents another molecule that is unchanged in the reaction.

As a result of this mechanism, ozone hole is formed (1) mainly over Antarctica (2) during spring time (September-October) and (3) replenishes after spring (November-December). Scanty depletion, however, also occurs (4) over Arctic and (5) a few pockets of non-polar regions. Almost entire depletion occurs through the mechanism initiated by reactive chlorine atoms. (6) Other mechanisms, though theoretically feasible, have been ruled out.

1. Why Ozone Depletion Occurs Mainly Over Antarctica?

The chain reactions (19.4 to 19.9) can theoretically occur all over the stratosphere. However, the ozone hole has been mainly observed in the stratosphere over Antarctica. In other parts of the stratosphere, two key reactions bind chlorine monoxide and chlorine atoms.

$$ClO^{\cdot} + NO_2 \rightarrow ClONO_2 \qquad (19.10)$$

$$Cl^{\cdot} + CH_4 \rightarrow CH_3^{\cdot} + HCl \qquad (19.11)$$

The presence of oxides of nitrogen in the stratosphere thus scavenges the chlorine monoxide radicals. The presence of methane scavenges chlorine atoms. The chlorine nitrate formed in reaction (19.10) and hydrogen chloride formed in reaction (19.11) thus act as natural reservoirs for reactive chlorine species. Once these are tied-up, the chain mechanism leading to ozone depletion is not initiated. Hence no ozone hole is created.

In Antarctica, however, the conditions are quite different. During winters, special type of clouds, called *Polar Stratospheric Clouds* (PSCs) are formed over Antarctica. These clouds are of two types. Type-I clouds, which contain mostly solidified nitric acid trihydrate ($HNO_3 \cdot 3H_2O$), form at about –77°C. Type-II clouds, which contain mostly ice, form at about –85°C. These clouds play an important role in ozone depletion.

On the surface of polar stratospheric clouds, chlorine nitrate, formed in reaction (19.10)—a reaction which also occurs over Antarctica—undergoes two important transformations. It gets hydrolysed,

$$ClONO_2 + H_2O \rightarrow HOCl + HNO_3 \tag{19.12}$$

and it reacts with hydrogen chloride formed in reaction (19.11).

$$ClONO_2 + HCl \rightarrow Cl_2 + HNO_3 \tag{19.13}$$

Reactions (19.12) and (19.13) convert an important reservoir of chlorine atoms into hypochlorous acid and molecular chlorine. Even under mild conditions, hypochlorous acid and molecular chlorine can be reconverted into reactive chlorine atoms, paving way for ozone depletion. These reactions also convert oxides of nitrogen—which are known to scavenge chlorine monoxide—into nitric acid. The latter becomes incorporated into ice, forming Type-I polar stratospheric clouds. Hence, the gaseous oxides of nitrogen are converted into a condensed phase. They are then not available to scavenge chlorine monoxide. The free chlorine monoxide propagates ozone depletion. Thus, the polar stratospheric clouds, on one hand reactivate chlorine radicals from a dormant sink, and on the other remove a scavenger of chlorine monoxide.

The species 'H_2O' in reaction (19.12) represents ice contained in the polar stratospheric clouds and not water vapours; the atmosphere of polar regions is devoid of water vapours. Likewise, 'HCl' in reaction (19·13) is not present in gaseous or solution state. It is present in solid state, dissolved in ice. Since 'H_2O' and 'HCl' are present in solid forms, reactions (19.12) and (19.13) are heterogeneous in nature. These do not occur under homogeneous conditions. In Antarctic region the surface for the heterogeneous reactions is provided by the polar stratospheric clouds. Such clouds are abundant over Antarctic skies. In other parts of the stratosphere no such surface is available for heterogeneous reactions. Therefore, the conditions for the catalytic destruction of ozone by chlorine radicals are established mainly over Antarctica.

2. Why Ozone Depletion Occurs During Spring Time?

Spring time in Antarctica lasts from early September to late October. During this period, sun begins to shine over Antarctic skies. The hypochlorous acid and molecular chlorine, formed in reactions (19.12) and (19.13) respectively, are photolysed by sunlight.

$$HOCl + h\nu \rightarrow {}^{\cdot}OH + Cl^{\cdot} \tag{19.14}$$

$$Cl_2 + h\nu \rightarrow 2Cl^{\cdot} \tag{19.15}$$

The reactive chlorine atoms are thus rendered free to deplete ozone through the chain mechanism culminating in reaction (19.9).

3. Why Ozone Hole is Replenished After Spring?

The earth's atmosphere is dynamic in nature. This means that the constituents of the atmosphere are in constant motion. For example, the ozone rich air in the stratosphere tends to circulate from high altitudes in the tropics towards lower altitudes in the polar region. This movement of stratospheric air is expected to 'fill-up' the ozone hole as soon as it is formed. Yet, more than 40% of stratospheric ozone over Antarctica remains depleted for almost two months of spring.

The local conditions of Antarctica inhibit the natural, dynamic character of the atmosphere to replenish the ozone hole. During winters, when polar stratospheric clouds are formed over Antarctica, stable wind patterns in the stratosphere, called the *Polar Vortex* encircle the continent. Polar vortex is a tight whirlpool of winds. It is so rigid that air within it is isolated from the sun and from warmer air of the temperate regions. Thus, as long as the polar vortex surrounds Antarctica, the ozone rich air from non-polar regions is prevented from 'filling-up' the ozone hole. The polar vortex remains intact throughout the Antarctic spring. Thence, the intensity of sunlight increases and the vortex breaks down (sometime in November). The ozone rich air from the temperate region rushes in and replenishes the ozone hole.

The mechanism of ozone hole formation and its subsequent replenishment is pictorially represented in Fig. 19.2.

4. Why Scanty Ozone Depletion Occurs Over Arctic Region?

We have seen that two geographical features—the polar stratospheric clouds and the polar vortex—play a central role in the chlorine-catalysed depletion of stratospheric ozone over Antarctica each spring. Polar stratospheric clouds and polar vortex also appear over the Arctic skies. The chlorine-catalysed chain mechanism (reactions 19.4 to 19.9), therefore, is eroding the stratospheric ozone layer over the Arctic too. However, the extent of depletion is quite less. A 5 to 6% ozone loss has been observed since 1970s (compared to 40% loss over Antarctica). The reason for the low level of ozone loss lies in the relatively warmer climate of the Arctic region as compared to the Antarctic zone.

The Antarctica is surrounded by ocean. Its temperature in winter drops to about –90°C. The Arctic region is in contact with habitated land masses. As a result, its winter temperature drops to about –80°C. The comparatively warmer climate of the Arctic zone has two consequences as far as ozone depletion is concerned.

Firstly, in Arctic region only Type-I polar stratospheric clouds are formed. Type-II polar stratospheric clouds, which form at about –85°C, are rarely seen over Arctic. In Antarctica, both Type-I and Type-II polar stratospheric clouds are prevalent. These clouds provide the surface for the heterogeneous reactions (19.12) and (19.13). Since lesser amounts of polar stratospheric clouds are created over Arctic region, these reactions proceed to a lesser extent. Hence, the chances of active chlorine species being released from the natural sinks ($ClONO_2$ and

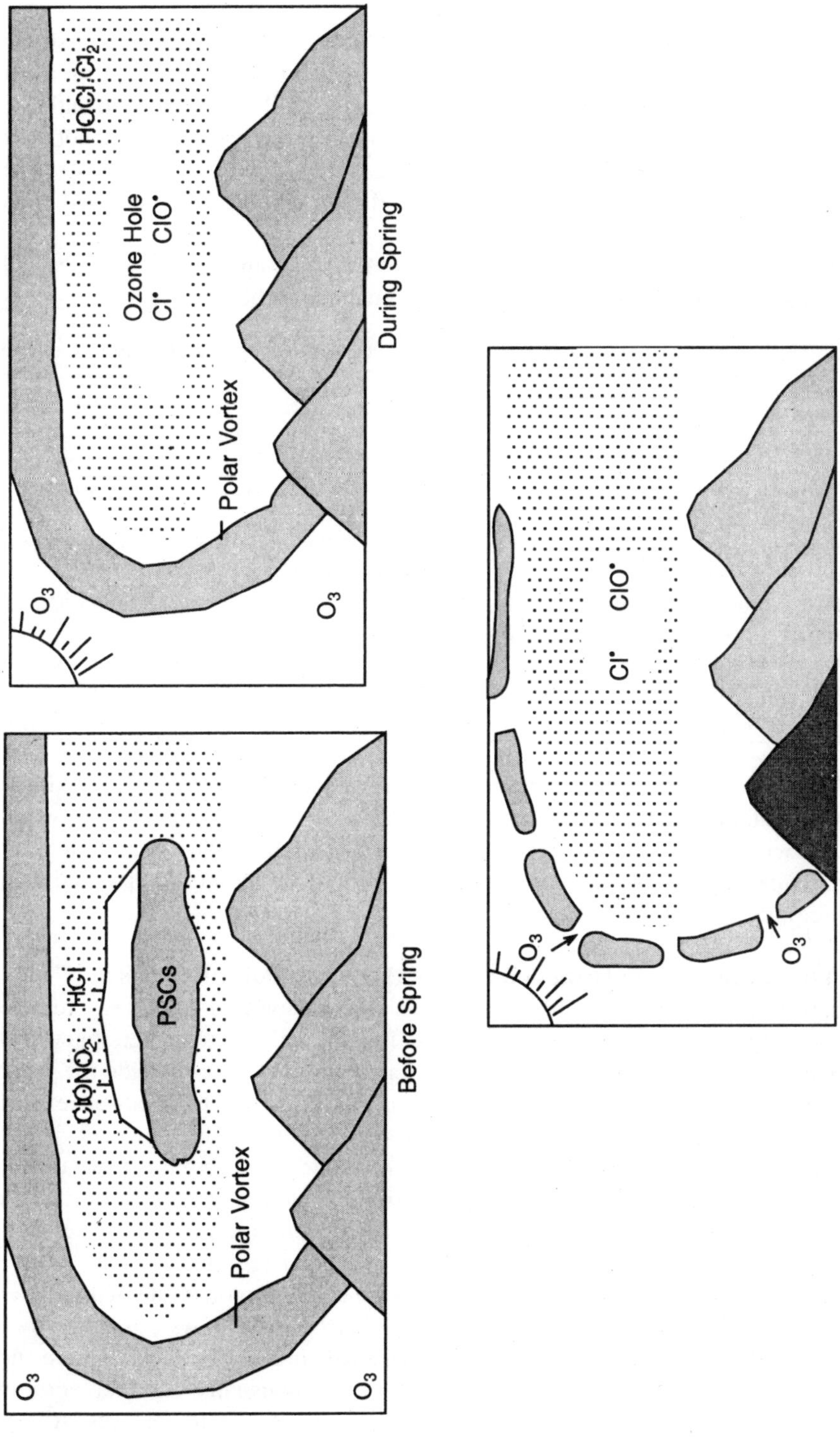

Fig. 19.2 Mechanism of ozone hole formation and its replenishment

HCl) are correspondingly less. Thus, ozone is depleted to a lesser extent over Arctic.

Secondly, the polar vortex encircling the Arctic is much weaker and short-lived than that over Antarctica. It begins to break up during spring season itself (late February). Thus, even before the chlorine-catalysed chain mechanism initiates, air from non-polar regions enters the Arctic zone. This air brings with it, not only ozone from temperate regions, but also oxides of nitrogen which scavenge the ozone depleting chlorine monoxide. Hence, the ozone hole is replenished even before it acquires a massive dimension.

A comparison of the geographical features of Arctic and Antarctic zones and the consequent levels of ozone depletion is made in Table 19.1.

Table 19.1 Local environment of Arctic and Antarctic regions that affect ozone depletion

Arctic	Antarctica
1. Winter temperatures drop to about –80°C	1. Winter temperatures drop to about –90°C
2. Polar stratospheric clounds are mostly Type-I	2. Polar stratospheric clouds are both Type-I and Type-II
3. Polar vortex breaks up during spring	3. Polar vortex breaks up after spring
4. Low concentrations of reactive chlorine atoms	4. High concentrations of reactive chlorine atoms
5. Significant ozone depletion	5. Massive ozone depletion

Thus, what is saving Arctic from a full-blown ozone hole is its milder winter. If an unusually cold Arctic winter occurs, a staggering ozone depletion would take place. This would prove to be more hazardous than the ozone hole over Antarctica, since the regions around North Pole are much more populated than the regions around South Pole.

5. Why Scanty Ozone Depletion Occurs Over Non-Polar Regions?

Reactions (19.12) and (19.13) are keys to ozone depletion since they release reactive chlorine atoms from its torpid sink. These reactions, being heterogeneous in nature, occur on the surface of polar stratospheric clouds. Outside the polar regions, these reactions normally do not occur because of the absence of a surface to carry the heterogeneous transformations through. However, it has been observed that sulfate aerosols in the stratosphere can provide a surface for these heterogeneous reactions, leading to ozone depletion in regions other than Arctic and Antarctic. Thus, an increase in the concentration of sulfate aerosols results in an increase in ozone depletion in non-polar regions.

During the year 1992, the average global concentration of ozone in non-polar regions was 2 to 3% lower than earlier years (1979–1991). As a result, ozone hole was observed over parts of Russia and Canada in Northern Hemisphere and Argentina and Chile in Southern Hemisphere. The unusually high depletion in these regions has been associated with high levels of sulfate aerosols. A year earlier, that is, in 1991, two volcanoes erupted: Mount Pinatubo in the Philippines and Mount Hudson in Chile. Large quantities of sulfur were injected into the stratosphere as a result of these eruptions. The sulfur was subsequently oxidised to sulfate aerosols, which were spread around the globe by stratospheric winds. These aerosols provided the surface for reactions (19.12) and (19.13), releasing the ozone depleting chlorine atoms.

6. Why Mechanisms Not Initiated by Chlorine Atoms have been Ruled Out?

The chlorine-catalysed chain mechanism of ozone depletion is supported by the following evidence.

(a) The large decrease in ozone concentration in the Antarctic stratosphere during spring season is accompanied by unusually high concentrations of chlorine atoms ($Cl^{\cdot}$), chlorinc monoxide ($ClO^{\cdot}$) and hypochlorous acid (HOCl), and low concentrations of oxides of nitrogen, relative to their observed values near equator. For example, within the ozone hole over the Antarctica, chlorine monoxide concentration upto 1300 ppt has been measured. Likewise, the maximum concentration of chlorine monoxide within the Arctic ozone hole is estimated to be upto 55 ppt. These observations support the hypothesis that ozone depletion is initiated by a reactive chlorine atom (reaction 19.4). It also supports the view that oxides of nitrogen act as scavengers for the ozone depleting chlorine monoxide species (reaction 19.10).

(b) On a typical day of Antarctic spring, the concentration of chlorine atoms within the ozone hole reaches a peak at noon, begins to decline during the evening and becomes undetected during the night. This observation supports the hypothesis that reactive chlorine atoms are generated by photolysis of hypochlorous acid and molecular chlorine (reactions 19.14 and 19.15). When the intensity of sunlight is maximum during noon, the photolysis reaction proceeds at a fast rate. When the intensity of sunlight declines in the evening, the rate of photochemical dissociation also slows down and hence lesser amounts of chlorine atoms are generated. In the absence of sunlight, the photolysis stops. Hence, no reactive chlorine atoms are detected at night.

(c) Laboratory studies show that heterogeneous reactions (19.12) and (19.13) can occur efficiently on surfaces that simulate polar stratospheric clouds. Thus, the rate of hydrolysis of chlorine nitrate (reaction 19.12), as well as that of reaction between chlorine nitrate and hydrogen chloride (reaction 19.13) is greatly enhanced in presence of ice. This observation supports the hypothesis that the heterogeneous reactions, which convert stable chlorine reservoirs into photochemically active forms of chlorine, occur in the polar regions over the surface of polar stratospheric clouds.

Laboratory studies also demonstrate that chlorine nitrate reacts with water and hydrogen chloride at –63°C on sulfate aerosols generated from 65% sulfuric acid. This observation supports the view that reactions (19.12) and (19.13) can occur very well over sulfate aerosols in non-polar regions. Besides the chlorine-catalysed reaction scheme (reactions 19.4 to 19.9), a few other mechanisms of ozone depletion have also been propounded. However, each of these suffers from drawbacks. Moreover, since these mechanisms are not supported by any concrete evidence, it is suspected that they are responsible for very little ozone depletion. These mechanisms are reviewed briefly.

(i) Ozone depletion due to shift in atmospheric dynamics

According to the dynamical theory of atmosphere, stratospheric air tends to circulate from higher altitudes in the tropics to lower altitudes in polar regions, carrying ozone along with it. This circulation depends on the temperature gradient between the tropics and the polar zones. Greater the temperature difference between the two regions, greater will be the circulation of stratospheric air. Before the spring season, the stratospheric air in Northern Hemisphere circulates all the way to North Pole. In the Southern Hemisphere, however, the circulation is impeded from reaching Antarctica due to the polar vortex. Therefore, the ozone hole is created over Antarctica. After the spring is over, the polar vortex breaks down and the ozone-rich air from the non-polar regions rushes in to replenish the ozone hole.

As a result of this natural pheonomenon, the concentration of ozone in non-polar regions decreases temporarily during the time following Antarctic spring. Ozone is a greenhouse gas (Chapter 18, p. 259). That is, it causes a warming of the atmosphere in its vicinity. When the concentration of ozone decreases, the temperature of the stratosphere over the non-polar regions also decreases. This means that the temperature difference between the non-polar and polar regions decreases. Due to this the circulation of stratospheric air weakens. In the years that follow, air carries lesser amounts of ozone to the Antarctica after each spring. In fact, a decrease in the stratospheric air circulation activity has been recorded since 1979 and since that period the size of the ozone hole over Antarctica has consistently increased.

Thus, according to dynamical theory of atmosphere, ozone is not depleted by a chemical process. In fact, it is not depleted at all. It is only redistributed by a geographical phenomenon. However, this theory fails to explain as to how the ozone hole, wherever it occurs, is embedded with above-normal concentrations of reactive chlorine atoms and chlorine monoxide.

(ii) Ozone depletion by oxides of nitrogen

Over the past few years, the atmospheric concentration of nitrous oxide has recorded a consistent increase. This has been attributed to large scale combustion of fossil fuels and enhanced use of nitrogenous fertilizers. Nitrous oxide, as well as other oxides of nitrogen, are also emitted by supersonic transport aircrafts.

Though nitrous oxide is quite inert, in stratosphere it is photochemically converted into more reactive nitric oxide.

$$N_2O + h\nu \rightarrow NO + N \tag{19.16}$$

The nitric oxide so formed then undergoes the following chain reactions.

$$NO + O_3 \rightarrow NO_2 + O_2 \tag{19.17}$$

$$O_3 + h\nu \rightarrow O_2 + O \tag{19.18}$$

$$NO_2 + O \rightarrow NO + O_2 \tag{19.19}$$

$$\text{Net reaction: } 2O_3 + h\nu \rightarrow 3O_2 \tag{19.20}$$

This mechanism has been ruled out because the oxides of nitrogen, rather then depleting ozone, act as scavengers for the ozone depleting reactive chlorine atoms.

(iii) Ozone depletion by hydroxyl radicals

The hydroxyl radicals in the stratosphere can also initiate ozone depletion by the following three mechanisms.

$$HO^{\cdot} + O_3 \rightarrow HO_2^{\cdot} + O_2 \quad (19.21)$$

$$O_3 + h\nu \rightarrow O_2 + O \quad (19.22)$$

$$HO_2^{\cdot} + O \rightarrow HO^{\cdot} + O_2 \quad (19.23)$$

$$\text{Net reaction: } 2O_3 + h\nu \rightarrow 3O_2 \quad (19.24)$$

$$O_3 + h\nu \rightarrow O_2 + O \quad (19.25)$$

$$HO^{\cdot} + O \rightarrow H^{\cdot} + O_2 \quad (19.26)$$

$$H^{\cdot} + O_3 \rightarrow HO^{\cdot} + O_2 \quad (19.27)$$

$$\text{Net reaction: } 2O_3 + h\nu \rightarrow 3O_2 \quad (19.28)$$

$$H\dot{O} + O_3 \rightarrow HO_2^{\cdot} + O_2 \quad (19.29)$$

$$HO_2^{\cdot} + O_3 + h\nu \rightarrow HO^{\cdot} + 2O_2 \quad (19.30)$$

$$\text{Net reaction: } 2O_3 + h\nu \rightarrow 3O_2 \quad (19.31)$$

The hydroxyl radicals ($\dot{O}H$)and the hydroxoperoxy radicals ($HO_2^{\cdot}$) are formed only where plenty of water vapours are present in the atmosphere. However, the concentration of water vapours in polar regions is insignificant. In fact, the hydroxyl and hydroperoxy radicals have not been detected in stratosphere over Arctic or Antarctic regions. Therefore, these three chain mechanisms have been ruled out.

Causes of Ozone Depletion

Since the chlorine-catalysed chain mechanism contributes to most of the ozone depletion, we must identify the source of reactive chlorine atoms in the stratosphere. These atoms are released by the photolysis of chlorofluorocarbons (CFCs). As their name implies, chlorofluorocarbons contain carbon, chlorine and fluorine. First synthesized in 1892, these chemicals were put to industrial use in 1928 (see Box). Besides their low cost of manufacture, three properties of chlorofluorocarbons were responsible for their wide industrial applications: Non-flammability, Non-toxicity and Non-reactivity.

Since the chlorofluorocarbons are non-flammable and non-toxic they are used—

1. As coolants for refrigerators and air-conditioners. Prior to chlorofluorocarbons, ammonia, sulfur dixoide, propane or other gases which are inflammable, explosive and/or toxic were used as coolants.
2. For making polystyrene containers common in food packaging.
3. For fire-fighting. Special types of chlorofluorocarbons, called *halons,* are used as fire extinguishers. Halons are bromine-containing derivatives of chlorofluorocarbons.
   Since the chlorofluorocarbons are unreactive, they are used—
4. As propellants for aerosol sprays. During the Second World War, chlorofluorocarbons were used to spray pesticides to clear unwanted vegetation and to fight malaria.
5. As blowing agents for foam plastics. They have been used to blow-up foam used in furniture, insulation, carpets and containers.
6. As solvents for removing grease from electronic equipments and for cleaning computer chips.
7. In polyurethane foam insulation that fills the walls and doors of refrigerators and other cooling units. The chlorofluorocarbons trapped within the foam contribute to its insulation value.

Ironically, the unreactive nature of the chlorofluorocarbons turns out to be their major drawback. Due to their inertness, these chemicals do not readily degrade in the troposphere. As a result, they eventually find their way into the stratosphere (Fig. 19.3). Here they are photochemically dissociated into reactive chlorine atoms. The latter catalyse the depletion of stratospheric ozone (reactions 19.4 to 19.9). It has been estimated that for every reactive chlorine atom generated in the stratosphere, 100,000 molecules of ozone are depleted.

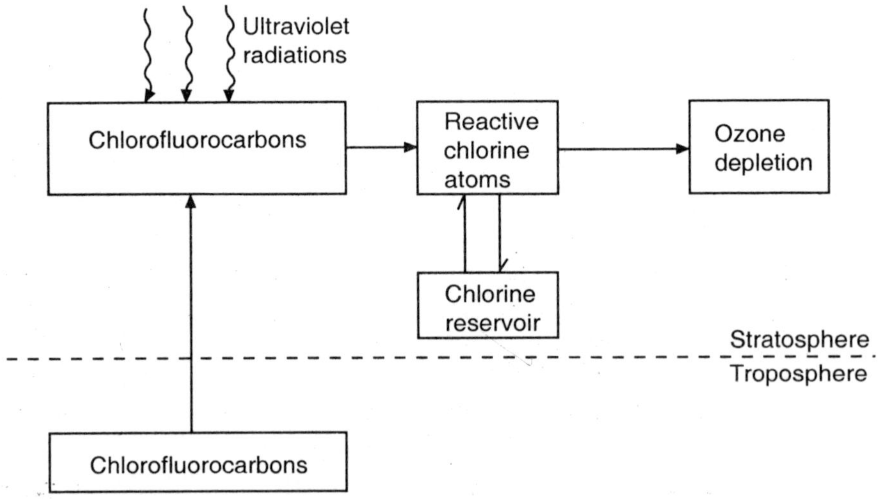

Fig. 19.3 Ozone depletion by chlorofluorocarbons

A Brief History of CFCs

1892–1907: The Belgian chemist Frederic Swarts describes the properties of synthetically produced groups of molecules in which chlorine and fluorine are grouped around the carbon atom—the CFCs.

1929: Thomas Midgley of General Motors has the compounds patented. Technical application begins.

1930: Midgley suggests that trichlorofluoromethane, termed CFC-11, and dichlorodifluoromethane, termed CFC-12, may be used as substitutes for the cooling agents, sulfur dioxide and ammonia, and has the substances patented. General Motors manufacture the "Frigidaire" refrigerator.

1932: The American Carl Iddings uses CFC-12 as a propellant for spray cans.

1936: CFC-11 and CFC-12 enter the market.

1942: The first aerosol insecticide is employed by the U.S. Army in Indo-China to protect soldiers against a bothersome insect plague.

1953/54: Aerosol production begins. CFCs are used in scent, hair and deodorant sprays, shaving creams, perfumes, glues, varnishes and paints.

The seventies: CFC application experiences a boom. The substances are now also used as cleansing agents in the electrical engineering and textiles industries, and as propellants for plastics. Halons are used in fire extinguishers.

1970: The British Scientist James E. Lovelock detects CFC-11 in the atmosphere.

1972: CFC manufacturers all over the world meet in Massachusetts at the *Seminar on the Ecology of Fluorocarbons*.

1974: Sherwood Roland and M.J. Molina, University of California in Irvine, publish their theory that CFCs could be destroying the protecting ozone layer of earth in the stratosphere in the scientific journal *Reviews of Geophysics and Space Physics*.

1978: In USA, a ban is imposed on CFC-11 and CFC-12 in aerosols. An exception in made in the case of pharmaceutical applications. In the European Community, a reduction of 30 percent in aerosols is achieved.

1985: The ozone hole over the Antarctica is discovered. The Vienna Convention on the protection of the ozone layer is agreed.

1987: The Montreal Protocol is adopted as a completion of the Vienna Convention. Forty six countries sign an agreement aimed at a step-by-step reduction of the production and use of fully halogenized CFCs to 50 percent of the 1986 level by 1999.

1988: The E.C. Decree on the introduction of the Montreal Protocol comes into force.

1989: The U.S. Congress passes a CFC taxation scheme: $1.37 per kilogram of CFC related to the *ozone depleting potential*.

The nineties: The internatioal consortia "PAFT" (*Programme of Alternative Fluorocarbon Toxicology Testing*) and "AFEAS" (*Alternative Fluorocarbon Environmental Acceptability Study*), both of which are funded by the CFC manufacturers, look for alternatives. The refrigerator industry starts to switch to HFC-134a as a cooling agent and employs a mixture of semi-halogenized

CFCs as foaming agents.

1990: The stipulations of the Montreal Protocol are tightened up in London. In accordance with its new version, a reduction of CFCs by 50 percent is to be achieved by 1995, while a 100 percent phase-out is to be completed by the year 2000. Article 5 countries are entitled to a grace period of further ten years. The establishment of the Multilateral Fund, which is funded by the industrialized nations, gives developing countries the opportunity to claim financial aid for a phase-out of CFCs.

1991: The CFC-halon ban decree is ratified in Germany. As from January 1, 1995, it is forbidden to produce or use CFCs in Germany.

1992: The East German Company Foron presents the first CFC and HFC-free refrigerator in which pure hydrocarbons are used for cooling and insulating purposes.

1994: In Europe, the employment of hydrocarbons becomes widespread in the refrigerator industry. USA and Japan continue to insist on using HFC-134a.

1995: Three scientists, viz., Professor Paul Crutzen, Professor Mario Molina and Professor E. Sherwood Rowland, awarded Nobel Prize in Chemistry for their pioneering work on ozone depletion by CFCs.

(Reproduced from *VATIS update*, June-July, 1995 with the permission of Asia and Pacific Centre for Transfer of Technology, New Delhi).

Consequences of Ozone Depletion

Ozone absorbs most of the ultraviolet rays from the sun, preventing them from reaching the earth. Reduced amounts of stratospheric ozone thus permit disproportionately larger amounts of ultraviolet radiations to penetrate the atmosphere. Solar ultraviolet radiations are composed of mainly UV-A and UV-B types. UV-A radiations have the wavelength in the range 240–280 nm, while for UV-B radiations, the wavelength varies in the range 280–320 nm. The UV-A radiations are absorbed both by oxygen (reaction 19.1) and ozone (reaction 19.3). However, the UV-B radiations are absorbed only by ozone. Exposure of the earth's environment to UV-B radiations results in more deleterious effects than exposure to UV-A radiations. Moreover, even if the ozone hole is formed, the UV-A radiations are abosrbed by oxygen cover and prevented from reaching the earth. The penetration of UV-B radiations through the atmosphere, due to ozone depletion, has the following environmental implications.

1. Human Health

UV-B radiations are sufficiently energetic to break apart important biological molecules, including proteins and DNA. If an overdose of UV-B rays gets through the ozone hole, it can blitz our exposed skin and eyes.

(a) Skin

Exposure to UV-B radiations causes the connective tissue of the skin to damage. As a result, the skin becomes finely wrinkled and loses its elasticity. If the exposure continues, the skin becomes coarsely furrowed and lumpy. Apart from the connective tissue, the cells in the skin's immune system also get damaged on prolonged exposure to UV-B radiations. These cells process foreign materials and organise responses to evict intruders, such as bacteria and viruses. Thus, a consistent exposure to UV-B radiations results in an increase in the incidence to herpes and leishmaniasis.

It has been estimated that a 1% decline in ozone layer could lead to a 4–6% increase in the incidence of *malignant melanoma,* a form of skin cancer. UV-B radiations have a tendency to destroy melamin-producing cells. Melamine is a brown pigment which acts as a natural sun-block by absorbing ultraviolet rays. Once the level of melamine comes down, the DNA molecules get damaged, leading to mutations which manifest themselves as melanoma. The symptoms of skin cancer include changes in the shape or size of pre-existing moles and formation of dark, irregular and spreading patches.

(b) Eyes

Eyes too are vulnerable to excessive UV-B exposure. The ocular pigment cells, which normally sharpen our sight by reducing reflected light inside the eye, can malfunction when exposed to ultraviolet light. Melanoma of eyes, blindness and even death can result. The UV-B radiations also damage the lens of the eye. The proteins of the eye are replaced naturally only very slowly. Hence, they are prone to accumulating damage. Over a period of time, the normally transparent arrays of proteins that focus light become cloudy, forming a *cataract*—one of the most preventable cause of blindness.

2. Crop Yields

More than two-third of the crop species are damaged by ultraviolet rays. An increased level of UV-B radiations, therefore, adversely affects crop yield. A 1% increase in UV-B radiations is estimated to cause a 1% decline in the yield of soya beans, peas and beans.

3. Phytoplankton

Phytoplankton, the tiny drifting plants that live near the surface of the sea, seeking light for photosynthesis, are vulnerable to UV-B damage. These organisms are the bottom link in the marine food chain and support almost all other forms of life in the sea. There is thus a potential for a major upheaval in the marine ecology, which could, in turn, affect human food sources.

A Cause of Concern?

Ozone depletion occurs largely over Antarctica which is not inhabited by humans. However, Antarctic region has a rich variety of polar animals. Have these animals been affected by ozone depletion? No. There are two reasons for the absence of deleterious environmental consequences originating from ozone depletion. Firstly, the sun is always very low on the horizon at high latitudes. Therefore, sunlight passes through the polar atmosphere at a large slant angle. As a result, low amounts of UV-B radiations reach Antarctic surface, even though ozone depletion is significant. Secondly, the skin of polar fauna is protected by heavy fur or thick feathers. Hence, the damage caused to biological molecules by UV-B radiations is prevented. Yet, their eyes may be vulnerable to increased UV radiations and in course of time these organs would certainly be affected.

The marine waters in the vicinity of Antarctica are the richest ecosystems on the earth. They sustain a wide variety of life by a marine food chain which initiates from phytoplankton. The latter are the microscopic free-floating algae. Fear has been expressed that large doses of UV-B radiations would kill phytoplankton, lesser amounts would retard their growth. If Antarctic phytoplanktons are affected, so will be krill, the tiny animals that feed upon them and then fish, seabirds, seals and whales.

However, it has been observed that ultraviolet radiations do not penetrate very deep into the water column. If the water is turbid or the sky cloudy, the radiations would not go beyond 5 meters. Therefore, the phytoplanktons would not be harmed. Moreover, it has also been observed that phytoplanktons have evolved mechanisms—called *photo-repair mechanisms*—to reverse the damage caused to DNA. They use the UV-A rays to fix what UV-B rays have damaged.

Nevertheless, in non-polar regions, if the ozone depletion occurs to a significant extent, the adverse consequeces arising from increased levels of UV-B radiations would be felt sooner than expected. Therefore, steps leading to abatement of ozone depletion can no logner be delayed.

Abatement of Ozone Depletion

The inert nature of the chlorofluorocarbons imparts them enormous stability. Their residence times in the atmosphere are very long. For example, two most commonly used chlorofluorocarbons, CFC-11 (CCl_3F) and CFC-12 (CCl_2F_2) persist in the environment for about 75 and 110 years, respectively. Thus, even if the release of these chemicals to the atmosphere were to stop with immediate effect, they would continue depleting ozone well into the next century. However, if their release is not stopped, these would eventually accumulate in the stratosphere to a level capable of seriously eroding the fragile ozone shield. Hence, the inescapable conclusion is that only a complete phase-out of the chlorofluorocarbons can prevent further deterioration of ozone layer. However, as we have seen, the chlororfluororcarbons have a wide range of industrial applications. Therefore, if these have to be phased out, their substitutes must be discovered.

Two sets of compounds are being developed as substitutes for chlorofluorocarbons. These are, the *hydrochlorofluorocarbons* (HCFCs) and the *hydroflurorcarbons* (HFCs). Together they are referred to as *alternative fluorocarbons* (AFCs). These compounds have the following advantages.

1. Their ozone depletion potential is far less than that of the chlorofluorocarbons. The relative ozone depletion potentials of chlorofluorocarbons and alternative fluorocarbons are compared in Table 19.2.

2. The alternative fluorocarbons carry hydrogen atoms that make them susceptible to degradation in the lower regions of atmosphere, unlike the fully

Table 19.2 Ozone depletion potential of chlorofluorocarbons, alternative fluorocarbons and related compounds

Code Number	Composition	Ozone Depletion Potential
	Chlorofluorocarbons (CFCs)	
CFC-11	CCl_3F	1.0
CFC-12	CCl_2F_2	1.0
CFC-113	CCl_2FCClF_2	0.8
	Alternative Fluorocarbons (AFCs)	
	Hydrochlorofluorocarbons (HCFCs)	
HCFC-22	$CHClF_2$	0.05
HCFC-123	$CHCl_2CF_3$	0.02
HCFC-124	$CHClFCF_3$	0.02
HCFC-141b	CH_3CCl_2F	0.11
HCFC-142b	CH_3CClF_2	0.06
	Hydrofluorocarbons (HFCs)	
HFC-22	CH_2F_2	0
HFC-125	CHF_2CF_3	0
HFC-134a	CH_2FCF_3	0
HFC-152a	CH_3CHF_2	0
	Others	
—	CCl_4	1.0
—	CH_3CCl_3	0.1

halogenated chlorofluorocarbons, whose long lives let them survive until they reach the stratospheric ozone layer.

3. The physical properties of alternative fluorocarbons are similar to those of chlorofluorocarbons, so that they could be substituted mostly without a major redesign or refitting of existing equipment.

The alternative fluorocarbons also possess the following disadvantages.

1. The ozone depletion potentials of hydrochlorofluorocarbons is not zero. No doubt, it is quite less than the depletion potential of chlorofluorocarbons, yet small amounts of ozone would contine to be depleted if hydrochlorofluorocarbons are used.

2. The ozone depletion potential of hydrofluorocarbons is zero. However, these chemicals possess greenhouse properties and, therefore, have a potential to increase global warming (Chapter 18).

3. Two compounds most likely to replace the chlorofluorocarbons are HCFC-123 ($CHCl_2CF_3$) and HFC-134a (CH_2FCF_3). These compounds have ozone depletion potentials of 0.02 and zero respectively. Their greenhouse potentials are also quite low. Yet, their degradation products are toxic in nature. The degradation pattern of HCFC-123 in the environment is depicted in Fig. 19.4. The products, trifluoroacetic acid and hydrogen fluoride are toxic in nature. The compound HFC-134a also releases trifluoroacetic acid on degradation, as shown in Fig. 19.5. Thus, by using alternative fluorocarbons, although the ozone layer would be safer, other environmental problems are likely to emerge.

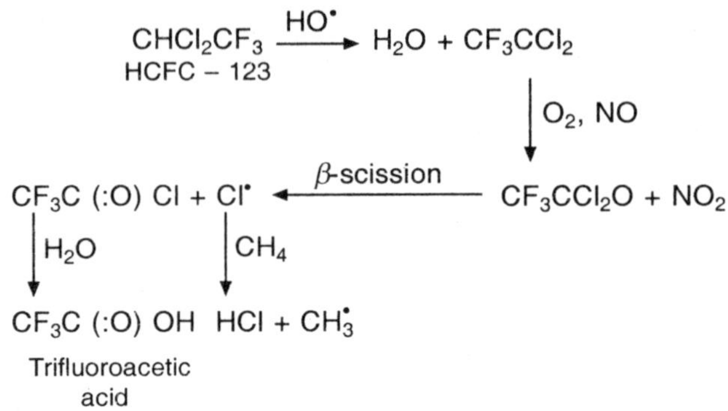

Fig. 19.4 Environmental degradation of HCFC-123

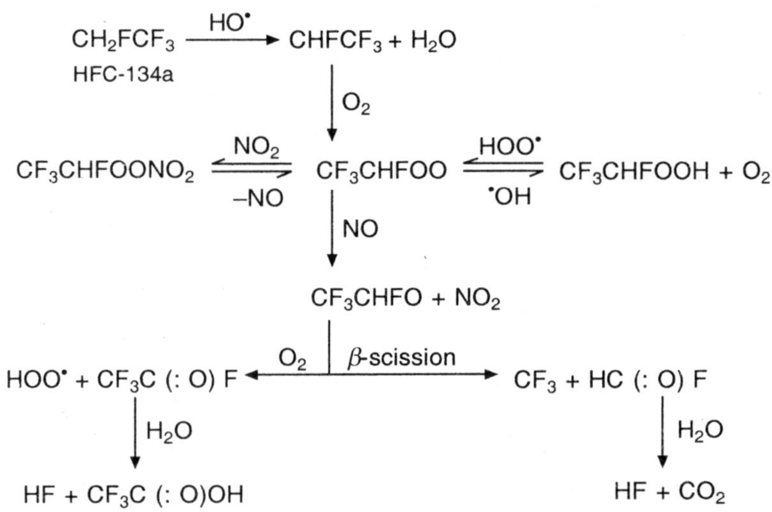

Fig. 19.5 Environmental degradation of HFC-134a

4. The cost of manufacturing alternative fluorocarbons is more than that of chlorofluorocarbons.

5. Certain related chemicals, such as carbon tetrachloride (CCl_4) and trichloroethane (CH_3CCl_3) possess ozone depleting potentials (Table 19.2). Carbon tetrachloride is used as a solvent, as a raw material for manufacture of chlorofluorocarbons and pesticides and as a fire extinguisher. Trichloroethane is used as a solvent. Since these chemicals cause ozone depletion, they too would have to be eventually phased out. However, there are no known substitutes for these compounds.

Thus, the phasing out of chlorofluorocarbons would ultimately turn out to be a two stage process: First from the chlorofluorocarbons to alternative fluorocarbons and the second from the alternative fluorocarbons to such compounds which have zero ozone depleting potential, zero greenhouse potential and low toxicity.

This is going to take a considerably long time. Till then individuals must gear-up to launch their own offensives against extra doses of UV-B radiations A universal weapon in this daily battle is sunglasses to protect the eyes. Sunglasses can cut-off a wide range of ultraviolet light. To protect the skin one needs to wear sun-blocks. These are the garments which are coated with reflective compounds like titanium dioxide. They reflect the impinging UV-B radiations away from the body and thus ward-off the ill effects associated with ozone depletion.

The Montreal Protocol

In September 1987, representatives of several countries met in Montreal and signed an agreement to limit the use of chlorofluorocarbons. India initially refused to sign the treaty for the following reasons.

1. India felt that the Protocol was discriminatory and biased in favour of developed countries. The developed nations consume about 85% of the chlorofluorocarbons; the developing nations only 15%. However, the Protocol demanded that the developing countries should freeze their consumption at the present level, while the developed countries may continue with chlorofluorocarbons for a longer time as their consumption is very high.

India consumes just 8000 tonnes of chlorofluorocarbons against the total world consumption of 10 lakh tonnes. Moreover, its *per capita* consumption of chlorofluorocarbons is just 10 g, compared to 3000 g in the USA. There is thus no justification that India be forced to freeze its consumption and USA allowed to carry it on.

2. Developing countries requested for free know-how to the technology for manufacturing alternative fluorocarbons. This request was turned down.

3. Developing countries suggested setting up of an international fund to finance their switch-over to alternative fluorocarbons. This suggestion was also turned down.

The Montreal Protocol was discussed at various forums. Since India was not a signatory to the Protocol, it was not invited at these discussions. However, in view of its rightful claims and its technological skill to manufacture chlorofluorocarbons and alternative fluorocarbons, India was accorded an observer's status in these meetings. Once this happened many non-signatory, developing countries looked upon India for leadership and guidance. Moreover, many signatory developing countries endorsed India's rightful claims. Finally, in June 1990, the Montreal Protocol was reviewed at a meeting in London. During this meeting, several clauses of the original Protocol were modfied so as to suit the interests of developing nations. The important amendments to the draft were—

1. The developed countries agreed to immediately set up a $240 m fund to be operated by the World Bank on the advice of the United Nations Environment Programme, to finance the developing countries switch-over to chlorofluorocarbons's substitutes.
2. Though the developed countries have to phase out chlorofluorocarbons

by the year 2000, developing countries have been allowed a 10 years grace period. That is, the developing countries can continue to consume chlorofluorocarbons till 2010. The other changes in the phase-out programme are highlighted in Table 19.3.

3. If the developed countries do not fulfil their obligations, the developing countries may withdraw from the Protocol.
4. A provision for a review of the implementation of the Protocol was incorporated.

Table 19.3 The phase-out programme of ozone depleting substances under the Montreal Protocol

Ozone Depleting Substances	Phase-out Programme	
	Original Protocol Montreal, 1987	Amendments London, 1990
Chlorofluorocarbons	50% reduction on 1989 production levels by 1999	Complete phase-out by 2000 for developed countries, and by 2010 for developing countries
Alternative fluorocarbons	No controls	Informal agreement to phase-out these chemicals by 2020–2040
Others	No controls	Carbon tetrachloride: 85% reduction by 1995; phase-out by 2000. Trichloroethane: 70% reduction by 2000; phase-out by 2005.

Since these amendments were quite favourable, India, subsequent to the London Meet, signed the Montreal Protocol. As a result, India was financed to carry out commercial production of alternative fluorocarbons by the Montreal Protocol Fund. Under this programme, the know-how for the production of HFC-134a has been developed at Indian Institute of Chemical Technology, Hyderabad.

It must be appreciated that the world moved fast to agree to phase out the chlorofluorocarbons once their dangers were demonstrated. This offers a hopeful precedent for international action on other threats to the global environment.

References

Cicerone, R.J. (1987). Changes in stratospheric ozone, *Science,* 237, 35–42.

Cicerone, R.J. (1994). Fires, atmospheric chemistry and the ozone layer, *Science,* **263,** 1243–1244.

Kerr, R.A. (1987). Halocarbons linked to ozone hole, *Science,* **236,** 1182–1183.

Kerr, R.A. (1993). Ozone takes a nose dive after the eruption of Mr. Pinatubo, *Science,* **260,** 490–491.

Lee, J. (1998). Our health in flux, *New Scientist (Inside Science),* December 12, p. 1–4.

Molina, M.J. and Rowland, F.S. (1974). Stratospheric sink for chlorofluoromethanes: Chlorine atom catalysed destruction of ozone, *Nature (London),* **249,** 810–812.

Molina, M.J., Tso, T.L., Molina, L.T. and Wang, F.C. –Y. (1987). Antarctic stratospheric chemistry of chlorine nitrate, hydrogen chloride, and ice: Release of active chlorine, *Science,* **238,** 1253–1257.

Roberts, L. (1989). Does the ozone hole threaten Antarctic life? *Science,* **244,** 288–289.

Sander, S.P., Friedl, R.R. and Yung, Y.L. (1989). Rate of formation of ClO dimer in the polar stratosphere: Implications for ozone loss, *Science,* **245,** 1095–1098.

Solomon, P.M., de Zafra, R., Parrish, A. and Barrett, J.W. (1984). Diurnal variations of stratospheric chlorine monoxide: A critical test of chlorine chemistry in the ozone layer, *Science,* **224,** 1210–1214.

Stolarski, R.S. (1988). The Antarctic ozone hole, *Sci. Amer.,* **258**(1), 20–26.

Tolbert, M.A., Rossi, M.J. and Golden, D.M. (1988). Antarctic ozone depletion chemistry: Reactions of N_2O_5 with H_2O and HCl on ice surfaces, *Science,* **240,** 1018–1021.

Tolbert, M.A. (1994). Sulfate aerosols and polar stratospheric cloud formation, *Science,* **264,** 527–528.

Zurer, P.S. (1988). Studies on ozone destruction expand beyond Antarctic, *Chem. Eng. News,* May 30, p. 16–25.

Zurer, P.S. (1993). Looming ban on production of CFCs, halons spurs switch to substitutes, *Chem. Eng. News,* November 15, p. 12–18.

CHAPTER 20

Suspended Matter

The term suspended matter refers to the suspension of finely divided solid particles or droplets of liquid in air. Such microscopic suspensions are also called *aerosols, particulate matter* or simply *particulates*. Their diameters lie in the range 0.001–100 μm (Table 20.1). Despite their extremely small size, particulates are known to cause several physical, chemical and physiological adversities. In fact, historically, particulate pollution is the oldest known air pollution problem. The problem is compounded by the fact that several other toxic components of the atmosphere become adhered to the particulates and produce synergistic effects that are more harmful than the individual components. A further cause of concern is that sometimes the original suspended matter may undergo atmospheric reactions which transform it into secondary particulates that are many times more obnoxious.

Table 20.1 Diameter range of some common particulates

Particulate Source	Diameter Range (μm)
Metallurgical refining	0.001–100
Carbon black	0.01–0.8
Combustion nuclei	0.01–1.10
Tobacco smoke	0.01–1.0
Smog	0.01–1.2
Sea salt nuclei	0.02–0.18
Aitken nuclei	0.05–0.15
Oil smoke	0.05–1.0
Paint pigment	0.10–10
Cement dust	0.50–100
Insecticide dust	0.80–10
Fly ash	0.80–100
Coal dust	1.00–100
Milled flour	1.00–100
Pollen	10–100

Types of Particulates

On the basis of their size and origin, suspended matter is classified into two types: Primary particulates and secondary particulates.

Primary Particulates

Primary particulates are those which are directly injected into the air by physical or chemical means. These are usually 1–20 μm in diameter, although in some cases the size may be as large as 100 μm. Examples of primary particulates from natural sources include pollen, soil dust and volcanic debris. Sea salt spray is also a natural primary particulate matter, although in this case the diameter is less than 1 μm. Anthropogenic primary particulates include coal and cement dust, fly ash and milled flour.

Secondary Particulates

Particulates produced in the atmosphere as a result of chemical reactions are called secondary particulates. These particulates are formed either in gas phase reactions or by heterogeneous reactions between contaminant gases and already existing primary particulates. The majority of secondary particulates possess a diameter of less than 1 μm.

The reactions leading to the formation of secondary particulates are largely catalysed by sunlight. The chief chemical reactants are sulfur oxides, nitrogen oxides, ammonia, water vapours and hydrocarbons. These constituents may enter the atmosphere from either natural sources or man-made sources. Natural phenomena contributing to the formation of secondary particulates include volcanoes and vapours from trees and decaying organic matter. Anthropogenic secondary particulates result from vehicular emissions, combustion processes and industrial operations. Photochemical smog is made up of secondary particulates.

Secondary particulates may be further subdivided into three types, based on sulfate, hydrocarbon and nitrate functional groups.

1. Sulfate particulates

Sulfate particulates constitute the largest fraction of the total tropospheric aerosol load. They are found everywhere, from the continents and the oceans to the polar regions.

Sulfate particulates include sulfuric acid droplets, ammonium sulfate and other metal sulfates. Ammonium sulfate is formed by the reaction between sulfuric acid and atmospheric ammonia. Other sulfate salts, like lead sulfate, are formed by the reaction between sulfuric acid and the corresponding metal oxide. Sulfuric acid itself is formed by the oxidation of atmospheric sulfur dioxide. There are four basic mechanisms which transform gaseous sulfur dioxide into sulfuric acid droplets.

(a) *Oxidation by oxygen atoms.* Sulfur dioxide may be oxidised by reactive oxygen atoms to sulfur trioxide in a 'three-body' reaction.

$$SO_2 + O + M \rightarrow SO_3 + M \tag{20.1}$$

where M is a molecule of nitrogen or oxygen (see equation 14.8, p. 219). Sulfur trioxide is hydrolysed to sulfuric acid by atmospheric water vapours (see equation 14.9, p. 219).

$$SO_3 + H_2O \rightarrow H_2SO_4 \tag{20.2}$$

The reactive oxygen atoms in reaction (20.1) are supplied by nitrogen dioxide, which photodissociates on absorbing sunlight in the wavelength range 300–360 nm.

$$NO_2 + h\nu \rightarrow NO + O \tag{20.3}$$

This mechanism is significant is producing sulfate aerosols in urban areas, where the concentration of nitrogen dioxide is quite high.

(b) *Direct photochemical oxidation.* In this mechanism sulfur dioxide absorbs sunlight in the wavelength range 240–320 nm and is transformed from the ground state (singlet) to the excited state (triplet).

$$SO_2 + h\nu \rightarrow SO_2^* \tag{20.4}$$

Excited sulfur dioxide combines with oxygen in a 'three-body' combination giving sulfate radicals.

$$SO_2^* + O_2 + M \rightarrow SO_4^{\cdot} + M \tag{20.5}$$

Sulfate radicals are converted to sulfur trioxide through reaction with sulfur dioxide,

$$SO_4^{\cdot} + SO_2 \rightarrow 2SO_3 \tag{20.6}$$

or molecular oxygen.

$$SO_4^{\cdot} + O_2 \rightarrow SO_3 + O_3 \tag{20.7}$$

Sulfur trioxide is subsequently hydrolysed to sulfuric acid droplets.

In rural areas, with relatively clean air, photochemical oxidation of sulfur dioxide is the primary mechanism of sulfate aerosol formation.

(c) *Catalytic oxidation.* In this mechanism, sulfur dioxide is first converted to sulfurous acid which is then oxidised to sulfuric acid by atmospheric oxygen.

$$SO_2 + H_2O \rightarrow H_2SO_3 \tag{20.8}$$

$$2H_2SO_3 + O_2 \rightarrow 2H_2SO_4 \tag{20.9}$$

The oxidation reaction (20.9) is catalysed by metal oxides. Ferric oxide is particularly effective and is able to carry the reaction (20.9) through even in the absence of sunlight. This accounts for the formation of sulfuric acid aerosols during night.

(d) *Free radical reactions.* In this mechanism sulfur dioxide is directly converted to sulfuric acid aerosols by hydroxyl radicals.

$$SO_2 + 2\dot{O}H + M \rightarrow H_2SO_4 + M \tag{20.10}$$

The hydroxyl radicals are formed from hydrocarbon-nitrogen dioxide-sulfur dioxide mixture. This mixture is a synergistic system. No two components react effectively, but the presence of all three produces hydroxyl radicals. This mechanism is largely operative in urban areas, especially during peak hours, when the concentrations of sulfur dioxide, nitrogen dioxide and hydrocarbons are quite high.

2. Hydrocarbon particulates

In urban areas the main sources of hydrocarbons are automobile exhaust, power generating stations and industrial effluents. In rural areas trees discharge terpene vapours to the atmosphere. The most important terpene emitted by vegetation is α-pinene (Fig. 20.1). Terpenes are unsaturated hydrocarbons, possessing olefin skeletons. These are easily oxidised by ozone to give peroxy radicals.

$$O_3 + \text{Olefin} \rightarrow \underset{\text{Peroxy radicals}}{RO_2^{\cdot}} + O \quad (20.11)$$

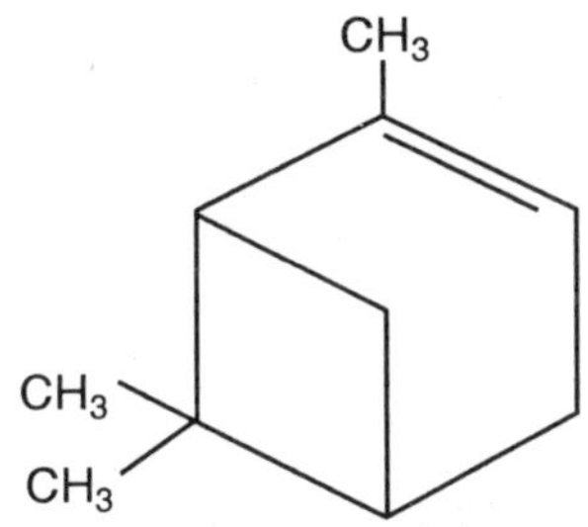

Fig. 20.1 α-**Pinene**

Peroxy radicals are the constituents of photochemical smog. In addition, aldehyde, ketone and epoxide particulates are produced by atmospheric oxidation of olefins.

3. Nitrate particulates

The key intermediary in the formation of nitrate particulates is nitric acid, which itself is produced from nitrogen dioxide by either of the following reactions.

$$4NO_2 + O_2 + 2H_2O \rightarrow 4HNO_3 \quad (20.12)$$

$$NO_2 + \dot{O}H + M \rightarrow HNO_3 + M \quad (20.13)$$

After the formation of nitric acid, nitrate aerosols are produced by its reaction with either gaseous constituents of the atmosphere,

$$NH_3 + HNO_3 \rightarrow NH_4NO_3 \quad (20.14)$$

or with metal oxide particulates.

$$PbO + 2HNO_3 \rightarrow Pb(NO_3)_2 + H_2O \quad (20.15)$$

In coastal areas, sea salt sprays are often converted into nitrate particulates.

$$NaCl + HNO_3 \rightarrow NaNO_3 + HCl \quad (20.16)$$

Sources of Particulates

Both the natural and man-made sources contribute to the airborne particulate matter.

Natural Sources

There are a variety of natural sources of atmospheric suspended matter. The important ones are discussed here.

1. Meteorites

Every day about 8 billion solid particles enter the atmosphere from the outer space. The majority of these are so small that they burn-up or disintegrate in their swift passage through the atmosphere. The larger ones, which number about 2000 per day, reach the surface of the earth in form of meteorite dust. Subsequently, they spread out from the peaks of mountains to the sediments of the oceans.

2. Volcanoes

Volcanic eruptions inject enormous amounts of particulate matter into the troposphere and stratosphere. The aerosols loading of some important volcanic eruptions are listed in Table 20.2.

Table 20.2 Aerosol load emitted by some important volcanoes

Volcanoes	Year of Eruption	Aerosol Load ($\times 10^{12}$g)
Tambora	1815	100
Krakatau	1883	50
Katmai	1912	20
Agung	1963	16–30
El Chichon	1982	12
Pinatubo	1991	30

The volcanic debris has a potential to travel several thousand kilometers from the site of eruption and damage the ecosystems of far flung places. Volcanoes also emit sulfur dioxide gas to the atmosphere. Subsequently, sulfur dioxide is converted into sulfate aerosols, increasing the particulate load.

The June 15, 1991 eruption of Mt. Pinatubo in Philippines discharged about 30×10^{12} g of suspended matter and about 20×10^{12} g of sulfur dioxide into the atmosphere. This is believed to be the largest aerosol perturbation to the atmosphere in the last century. The bulk of the suspended matter was concentrated at a height of 20–27 km above the earth's surface. However, it did not remain stationary over the site of eruption. Rather, it began encircling the globe and made its presence felt at various places.

3. Vegetation

Spores and pollens, being very light, are found in both troposphere and stratosphere. The terpenes, especially α-pinene (Fig. 20.1), is discharged from forest trees as minute aerosols. It causes a blue haze in the atmosphere due to dispersion of sunlight.

4. Storms

Duststorms and sandstorms lift particulates—both large and small—and cause them to spread out over vast areas. The turbulent conditions created by storms interfere with the natural settling phenomena of particulates. Thus, the particulates remain suspended in the air for a longer time.

5. Sea salt sprays

Ocean waves inject salt laden water droplets into the air. When the water evaporates, it leaves behind the salt particles in the atmosphere. The atmospheric load of sea salt sprays is especially high in the coastal areas.

6. Forest fires

The frequent occurrence of forest fires in yet another important natural source of particulates. In September 1997, the haze caused by forest fires in Indonesia was so severe that the authorities declared a state of emergency that lasted for ten days and prompted the closure of schools and offices. In May 1998, forest fires in Mexico created a similar haze by the dispersal of particulates.

The suspended matter emitted during the forest fire episodes consists of polynuclear aromotic hydrocarbons (for example napthalene, phenanthrene, anthracene, pyrene) and salts of calcium, iron, zinc and magnesium. In addition, gaseous pollutants like sulfur oxides, nitrogen oxides and ozone are also emitted during forest fires. However, these gases diffuse away rapidly from the source and do not represent any hazard other than to fire-fighters and those in close proximity to fire. Particulates, on the other hand, do not diffuse so easily, causing problems at the regional level.

Man-Made Sources

The amount of particulates discharged to the atmosphere by man-made sources is comparable to that emitted by natural sources. The important anthropogenic sources of particulates are now discussed.

1. Vehicular exhaust

Transportation systems contribute only about 1% of the total mass of particulates. However, their impact is far more than natural dust because the transport related particulates are very minute. They, therefore, take a longer time to settle down. Moreover, they are easily carried away by air and, therefore, spread from urban to rural areas. In addition, as we shall see later, the smaller particles cause greater adversities.

Particulates emitted from automobile exhaust consist of soot, lead compounds, crankcase oil and hydrocarbons. These interact with other air pollutants, especially nitrogen oxides, to produce secondary particulates which, in turn, produce photochemical smog. It has been observed that diesel engines emit more particulate matter than gasoline engines.

2. Combustion processes

Combustion processes produce particulates having a diverse chemical composition and a wide range of size. Sulfur oxides and nitrogen oxides are the main components of combustion processes. They are converted into sulfate and nitrate aerosols, respectively. Incomplete combustion of fossil fuels produce soot particles. The latter have a high potential to adsorb toxic gases and metallic oxides. Several industrial combustion processes produce fly ash, which is a mixture of the oxides of silicon, aluminium, calcium and phosphorus.

3. Metallurgical operations

Extraction processes emit particulates in form of metallic dust and metal oxides. These particulates provide catalytic surface for unwanted atmospheric reactions.

Fate of Particulates in the Environment

We have seen that both the natural and man-made sources contribute to the airborne particulate concentration. Once in the atmosphere, the particulates spread out far from the land of their origin. For example, dust from Sahara desert in Africa is known to reach Caribbean in central America. However, the sea salt sprays do not spread out very far inland or to high altitudes. They absorb water droplets from the humid sea air. Their weight increases and therefore they settle down rapidly under the influence of gravity.

The main sink for all the atmospheric particulates is precipitation, that is, settling down under the influence of gravity. The rate of precipitation follows the Stokes law equation,

$$V_0 = gd^2(\rho_1 - \rho_2)\ 18\eta \tag{20.17}$$

In equation (20.17), V_0 is the velocity by which a particulate of diameter d and density ρ_1 settles down through an air column of density ρ_2 and viscosity η (g is the acceleration due to gravity).

It is evident from equation (20.17), that the particles having large diameters and high densities settle down rapidly under the influence of gravity. Conversely, small, light particles take longer periods to precipitate. It has been calculated that particulates having diameters around 0.1 μm and density close to 1 g cm^{-3}, settle down with a velocity of about 10^{-6} ms^{-1}. This corresponds to a descent of about 32 m in one year.

However, it is not always possible to assess the rate of particulate descent very accurately. The reason being that Stoke's law is not strictly applicable to particulates. This law is ideally valid for spherical bodies. The particulates often possess irregular shapes. Moreover, wind movement, turbulence and thermal conditions lower the rate of precipitation of particulates. On the other hand rainfall increases the rate of precipitation. Both the small and large particulates are quickly washed out of the atmosphere by rainfall.

Another factor which increases the rate of precipitation is the condensation of particulates. During their random motion in the atmosphere, the particulates collide with each other and condense into larger particles which settle down relatively rapidly. Some of the particles disappear from the air by impinging on objects like trees and buildings.

Analysis of Particulates

In order to assess the extent of air pollution by particulates, one of the most widely accepted method is to determine the amount of dust that settles from the atmosphere or that is washed from the air by rainfall. The determination of dustfall is made gravimetrically on samples that have been collected for a given period of time.

A convenient equipment for collecting dustfall is shown in Fig. 20.2. It consists of a perforated, polished aluminium box containing a polyethylene cup for collecting dust. The vents in the box are provided for the drainage of rain water.

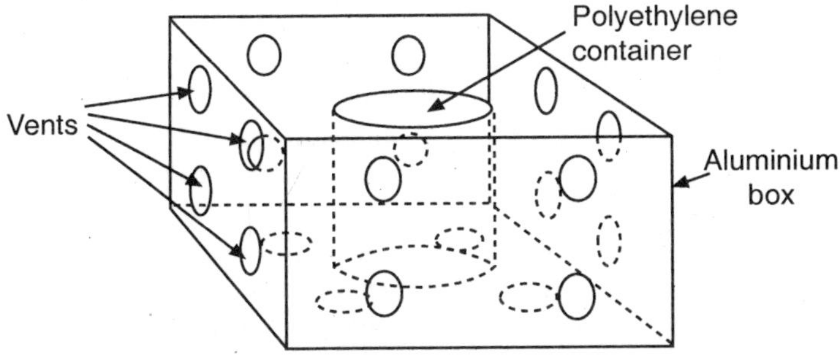

Fig. 20.2 Gravimetric estimation of dustfall

In areas where wind velocity is high, the inside of the polyethylene container is coated with grease. The dust sticks to the grease and is not lost by wind action. In very cold places, a volatile anti-freeze like isopropanol is added to the dust collector, so as to avoid the freezing of rainwater or moisture. About 2 mg of quaternary ammonium chloride is added as an algicide. This ensures that, over a period of time, algae blooms will not crop up in the container.

The aluminium box is placed in an open place. After one month, the polyethylene container is removed and exposed to an infrared lamp. This treatment eliminates any moisture which may have adhered to the collector. The polyethylene cup is then heated to 110°C for one hour. The weight of the particulates in the container is determined. The amount of dust, X, in milligrams per square centimeter per month is given by,

$$X = \frac{a}{\pi b^2} \tag{20.18}$$

where a is the weight of the dust in milligrams and b is the radius of the container in centimeters.

It is sometimes advantageous to differentiate between water-soluble and water-insoluble particulates. In order to achieve this, the total dust collected in the polyethylene container is washed with a jet of hot, distilled water. The washings are quantitatively filtered in an evaporating dish. Subsequently, water is evaporated from the filtrate by heating. The sample is dried by further heating at 110°C for one hour. The dried sample is weighed. The amount of water-insoluble dust, Y, is calculated from the difference,

$$Y = a - c \tag{20.19}$$

where c is the weight of water-soluble particulates.

Effect of Particulates

The suspended matter in the atmosphere produces a host of adverse effects. It aggravates the deleterious effects of other air pollutants. The problem is further aggravated by the fact that the particulates sometimes travel across national boundaries, eliciting harmful effects many thousand kilometers from their place of origin.

1. Human Health

Particulates are known to increase morbidity and mortality rates. As a result, a standard of 75 ppm has been established as an acceptable upper limit for the annual average of total suspended particulates in the atmosphere in any given region. For a 24 h period, the maximum tolerable limit of particulates has been fixed at 260 ppm.

The ultrafine small particulates are more harmful to human health than are the large particulates. The particulate matter having a diameter less than 10 μm, called PM_{10}, is especially hazardous. The PM_{10} s are capable of bypassing the body's respiratory filters and penetrating deep into the lungs. Once in the lungs, they cause greater damage than gaseous pollutants, since the latter are more easily expelled. Chronic respiratory diseases, such as bronchitis, bronchial asthma and pulmonary edema are caused by deposition of PM_{10} s in the lungs.

The PM_{10} s are harmful in yet another way. Because of their small size, they possess large surface area. As a result, other contaminants become adsorbed on them and are carried to the respiratory tract. The respiratory problems, therefore, become aggravated. For example, it has been observed that physiologically injurious effects of sulfur dioxide are more severe in the presence of particulates. The synergistic effect of sulfur dioxide-particulates combination was noted during London smog of December 5–9, 1952, when an estimated 4000 people died mainly due to respiratory problems.

At least eight people died during the September 1997 episode of forest fires in Indonesia and many more suffered upper respiratory tract infections, asthma, bronchitis and breathlessness. These ailments were traced to the presence of PM_{10}s in the haze. The most seriously affected were the elderly and those already suffering from respiratory and cardiovascular diseases.

The PM_{10} s in the nasal cavity and the tracheobronchial section may also be swallowed and passed to the gastrointestinal tract. Toxic substances adsorbed on the particulates may be transferred from any of these reservoirs into the blood stream and then transmitted to other organs via the circulatory system. Long term effects on organs other than respiratory system then follow. For example, it has been found that PM_{10} s from wood smoke, containing aldehydes and polycyclic aromatic hydrocarbons, have the potential to suppress the body's immune system. The preschoolers are especially susceptible to this type of hazard.

Women exposed to high levels of PM_{10} s have been observed to give birth to children with smaller heads and bodies. Such children are slow to learn and run an increased risk of cancer. PM_{10} s, particularly those coated with polycyclic aromatic hydrocarbons, damage the DNA of the growing fetus. The latter has no repair mechanism to reverse the damage.

Most of the secondary particulates fall in the PM_{10} range. Hence, it may be safely generalized that secondary particulates are more hazardous to human health than are the primary particulates.

2. Visibility

Visibility or visible range refers to the distance at which an object can be just perceived against the horizon sky. The airborne particulates scatter the sunlight and reduce the light flux. As a result, the contrast between the objects and their background gets diminished and visibility becomes impaired. The fraction of light scattered by any one particle is called the *single scattering albedo,* w_0.

The single scattering albedo depends upon the diameter of the particulates. The aerosols having the size range 0.1–1.0 μm have a high value of w_0, since their diameters are comparable to the wavelength of visible light. Hence, they cause a significant reduction in visibility.

On an average, a typical urban area has a particulate concentration of 100 ppm. Because of such heavy load of particulates, one out of every four photons of visible light, vertically traversing an urban atmosphere, encounters an aerosol before reaching the ground, while a photon horizontally crossing a city may travel only about 10 km before intercepting a particle. Hence. the visibility in urban areas in only 8–10 km. On the other hand, in rural areas the particulate level is just about 20 ppm. The visibility range in rural areas is of the order of 50–60 km.

3. Odour

Particulates are involved in the transmission of unpleasant odours. Some aerosols are volatile and therefore their odour spreads from one place to another. Examples of such aerosols include hair sprays, paints and deodorants. Some aerosols are non-volatile. They themselves do not possess any odour, but they serve as a transport medium for volatile matter which possesses odour. For example, odours from incinerators, open dump fires and food products are transmitted by particulates. If particulate matter is removed from diesel exhaust, the odour of the fumes is significantly reduced.

4. Materials

Particulate matter has damaging effects on buildings, metallic structures and fabrics. The corrosiveness of particulates is due either to their own chemical activity or to their ability to accelerate the obnoxious effects of other gaseous pollutants like the oxides of sulfur and nitrogen.

The acidic aerosols, such as sulfate and nitrate, attack limestone, sandstone, cement and other masonry. They, therefore, accelerate the deterioration of buildings. Soot deposited on urban buildings imparts an unaesthetic appearance. Airborne particulates also hasten the corrosion of metals. Even the structures made of steel are prone to rapid disintegration in presence of particulates. They also catalyse the erosion of metal coatings and painted surfaces. The deposition of grime, dust and soot on fabrics lowers their texture and life span. The cleaning requires a large quantity of solvents and detergents, which further add to the pollution cycle.

5. Global Climate

As shown in Fig. 20.3, the particulate matter is largely concentrated in the first two kilometer region above the earth's surface. Of these, the bulk consists of secondary particulates. The secondary particulates, especially the sulfate aerosols, can efficiently scatter the solar radiation back into the space. Solar rays heat up the earth's surface. If they are scattered back into the space, the atmosphere cools down. The secondary particulates thus show anti-greenhouse characteristics.

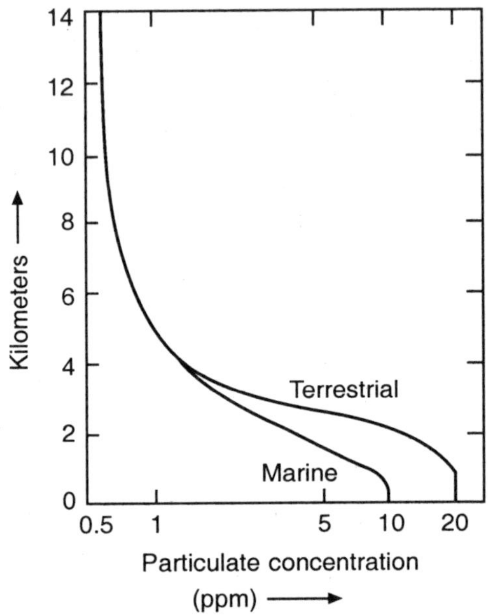

Fig. 20.3 Particulate concentration as a function of altitude

The sulfate aerosols produce the cooling effect by yet another phenomenon. They act as *cloud condensation nuclei,* that is, they trigger cloud formation. Greater cloud cover means greater rainfall and cooler atmosphere.

The primary dust particles are largely concentrated in the stratosphere. The reason being that the main emitter of dust are the volcanoes. The dust from the volcanic eruptions is ejected with such a force that it accumulates in the stratosphere. The dust particles in the stratosphere can both scatter and absorb solar radiation. We have seen that scattering effect produces atmospheric cooling. The absorption of solar radiation does the same. The dust particles in the stratosphere absorb heat energy from the sun's rays. The stratosphere, no doubt, becomes warmer, but a curtailed heat flux reaches the earth's surface, causing a drop in temperature of the troposphere.

The net effect of sulfate aerosols and dust particles ejected by Mt. Pinatubo eruption has been a lowering of average atmospheric temperatures from mid-1992 onwards. Wherever the volcanic plume traversed, the mercury plummeted. A similar drop in the mean atmospheric temperature was observed in 1964, the year following the eruption of Mt. Agung in Bali. The 1815 eruption of Mt. Tambora in Indonesia was followed by two extremely cold years. In fact, climatologists call 1816 as the *year without a summer*!

6. Ozone Depletion

We have stated that most of the sulfate particulates are concentrated in the troposphere. However, some of these reach the stratosphere. There, they catalyse ozone depletion.

As mentioned in Chapter 19, ozone in the stratosphere filters the harmful ultraviolet rays, called UV-B rays, and prevents them from reaching the earth's surface. If the atmospheric ozone level falls, the UV-B radiations would penetrate through the ozone hole, to the earth's surface, causing skin burns and skin cancer, cataract and blindness, damage to body's immune system and fall in crop productivity.

Before the eruption of Mt. Pinatubo, ozone hole had appeared over Antarctica and, to some extent, over Arctic region. In these regions special types of clouds, called *polar stratospheric clouds,* provide surface for the atmospheric reactions that lead to ozone depletion.

Polar stratospheric clouds are not formed in non-polar regions. Hence, it was assumed that the harmful effects of ozone loss would be restricted to polar regions alone. Here too the threat was not considered very serious. Neither are these cold deserts inhabited, nor do they support any plant life. The marine animals are protected from the UV-B rays by sea water. The land animals too are not affected by UV-B rays because nature has provided them with thick furs or feathers. Although these coverings were meant to sustain the polar fauna in extremely cold and hostile environment, they are now affording protection against UV-B radiations as well.

It has been observed that, like polar stratospheric clouds, sulfate aerosols can also provide surface for the ozone depleting reactions. Unfortunately, sulfate aerosols are found even over densely inhabited tropical regions. A 6–8% loss of stratospheric ozone was recorded over the Equator following Mt. Pinatubo eruption. If the concentration of sulfate aerosols increases, the ozone level would fall further. Nobody would be safe in such an event because nature has not provided us with means to dodge the harmful ultraviolet rays.

Control Measures for Particulate Pollution

Electrostatic precipitators are commonly used to control the level of particulates in the atmosphere. An electrostatic precipitator is made up of a series of flat, parallel collector plates spaced 15 to 30 cm apart with a wire discharge electrode mounted between each pair of plates (Fig. 20.4). A strong field is put across the wires and plates. The particulate-laden flue gas from a combustion source passes between the plates. An electirc discharge ionizes the constituents of flue gases, such as carbon dioxide, nitrogen oxides and sulfur oxides. These ions then impart electrostatic charges to the particulates. The charged particulates are pulled by the electrostatic field force to the oppositely charged collecting electrodes, which are periodically shaken to dislodge the accumulated particulate matter. The particulate load falls into a hopper beneath the electrodes, from where it is removed and transported to a landfill.

Collection efficiencies of 99% or greater have been achieved by electrostatic precipitators. Efficiency increases with the time the flue gas spends in the precipitator and with the strength of the electric field. Particulates with diameters greater than 0.005 μm can be collected by this technique. The electrostatic precipitators are reasonably compact and have a low maintenance cost.

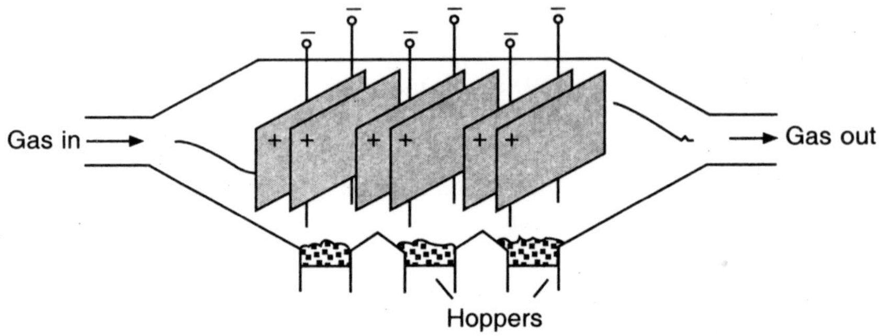

Fig. 20.4 An electrostatic precipitator

References

Baldwin, R., Pollack, J.B., Summers, A., Toon, O.B., Sagan, C. and Van Camp, W. (1976). Stratospheric aerosols and climate change, *Nature (London),* **263,** 551–555.

Charlson, R.J., Schwartz, S.E., Hales, J.M., Cess, R.D., Coakley, J.A., Hansen, J.E. and Hofmann, D.J. (1992). Climate forcing by anthropogenic aerosols, *Science,* **255,** 423–430.

Edwards, R. (1996). Smog blights babies in the womb, *New Scientist,* October 19, p. 4.

Ellsaesser, H.W. (1972). Turbidity of the atmosphere: Source of its background variation with the season, *Science,* **176,** 814–815.

Fennelly, P.F. (1976). The origin and influence of airborne particulates, *Amer. Sci.,* **64,** 46–56.

Lee, R.E. (1972). The size of suspended particulate matter in air, *Science,* **178,** 567–575.

Maxwell, K.E. (1973), *'Environment of Life',* Dickenson Publishing Co., Inc., California, p. 155–158.

McCormick, M.P., Thomason, L.W. and Trepte, C.R. (1995). Atmospheric effects of the Mt. Pinatubo eruption, *Nature (London),* **373,** 399–400.

Radojevic, M. (1998). Burning issues, *Chem. Brit.,* **34,** 38–42.

Sassen, K. (1992). Evidence for liquid-phase cirrus cloud formation from volcanic aerosols: Climatic implications, *Science,* **257,** 516–518.

Schneider, S.H. and Mass, C. (1975). Volcanic dust, sunspots and temperature trends, *Science,* **190,** 741–746 (1975).

Shaw, R.W. (1987). Air pollution by particles, *Sci. Amer.,* **257**(2), 84–91.

Stone, R. (1995). Environmental toxicants under scrutiny at Baltimore meeting, *Science,* **267,** 1770–1771.

Toon, O.B. and Pollack, J.B. (1980). Atmospheric aerosols and climate, *Amer. Sci.,* **68,** 268–278.

Toumi, R., Bekki, S. and Law, K.S. (1994). Indirect influence of ozone depletion on climate forcing by clouds, *Nature (London),* **372,** 348–351.

West, P.W. (1976) in, *'Air Pollution',* Vol. III, 3rd Edn., Stern, A.C. (Ed.), Academic Press, New York, p. 51–97.

Part V

Water Pollutants

CHAPTER 21

Classification of Water Pollutants

Water pollution is defined as the addition to water of an excess material or heat that is harmful to the living organisms or which impairs the beneficial uses of water. Pollution makes water physically impure, foul or filthy. It changes the natural qualities of water, so that it becomes unsuitable for the uses to which it is normally put to. Pollutants may be present in water either as suspended particles or as dissolved compounds or both. The dissolved compounds are more difficult to remove and hence these taint the water resources more seriously than the suspended impurties.

Unique Characteristics of Water

Water has a number of properties that are of critical importance to life and the environment. Water is the only naturally occurring inorganic liquid. It is also the only chemical on this planet that occurs naturally in all three states of matter—solid, liquid and gaseous.

Water and the Living Environment

The origin of water preceded the evolution of life. In fact, the reactions that make up life, such as the synthesis of proteins and nucleotides, occurred in the aqueous medium. It is for this reason that water forms a part of biological structures. For example, water plays an important role in maintaining protein structure. Water molecules are bound to the protein chains at varied locations through hydrogen bonding. They hold up the proteins in their peculiar folded conformations. Without the folded conformations, the proteins would have been unable to perform their specific functions.

An adult human has a water content of 65–70%. Table 21.1 shows that a normal adult person consumes two and a half liters of water every day and loses an equal amount. The daily gain of 350 ml water, due to glucose oxidation, results from the reaction,

$$C_6H_{12}O_6 + 6O_2 \rightarrow 6CO_2 + 6H_2O \quad (21.1)$$

This reaction liberates about 8000 KJ of energy which is utilized by the body for performing essential work. Had this energy been dissipated as heat, the body temperature would have risen by 26° and would have resulted in death.

Table 21.1 Daily balance sheet of water of an adult

Gain (ml)	Loss (ml)
Drinks: 1300	Respiration: 400
Food: 850	Evaporation: 500
Glucose oxidation: 350	Excretion: 1600
Total: 2500	Total: 2500

Plants derive their energy from the reverse reaction, that is,

$$6CO_2 + 6H_2O \rightarrow C_6H_{12}O_6 + 6O_2 \qquad (21.2)$$

Carbon dioxide and water combine to form glucose and oxygen in photosynthesis. This process takes place in the presence of solar energy. The levels of carbon dioxide and water in the biosphere are maintained by reaction (21.2)

The large surface tension of water (73 dynes cm^{-1} at 20°C) and its ability to wet surfaces are the basis for the capillary action, which carries water to the leaves of plants and trees. Water has a maximum density at 4°C, a little above its freezing point. Thus, ice has a lower density than liquid water. It, therefore, floats on the water surfae and freezing of aquatic systems occurs from surface downwards. Unlike, the solid state of other substances is more dense than the liquid state. Had water behaved in this manner, the aquatic systems would have frozen from bottom up. As a result, the aquatic life in oceans, rivers or lakes would have perished in the winters.

Water and the Non-living Environment

Water has a high heat capacity per unit weight. This means that it can absorb relatively large quantities of heat without large changes in temperature. Thus, during summers a large body of water, such as a lake or an ocean, absorbs more heat from the air than the land. An area in the vicinity of the water body, therefore, would have a lower temperature than a distant region. During winters, the temperature of the air drops below that of the water and the aquatic system gives up its stored energy as heat. This effect tends to warm up the air. As a result, the areas surrounding a lake or an ocean have milder winters. Moreover, the high heat capacity of water enables the oceans to store considerable quantities of thermal energy. The oceans are thus able to act as large thermostats from which heat energy is carried to the cooler regions by waves or currents.

Water has a very low coefficient of compressibility (5 pa^{-1}) at ordinary temperatures and pressures. However, if water would have been completely incompressible, then the level of the sea would be higher by 40 meters, reducing the inhabitable area by 5%.

The high dielectric constant of water makes it an excellent solvent; it is often referred to as the *universal solvent*. As a result, natural water is never pure, but a solution of substances which come into contact with it. When the number and/ or concentration of the chemicals entering an aquatic body becomes so large that the natural qualities of water are altered, we say that the water has become polluted. The dissolved impurities are not the only ones which make the water stained. The suspended matter is also hazardous.

When the aquatic system becomes polluted it becomes mandatory to purify water, so that its natural qualities are restored. However, before water can be purified, the type and nature of the pollutants must be identified.

The Different Types of Pollutants

Water pollutants are classified into four broad categories: Chemical, physical, physiological and biological (Table 21.2). The introduction of these contaminants into the aquatic systems is likely to cause health hazards, harm to ecology, damage to structures or amenities and interference with legitimate uses of water.

Table 21.2 Classification of water pollutants

Chemical	Physical	Physiological	Biological
Organic pollutants	Colour	Taste	Weeds
Inorganic pollutants	Turbidity	Odour	Algae
	Suspended matter		Viruses
	Froth		Bacteria
	Radioactivity		Protozoa
	Thermal		Parasitic worms

Chemical Pollutants

Chemical contaminants are subdivided into organic pollutants and inorganic pollutants, the former being more threatening to the environment.

Organic pollutants

Organic pollution is the most common form of water pollution. It is caused by the naturally occurring compounds like proteins, fats, carbohydrates etc., as well as by synthetic compounds like dyes, pesticides, herbicides etc. The synthetic organic derivatives cause more harm to the environment than the naturally occurring ones. The organic pollutants originate mainly from domestic sewage, urban runoff, industrial effluents and farm wastes. Some important organic pollutants, along with the sources of their origin, are listed in Table 21.3.

When an organic polluting load is discharged into a water body, it is gradually biodegraded by aerobic bacteria. The complex, hazardous, organic compounds are broken down into simple, harmless, inorganic components and water is said to have undergone *Self purification.* Dissolved oxygen is consumed in the biochemical reactions leading to self purification of water.

Some organic compounds like soaps and aliphatic derivatives are easily biodegraded. When discharged into the aquatic systems, these compounds are rapidly eliminated by microorganisms. Low concentrations of dissolved oxygen are used up in these reactions. However, some organic compounds like detergents and aromatic derivatives are resistant to microbial decomposition. Their biodegradation occurs extremely slowly, consuming massive amounts of dissolved oxygen. For such reactions, the rate of oxygen consumption is greater than the

Table 21.3 Some common organic pollutants

Pollutants	Representative Examples	Sources
Proteins	Gelatin, keratin, casein	Food processing, cannery waste, slaughter house waste, tannery waste
Fats	Glycerol palmitate, glycerol stearate	Fat refining, wool scouring waste, edible oil waste
Carbohydrates	Starch, cellulose	Paper and pulp industry, textile industry
Resins	Amber, rosin	Paint manufacture, textile industry, lacquer industry
Soaps	Sodium palmitate, potassium stearate	Laundry waste, textile waste
Detergents	Sodium lauryl sulfate, polyglycol ether	Laundry waste, textile waste
Dyes	Fluorescein, magenta	Dyeing and printing textile, paper, leather
Agrochemicals	DDT, 2, 4-D	Agricultural runoff

rate of oxygen replenishment. Under these circumstances, depletion of dissolved oxygen occurs, resulting in severe consequences for the aquatic biota.

The amount of oxygen required to oxidise a given quantitiy of an organic material is called *biochemical oxygen demand* or *BOD*. If water contains low level of organic pollutants, a relatively small amount of oxygen is consumed during the self purification phenomena. The biochemical oxygen demand is said to be low and such a watercourse is rapidly restored. If the pollution load is massive, large quantities of oxygen are consumed during self purification and the biochemical oxygen demand of water is said to be high. Biochemical oxygen demand is experimentally determined by incubating a sample of water for 5 days at 20°C. At the end of this time, the amount of oxygen which has been consumed is taken as the biochemical oxygen demand.

If the organic pollution load is small and dilution by well oxygenated water is high, the aerobic bacteria are able to break down the organic matter completely. The self purification that occurs is because of the following reactions:

$$\mathrm{C} \rightarrow \mathrm{CO_2} + \text{Carbonates} + \text{Bicarbonates} \tag{21.3}$$

$$\mathrm{H} \rightarrow \mathrm{H_2O} \tag{21.4}$$

$$\mathrm{N} \rightarrow \mathrm{NH_3} \tag{21.5}$$

$$\mathrm{S} \rightarrow \text{Sulfates} \tag{21.6}$$

$$\mathrm{P} \rightarrow \text{Phosphates} \tag{21.7}$$

If the oxygen demand of sewage effluents is studied over a number of days it is found that oxidation proceeds quite rapidly for a fortnight and then slows down. During the first phase, reactions (21.3–21.7) occur at a rapid pace. During the second phase, ammonia, formed in reaction (21.5), is oxidised to nitrites and nitrates.

$$NH_3 \rightarrow HNO_2 \rightarrow HNO_3 \tag{21.8}$$

Reaction (21.8), called *nitrification* or *nitrosification reaction*, occurs relatively slowly.

The rate of biochemical oxidation of organic matter by microorganisms is directly proportional to the concentration of unoxidised organic material,

$$-\frac{dc}{dt} = K_1 C \tag{21.9}$$

where K_1 is the rate constant for the oxidation reaction. It is commonly referred to as the *deoxygenation constant.* On rearranging and integrating, equation (21.9) becomes,

$$-\int_{C0}^{C} \frac{dc}{c} = K_1 \int_{t=0}^{t=t} dt \tag{21.10}$$

$$C = C_0\ e^{-K_1 t} \tag{21.11}$$

where C_0 is the concentration of the pollutant when it is discharged to the watercourse ($t = 0$) and C is the concentration at time t (measured in days).

Requirements of modern, industrialized society are such that massive quantities of organic load are discharged into natural waters. A part of the pollutants is decomposed in accordance with reactions (21.3–21.8). However, because of the high biochemical oxygen demand of the contaminants, the dissolved oxygen is rapidly exhausted. Under these circumstances, putrefaction results. Aquatic life perishes. The aerobic bacteria also die. The components of organic pollution decompose into an entirely different set of compounds.

$$\text{C} \rightarrow \text{Organic acids} \tag{21.12}$$

$$\text{N} \rightarrow \text{Organic amines} \tag{21.13}$$

$$\text{S} \rightarrow H_2S + \text{Organic sulfur compounds} \tag{21.14}$$

$$\text{P} \rightarrow PH_3 + \text{Organic phosphorus compounds} \tag{21.15}$$

These decomposition end products are not harmless. As their concentrations increase, the biochemical oxygen demand further increases. If these conditions continue to prevail, self purification of water becomes impossible. However, nature has a back-up system to counter these conditions. A whole new set of microorganisms, called anaerobic bacteria, takes over. These bacteria thrive in the absence of oxygen. They extract oxygen from oxygen-containing compounds like sulfates, nitrates, oxides etc. and oxidise the organic matter to harmless end products shown in reactions (21.3–21.7). During the time anaerobic bacteria are acting upon the organic load, sufficient oxygen becomes dissolved in water again and aerobic conditions return.

There are a large number of industrially important organic compounds, such as pesticides, herbicides and detergents, which are not only resistant to biodegradation, but are also toxic in nature. Before these are acted upon by microorganisms they manifest their toxic effects on the aquatic living community. Thus, such compounds elicit a wide range of environmental hazards. One cannot rely on the self purification phenomena for the elimination of these compounds and, therefore, proper control measures have to be instituted for their removal.

Inorganic pollutants

Some common classes of inorganic pollutants are listed in Table 21.4. These pollutants largely originate from industrial wastes. Inorganic acids and alkalis can do extensive damage to a water body by breaking down its natural buffer system and altering its normal pH value. Changes in pH have a profound effect on aquatic organisms, especially fish. Freshwater fish usually thrive in the pH range 5.0–8.5. Under more acidic and alkaline conditions, the fish population tends to fall. In particular, fish are liable to be adversely affected if suddenly transferred from an alkaline watercourse to an acidic stream or *vice versa*.

Table 21.4 Some common inorganic pollutants

Pollutants	Representative Examples	Sources
Acids	Sulfuric acid, phosphoric acid	Mine runoff, wool scouring waste, iron pickle liquor
Alkalis	Caustic soda, lime	Tannery waste, cotton processing waste
Cations	Mercury(II), lead(II)	Metallurgical operations
Anions	Sulfides, cyanides	Plating waste, gas liquor, mine runoff

Acids are also objectionable on account of their corrosive nature. If the pH of the watercourse falls below 5.0, the metal or concrete structures, pumps and other equipments installed in water become corroded. Acids also generate hydrogen sulfide gas from sludge deposits. Both acids and alkalis destroy bacteria and so inhibit or prevent the self purification of a stream.

Self purification is also hindered by the presence of heavy metal ions in water. Moreover, these ions destroy algae and bring about the death of fish by coagulating the mucus around the gills. Anions, such as cyanide, fluoride and sulfide, are extremely toxic, especially if present in large amounts. In recent years a good number of cases pertaining to fish mortality have been attributed to the presence of these anions in water.

Physical Pollutants

The onslaught on the natural qualities of water occurs not only when chemical constituents are introduced in the aquatic systems, but also when the physical parameters of water are drastically altered. No doubt, physical pollution is less damaging than chemical pollution, yet it cannot be altogether ignored. In most cases physical pollution originates as a secondary consequence of chemical pollution.

Colour

Many trade wastes discharging into watercourses have a distinct colour which they may impart to the water. The colour is due, in most cases, to organic dyes. For example, fluorescene imparts its characteristic yellow-green colour to water at as low a concentration as 0.025 ppm. Likewise, the dye magenta shows a distinct reddish colour at 0.020 ppm.

Interaction between two different trade wastes in a stream or between a trade effluent and substances present naturally in stream can also produce intense colourations. For example, when effluents from mine runoff, rich in iron content, enter an aquatic system containing hard water, a reddish brown opalescence develops. This colouration is due to ferric hydroxide which is formed by the interaction of ferric ions present in mine runoff and bicarbonate ions present in hard water.

$$Fe^{3+} + 3HCO_3^- \rightarrow Fe(OH)_3 + 3CO_2 \qquad (21.16)$$

Similarly, chromium compounds impart a yellow or orange colouration on interaction with constituents already present in water.

Organic dyes and compounds of iron and chromium are generally toxic in nature. However, colour is not always harmful. River water is often coloured green because of the presence of humic acid of vegetable origin. Humic acid is not toxic. Nevertheless, coloured water presents an unaesthetic sight.

Turbidity

A distinct physical characteristic of sewage and industrial waste water is their degree of turbidity which is caused by the presence of colloidal matter. Colloidal particles do not settle on standing and, therefore, impart cloudiness to water.

As a general rule, greater the degree of pollution, more pronounced is the turbidity. Therefore, often the degree of turbidity is taken to be an approximate measure of the intensity of pollution. However, in this connection two reservations must be made. Firstly, turbidity in natural watercourses might be due to the presence of small amounts of some inert and relatively harmless minerals like clay particles. Such water is considered satisfactory for most purposes. Secondly, absence of turbidity does not necessarily imply that a stream is unpolluted, for clear water might be heavily contaminated with dissolved impurities.

Colloidal particles which cause turbidity are generally non-toxic. However, they prevent the penetration of sunlight and reduce the extent of photosynthesis. Turbidity in water used for domestic purposes is very objectionable, particularly if it is due to the presence of hydrated iron or manganese oxides, in quantities greater than 1.0 ppm.

Suspended matter

Coarse, insoluble matter suspended in water is the most common form of physical pollutant. All natural water resources, even those which are relatively unpolluted, contain suspended matter in form of silt, sand, clay and pulverized rocks. Suspended matter of natural origin is largely harmless, but watercourses receiving suspended solids from sewage and industrial wastes suffer the consequences of pollution.

Suspended matter in sewage effluents, dairy wastes and coal washery residues is mainly orgnic in nature. On the other hand, wastes from china-clay works and stone quarrying contain chiefly inorganic suspended solids. The particulates tend to cause silting of a watercourse, especially if the stream is slow moving. They have an abrasive action on pumps and other metallic structures installed in water. Moreover, they cause mechanical injury to the fish by choking their gills. Further, they reduce the penetration of sunlight in water and curtail photosynthesis. The self purification phenomenon is also retarded in presence of suspended matter. Lastly, suspended solids are unsightly, as well as a nuisance aesthetically.

Froth

Froth or foam consists of a dispersion of gas bubbles in water. The formation of foam is a physical phenomenon, arising due to the lowering of surface tension of water by impurities, such as soaps, detergents and saponins. The tendency to foam is greatest in relatively clean waters and diminishes as the amount of pollution increases. Foam persists for a longer time if the atmospheric humidity is high and wind velocity is low. Froth curtails photosynthesis by reducing the penetration of sunlight. It is also aesthetically objectionable.

Radioactivity

Many radioactive elements like uranium, thorium and actinium, have always existed in the earth's crust. Granite rocks are particularly rich in radioactive materials. A part of these radionucleides enter the natural watercourses. However, the background radiation of water receiving radioisotopes from natural sources is very low. The danger lies in any large scale increase in radiation due to the development of nuclear fission processes and also due to the increased use of man-made sources of radioactivity in medicine, industry and research.

When radionucleides are taken up by aquatic organisms, exchange takes place with chemically similar elements already present in the organism's body. For example, radioactive strontium-90 and radium-226 replace calcium. Thus, strontium-90 becomes concentrated in scales and bones of fish. Phosphorus-32 is readily absorbed by bones of aquatic animals. Water weeds tend to assimilate strontium-90 and radium-226.

Once these isotopes are accumulated in the body, they start emitting radiations. For example, strontium-90 is a β-emitter.

$$^{90}_{38}\text{Sr} \rightarrow \, ^{90}_{39}\text{Y} + \, ^{0}_{-1}\beta \tag{21.17}$$

Radium-226 is an α-emitter.

$$^{226}_{88}\text{Ra} \rightarrow \, ^{222}_{86}\text{Rn} + \, ^{4}_{2}\text{He} \tag{21.18}$$

The radiations emanating from these elements cause a multitude of hazardous effects. Since the aquatic flora and fauna can concentrate radioactivity, the possibility exists that those used as food by man could bioaccumulate dangerous amounts of radiations in the human body.

Thermal pollution

Thermal or temperature pollution refers to the discharge of heated water from electricity generating plants into watercourses. The abrupt change in the temperature of natural water has severe environmental consequences. Of all forms of physical pollutants, thermal pollution is the most important. Hence, it is detailed in a separate section. (p. 315)

Physiological Pollutants

Physiological pollution is always a secondary phenomenon, arising from the presence of trace quantities of chemicals. It manifests itself when natural water no longer remains tasteless or odourless. It is in the event of shortage of water supply or lack of purification facilities that the problem of taste and odour assumes considerable importance.

Taste

Industrial wastes contain many chemical compounds which impart characteristic and unpleasant taste to water, even when present in extremely small quantities. For example, the presence of synthetic detergents to the extent of 1.0 ppm imparts a mouldy or mildewy taste to water. Inky taste arises due to ferrous ions present at as low a concentration as 0.1 ppm in water. Phenolic taste in water is due to the presence of 0.002 ppm or more of phenols. When water containing phenols is chlorinated, the taste becomes more intense and objectionable due to the formation of chlorophenols. Wastes from explosive industry are rich in nitro derivatives of aromatic compounds, which impart a bitter almond taste to water.

Natural waters containing taste-producing substances may curtail the value of fisheries by imparting unpleasant tastes or 'taints' to fish. However, the objectionable taste may be eradicated by transferring the fish to clean waters for some time.

Odour

Under the influence of microorganisms, the proteins undergo putrefaction giving intermediates having offensive odours. For example, sewage possesses unpleasant odour due to the decomposition of proteins into heterocyclic amines—indole and skatole. Other amines impart a fishy odour to natural waters. Putrid smell is because of the decomposition of proteins into organo-sulfur compounds like methyl or ethyl mercaptans. Organo-phosphorus compounds impart a fishy odour. The presence of aldehydes and cyano derivatives give an odour of bitter almonds.

Certain microorganisms have also been shown to be responsible for odours in water supply. For example, the protozoa *Dinobryon* imparts a fishy odour to water, while the algae *Anabaena* gives a strong grassy odour. Such odours are due to the liberation of essential oils by these microorganisms. In unpolluted waters, the earthy odour is because of the presence of humus.

Biological Pollutants

Those aquatic organisms which either multiply excessively or are otherwise undesirable, harmful or injurious are classified as biological pollutants. These pollutants generally arise as a secondary result of pollution by sewage or trade wastes.

Weeds

Excessive growth of freshwater weeds like chestnut is because of the discharge of untreated sewage or effluents into water. The weeds cause fish mortality and serve as breeding grounds for mosquitoes. During summers these decompose with the evolution of hydrogen sulfide, thus imparting offensive odour to water.

Algae

Algae multiply at an accelerated pace, causing odours in water. When they die, their cells decompose, encouraging the growth of bacteria and fungi which thrive on dead cell constituents. Moreover, the decomposition of algae cells consumes oxygen. The depletion of dissolved oxygen causes eutrophic conditions in closed bodies of water like lakes (Chapter 29, p. 409). High density of algae may also produce toxins which are lethal to animals. For example, *Prymnesium parvum* produces a toxin which is harmful to the fish.

Bacteria

The faecal contamination of water can introduce a variety of pathogenic bacteria into waterways. *Salmonella* can cause acute gastro-entritis with diarrhoea, fever and vomiting. These in rivers can survive for long distances downstream of their source. *E. coli* produces diarrhoea in children. *S. typhi* causes typhoid, while *Vibrio cholerae* causes cholera. The latter disease is largely under control in countries where modern sewage treatment facilities are available.

Viruses

Viruses are inducted into natural waters through faeces. Although these organisms are present in water in much lower numbers than bacteria, yet they pose a considerable health hazard because they possess a greater resistance to disinfection. Many of the common infections of man are virus diseases. Influenza, small pox, poliomyelitis and yellow fever are some of the common ailments caused by viruses. Hepatitis A, a form of jaundice, is also transmitted by virus-contaminated water. The disease is endemic during floods. Entritis is caused by ECHO virus and its outbreak may develop very rapidly if water supply becomes contaminated with untreated sewage.

Protozoa

Faecal contamination of drinking water is the major source of transmission of pathogenic protozoa. Amoebic dysentery is caused by the parasitic protozoa,

Entamoeba histolytica. It is endemic in the tropical and subtropical regions. Another parasitic protozoa, *Naegleria fowleri* causes amoebic meningoencephalitis, a fatal disease.

Worms

Common parasitic worms present in water are roundworms, tapeworms and flukes. The ova of these parasites are passed out in the faeces. They are resistant to sewage treatment processes and multiply at a fast rate. Their lifecycles involve an intermediate host. In case of tapeworm the intermediate host is the cattle. It becomes infected by grazing in pastures sprayed with sewage sludge or by drinking water contaminated by sewage.

Thermal Pollution

The discharge of waste heat into natural waters is termed as thermal pollution. It is one of the prominent causes of environmental damage.

Causes

One of the characteristic properties of water is its high specific heat. By absorbing and releasing heat slowly, water serves as an excellent medium for heat exchange. Hence, water has been used for many heating and cooling purposes. The major use of water as a coolant is in the electric power industry. The steam electric generating unit receives exhaust steam from the turbine, which is subsequently cooled to condense it to water. The cooling of exhaust steam requires large quantities of water, which is drawn from a river or estuary adjacent to the electric power plant and made to circulate through the unit condenser. When water is finally discharged into the stream, its temperature is generally 10–16°C higher than the initial temperature, thus causing a progressive warming of the water body. This phenomenon is called thermal pollution.

Consequences

A good amount of information is available on how temperature affects the life process of fish. Most of the effects stem from impact of temperature on the rate of metabolism, which is speeded up by heat in accordance with the van't Hoff principle that rate of chemical reaction increases with rising temperature. In general, the metabolic rate doubles with each increase of 10°C.

1. Life Span

All organisms have temperature limits to their survivial ability, but for most acclimation is possible to some extent if the temperature change is gradual. A sudden change in temperature, however, is generally lethal. Hence a rapid change in temperature or sudden transference of fish to warm water, may result in death at a temperature well below that regarded as lethal for the species.

Temperature also affects the longevity of fish. High temperature may lead to a short but exhaustive spell of rapid growth. Daphnia can live for 108 days at 8°C, but its life span at 28°C is only 29 days. The water flea moina has a life span of 14 and 5 days at 13°C and 33°C, respectively.

2. *Respiration*

An increase in temperature causes not only disappearance of dissolved oxygen, due to the lower solubility of oxygen at higher temperatures, but also an increase in the rate of respiration of fish. The oxygen consumption for most fish increases four-fold as the temperature of water is raised to the maximum at which they could survive. At higher temperatures, the haemoglobin in the blood of the fish has a reduced affinity for oxygen and, therefore, becomes less efficient in delivering oxygen to the tissue. The combination of increased need for oxygen and reduced efficiency in obtaining it at higher temperatures can put severe stress even on fish that are normally capable of living in a limited supply of oxygen. For example, the hardy carp, which at a water temperature of 1°C can survive at an oxygen concentration of 0.5 ppm, needs a minimum of 1.5 ppm when the temperature is raised to 35°C.

3. *Reproduction*

Temperature profoundly affects the reproduction of aquatic animals. An increase in temperature induces seasonal development of gonads and triggers females to deposit eggs. A rise in temperature also reduces the time taken by the fish's eggs to hatch. For example, fertilized eggs of Atlantic salmon hatch in 114 days in water at 2°C, but the period is shortened to 90 days at 7°C. Trout eggs hatch in 165 days at 3°C and in 32 days at 12°C. Excessive temperatures, however, retard the normal development of eggs. Further, there are critical temperatures above which fish will not reproduce. Brown trout and sunfish do not reproduce above 11°C and 22°C, respectively.

4. *Heart Beat*

Experiments with crayfish reveal tbat its heart beat increases from 30 beats per minute at a water temperature of 4°C to 125 beats per minute at 22°C and then decreases to 65 beats per minute as the temperature approaches 35°C, the lethal temperature for this crustacean. The final decrease in heart beat shows the weakening of the fish under thermal stress.

5. *Swimming*

In general, the fish tend to raise their swimming speed and show more spontaneous movement as the temperature rises. Stockeye salmon cruises twice as fast in water at 16°C as it does in water at 2°C. Above 16°C, however, its speed decreases. The decrease in speed at higher temperatures affect their ability to get food. They slow down in pursuing minnows and at still higher temperatures are almost incapable of catching minnows.

6. *Food Assimilation*

The temperature of water has a pronounced effect on the assimilation of food in fish. It has been established that digestion is fast at higher temperatures. Tracer experiments with young carp have revealed that at 10°C the food takes 18 h to pass through the alimentary canal, while at 26°C it takes only 4–5 h. Since at higher temperatures, the metabolic activity of the fish increases, a substantial portion of the food intake in warm waters is utilized for merely maintaining the body functions.

7. Morphology

Temperature differences during growth and development induce morphological changes in aquatic organisms. A variety of fish, when subjected to thermal pollution, show agape jaws, eye abnormalities and twinning. Above 30°C the incidence of spinal deformation is high.

8. Synergism

The toxicity due to insecticides, detergents, paper mill effluents and other chemicals increases with rise in temperature. For example, the lethal effects of 5 and 10 ppm zinc sulfate to rainbow trout are roughly doubled on raising the temperature from 13.5 to 21.5°C. At 1°C carp is able to tolerate a carbon dioxide concentration of 120 ppm, but at 36°C a concentration of 60 ppm is lethal. In the presence of cadmium salts, the LC_{50} for *Galaxias meculatus* at 21°C is 5–7 times lower than that at 15°C.

Abatement

It is becoming increasingly evident that we cannot continue to inject waste heat into water without causing serious environmental problems. Two mechanical means are employed for abatement of thermal pollution. These are: The wet cooling towers and the dry cooling towers (Fig. 21.1).

In the wet cooling method, hot water is sprayed over baffles in a huge hyperbola shaped tower. Cool air rising through the tower carries waste heat and the cooled water is either discharged into the stream or recycled. Two problems are associated with wet tower technique. Firstly, a substantial amount of water is lost through evaporation. This not only vitiates the benefit of recycling, but also concentrates whatever pollutants the effluent may contain. Secondly, in cooler regions, there is extensive fog formation in the vicinity of the wet cooling tower, causing damage to vegetation.

In a dry cooling tower the heated effluent is contained in a system of pipes. Air is passed over the pipes by a large fan facilitating heat exchange by radiation and convection. Water loss and fog formation are controlled by this method, but installation and maintenance costs are much higher as compared to those of a wet cooling tower.

Applications

Since both the abatement techniques suffer from one or more drawbacks, it is desirable to take advantage of some positive aspects of thermal loading. Warm water outfalls in winter attract fish because they extend the feeding period and forestall ice formation, which often prevents adequate oxygenation of the water beneath. The artificial rearing of salmon and catfish in winter is carried out in cages exposed to thermal loading.

By combining waste water from a power plant and effluents from a sewage treatment plant, the heat and nutrients, considered a nuisance in most environments,

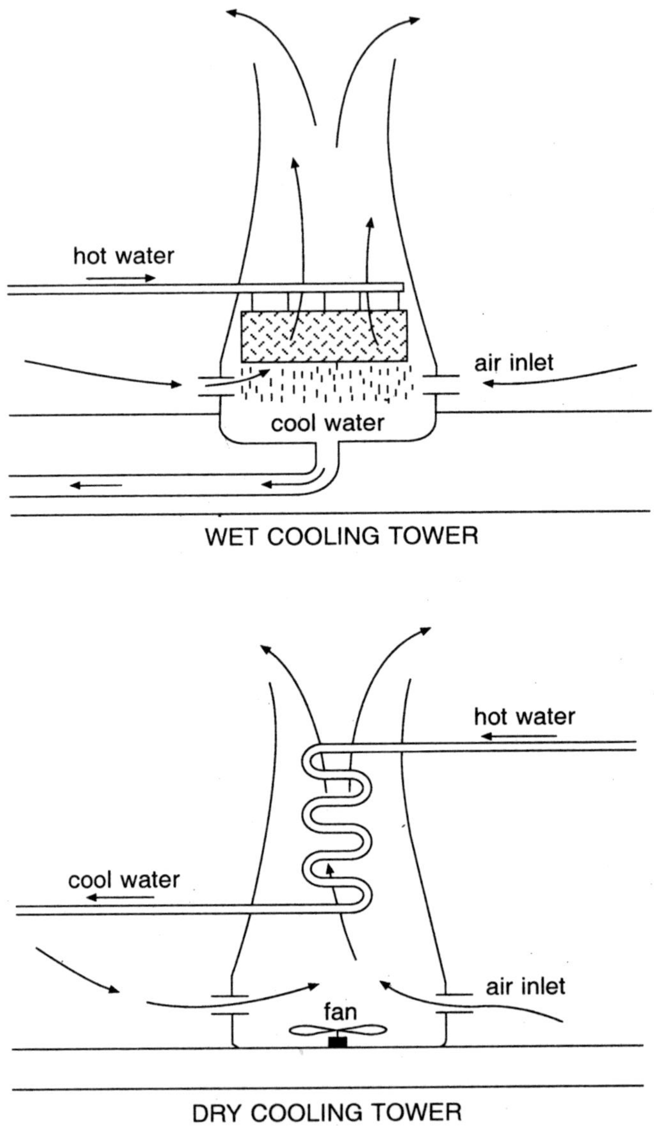

Fig. 21.1 Cooling towers

can be put to good use. By supplying heat-tolerant algae as primary producers and introducing warm-water fish to harvest the algae, water would have time to cool, have its nutrient content sharply reduced, and its algae harvested in the form of useful protein, before being returned to the natural environment.

It has been established that growth of some crops is accelerated by irrigating with warm water. Hence, heated water may be put to good use by applying it to such crop. Frost protection can also be achieved by spraying the trees with warm water. Waste heat is also being used in desalination plants to aid in the evaporation process.

References

Clark, J.R. (1969). Thermal pollution and aquatic life, *Sci. Amer.,* **220**(3), 19–27.

Franks, F. (1976). Water—the unique chemical, *Chem. Brit.,* **12,** 278–280.

Hites, R.A. and Biemann, K. (1972). Water pollution: Organic compounds in Charles River, Boston, *Science,* **178,** 158–160.

Holum, J.R. (1977), *'Topics and Terms in Environmental' Problems,* John Wiley & Sons, New York, p. 653–665.

Jones, M.M., Johnston, D.O., Netterville, J.T., Wood, J.L. and Joesten, M.D. (1987), *'Chemistry and Society',* 5th Edn., Saunders College Publishing, Philadelphia, p. 429–461.

Klein, L. (1962), *'River Pollution II. Causes and Effects',* Butterworths, London, p. 22–109.

Marx, J.L. (1974). Drinking water: Another source of carcinogens? *Science,* **186,** 809–811.

Mason, C.F. (1981), *'Biology of Freshwater Pollution',* Longman, London, p. 14–62.

Moore, J.W. and Moore, E.A. (1976), *'Environmental Chemistry',* Academic Press, New York, p. 355–374.

Sodhi, G.S. (1988). Biological aspects of thermal pollution, *Chemistry Education,* **5,** 26–28.

CHAPTER 22

Heavy Metals

Among the natural substances that man concentrates in his immediate environment, metals are the most ubiquitous. Metals can be neither degraded nor metabolized; they are an example of ultimate persistence. Metals enter the living organisms either as inorganic salts or as organometallic derivatives. Amongst the heavy metals, mercury and lead are the ones that have most severely affected the environment. These metals are present in all components of the environment—atmosphere, land and aquatic systems. However, we have considered environmental implications of heavy metals in the section of water pollutants because a significant portion of these centaminants present in soil is ultimately leached into water bodies. Likewise, the atmospheric components of heavy metals are, in large part, rained down into natural watercourses. Moreover, mercury in the aquatic environment is transferred into physico-chemical forms that are highly toxic and hazardous to humans.

Mercury

Mercury is known to be toxic at such low levels that it leaves but a small margin of safety to humans. It is, therefore, pertinent to ascertain how mercury is being redistributed and concentrated in the environment and in what forms and compounds it may be harmful to life.

Sources of Mercury

Mercury is discharged into the aquatic systems both by natural sources and man-made sources, the consequences of the latter being more deleterious. The natural sources of mercury are scanty and scattered. However, certain locations, such as Almaden (Spain), Tuscany (Italy), California (USA) and Crimea (Russia) have abundant concentrations of mercury and are, therefore, sites for mining operations. The flora and fauna in the immediate vicinity of these sites sustain a significantly high concentration of mercury.

Mercury in the form of cinnabar (HgS) or in metallic state is embedded in igneous rocks, in particular, basalt and granite. These rocks contain about 0.080 ppm mercury content. Each year about 800 tonnes of mercury are released to the environment through weathering of rocks.

Volcanic action is yet another natural source of contamination of environment by mercury. Volcanic lava is always rich in the constituents of igneous rocks.

Thus, the mercury content of Lake Ontake (Japan) has shown a steady rise because of the repeated volcanic eruptions of the nearby Mount Kisoontake.

The natural contribution of mercury to the aquatic systems is minor and stable. Except for locations near rich deposits of mercury-bearing minerals, the only concentrated sources of mercury in the aquatic systems are those resulting from human activities. Mercury content of the aquatic systems has elevated due to progressive urbanization and industrialization.

The combination of physico-chemical properties, such as liquidity at room temperature, uniform volume expansion over a wide temperature range, high surface tension and non-wettability to glass surfaces makes mercury uniquely useful for measuring devices such as thermometers, barometers and manometers. The low electrical resistivity and high thermal conductivity make mercury an excellent electrical conductor and coolant, respectively. Because of its capacity to absorb neutrons, mercury is employed as a shield against atomic radiation.

Mercurials have found widespread uses as insecticides, fungicides, bactericides and pharmaceuticals. Due to their stereospecific nature, several compounds have been used as catalysts, especially in the manufacture of synthetic polymers. Mercury amalgamated with silver, tin, cadmium and copper is used to make dental fillings.

The large hydrogen overpotential of mercury permits its use as a cathode in electrochemical operations. In fact, the chlor-alkali process is the single largest source of mercury in the natural waters. For every tonnes of chlorine produced by the process, 0.1–0.2 kg of mercury is discharged to the aquatic systems. Even in cases where the chlor-alkali plants have been shut down, the amount of mercury that has already settled in the river beds continues to be a cause of concern.

Biomethylation of Mercury

Any natural phenomenon which generates virulent forms of contaminants has serious consequences. This is particularly true in case of mercury. In its elemental form or as inorganic salts, mercury is relatively less toxic. However, when mercury in these forms is discharged into natural waterways, it is converted into highly toxic and malicious methylmercury ion, CH_3Hg^+. Since chloride ions are present in every aquatic system, the methylmercury ion is transformed into methylmercury chloride, CH_3HgCl.

The conversion of elemental or inorganic mercury into methylmercury chloride has been recognized as a critical step in the environmental behaviour of this heavy metal. It provides an explanation for the fact that methylmercury chloride is found in fish even if all known sources of mercury in the aquatic system are in form of elemental or inorganic mercury.

The methylation of mercury in water bodies is brought about by a set of microorganisms called *methanogenic bacteria*. That is why the process is called biomethylation. Higher levels of methylmercury chloride are found in deep, rather than shallow sediments, indicating that biomethylation largely occurs under anaerobic conditions. The conversion of mercury into methylated form is fostered by the enrichment of water with organic impurities that permit the growth of methanogenic bacteria. Biomethylation proceeds most effectively in the pH range 5.5–6.5.

Bioamplification of Mercury

Certain substances, of which methylmercury chloride is an excellent example, tend to concentrate among the food chains. Being more soluble in the tissues of simple organisms than in surrounding water, these substances penetrate inside the bodies of simple organisms. Subsequently, they become a concentrated component of the more complex species which feed on the simpler ones. This process is called bioamplification. As a result of bioamplification, methylmercury chloride concentrates in the tissues of aquatic organisms, mainly fish.

In aqueous solution, methylmercury chloride ionizes as follows.

$$CH_3HgCl \rightleftharpoons CH_3Hg^+ + Cl^- \qquad (22.1)$$

However, the equilibrium constant for the reaction (22.1) is only $10^{-5.45}$, indicating that methylmercury chloride is weakly dissociated. The undissociated methylmercury chloride is apparently readily transported across the body membranes and collects in the fatty tissues, from where the chances of its expulsion are quite remote. Any factor which shifts the equilibrium (22.1) to the left, increases the chances of bioamplification.

As shown in Fig. 22.1, the factors determining the extent of bioamplification are classified into two broad types: Those depending on the quality of water and those depending on the characteristics of the fish.

Bioamplification—Factors Depending On

Water Quality	Fish Characteristics
1. pH	1. Age
2. Temperature	2. Size
3. Hardness	3. Weight
4. Salinity	4. Sex
5. Contaminants	5. Species

Fig. 22.1 Factors determining bioamplification

Taking the former factors first, it is observed that the uptake of methylmercury chloride is favoured at low pH. As the pH of the water body exceeds 5, methylmercury chloride is transformed to methylmercury hydroxide, CH_3HgOH. The latter compound is insoluble in the lipid constituent of fatty tissues and hence the chances of its bioamplification are considerably reduced. Thus, fish in basic or neutral waters contain smaller mercury content. It was noted that fish taken from Lake Ontake (Japan) at pH 4.2 contained 0.61–0.68 ppm mercury. When the pH of the lake had risen to 6.7, the mercury level in the fish decreased to 0.15–0.24 ppm.

Bioamplification of methylmercury chloride in fish increases with temperature. Thus, *Daphnia magma* accumulates 48.2% and 20% methylmercury chloride at 18°C and 10°C, respectively.

The extent of bioamplification increases with salinity. With an increase in the concentration of chloride ions, the equilibrium (22.1) shifts to the left, increasing the concentration of undissociated methylmercury chloride and thus enhancing the rate of its uptake. The presence of calcium and magnesium ions in water removes the chloride ions, shifting the equilibrium (22.1) to the right and thus reducing the extent of bioamplification. Thus, the relative efficiency of methylmercury chloride uptake in soft water is more than double to that measured in hard water.

Organic contaminants, in general, do not affect the rate of accumulation. However, certain chelating ligands such as cysteine, citrate and ethylenediamine tetraacetic acid reduce the uptake of mercury by removing it from water as complexes.

Turning to the factors depending on the characteristics of the fish, it is observed that the bioamplification increases with the age, size and weight of the fish. There exists no definite correlation between sex of the species and the extent of bioamplification. Nevertheless, female American smelts are known to accumulate higher concentrations of mercury than males of the same age.

The degree of bioamplification also depends on the type of fish. Species like wall-eye, bass, or trout which occupy a higher trophic level have greater amplification factors than blue-gill, carp or catfish, which occupy a lower level on the trophic scale. Thus, at each higher trophic level on the food chain, mercury becomes more concentrated. Since humans are usually at the top of the food chain, high concentrations of mercury may occur in human diets, with tragic results, such as Minamata disease (see box).

Minamata Disease

One of the most visible tragedies caused by water pollution is Minamata disease. Originating about four decades ago in the Japanese fishing port of Minamata, the disease remains the most disastrous case of environmental mercury poisoning, and perhaps the most illustrative of the complicated ecological pathway of this heavy metal. The cause of the disease was contamination of water with mercury- rich effluents discharged into the Bay by Chisso Chemical Company. By 1963, when it was determined that the illness symbolized a case of methylmercury poisoning, caused by eating contaminated fish from the Bay, the disease had claimed about 50 lives and permanently paralysed over 700 people. Numerous newborns sustained congenital defects. Although the outbreak of Minamata disease was a tragedy of high order, nevertheless it exposed two important facts. Firstly, the Chisso plant used mercuric chloride catalyst in acetaldehyde production and, therefore, only the comparatively non-toxic inorganic mercury was released in the industrial effluents. However, the sediments from the Minamata Bay were found to be rich in methylmercury chloride. It thus became evident that inorganic mercury underwent biomethylation in aquatic systems. Secondly, the human deaths, in and around Minamata, coincided with a large scale fish mortality and, therefore, it was inferred that methylmercury chloride bacame concentrated in the body of the fish through the process of bioamplification.

Mercury Poisoning

Both inorganic mercury and methylmercury are poisonous in nature. Yet, methylmercury elicits more severe effects. The reaons for the greater toxic effects of methylmercury are manifold. Firstly, the methylmercury is amphilitic in nature and, therefore, interacts favourably with the hypophilic cell structures, especially the cell membrane. Secondly, methylmercury can penetrate the membrane separating the blood stream from the brain and those in placenta through which the nutrients must pass to an unborn fetus. The latter was apparently the cause of congenital birth defects in Minamata disaster. Thirdly, in methylmercury, the metal ion is firmly bound to the carbon so that the molecule is not easily broken down and maintains its destructive action for weeks or months. In contrast, the injury caused by inorganic mercury is almost invariably reversible.

Mercury poisoning manifests itself in several forms

1. Enzyme inhibition

Mercury reacts with a large number of enzymes, inhibiting catalysis of general metabolic reactions. Thus, mercury is often considered as an enzyme inhibitor. The chemical basis of this effect results from the mercury's strong affinity for the sulfhydryl (– SH) group in the enzymes.

$$CH_3Hg^+ + \text{enzyme–SH} \rightarrow CH_3Hg\text{–S–enzyme} + H^+ \qquad (22.2)$$

2. Cellular dysfunction

Mercury also reacts with sulfhydryl function in proteins. Bound to the proteins in a cell membrane, the mercury may alter the distribution of ions, change electrode potentials and thus interfere with the movement of fluids across the membrane. The binding of mercury to proteins also disturbs the normal operation of constituents, such as mitochondria and lysosomes within the cell.

3. Neurological disorders

The brain is the target organ in methylmercury poisoning, where essentially irreversible damage is done to cells of the nervous system. The first symptoms are loss of sensation at the extremities of hands and feet and areas around mouth (paresthesia). As the body burden of mercury increases, these symptoms are followed by loss of coordination in gait (ataxia), slurred speech (dysphasia) and loss of hearing. Severe poisoning can cause blindness, coma and death. Estimates of the body burden at the onset of various symptoms are given in Table 22.1.

Table 22.1 Body burden at the onset of symptoms of methylmercury poisoning

Symptoms	Body Burden (ppm)
Paresthesia	0.50
Ataxia	1.00
Dysphasia	1.75
Deafness	3.50
Death	4.00

4. Erethism

When a person is exposed to relatively mild doses of mercury, the neurological disorders manifest themselves as behavioural problems.

The psychic disorders that result from mercury poisoning are termed as erethism. These are characterized by self-consciousness, timidity, embarrassment with insufficient reasons, anxiety, indecision, lack of concentration, depression, despondency, resentment of criticism, irritability and excitability. Sometimes these symptoms cause a complete change in personality.

5. Teratogenic effects

Methylmercury chloride readily penetrates the fetus through the placenta. The concentration of mercury in the blood and the brain of the fetus is about 20% higher than in the mother. Infants whose mothers are exposed to large amounts of methylmercury chloride are liable to be afflicted with mental retardation, cerebral palsy and convulsions.

Analytical Procedures for Estimation of Environmental Mercury

The reduction-aeration-flamless atomic absorption method is quite sensitive for analysing the mercury content in a variety of samples after appropriate pre-treatment.

For determining mercury in water, rocks or soil, the sample is digested with sulfuric acid or nitric acid. Any organic material that might be present is oxidised with a mixture of hydrogen peroxide and potassium permanganate. A sodium chloride-hydroxylamine mixture is added to destroy excess of oxidising agents, while retaining mercury in the mercuric (Hg^{2+}) form. Finally, a solution of stannous chloride is added to reduce mercuric ions to elemental mercury.

$$Sn^{2+} + Hg^{2+} \rightarrow Sn^{4+} + Hg \qquad (22.3)$$

The elemental mercury is volatilized by passing a stream of air at a rate of about 2 liters min^{-1}. The air stream is directed to an atomic absorption spectrophotometer. Intensity of the absorbance at 253.7 nm is directly proportional to the concentration of mercury in the air stream which, in turn, is proportional to the total mercury concentration in the sample.

For analysis of mercury in tissues, biological fluids and foodstuff, the sample is digested with sulfuric acid and potassium permanganate in a closed vessel. After sodium chloride-hydroxylamine treatment, the inorganic mercuric ions are reduced to elemental form by stannous chloride. If the sample contains mercury as methylmercury the procedure has to be modified. Methylmercury is not reduced by stannous chloride. However, a mixture of stannous chloride and cadmium chloride is effective in converting methylated mercury into elemental form. The elemental mercury is carried to the atomic absorption spectrophotometer by an air stream and determined as already described.

Antidotes to Mercury Poisoning

As a general safeguard against mercury poisoning, it has been recommended that the weekly intake of mercury per person should be limited to 0.3 ppm, of which no more than 0.2 ppm should be in form of methylmercury. The permissible upper limit of mercury in foods is 0.5 ppm. The level of mercury in blood should not be more than 1.0 ppm. Drinking water should have a maximum of 0.005 ppm mercury content.

When these limits are exceeded, the symptoms of mercury poisoning set in. As an antidote to poisoning effects, certain mercury-binding chemotherapeutic agents are administered to the patient. These agents tie-up mercury in form of complexes which are rapidly excreted from the body.

Some of the common chemotherapeutic agents used for the treatment of mercury poisoning are depicted in Fig. 22.2. Of these, N-acetyl-D, L-penicillamine and dimercaptosuccinic acid are most promising against methylmercury poisoning. These penetrate through the cell membrane and convert the intracellular methylmercury into a mobile complex that readily diffuses out of the cell. It is then passed out of the body as urinary or faecal excretions.

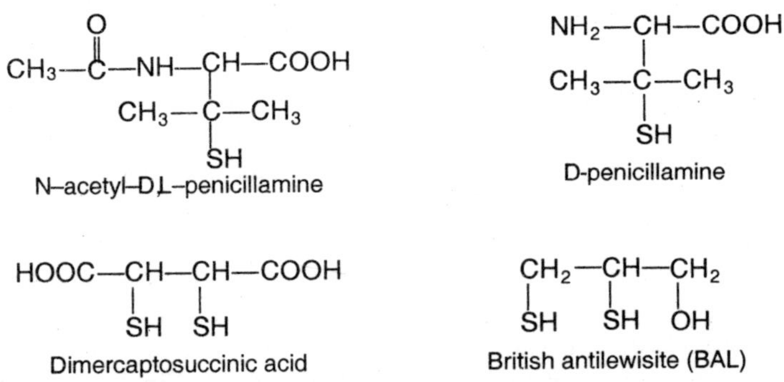

Fig. 22.2 Chemotherapeutic agents for treatment of mercury poisoning

British antilewisite or BAL is very effective in treatment of inorganic mercury poisoning. However, it is not used in cases of methylmercury poisoning. The reason being that the complex which it forms with methylmercury crosses the blood-brain barrier very rapidly and becomes concentrated in the brain, rather than being excreted out of the body.

Abatement Procedures against Mercury Poisoning

Once it is realised that mercury released by human activities has become an environmental problem, necessary control measures must be instituted. Toxic mercurials in industry and agriculture should be replaced by less toxic substitutes. In fact, a large number of mercury-based biocides have been phased out. Mercury-amalgam dental fillings too have been banned.

It is also necessary to prevent release of mercury and its compounds as waste by instituting recovery processes. For example, the discharge of mercury from chlor-alkali plants has appreciably reduced following the use of diaphragm cells. Advnaced waste water treatment has also contributed to the reduction of mercury pollution.

Unfortunately, the prevention of further mercury discharges is only a small portion of the abatement problem. Bodies of water whose underlying sediments are already overburdened with mercury shall continue to release the metal for a considerable time. In the meantime, fish will continue to accumulate large concentrations of mercury and humans will have to be careful of how much fish they should consume.

Attempts to dredge up the mercury-containing sediments apparently makes the problem worse instead of better. Dredging stirs up the bottom of the stream and temporarily increases the dissolved mercury level. When this method was tried in Minamata, the mercury content of fish increased.

Suggestions that sediments be covered preferably with sulfide-rich minerals have also been tried. Sediments of Jiyun River (China) were covered with 1 mm ferrous sulfide and 1 cm sand. The release of mercury from the sediments was substantially reduced. However, covering methods can only be seriously considered where the mercury contamination occurs over a limited area that can be pinpointed.

At present, perhaps the most effective control measure shall be to give up the contaminated water bodies for food production until they are able to cleanse themselves and to institute a widespread and effective monitoring network so as the prevent the creation of new ones.

Lead

Lead is one of the oldest metals known to man. Since medieval times it has found wide range applications and is, therefore, prevalent in natural environment. However, lead is a dangerous and insidiously deceptive poison. Children are the most common victims of lead poisoning.

Sources of Lead

The most important natural source of lead in the environment is the weathering of rocks. The lead content of the rocks is eroded by fast moving rivers and carried to long distances. Wind-blown dusts, forest fires, volcanic eruptions and sea salt sprays are some other natural phenomena leading to the distribution and concentration of lead in the environment.

As in the case of mercury, the man-made sources of lead have played a greater havoc with the environment. Lead has been mined and worked by man for millennia. Its ductility, high strength, high density, impermeability to radiation and other properties make it a very useful metal.

Lead is used in acid storage batteries, ammunition, solder and casting materials. Lead-coated steel sheets have sound attenuation properties and are commonly used in construction of appartments. Lead-asbestos anti-vibration pads are employed in foundations for buildings exposed to vibrations from nearby rail or road traffic. Leaded-porcelain enamel is used as a protection against radiation in nuclear powered reactors. Collapsible tubes, caulking materials and corrosive liquid containers are also made from lead. Anti-corrosive and highway traffic safety paints incorporate lead salts.

Most of the lead poisoning incidents in children below the age of five are caused by eating flecks of lead-containing paints from the walls of deteriorating buildings. This behaviour is termed *pica.* Lead monoxide, also called *litharge,* is an orange-yellow pigment that is used in glazing pottery. When acidic foods, like fruit juices and pickles, are stored in improperly glazed earthenwares, lead slowly leaches from the glaze and becomes a component of the diet. Lead enters drinking water from old conduit water pipes.

Tetraethyl lead, $(C_2H_5)_4Pb$, is an important gasoline additive. Though an extremely poisonous substance, it possesses anti-knocking characteristics. During combustion of gasoline, it is converted into lead oxide. The latter is subsequently reduced to elemental lead that gets deposited in the combustion chamber. The deposit causes slow corrosion of the engine. In order to avoid the degradation of the engine, a mixture of ethylene dichloride and ethylene dibromide is added to gasoline. During combustion, lead(II) chloride and lead(II) bromide are produced. These halides are not only resistant to reduction but are also volatile and are, therefore, emitted with the exhaust. The automobile engine is saved, but the atmosphere is despoiled!

Lead Poisoning

Lead poisoning, also called *plumbism* or *saturnism*, has proved to be more dangerous to the children than the adults. The following reasons justify this observation.

Firstly, once lead enters the human body it tends to concentrate in the bones. It remains in the bones in a relatively inert form, causing no ill effects. However, when the body feels shortage of an essential element like calcium or phosphorus, the blood leaches out these minerals from the bones and supplies it to the relevant organ. In this process lead too becomes labile and enters the blood stream. It then concentrates in the tissue where it elicits toxic effects. Since the metabolic rates are higher in children than in adults, the chances of lead being leached out of the bones are more in case of infants.

Secondly, due to the higher metabolic rates in children, the ratio between the quantity of lead consumed through food to the body weight is more in case of infants. This means that the body burden of lead is more in case of children than in adults.

Thirdly, the opportunity for the oral intake of significant amounts of lead from non-food sources, for example, from lead-bearing dust, soil, painted walls, toys and plasters, is considerably higher in small children. The habit of oral exploration, sucking fingers, mouthing and chewing objects is a natural activity amongst the toddlers. These habits contribute to the body's lead burden.

Plumbism manifests itself in several forms.

1. Enzyme inhibition

Like mercury, lead too coordinates with the sulfhydryl function in enzymes. This causes an inhibition of enzymatic activity. The clearest manifestation of the inhibitory effect of lead is the disturbance in the biosynthesis of heme. The latter is the iron-containing biomolecule that is a precursor to haemoglobin, the oxygen-carrying pigment of the red blood cells. Heme is also an essential constituent of the other respiratory pigments, the cytochromes, which play key roles in energy metabolism.

The decrease in heme biosynthesis leads at first to a decrease in the life span of red cells and later to a decrease in the number of red cells. To compensate the shortage, the blood forming tissue steps up its production of red cells. However, the rate of production is less than the rate of disappearance. Therefore, in a person suffering from lead poisoning, immature red cells, called *reticulocytes* and *basophilic stippled cells*, appear in circulation, causing anaemia.

2. Kidney impairment

Acute lead poisoning causes impairment of kidney functions. Lead tends to concentrate in the proximal renal tubules lining. There is an increased loss of amino acids, glucose and phosphate in the urine because the damaged tubular cells fail to reabsorb these substances as completely as the normal tubular cells do. The energy storage system of the body suffers. Exposure to milder doses of lead cause *chronic nephritis,* a disease characterized by scarring and shrinking of kidney tissue.

3. Neurological disorders

Central nervous system effects are the most serious manifestation of lead poisoning. Continuous exposure to lead over prolonged periods causes *encephalopathy* or brain damage by two mechanisms. Firstly, the blood capillaries reaching the brain start leaking and cause edema, that is, swelling of the brain. There is not enough space to accomodate a swelled brain within the periphery of the rigid skull. Therefore, the brain cells get damaged. Secondly, the cells constituting the motor nerve extremities are damaged and the conduction of nerve impulse is impaired.

Symptoms of encephalopathy initiate with the onset of vomiting, ataxia, periods of alternative stupor and hyperirritability. Subsequently, they culminate in delirium, convulsions, coma and death. For reasons already cited, children are more sensitive to neurological damage than the adults. In adults, most of the neurotoxic effects are reversed once the exposure to lead ceases. In children they are not.

4. Behavioural problems

When a person is exposed to milder doses of lead, the neurological damage manifests itself in form of behavioural problems. In such cases the symptoms include excitement, restlessness, insomnia, nightmares, impairment of memory and loss of concentration. In adults, these effects are reversed once the exposure to lead stops.

The picture with children, however, is very different. In a few cases, childhood attacks of symptomatic lead poisoning can produce in a child hostile, aggressive and destructive behavioural patterns so that there may eventually be a damand for institutionalization. In larger number of cases, children suffer because of dulling of mentation and chronic hyperkinesis. The former is followed by educational abnormalities; the latter by an impulsive behaviour, with a tendency to violence.

5. Reproductive problems

Women employed in pottery works, where exposure to lead is high, are more liable to be sterile than those in general population. If they become pregnant, the pregnancy is more likely to result in miscarriage or stillbirth. If the child is born living, there is a higher infant mortality. Plumbism in father also affects the survival, vigour and fertilization of the offspring.

6. Teratogenic effects

Children born to women who are exposed to lead during pregnancy suffer from convulsions and a form of macrocephaly characterized by a square shaped head. It has also been observed that children born to women who suffered from plumbism during their childhood, may also show birth defects. This means that a girl growing up in a lead-polluted environment might, years later, pass that lead to her offspring. During adolescence, the lead becomes concentrated in the girl's bones. During pregnancy, calcium is removed from a woman's bones and passed to the blood stream. It then enters the placenta to help build the developing embryo's skeleton. In this process lead also comes out of the bones, crosses the placenta and enters the fetus, where it elicits teratogenic effects.

Analytical Procedures for Estimation of Environmental Lead

Though there are several techniques for analysing environmental lead, the spectrophotometric method is the most popular. In this method, lead is extracted from the environmental sample as a coloured lead-dithizonate complex. The colour intensity of the complex reflects the concentration of lead.

If lead is to be determined in blood, tissue, soil or vegetation, the sample is digested in a mixture of 80% nitric acid, HNO_3 and 20% perchloric acid, $HClO_4$. Perchloric acid serves to oxidise organic constituents of the sample to carbon dioxide and water, while nitric acid converts the lead content into soluble lead nitrate. If the atmospheric content of lead is to be determined, then air is passed through filter materials, such as fibre glass, asbestos or porous plastic. The residue left on the filter is digested in nitric acid-perchloric acid mixture. If lead is to be determined in water, a measured quantity of the sample is evaporated to dryness. The residue which is left out is digested in nitric acid-perchloric acid mixture.

The pH of the digested solution is adjusted to 8.5–9.0 by adding a buffer. Next, a mixture of ammonium citrate, potassium cyanide and hydroxylamine is added. Ammonium citrate prevents the precipitation of alkaline earth and transition metal hydroxides and phosphates. Potassium cyanide is a masking agent to prevent the extraction of metals, such as copper and zinc. Hydroxylamine is added to prevent the oxidation of dithizone.

A solution of dithizone (Fig. 22.3) in an organic solvent like carbon tetrachloride is then added. The lead content of the sample is converted into a red lead-dithizonate complex (Fig. 22.3). This complex, being soluble in carbon tetrachloride, passes into the organic phase. The aqueous phase is discarded. The absorbance of the organic phase is measured at 510 nm. Following the Lambert-Beer's law, the intensity of absorbance is proportional to the concentration of lead-dithizonate which, in turn, is proportional to the concentration of lead in the environmental sample.

Antidotes to Lead Poisoning

The permissible limit of lead in food is 2 ppm. Lead content of blood should not exceed 0.25 ppm. Drinking water should not have more than 0.05 ppm lead content. Beyond these limits the symptoms of lead poisoning begin to manifest themselves.

As in the case of mercury, the treatment of lead poisoning is carried out with

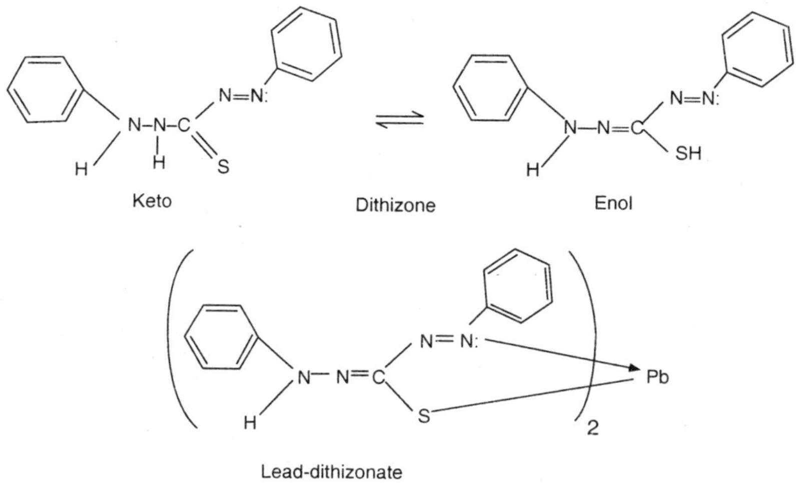

Fig. 22.3 Structures of dithizone and lead-dithizonate

potent compounds, called chemotherapeutic agents. These agents tend to sequester lead from the body tissues and bind it in form of soluble complexes. In soluble form, lead is excreted through the kidney. Initially two agents are administered by injection: British antilewisite (BAL, Fig. 22.2) and ethylenediamine tetra-acetic acid (Fig. 28.3, p. 391). With these complexing reagents, very high tissue levels of lead are rapidly reduced to levels approaching normal and the adverse metabolic effects are suppressed. After the lead level has been reduced, another chemotherapeutic agent, D-pencillamine (Fig. 22.2) may be administered orally as a follow up therapy.

Abatement Procedures against Lead Poisoning

In order to prevent the release of lead to the environment, control measures have to be instituted at every stage of lead processing: Mining, extraction and manufacture of commodities.

The practice of transporting lead ore in open railroad cars or trucks is being discontinued. The cars are covered with a protective layer of latex which is easily removed during extraction. Within the extraction units, the ore is transferred in enclosed belt conveyors The lead smelting and refining plants are located in isolated regions. The area around the smelter is brick-paved, so as to prevent the absorption of lead residues by soil or vegetation.

The most commonly used technique to reduce airborne lead emissions is the electrostatic precipitator. The details of this technique are outlined in Chapter 20, p. 301. The lead content of natural water systems is removed by adding lime or dolomite. Lead gets precipitated as insoluble hydroxide or carbonate. The insoluble compounds settle down through sedimentation and become a part of the sediments.

Lead-based commodities are being replaced by more environmental-friendly materials. The use of titanium dioxide in place of lead pigments in paints has brought down cases of pica. Unleaded petrol is gradually replacing the conventional leaded gasoline. Reformulation of gasoline to produce equivalent octane rating in the absence of tetraethyl lead is accomplished by increasing the aromatic hydrocarbon content of gasoline feedstocks and by installing catalytic converters in the automobiles.

In the new townships, emphasis is being placed on urban planning strategies to reduce lead pollution. Major sources of lead in the urban environment originate from automobile exhaust emissions and from industrial stacks. Accordingly, planning to assure minimum human exposure to high lead concentrations requires locating main traffic arteries, automobile parking complexes and industrial sites in such pockets where the prevailing winds and atmospheric transport directions are away from the areas of high density dwellers. This strategy may seem to be of academic interest only, yet it highlights the increasing importance of pollution control measures.

Inference

Though all heavy metals are cumulative environmental pollutants, mercury and lead are especially dangerous because they possess extremely high neurotoxic characteristics. Even below their clinical thresholds, these metals produce short term and long-delayed effects in behaviour, intelligence and central nervous system.

The chemotherapeutic agents may remove the heavy metals from the body quite efficiently, yet they cannot completely reverse the neurotoxic effects. This is especially true of children below the age of five years. Thus, patients treated for heavy metal exposure continue to experience recurring episodes of poisoning. This means that treatment alone is not sufficient. The heavy metal poisoning must be prevented through control measures.

Since the beginning of heavy metal mining, several million tonnes of mercury and lead have been taken from beneath the earth where they were buried in an inert form and subsequently released into the environment. The prevention of further discharge is only a small portion of the abatement problem, for there is no known terrestrial process that will lock-up the already released metals in an inert form.

References

Bryce-Smith, D. (1972). Behavioural effects of lead and other heavy metal pollutants, *Chem. Brit.*, **8,** 240–243.

Chisolm, J.J. (1971). Lead poisoning, *Sci. Amer.,* **224**(2), 15–23.

Chow, T.J. (1973). Our daily lead, *Chem. Brit.*, **9,** 258–263.

Day, M. (1996). Past lead pollution poisons unborn children, *New Scientist,* June 22, p. 4.

Gavis, J. and Ferguson, J.F. (1972). The cycling of mercury through the environment, *Water Res.*, **6,** 989–1008.

Goldwater, L.J. (1971). Mercury in the environment, *Sci. Amer.,* **224**(5), 15–21.

Jannett, J.C., Wixson, B.G. and Lowsley, I.H. (1979) in, *'Lead in the Environment'*, Boggess, W.R. and Wixson, B.G. (Eds.), Castle House Publications Ltd., Texas, p. 207–219.

Klein, D.H. (1972). Some general and analytical aspects of environmental mercury contamination, *J. Chem. Ed.*, **49,** 7–10.

Krishnamurthy, S. (1992). Biomethylation and environmental transport of metals, *J. Chem. Ed.*, **69,** 347–350.

Lanphear, B.P. (1998). The paradox of lead poisoning prevention, *Science,* **281,** 1617–18.

Maxwell, K.E. (1973), *'Environment of Life'*, Dickenson Publishing Co., Inc., California, p. 40–56.

Moore, J.W. and Ramamoorthy, S. (1984), *'Heavy Metals in Natural Waters',* Springer-Verlag, New York, p. 100–160.

Rabenstein, D.L. (1978). The chemistry of methylmercury toxicology, *J. Chem. Ed.,* **55,** 292–296.

Ratcliffe, J.M. (1981), *'Lead in Man and the Environment',* Ellis Horwood Ltd., Chichester, p. 51–95.

Sodhi, G.S. (1987). Biomethylation and bioamplification of mercury in aquatic environment, *Chemistry Education,* **4**(2), 31–34.

Wessel, M.A. and Dominski, A. (1977). Our children's daily lead, *Amer. Sci.,* **65,** 294–298.

CHAPTER 23

Soaps and Detergents

Soaps are the sodium or potassium salts of fatty acids. They are manufactured by the saponification of fats and oils. Synthetic detergents or *syndets* were developed as soap substitutes in an economy which was running short of edible fats and oils. A common class of detergents is the alkylbenzene sulfonates. Both the soaps and the detergents maintain an optimum hydrophobic-hydrophilic balance in their structures. The fatty acid chain in soaps and the alkylbenzene group in detergents are hydrophobic or water repelling in character. The carboxylate unit in the former and the sulfonate function in the latter are hydrophilic or water attracting in nature. Organic compounds whose molecules have both hydrophobic and hydrophilic groups lower the surface tension of water to aid in wetting of clothes, dishes or hard surfaces and in emulsifying and dispersing soil particles. Such compounds are called surface active reagents or *surfactants*. Because of their cleansing properties, the surfactants have become an inseparable part of the modern society. However, surfactants also produce a host of environmental problems which must be taken into account when the industrial importance of these products is being assessed.

The Need

It might seem trivial to state the utility of soaps and detergents, for even a layman is well versed with their cleansing properties. However, in industrial and institutional markets, many surfactant materials find their way to more than just cleaning jobs. This is not surprising since by varying the organic substituents on the hydrophobic functions of the surfactants, a multitude of properties and products may be produced.

Thus, the surface active agents are widely used to separate oil from water in enhanced oil recovery techniques. In road construction, detergents are used as emulsifiers. A detergent-induced emulsion of tar and water ensures quick application of tar on the road. As emulsifiers they are also introduced in agricultural products such as fertilizers, pesticides and insecticides.

Most soaps and detergents possess germicidal and bactericidal properties. For this reason, they are often incorporated in the cosmetic products. Surfactants are also used as fabric softeners. Some varieties of detergents, based on polyoxyethylene derivatives, find applications as deinking agents and as penetrants. Others are used for preparing coatings for metal protection and corrosion inhibition. The quaternary nitrogen-based detergents are used as sensitizers and floating agents. They also possess anti-static properties and are, therefore, used to remove the static charge associated with synthetic fibres. Detergents based on sucrose ester structures are non-toxic, tasteless and edible. They are used in toothpastes, in food industry and in pharmaceutical formulations.

The Classification

Surfactants fall into four major categories, depending on their ionic activity. These are: Anionic, cationic, non-ionic and amphoteric. The last three categories include only synthetic detergents. The anionic surfactant category, however, includes both soaps and detergents. The strucutures of some representative examples from each category are depicted in Fig. 23.1.

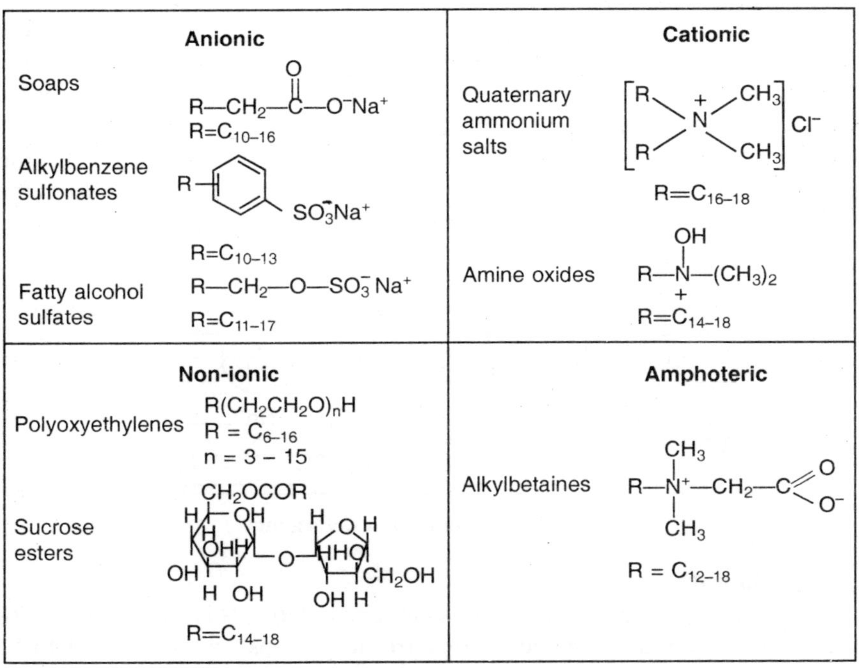

Fig. 23.1 Classification of surfactants

The most widely used variety of surfactants belongs to the anionic class. Besides the conventional powdered grades, they are also available in liquid and spray-dried formulations. Toilet soap bars, heavy-duty laundry detergents and shampoos are the major end products containing anionic surfactants. Cationic surfactants may be formed from nitriles, amines or quaternary nitrogen bases. At least one hydrophobic group is attached directly to a positively charged nitrogen atom. Cationic detergents are being increasingly used as fabric softeners.

The most common non-ionic type of surface active agents are polyoxyethylene derivatives. These are popular ingredients for heavy-duty laundry formulations. The non-ionic surfactants are versatile industrial products in the sense that by varying the number of ethylene oxide groups, their properties may be modified from a water-soluble to oil-soluble products. Amphoteric surfactants contain both cationic and anionic groups, These are the most expensive variety of detergents and are mainly used in shampoos and bubble baths.

The Characteristics

The characteristics of surfactants which have a direct bearing on environment include their mode of action, their biodegradability and the nature of additives incorporated in their formulations.

Mode of Action

Soaps and detergents have similar mechanisms of cleansing action. Both contain a non-polar, hydrophobic group and a polar, hydrophilic group. When these substances are placed in a polar solvent like water, their molecules reorient themselves on the surface so that the hydrophilic part is turned towards the solvent, while the hydrophobic end protrudes into the air in the interior of the bubble (Fig. 23.2). This phenomenon lowers the surface tension of water so that it penetrates the surface and interstices of the object being cleaned. If the binding of the detergent to the substance being cleaned is greater than the binding of the soiling agent, the latter will then be displaced into the aqueous phase.

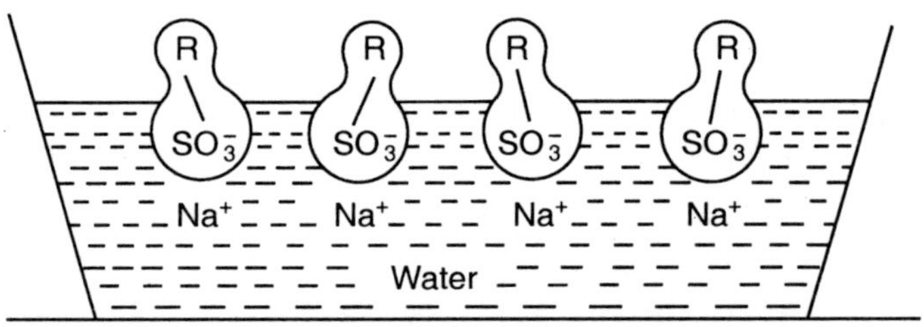

Fig. 23.2 Mechanism of detergency

Biodegradability

Although the cleansing mechanism of soaps and detergents is similar, their biochemical and chemical characteristics are not. Soaps are considered to be environmental-friendly products. They are 100% biodegradable. This means that when soaps are discharged into aquatic systems, they are acted upon by microorganisms in presence of oxygen dissolved in water, and are completely broken down into harmless, inorganic compounds like carbon dioxide, water, carbonates and bicarbonates. However, soaps have two disadvantages. Firstly, they do not function in hard water. The calcium(II) and magnesium(II) ions present in hard water intereact with soap to produce slimy precipitates.

$$2RCOO^-Na^+ + Ca^{2+} \rightarrow (RCOO)_2Ca + 2Na^+ \quad (23.1)$$

$$2RCOO^-Na^+ + Mg^{2+} \rightarrow (RCOO)_2Mg + 2Na \quad (23.1)$$

The precipitates adhere to the fabrics and stain them. Secondly, soaps cannot be used for industrial purposes in acidic medium. At low pH, the soap is precipitated as fatty acid.

$$RCOO^-Na^+ + H^+ \rightarrow RCOOH + Na^+ \qquad (23.3)$$

These disadvantages of soap paved the way for a large scale emergence of detergents. The detergents can function in hard water and also in acidic solutions. However, detergents are not as environmental-friendly as soaps. They are not completely biodegradable. That is, microorganisms cannot completely convert them into harmless end products. Once discharged into the streams, a part of these chemicals always persists, producing environmental problems.

It has been observed that detergents having a branched chain hydrophobic group are less biodegradable than those having a linear carbon chain (Fig. 23.3). The reason being that the microorganisms first nibble away the branched substituents and only then severe the main chain. Thus, the branched chain alkylbenzene sulfonates are only 50–60% biodegradable, while the unbranched analogs, the linear alkylbenzene sulfonates, are 90–95% biodegradable. Likewise, in the fatty alcohol sulfate category of detergents, the branched chain, Tergitol 7, is less biodegradable than the straight chain Teepol.

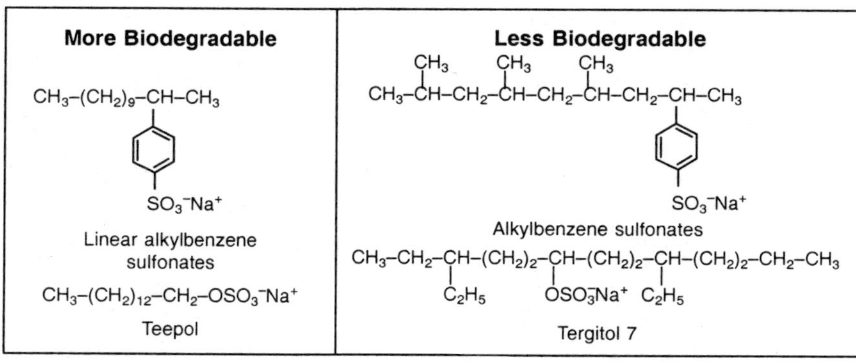

Fig. 23.3 Biodegradability of detergents

Additives

Important additives present in surfactant formulations are listed in Table 23.1, along with their significance and examples. From an environmental point of view, the most important additives are the builders.

Although detergents are not precipitated in hard water, their cleansing efficiency is curtailed markedly by the presence of calcium(II) and magnesium(II) ions. Builders are added to syndets to avoid this problem. The most popular/useful builders are, sodium tripolyphosphate, $Na_5P_3O_{10}$; potassium pyrophosphate, $K_4P_2O_7$; and sodium trimetaphosphate, $Na_3P_3O_9$. Their structures are presented in Fig. 23.4.

The builder's primary role in a detergent is as a sequestering agent which ties up calcium(II) and magnesium(II) ions that are present in hard water.

$$K_4P_2O_7 + 2Mg^{2+} \rightarrow Mg_2\,(P_2O_7) + 4K^+ \qquad (23.4)$$

However, builders perform other functions as well. Builders are typically added to a detergent formulation to extend or improve its cleansing performance across a wide range of use conditions. Therefore, in some formulations as much as

Table 23.1 Surfactant additives

Additives	Functions	Examples
Builders	To make cleansing action possible in hard water and to improve cleansing efficacy	Polyphosphates
Optical brighteners	To counteract yellowish appearance of clothes by their blue fluoresce	Ultramarines
Bleaching agents	To eradicate stains	Perborates
Enzymes	To hydrolyse and remove starch-based and protein-based stains	Amylase
Anti-redeposition agents	To prevent redeposition of soil on clothes	Sodium carboxy-methyl cellulose
Fabric softeners	To give clothes a soft feel	Quaternary ammonium salts
Hydrotropes	To hold less soluble surfactants and additives in aqueous solution	Sodium xylene sulfonate

40% of the total weight is in form of builders. The combination of builders and surfactants exhibits a synergistic effect which boosts the cleaning efficacy of the mixture as compared to an equal amount of either compound alone. Builders also make the wash solution alkaline, which is necessary for the effective removal of soil particles. Further, they buffer against pH changes that may occur during washing. Lastly, builders help the surfactant molecules in a concentrated wash solution to aggregate in groups called *micelles.* The micelles dissolve oil and grease particles and thereby remove them from the object desired to be cleaned.

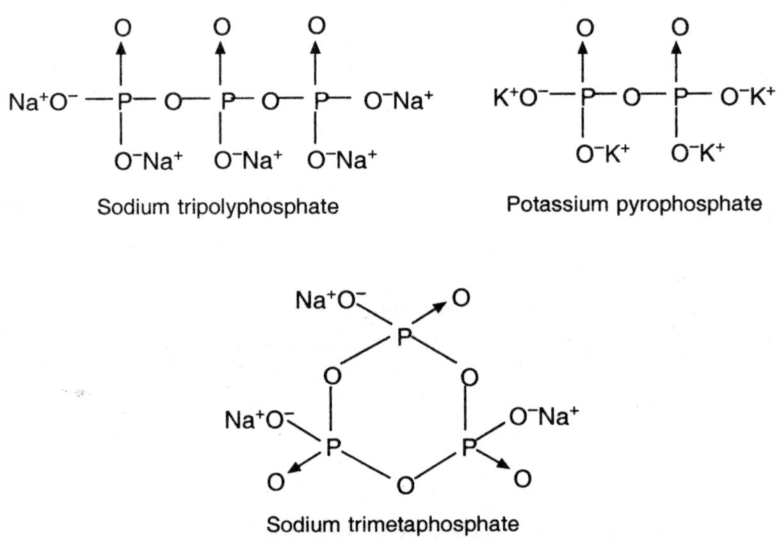

Fig. 23.4 Structures of some common builders

However, phosphate-based builders also cause environmental problems. They cause extensive colonies of algae to bloom in lakes, resulting in eutrophication. Therefore, some formulations incorporate non-phosphate builders.

Environmental Impacts of Soaps and Detergents

Some important environmental problems associated with the growing use of surfactants are described below.

1. Eutrophication

Detergents containing polyphosphate builders have been under indictment for some time as a significant cause of the eutrophication of lakes. For the same reason, phosphate-based fertilizers are being censured. Phosphates are believed to contribute to the unwanted growth of algae that often choke bodies of water and that, in their decay, can exhaust the oxygen supply, so that the lake becomes uninhabitable for fish and less attractive for recreation. The causes, consequences and mitigation of eutrophication are detailed in Chapter 29 p. 409.

2. Foam

It is now generally accepted that soaps and detergents are responsible for the present day foaming troubles in rivers. Often when a new product enters the market, its free samples are generously distributed and the inevitable result is a big increase in froth at the local sewage works and in the river receiving sewage works effluents. Although there is no correlation between foam stability and detergency, some of the most widely used surfactants produce foams that may persist for days or even weeks. Scenic rivers may become blanketed with foam and develop an unaesthetic appearance. At times the blanket is so high that it is nicknamed *foam iceberg*. Being light, the foam is easily blown to distant places.

Under normal circumstances, natural bodies of water absorb oxygen from the air. The rate at which the oxygen is absorbed by water from air is called *exchange coefficient*. Due to the presence of heavy foam on the surface of the aquatic system receiving effluents rich in detergents, the reaeration of water is curtailed. The presence of syndets lowers the exchange coefficient of oxygen by as much as 20%. Aquatic life of every sort is suffocated or destroyed.

3. Toxicity to Aquatic Organisms

Water plants and animals are adversely affected by synthetic detergents. Rooted plants like *Ranunculus aquatilis* are unable to grow if the concentration of syndets in water exceeds 2.5 ppm. A concentration of 3 ppm of common household detergents causes 50% mortality of trout fish in 12 weeks. Soaps are less toxic. They cause fish mortality only when their level exceeds 10 ppm. Moreover, it has been observed that the toxicity of soaps is greatly reduced in hard water due to their precipitation as insoluble calcium or magnesium salts (reactions 23.1–23.2). However, the syndets show the same toxicity in pure water as in hard water (Table 23.2).

4. Human Health

Some questions have been raised about the adverse health effects from enzyme additives present in detergent formulations. The enzymes are added to dissolve starch- or protein-based stains on clothes. However, enzymes attack proteins not only on soiled clothes but wherever they are. The workers in the detergent industry

Table 23.2 Toxicity of surfactants to fish

Type	Mimimum lethal concentration (ppm)	
	Pure water	Hard water
Soaps		
Sodium palmitate	10–12	900–1000
Sodium stearate	10–12	250–300
Detergents		
Sodium lauryl Sulfate (anionic)	6–7	6–7
Polyglycol ether (non-ionic)	2–3	2–3

are continually exposed to low levels of enzymes which attack the proteins residing in soft lung tissue. An allergic sensitization is induced and the workers suffer from respiratory ailments.

Another occupational hazard of surfactants results from their alkaline nature. Detergent formulations normally possess a pH of 9–10.5. Highly alkaline solutions can cause gel formation in protein tissue. Thus, workers in detergent industry often complain of skin problems. However, it must be stated that the exposure level and the frequency of occurrence of these occupational hazards vary from person to person.

5. Objectionable Taste of Potable Water

The presence of syndets in river waters used for potable supplies imparts objectionable taste to water. However, there is a considerable variation amongst individuals in their sense of taste. As little as 1 ppm of an alkylbenzene sulfonate detergent prompts complaints of mouldy or mildewy taste from many persons. Others are not so sensitive. When detergents are present to an extent of 1 ppm, their objectionable taste may be eradicated by chlorination of water. However, when the detergent concentration is in the 10–12 ppm range, the objectionable taste continues to persist even after chlorination. Below 0.5 ppm, detergents do not impart taste to drinking water.

6. Reduced Efficiency of Water Purification

During water purification—

(a) The organic pollutants are oxidised to harmless materials by microorganisms in presence of dissolved oxygen.

Detergents in sewage reduce this purification phenomenon firstly, because a good number of microorganisms are killed due to the toxicity of syndets and secondly, because the reaeration of water is curtailed due to froth formation.

(b) The greasy and oily impurities, being immiscible with water, are skimmed off.

Syndets emulsify grease and oil droplets and mix them thoroughly with water, making skimming operation impossible.

(c) The suspended and colloidal impurities tend to undergo sedimentation, that is, they settle down under the influence of gravity.

In presence of syndets, the sedimented impurities become attached to the froth bubbles and are dispersed to the surface.

All these factors reduce the efficacy of water treatment techniques.

Abatement Procedures for Soaps and Detergents Pollution

How is it that products which are supposed to help us improve our cleanliness and hygiene are considered to be pollutants? A part of the problem stems from the fact that syndets are only partially degraded by microorganisms. Hence, once discharged into a stream, syndets elicit toxicity, impart objectionable taste to water and reduce the efficiency of water purification techniques.

It must, however, be emphasized that the degree of biodegradability of all detergent formulations is not the same; some are more biodegradable than others. As has already been stated, the linear alkylbenzene sulfonate detergents are more easily degraded than branched chain alkylbenzene sulfonates (Fig. 23.3). Even amongst the alkylbenzene sulfonates the resistance to breakdown varies. An important factor controlling the degradation of these detergents is the distance between the sulfonate group and the end of the alkyl chain; the greater this distance, the faster is the decomposition by biochemical action. Moreover, the sucrose ester types of detergents are easily broken down by biological oxidation. It is, therefore, possible to structurally manipulate environmental-friendly syndets.

All types of surfactants, including soaps, produce foam. Yet, the foaming problem is largely attributed to anionic surfactants for these are the most widely used cleansing agents. Foam may be eliminated by adding cationic detergents to the effluents rich in anionic surfactants. The two counter ions interact to give an insoluble precipitate, which settles down under the influence of gravity. Overdosage of cationic detergents should be avoided for these are corrosive in nature. It has also been found that exposure of sewage to radiation from cobalt-60 degrades the anionic detergents and thereby eliminates foaming troubles. No residual radioactivity is left. However, the cost of this procedure is high.

The occupational hazards associated with detergents can be avoided by monitoring the formulation of these products. For example, the skin problems arise only when the pH of the detergent solution exceeds 11. Present day products give a maximum pH of 10.5. The enzymes, which cause allergic reactions amongst workers in detergent industry, are added in amounts much below their maximum allowable concentrations.

The use of phosphate builders in detergents has been debated, and in many places restricted, over the past 20 years. Phosphates are popular builders because of their effectiveness, reasonable costs and safety for use with washing machines and fabrics. The problem is that they nourish algae. The excessive growth of algae in lakes causes eutrophication.

One way to get around the problem is not to use any builders in detergent formulations. Such formulations, however, possess poor cleaning performance. A more reasonable approach, therefore, is to seek alternative builders. Some promising substances which can replace phosphates are depicted in Fig. 23.5.

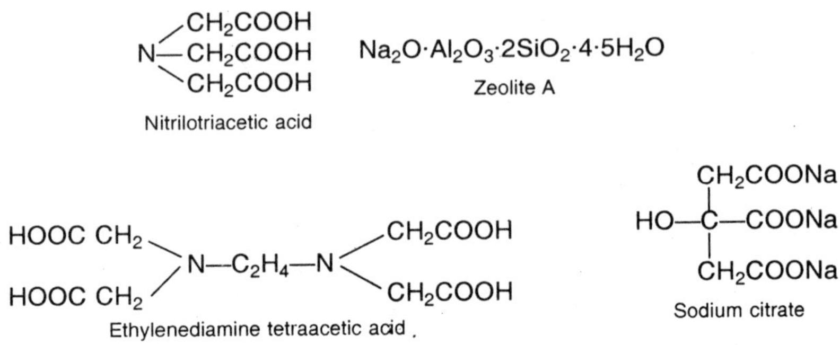

Fig. 23.5 Non-phosphate builders

Unfortunately, none of these is comparable to phosphates, as far as the detergency is concerned.

Nitrilotriacetic acid possesses excellent sequestration ability towards calcium(II) and magnesium(II) ions present in hard water. Therefore, at one time it was deemed to be a sure substitute for phosphate builders. Later, however, the compound was tested to be teratogenic. When discharged in water, it loosens heavy metals like mercury and cadmium, present in the sediments, through chelation. In chelated form, the metals may be transmitted across the placental barrier to fetus, thus increasing the likelihood of birth defects. Nitrilotriacetic acid is normally degraded rapidly under aerobic conditions. However, it does not degrade in anaerobic environments that are commonly found in septic tanks. Since the heavy metals occur widely in many aquatic systems, there is apparently some cause of concern.

Zeolite A is considered to be another substitute for phosphate builders. It removes calcium(II) ions from hard water by accommodating them in its pores. It is also effective in removing manganese and iron ions that can stain the fabrics. However, it cannot remove magnesium(II) ions, since the hydrated radius of magnesium (II) ion is greater than the pore size of zeolite A. Another drawback of zeolite A is that it is insoluble in water and, therefore, cannot be used in liquid surfactant formulations.

Sodium citrate is inferior in sequestrating effectiveness to and more costly that phosphate builders. Ethylenediamine tetraacetic acid possesses excellent sequestrating ability but its cost performance limits its usage. A combination of sodium carbonate and sodium silicate has also been tried as a builder. Sodium carbonate causes calcium(II) and magnesium(II) ions to precipitate as carbonates, while sodium silicate provides for buffering in wash water. However, the precipitated calcium and magnesium carbonates tend to redeposit on fabrics and leave a 'hard' feel on the clothes. The anti-redeposition agents are ineffective with carbonate precipitates.

We are back to square one. No other builder matches in its performance to phosphates. Yet, the silver lining is that a long term study on the problem of eutrophication has recently concluded that phosphate builders are not the environmental villains they have been painted to be. Algal problems are related to a combination of nutrients and not to any one nutrient. Phosphate certainly causes eutrophication, but only when other nutrients like carbon, nitrogen and potassium are also present. The critical phosphate concentration for causing eutrophication is also dependent on the concentrations of toxicants like heavy metal ions, pesticides and herbicides. If these toxicants are absent then organisms such as water fleas and daphnia begin to flourish in water bodies. These organisms graze on algae and prevent eutrophication, even if phosphate is present in abundance.

Thus, the conclusion is that reducing phosphate concentration is not enough to gain clear water. Therefore, there is no environmental reason to exclude phosphate builders from surfactant formulations.

References

Ainsworth, S.J. (1994). Soaps and detergents, *Chem. Eng. News,* January 24, p. 34–59.

Bailey, R.A., Clarke, H.M., Ferris, J.P., Krause, S. and Strong, R.L. (1978), *'Chemistry of the Environment',* Academic press, New York, p. 134–146.

Cheneval, J.P. (1993). The toxicity of detergents, *The World Scientist,* September, p. 56–65.

Deming, H.G. (1975), *'Water—The Fountain of Opportunity',* Oxford University Press, New York, p. 227–229.

Feierstein, G. and Morgenthaler, W. (1983) in, *'Riegel's Handbook of Industrial Chemistry',* 8th Edn., Kent, J.A. (Ed.), Van Nostrand Reinhold Co., New York, p. 450–487.

Greek, B.F. (1988). Detergent components become increasingly diverse, complex, *Chem. Eng. News,* January 25, p. 21–53.

Hammond, A.I. (1971). Phosphate replacements; Problems with washday miracle, *Science,* **172,** 361–363.

Layman, P. (1995). Studies help rehabilitate detergent phosphates, *Chem. Eng. News,* February 13, p. 21–23.

Stinson, S.C. (1987). Consumer preferences spur innovation in detergents, *Chem. Eng. News,* January 26, p. 21–46.

CHAPTER 24

Paper Mills

Paper is defined as the sheet material made by felting together the vegetable cellulose fibres which are contained in almost all plant materials. The fibres are first converted into pulp and subsequently treated with a myriad of speciality chemicals which serve as strengtheners, coatings and impregnations. The paper industry, therefore, employs a vast amount of natural resources, equipment, energy and synthetic materials. Obviously, it produces environmental problems too. As paper is ubiquitous, it is generally taken for granted. However, in recent years several pressure groups are targeting the paper industry to change its production methods, its chemical ingredients and its energy consumption so as to reduce its impact on the environment.

Paper Manufacture

Paper manufacture involves the following five stages.

Stage-I: Raw Materials

Raw materials for making paper are grouped in two broad categories: Fibrous and non-fibrous.

Fibrous raw materials

The primary source of fibrous raw materials for paper making is wood, the supporting and conducting tissue for the trees. Wood is composed of strong, relatively thick-walled, long cells. These, when separated from each other, are fine fibres, suitable for paper making.

Chemically, the cell wall tissue of wood is made up of two groups of polymers: The polysaccharides and the lignin. The polysaccharides constitute 70–80% of the woody tissue, with lignin making up the remainder. The polysaccharides of wood are collectively known as *holocellulose.* The major constituent of holocellulose is cellulose, a high molecular weight, linear polymer of condensed glucose units (Fig. 24.1). The minor components of helocellulose include linear and branched chain condensed polymers of pentose and hexose sugars. Lignin is a complex polymer of condensed phenylpropane monomeric units (Fig. 24.2). It serves as the adhesive material of wood, cementing the fibres together to form the anatomical structure of wood. Lignin is susceptible to degradation. Therefore, in paper making it is removed from wood, leaving the separated cellulose fibres in the form of pulp.

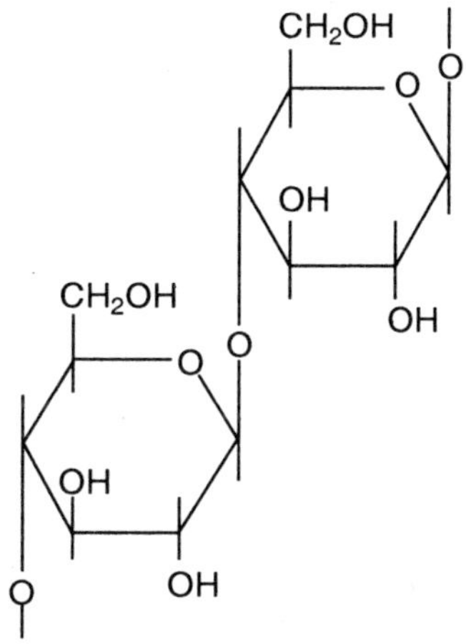

Fig. 24.1 Cellulose

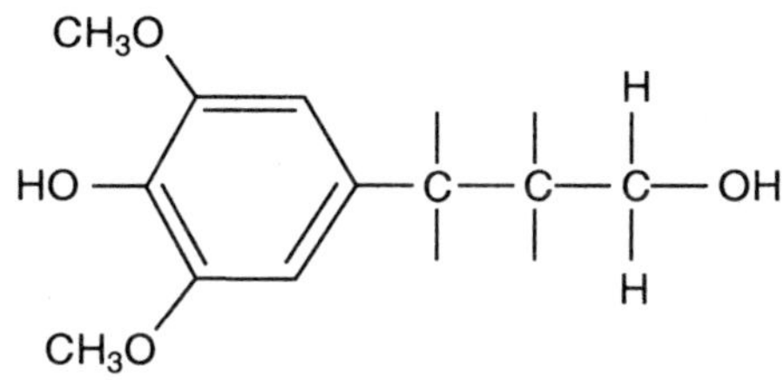

Fig. 24.2 A monomeric unit of lignin

Apart from wood, cereal straws, hemp, jute and bamboo may also be used as sources of fibres in paper manufacture. In order to increase the strength and tear resistance of paper, a small percentage of synthetic fibres like rayon or nylon are added.

Non-fibrous raw materials

For making paper for high grade applications, chlorine and chlorine dioxide are the common *bleaching agents. Sizing agents* like rosin are used for the purpose of decreasing the water absorption characteristics of the paper. This water repellency is a useful property for printing with water-based inks. Rosin is generally added as soluble sodium salt and then precipitated by introducing a solution of aluminium sulfate—the so called *paper-maker's alum.*

In order to provide opacity, the wood fibres are mixed with *fillers*, like calcium carbonate and pigments, like titanium dioxide. *Coating materials* like wax or latex are added not only to increase the water resistance of paper, but also to impart a glossy appearance. In this form the paper is commonly used for printing magazines.

Stage-II: Pulping

The purpose of pulping is to separate out the cellulose fibres of the wood from unwanted lignin. This must be done in such a way that the cellulose fibres are undamaged or are damaged to a minimum extent. Pulp produced from hardwood sources like deciduous trees tends to have shorter, weaker fibres as compared to that obtained from softwood sources like coniferous trees.

Wood is converted into pulp by mechanical, chemical or semi-chemical processes.

Mechanical pulping

Mechanical pulping consists of subjecting wood to the action of a grindstone in a stream of water. Previously, natural sandstone was used for grinding. However, modern grindmills employ either silicon carbide or aluminium oxide grits.

Mechanical pulp is quite weak. As a result, paper made from groundwood pulp soon loses strength and turns yellow. Therefore, mechanical pulp is used in relatively impermanent papers, such as newsprints. The advantage of this type of pulp is that it possesses excellent printing qualities and ink absorption characteristics. Moreover, it is produced inexpensively and utilizes the entire wood, giving essentially 100% yield.

Chemical pulping

Chemical pulping involves the removal of lignin by chemical treatment and separating out of cellulose fibres. Generally, the pulp obtained by chemical methods is tan to dark brown in colour. In this form, it may be used for making grocery bags, corrugated containers and wrapping sheets. However, if white paper is required, a suitable bleaching agent is added to the fibres liberated chemically from lignin.

Chemical pulping methods are carried out either under acidic conditions or under alkaline conditions. The most significant chemical-related change in recent years has been a move from acid to alkaline paper making. Alkaline paper is valued more because it lasts longer and is less susceptible to yellowing.

The acid process—also called the *sulfite process*—is based on the fact that lignin reacts with bisulfite ions, HSO_3^-, to form lignosulfonates that are soluble in water and hence can be drained off. Sodium, magnesium or ammonium bisulfites are generally used. The wood chips are pressure-heated with bisulfite for about 6h, at a temperature of 135–165°C, while the pH of the ingredients is maintained at 2.7–3.1. The lignosulfonates are removed by filtration.

The alkaline process initially used sodium hydroxide for digestion. The resin acids were converted into soluble sodium salts, which were filtered off. However, the amount of alkali had to be carefully monitored. If sodium hydroxide was present in excess, it degraded the cellulose fibres too, resulting in weak pulp.

This problem was overcome by replacing a part of sodium hydroxide by sodium sulfate. The modified method is called *sulfate or Kraft process.* In this process, the wood chips are digested under pressure with a mixture of sodium hydroxide and sodium sulfate. A part of sodium sulfate is reduced to sodium sulfide by the organic constituents present in wood. Likewise, some of the sodium hydroxide is converted into sodium carbonate. Sodium hydroxide is regenerated by treatment of residual liquor with slaked lime.

$$Na_2CO_3 + Ca\,(OH)_2 \rightarrow 2NaOH + CaCO_3 \qquad (24.1)$$

The spent liquor now contains a mixture of sodium sulfide and sodium hydroxide. To this sodium sulfate is added and the liquor is reused for digestion. This operation has several advantages. Firstly, sodium sulfide speeds up saponification of resin acids and accelerates pulping without damaging cellulose fibres. Secondly, the expensive sodium salts are conserved for reuse. Thridly, the pollution problem is alleviated to quite some extent. Last, but far from least, the pulp produced has an exceptionally good strength (Kraft in German means strength) and, therefore, can be used for high grade applications.

Semi-chemical pulping

The semi-chemical pulping method, as the name suggests, is a combination of chemical and mechanical procedures. In this technique, firstly a mild chemical treatment is given to affect delignification and then a mechanical treatment is given to segregate the fibres. The wood is digested with sodium sulfite, buffered with sodium bicarbonate or sodium carbonate to maintain a slightly alkaline pH. The digestion is followed by groundmill treatment, as described in mechanical pulping.

Stage-III: Stock Preparation

Stock preparation involves two operations: Furnishing and beating.

Furnishing

During this step, the pulp is dispersed in water using a washing machine type of equipment called *hydrapulper*. The pulp, at a consistency of 4%, is suspended in water maintained at about 60°C and agitated for about 1h. This operation ensures the segregation of individual fibres of cellulose. The suspension of fibres is called *paper slurry*.

Beating

Beating—also called *refining*—is meant to enhance the bonding capacity of fibres amongst themselves. As the fibres are brought into close contact, a large number of hydrogen bonds are formed between their surfaces. Beating is carried out by pressing the paper slurry between the two rotating surfaces of a *beater*. The greater the extent of beating, the stronger is the quality of paper.

After beating, the fibres are resuspended in the hydrapulper and such additives as sizing agents, fillers and pigments are added to the paper slurry.

Stage-IV: Paper Making Proper

The paper slurry is fed into a head box from where it is passed to a rotating loop of wire guaze. The head box controls the rate at which the slurry is introduced on the wire gauze. The slurry is pressed by table rolls to drain off water. As the slurry moves further, vacuum suction rolls remove much of the residual water. The fibre mat then passes through a series of steam-heated cylinders which bring about the final drying of the paper web. Finally, the web is pressed by a series of heavy rollers to produce a compact sheet. This step is called *calendering*.

If a glossy surface is required, the web is subjected to *super-calendering*. In this operation the fibre mat is alternately pressed by steel and hard-rubber rollers. Since the rollers are made from different materials, the paper sheet is subjected to a varying degree of friction. This differential friction has a polishing effect and produces a fairly high degree of glazing.

Stage-V: Finishing

Finishing step includes such surface treatments as coating, impregnation and lamination of paper. These operations confer special properties to paper, such as glossy appearance and water repellency. Above all, they increase the strength of the paper, so that it can withstand corrosive environments.

Environmental Implications of Paper Mills

Paper mills are responsible for a host of environmental problems. The important ones are listed below.

1. Water Pollution

By far, the greatest environmental problem associated with paper mills is that of water pollution. The discharge of spent water from pulp and paper making introduces two broad categories of pollutants into the natural aquatic systems. These are: Suspended matter and dissolved organic substances.

The suspended matter present in paper mills is in form of wood particulates, fibre debris and filling and coating materials, such as china-clay, kaolin and calcium carbonate. These suspended particles are detrimental in that they can form bottom deposits in receiving streams. They are not only inimical to aquatic life, but also unsightly. On decomposition they produce odoriferous components. They also render a stream opalescent, as a consequence of which the penetration of sunlight into the water is greatly reduced. This, in turn, reduces photosynthesis and thus adversely affects the aquatic life forms.

The dissolved organic compounds present in spent liquors too cause a variety of effects on streams. These pollutants are slowly acted upon by microorganisms and are transformed into simple, harmless, inorganic compounds. But in this transformation oxygen, dissolved in water, is consumed and some of the aquatic organisms are choked to death. Some of the dissolved compounds like polysaccharides and proteins stimulate the growth of slime organisms, causing biological imbalance to occur. Others, such as lignins and tannins, discolour water and impart to it an unaesthetic appearance.

2. Air Pollution

The delignification of wood by alkalis produces alkyl sulfides, like dimethyl sulfide, and mercaptans, like methyl mercaptan. These products are volatile and, therefore, escape to the atmosphere. They are not only toxic, but also possess offensive odour. However, as far as pulping is concerned, the air pollution problem is a relatively minor one, being restricted to the areas in the immediate vicinity of the mills.

3. Land Pollution

During paper manufacturing, huge quantities of sludge are produced. These contain mainly fine and coarse wood debris. Small amounts of residual additives like calcium carbonate, china-clay and titanium dioxide may also be present. The sludge is first dried and then disposed off in landfills. Solid waste disposal problems arise not only at the paper processing stage, but also at the postconsumer stage. *Postconsumer wastes* are those which are generated by end users, that is, by the consumers. Paper and paper products are the largest postconsumer sources of solid wastes. These wastes too need disposal sites.

4. Toxicity

Chlorine, the commonly used pulp-bleaching agent, is being targeted by environmental groups. No doubt, chlorine has excellent bleaching ability, but once it is discharged in aquatic systems, it interacts with other industrial effluents to produce a host of chlorinated organics, the most malicious member being 2, 3, 7, 8-tetrachlorodibenzo-p-dioxin (Fig. 24.3). The latter is persistent in the environment for prolonged periods, has a tendency to bioaccumulate in the food chains and elicits toxic effects to humans (see also Chapter 31, p. 437).

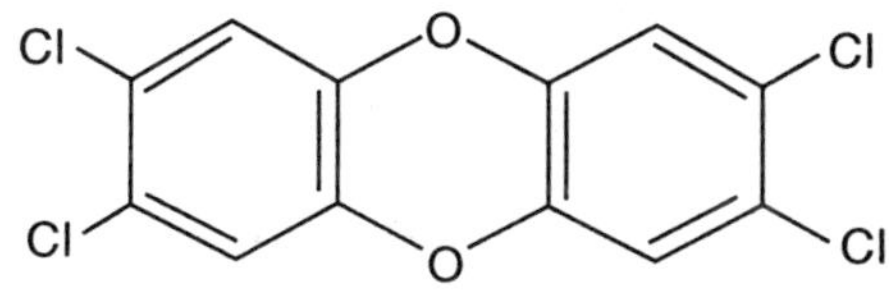

Fig. 24.3 2, 3, 7, 8-tetrachlorodibenzo-p-dioxin

The liquor left out after pulping is rich in sodium salts of resin acids and higher mercaptans. When present in high concentrations, these derivatives are toxic to fish and other aquatic organisms.

5. Ecological Degradation

The major raw material for paper manufacture is wood, a renewable resource. However, acquisition of wood for pulp making has visible negative impacts on the landscape in form of construction of road access and typical forest cutting practices. When the deforested area occupies an extensive portion of a watershed, it increases the maximum and decreases the minimum flow of streams draining the watershed. This, in turn, increases the likelihood of floods and droughts in the areas served by these streams.

Abatement of Paper Mill Pollution

The environmental degradation resulting from paper manufacture may be curtailed by the following measures.

1. Sedimentation Tanks

In order to reduce the problem of water pollution, the suspended solids in paper mill effluents are removed by sedimentation, that is, they are allowed to settle under the influence of gravity and then removed from the base of settling tanks. The fillers, pigments and coating materials might be isolated from the settled particulates and reused. When the suspended particles are too light, they keep floating on the surface. In such instances, coagulants, such as alum or silica, are added to speed up their rate of settlement.

2. Stabilization Basins

The problem of water pollution is also obviated by removing the dissolved organic contaminants in the paper mill effluents, in special type of tanks called stabilization basins. Here, the organic constituents of the effluents are biologically oxidised in the presence of microorganisms and oxygen. The basins may be of naturally aerated or mechanically aerated varieties. The microbes multiply in the oxygen-rich environment and oxidise the complex, hazardous, organic compounds to simple, harmless, inorganic compounds.

However, the efficiency of the stabilization basins is reduced in the presence of suspended solids. The reason being that a part of these solids, over a period of time, gets converted into complex organics and thus increase the pollution load. Therefore, the inlet to the basin is provided with a trap to capture the particulate matter.

3. Recycling Water

The problem of water pollution is further curtailed by recycling the water used in processing of paper. Large quantities of water are employed in the production of pulp and paper products. However, water is not discharged to the streams after each cycle. It is recycled. This ensures savings in heat, fibre, chemicals and water requirements. Of course, it also reduces the discharge of noxious chemicals to the natural aquatic systems.

4. Alternative Bleaching Agents

A ban on chlorine-based bleaching agents is being proposed to prevent the formation of toxic, chlorinated organics. Therefore, other pulp-bleaching agents are gaining ground, the chief ones being hydrogen peroxide, ozone and oxygen. These agents do not leave any unwanted residues and, therefore, are considered environmentally favourable.

However, the shift from chlorine shall result in an economic problem. Both chlorine and sodium hydroxide are produced by the electrolysis of brine. The paper industry buys both the chemicals—chlorine for bleaching and sodium hydroxide for pulping. It, therefore, receives appreciable price incentives. When the ban on the use of chlorine comes into effect, the manufacturers would be buying only sodium hydroxide and thus lose the price incentives. Moreover, it would create an imbalance within the chlor-alkali industry.

5. Processing Spent Liquor

The land pollution problem is substantially reduced by utilizing spent liquor, rather than dumping it. The spent liquor may be concentrated by evaporation and then used as a fuel, as a road binder or as an adhesive for making briquettes. It may also be mixed with phosphate or ammonium salts and sprayed as fertilizer.

However, such approaches are costly. A better option is to process the liquor into useful chemicals. Not only does this solve the disposal problem, but it also helps defray the cost of processing the liquors. Vanillin, a food flavouring agent, has been commercially obtained from the spent liquor. Torula yeast has also been produced from the pulping spent liquor. This yeast is used as a protein concentrate and a source of vitamin B. Another product made from waste liquor is ethyl alcohol. Turpentine oil, a widely used solvent, and thinner are also extracted from pulp waste.

6. Recycling Paper

In order to curtail ecological degradation, it is advisable to recycle waste paper. The used paper is treated with sodium hydroxide and steam and then bleached to produce a pulp which is of sufficiently good quality to replace virgin pulp from wood. Recycling not only obliterates the postconsumer trash problem, but also reduces the extent of deforestation, and hence conserves the natural resources—and the ecology.

References

Deming, H.G. (1975), *'Water—The Fountain of Opportunity'*, Oxford University Press, New York, p. 220–223.

Gehm, H.W. (1965) in, *'Industrial Wastewater Control'*, Gurnham, C.F. (Ed.), Academic Press, New York, p. 357–373.

Hocking, M.B. (1991). Paper versus polystyrene: A complex choice, *Science*, **251**, 504–505.

Imhoff, K. and Fair, G.M. (1956), *'Sewage Treatment'*, 2nd Edn., John Wiley & Sons, New York, p. 265–266.

John, E.C. and Strauss, R.W. (1983) in, *'Riegel's Handbook of Industrial Chemistry'*, 8th Edn., Kent, J.A. (Ed.), Van Nostrand Reinhold Co., New York, p. 519–576.

Moore, J.W. and Moore, E.A. (1976), *'Environmental Chemistry'*, Academic Press, New York, p. 406–410.

Pankhurst, K.G.A. (1965). Chemistry in paper industry, *Chem. Brit.*, **1**, 563–568.

Thayer, A.M. (1993). Paper chemicals, *Chem. Eng. News*, November 1, p. 28–41.

CHAPTER 25

Water Treatment

If small amounts of biodegradable pollutants are present in the natural aquatic systems, they are easily decomposed by microorganisms into harmless materials and the water is said to have undergone *self purification*. There are several other mechanisms by which nature tends to purify water during the course of hydrologic cycle. When water crystallizes as ice or evaporates as vapours, it gets rid of the dissolved compounds. In slow moving streams, the coarse particles undergo sedimentation, that is, they settle down under the influence of gravity and water gets rid of suspended matter. Filtration through sand and porous rocks rids water of algae and weeds. Aeration of groundwater, as it trickles over rock surfaces, allows volatile impurities to be released to the atmosphere. Before the advent of exploding human population and the industrial revolution, natural purification phenomena were quite adequate to provide sufficient water of high purity. However, human activity has pushed the natural purification processes beyond their limits. Therefore, a series of purification operations are required to restore the natural characteristics and life sustaining qualities of water.

Water Purification

Water is purified in the following four stages:

(I) Preliminary treatment, which involves screening, for the removal of large objects and skimming, for the removal of grit.

(II) Primary treatment, where the suspended solids and colloidal matter are first allowed to settle down and then removed.

(III) Secondary treatment, where the organic pullutants, both suspended and dissolved, are oxidised in the presence of microorganisms.

(IV) Tertiary treatment, which involves the use of a number of techniques for removing non-biodegradable pollutants, ionic constituents and trace impurities.

These treatment stages are now described in details.

Stage-I Preliminary Treatment

The preliminary stage of water treatment involves the removal of two types of impurities: Firstly, those which are suspended in water as coarse matter and secondly, those which float or can be made to float on the surface. Suspended matter is eliminated by screening, while the floating matter is eliminated by skimming. In some purification plants, screening is followed by skimming; in others the operation is carried out in the reverse order.

Screening

The impure water is made to pass through screens made up of rows of iron bars with a spacing of 1–2 inches. The spacings are narrow enough to hold back sticks, rags, wood pieces, papers and polythene items. The iron bars are periodically cleaned to prevent clogging.

Skimming

Oil, grease and other impurities which are lighter than water rise to the surface and can be removed by mechanical skimming. The skimming technique can also be extended for the removal of impurities heavier than water by blowing compressed air into the polluted water. The air bubbles that are formed, attach themselves to the grit particles and lift them to the surface, from where they are skimmed off. The efficiency of skimming technique can be increased by adding *collecting* and *activating agents* that form a thin film of suspended particles to which air bubbles can attach themselves. Potassium xanthates serve as excellent collecting and activating agents. In addition, *foaming agents* like oil or glue may also be added to the skimming tanks. These agents stabilize the froth for a longer spell of time, so that the grit particles do not get detached from the bubbles before they can be skimmed off.

Stage-II Primary Treatment

The primary treatment stage of water purification involves settling of the colloidal or suspended matter as sludge. The two common methods used to settle out impurities are sedimentation and flocculation.

Sedimentation

Most of the solids suspended in polluted water are too fine to be screened out and too heavy to be skimmed off. These may, however, be removed by allowing them to settle down under the influence of gravity. This technique is called sedimentation. A sedimentation tank is generally shallow and radial shaped.

Polluted water is retained in the tank for 1–3 h, during which period the impurities settle down. The sludge which settles out is prone to undergo putrefaction. It should, therefore, not be left in the sedimentation tank for long. The tanks are equipped with mechanical gears to remove the sludge at periodic intervals. In order to reduce the accumulation of solids on exposed surfaces, other than the tank bottom, all troughs, beams and other structural components of the sedimentation tank slope steeply away from the horizontal axis.

Flocculation

Flocculation or chemical precipitation is meant to rid water of the extremely fine particulates. Ordinarily, these particulates take a very long time to settle down under the influence of gravity. The method of plain sedimentation is, therefore, not effective for their removal. Their rates of settlement can, however, be considerably enhanced by the addition of flocculating agents, such as potassium aluminium sulfate (potash alum), ferrous sulfate or ferric chloride.

$$K_2SO_4Al_2(SO_4)_3 \cdot 24H_2O \rightarrow 2K^+ + 2Al^{3+} + 4SO_4^{2-} + 24H_2O \quad (25.1)$$

$$FeSO_4 \cdot 7H_2O \rightarrow Fe^{2+} + SO_4^{2-} + 7H_2O \quad (25.2)$$

$$FeCl_3 \rightarrow Fe^{3+} + Cl^- \quad (25.3)$$

The fine, suspended particles in polluted water bear either a positive or a negative charge. The similarly charged particles repel each other and, therefore, have no tendency to coalesce. Hence, they remain suspended for a long time. When flocculating agents are added, the positive charges on the particulates are neutralized by SO_4^{2-} or Cl^- ions, while the negative charges are neutralized by Al^{3+}, Fe^{2+} or Fe^{3+} ions. Once the adhering charges are nullified, the particles come into contact with each other and coalesce. Their weight increases and they rapidly settle down.

The residual metal ions are transformed into hydrated oxides.

$$2Al^{3+} + 6OH^- + (x-3)\,H_2O \rightarrow Al_2O_3 \cdot xH_2O \quad (25.4)$$

$$2Fe^{3+} + 6OH^- + (x-3)\,H_2O \rightarrow Fe_2O_3 \cdot xH_2O \quad (25.5)$$

The hydrated aluminium and iron oxides are insoluble and tend to settle down rapidly. As they undergo sedimentation, they enmesh or adsorb the uncharged suspended particles and carry them to the bottom of the settling tank.

Reactions (25.4) and (25.5) occur only in alkaline conditions. If the polluted water is acidic, lime is added to adjust the pH in the alkaline range. The extra lime is converted into insoluble carbonate by the following steps.

$$Ca(OH)_2 \rightarrow Ca^{2+} + 2OH^- \quad (25.6)$$

$$OH^- + CO_2 \rightarrow HCO_3^- \quad (25.7)$$

$$HCO_3^- + OH^- \rightarrow CO_3^{2-} + H_2O \quad (25.8)$$

$$Ca^{2+} + CO_3^{2-} \rightarrow CaCO_3 \quad (25.9)$$

Calcium carbonate formed in reaction (25.9) enmeshes and removes the suspended particles in the same way as the hydrated aluminium and iron oxides do.

Stage-III Secondary Treatment

Secondary treatment of water involves the oxidation of dissolved and colloidal organic compounds in the presence of microorganisms. As in the case of self purification phenomena, the complex, hazardous organic compounds are biodegraded into simple, harmless, inorganic compounds (see Chapter 21, reactions 21.3–21.8, p. 308). Sufficient quantity of dissolved oxygen is a necessary condition for this process. The aerated conditions are obtained usually either by trickling filters or activated sludge tanks. In warmer regions the oxidation ponds may also be used.

Trickling filters

Trickling filters, also called *percolating filters* or *sprinkling filters,* consist of circular or rectangular beds, usually 3–5 cm at the top and 10–15 cm at the base with a depth of 1–3 m. They are provided with brick or concrete walls. The beds are packed with clinker, slag, stones, coal, gravel, etc. Such a medium is intended to serve as a habitat of bacteria, fungi and other macro- and microorganisms which play an important role in water purification.

The material constituting the medium should

1. be durable
2. be resistant to the action of pollutants
3. have a high ratio of surface area to weight
4. be rough enough to encourage the formation of biological film
5. have a large percentage of voids to permit the passage of air.

Modern filters incorporate synthetic plastic media which have a greater efficiency, better uniformity and larger surface area as compared to the conventional media.

A part of the sludge from the primary settling tank (Stage-II) is applied to the beds from above. Then polluted water is passed over the beds. Initially, water seeps through the beds practically unchanged. However, gradually a biotic community is established as a gelatinous layer on the surface of the beds. This layer contains bacteria, fungi, protozoa, etc. When water trickles through the biological layer, the microorganisms, in presence of dissolved oxygen, breakdown the putrescible organic impurities into carbon dioxide, water, nitrates, sulfates, phosphates and a brown, insoluble humus material. During the percolating filter treatment, the microorganisms already present in water are also retained on the beds and they supplement the purification phenomena.

Percolating filters are not 'filters' in the real sense of the term. Purification is not achieved due to the physical process of straining out the impurities but is largely brought about by contact with biological agents.

Although trickling filter technique is quite efficient and widely used, it suffers from a number of drawbacks. Firstly, an adequate period of time is required for the establishment of a well-balanced biological film before efficient purification can be achieved. Maturing of filters takes place more easily in summers than in winters and hence it is advisable not to start fresh beds during colder months. Secondly, the efficiency of trickling filters falls in winter. The percolating beds contain some macroorganisms like rotifers, nematodes, larvae etc. which feed on the biological film. These grazing organisms are useful scavengers in the sense that they prevent the filters from becoming clogged by an excessive accumulation of biological matter. During winters, however, the activities of these organisms are at a low ebb and, therefore, the filter beds become clogged. This phenomenon is called *ponding*. With the return of summer season, the macroorganisms once again become active and relieve the filter of excessive load.

Thirdly, a certain amount of unpleasant odour is associated with percolating filters. Fourthly, files and midges are present in large numbers around the beds. It is, therefore, suggested that the trickling filters should be located at some distance from the residential areas. Some of the problems associated with trickling filters are avoided by following the activated sludge method.

Activated sludge method

When water containing organic pollution load is aerated, some amount of sludge settles down. Soon microorganisms inhabit the sludge. When a fresh batch of polluted water is aerated over this sludge, it gets purified more efficiently than the previous one. More sludge accumulates and a greater density of microorganisms settle on it. Thus, after each purification slot, the sludge becomes more activated. The aeration time is, therefore, decreased after each slot because purification is achieved in shorter periods.

If sterile sludge is used, the water remains polluted even after prolonged periods of aeration. This shows that purification is achieved by the microorganisms inhabiting the sludge. As in the case of trickling filters, the microorganisms originally present in water also enter the sludge and help in the purification of fresh batches.

For efficient operation of this method, it is necessary that sufficient amount of activated sludge, containing biologically active microorganisms, must be present. Further, there should be sufficient turbulence to produce requisite aeration and also to bring larger volumes of water in contact with the sludge.

Activated sludge method does not suffer in efficiency because of seasonal variations; it is as efficient in winters as in summers. No odour emanates from an activated sludge plant. The problem of flies is also non-existent. No macroorganisms have to be inducted in the sludge. However, this method suffers from the drawback that when a massive quantity of sludge accumulates, a part of it has to be mechanically removed.

Oxidation ponds

Oxidation ponds are used in warmer climates to purify polluted water through an interaction between bactria and algae. The ponds are shallow lagoons, with an average depth of 1 m. Polluted water is made to flow through the ponds at a slow speed. The bacteria present in the pond decompose the biodegradable organic matter. Dissolved oxygen is consumed in this process and carbon dioxide, nitrates and phosphates are generated. While nitrates and phosphates are consumed by algae as food material, carbon dioxide is utilized in photosynthesis. The latter phenomenon releases oxygen. Thus, the dissolved oxygen which was consumed in microbial degradation of organic load is restored and the cycle restarts. A flow diagram showing the working of an oxidation pond is shown in Fig. 25.1.

Sludge disposal

The solid matter that settles down during the primary and secondary stages of water treatment is called sludge. The disposal of sludge is considered to be a troublesome task firstly, because the sludge is quite bulky due to its high water content and secondly, because it is putrescible due to its high organic matter

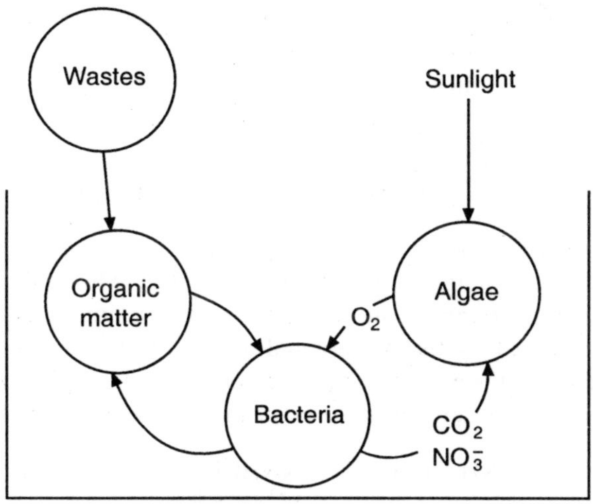

Fig. 25.1 Mechanism of water purification in an oxidation pond

content. Therefore, the water purification plants have a separate unit for the pre-treatment and disposal of sludge.

The pre-treatment of sludge is called *sludge digestion.* The digestion of sludge takes place in two steps. In the first step, the sludge is heated in a closed tank to a temperature of 27–35°C for 7–30 days. The microorganisms present in the sludge initiate its degradation. Oxygen, which is present in low concentration in the sludge, is quickly exhausted and, therefore, most of the decomposition takes place anaerobically. The materials initially attacked by the microbes are the easily available food substances like sugars, starch, cellulose and nitrates. Later on, more resistant materials, such as proteins and fatty acids, are attacked. Large volumes of a gas, called *sludge gas,* evolve during anaerobic biodegradation of sludge. In the second step, the residual solid matter is further digested in open tanks for 20–60 days at about the same temperatures as in the first step. By this treatment the sludge consolidates and dries up.

Thus, the pre-treatment of sludge results in two products: Dried residual solid and sludge gas. The dried solid may be directly dumped into the sea. It may also be used as a landfill. If it is devoid of toxic compounds, it may be used as a fertilizer. Sludge gas is a mixture of 65–80% methane and 20–35% carbon dioxide. It is combustible and liberates recoverable energy when used in water boilers or gas engines. Most of the time, however, it is used to heat the sludge digestion tanks.

Stage-IV: Tertiary Treatment

Tertiary treatment of water encompasses a set of techniques which remove the impurities that remain even after the first three stages of purification. These are mainly the soluble inorganic impurities. Depending upon the nature of pollution, one or more of these techniques may be used in this final stage of water treatment.

Chlorination

Chlorination refers to the addition to impure water of gaseous chlorine or compounds containing active chlorine, such as bleaching powder, CaO (OCl) or sodium hypochlorite, NaOCl. In aqueous solution, these compounds generate hypochlorous acid, which acts as a purifying agent.

$$Cl_2 + H_2O \rightarrow 2HOCl \tag{25.10}$$

$$NaOCl + H_2O \rightarrow NaOH + HOCl \tag{25.11}$$

Chlorination purifies water in three ways.

1. It disinfects water of pathogenic bacteria. Although almost 90% of microorganisms are removed during the secondary treatment stage, the remaining 10% are eliminated by the bactericidal action of chlorine. As a disinfectant, chlorine inactivates the enzymes that are essential for the life processes of bacteria.

When ammonia or amines are present in water, chloramines are generated.

$$RNH_2 + HOCl \rightarrow \underset{\text{Chloramines}}{RNHCl} + H_2O \tag{25.12}$$

Chloramines too are disinfectants. Although their ability to kill microorganisms is less than that of chlorine, yet chloramines persist for a longer time in water and, therefore, their purifying ability is more lasting.

2. Chlorination destroys or controls the growth of undesirable algae in the water treatment plants. Algae grow at an accelerated pace and clog conduits in the water purification units. Passage of chlorine through these conduits removes the unwanted algal growth.

3. Chlorination eliminates the odours associated with anaerobic decomposition of organic matter in water. Amongst the degradation products, hydrogen sulfide is the most common. It is destroyed by chlorination, preventing its foul odour being released to the air.

$$H_2S + HOCl \rightarrow H_2O + HCl + S \tag{25.13}$$

Although chlorination is a standard method of water treatment, it suffers from two drawbacks. Firstly, chlorine interacts with those organic impurities which persist in water even after the secondary stage of water treatment. A set of compounds, called *disinfection by-products,* are produced by such interactions. Some of these by-products like chloroform and carbon tetrachloride are carcinogenic. Thus, increasing use of chlorination for water purification has been linked to elevated rates of bladder and rectal cancers. It has, therefore, been suggested that the amount of chlorine being added to water must be carefully monitored, so that the threat of pathogens must be balanced against the statistical risk of cancer. Secondly, though chlorine acts efficiently against pathogenic bacteria, it is incapable of destroying most viruses.

Wet oxidation

Impurities present in water may be oxidised to harmless materials using a suitable oxidising agent like hydrogen peroxide. The oxidising ability of hydrogen peroxide is catalysed by ultraviolet light, under whose influence it is rapidly decomposed to hydroxyl radicals.

$$H_2O_2 + h\nu \rightarrow 2\dot{O}H \qquad (25.14)$$

The hydroxyl radicals immediately attack inorganic and organic pollutants. The sulfur-containing organic compounds are oxidised to sulfates, phosphorus-containing compounds to phosphates, cyanides to cyanates and halogen derivatives to halides. Often a number of intermediates, called *hydroxylated derivatives,* are formed. Even this transformation is beneficial because the hydroxylated derivatives are more biodegradable than the non-oxidised original substances.

By using the ultraviolet-catalysed wet oxidation process, no solid or liquid waste remains after purification. The gaseous products that are emitted are mainly oxygen from the decomposition of excess hydrogen peroxide,

$$2H_2O_2 \rightarrow 2H_2O + O_2 \qquad (25.15)$$

or carbon dioxide from the oxidation of organic compounds. Both the gases are non-hazardous. Another advantage of the wet oxidation method is that it is cost-effective.

Adsorption

A number of compounds which are resistant to biodegradation persist in water even after the primary and secondary treatment stages. Many of these compounds are carcinogenic and toxic. These may be eliminated by passing water over a bed of granulated, active charcoal. A wide variety of compounds, both in the dissolved and suspended states, are adsorbed by charcoal and thereby removed from water. When charcoal becomes saturated with impurities, it is heated in vacuum to about 1500°C and the adsorbates are driven off. The active sites on charcoal once again become bare and hence the adsorbent may be reused.

Charcoal is also effective in removing small amounts of dyes which impart undesirable colour to water. In addition, those derivatives which produce objectionable tastes and odours may also be adsorbed on charcoal and eliminated from water.

Reverse osmosis

The term osmosis refers to a process whereby relatively pure water flows into a concentrated solution through a semi-permeable membrane. This process is spontaneous because when a concentrated solution becomes diluted, the entropy of the system increases. The minimum pressure required to counteract osmosis is called *osmotic pressure,* π,

$$\pi = CRT \qquad (25.16)$$

where C is the concentration of the solution in moles per liter, R is the gas constant and T is the absolute temperature.

If pressure is increased beyond the value of osmotic pressure, water begins to flow in the reverse direction. This process is called reverse osmosis or *ultrafiltration.* Impure water, containing dissolved solutes, is placed above a semi-permeable membrane and subjected to high pressures. Pure water flows down the semi-permeable membrane. The solute molecules, whether ionic or covalent, generally have larger diameters than the pores of semi-permeable membrane. Hence they are retained by the membrane, while pure water filters off. The working of reverse osmosis is shown in Fig. 25.2.

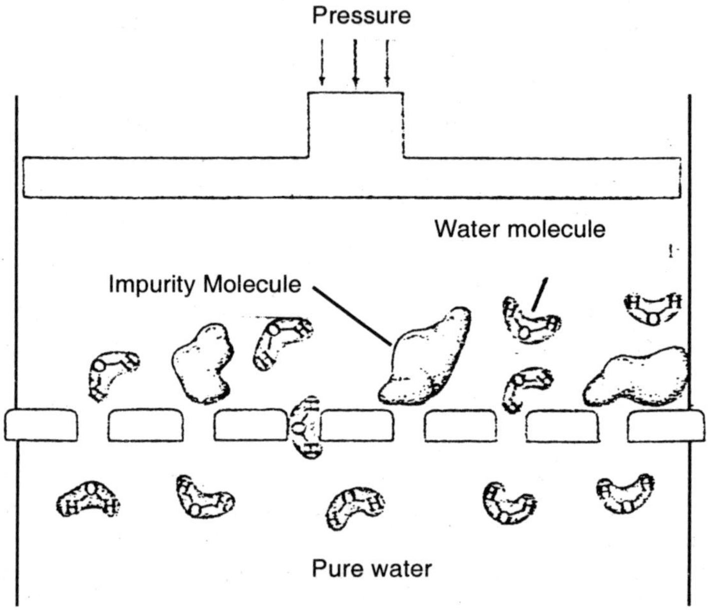

Fig. 25.2 Principle of reverse osmosis

There is a limit to which the mechanical pressure can be put on the solution. Application of extremely high pressure tends to tear the semi-permeable membrane. Three membrane materials are commonly used in reverse osmosis: Cellulose acetate, cellulose triacetate and polyamide. Over a period of time the pores of membrane become clogged. The membrane is then flushed with a 1% detergent solution. This cleans away small solid particles trapped in the pores. However, insoluble inorganic compounds like calcium carbonate and iron oxide are not washed out by detergents. These contaminants are removed by flushing the membrane with a solution of citric acid.

Electrodialysis

Electrodialysis technique is used to purify brackish water, that is, water containing large concentration of ionic impurities. It may also be used for the desalination of sea water for irrigation and drinking purposes.

The equipment for electrodialysis consists of an electrolytic cell divided into three compartments, using two semi-permeable membranes (Fig. 25.3). The orifices on the membrane near the cathode are coated with negatively charged ions. This membrane, therefore, repels the anions present in impure water, but allows the cations to permeate through it. Likewise, the orifices on the membrane near the anode are coated with positively charged ions. This membrane repels the cations but allows the anionic constituents of impure water to pass through it.

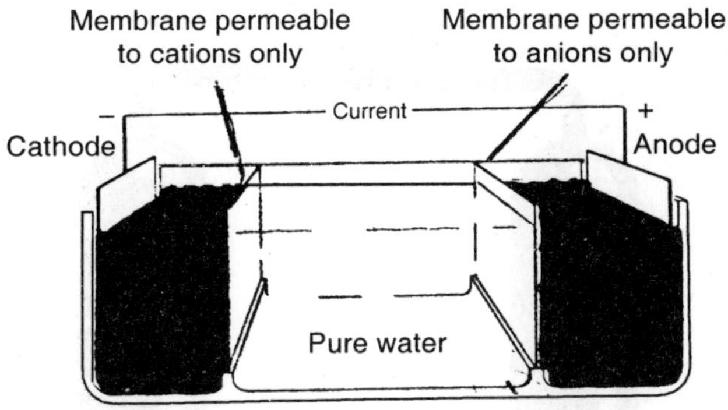

Fig. 25.3 Principle of electrodialysis

When current is induced, the cations present in impure water pass through the cation-permeable membrane and discharge at the cathode. Similarly, the anions have their way through the anion-permeable membrane and discharge at the anode. The central compartment of the cell thus becomes devoid of the ionic impurities.

Electrodialysis technique is ineffective when water contains a large amount of organic impurities. The organic molecules are generally larger in size than the inorganic ions. Therefore, the former become clogged in the orifices of the semi-permeable membranes. As a result, the passageway of the ions becomes blocked and the membranes have to be removed and washed with a 1% detergent solution to clear the orifices. However, the efficiency of the process deteriorates after each cleaning operation.

Ion exchange

The ion exchange technique is used for the purification of brackish water. Its equipment consists of a cation exchanger coupled to an anion exchanger (Fig. 25.4). Each exchanger incorporates a resin, which is a high moleculer weight polymeric material containing an ionic functional group. The functional group can exchange one ion for the other as water containing ionic impurities is passed over it.

Impure water is first passed through the cation exchanger. The resin of this exchanger generally contains ionic sulfonic acid group. The cations, M^+, present in impure water exchange with the hydrogen ion, H^+, of the resin.

$$\text{Resin/SO}_3^-\text{H}^+ + \text{M}^+ \rightarrow \text{Resin/SO}_3^-\text{M}^+ + \text{H}^+ \qquad (25.17)$$

Water is then passed through the anion exchanger. The resin of this exchanger generally contains ionic, quaternary ammonium hydroxide group. The anions, X^-, present in impure water exchange with the hydroxide ions, OH^-, of the resin.

$$\text{Resin/N}^+\text{OH}^- + \text{X}^- \rightarrow \text{Resin/N}^+\text{X}^- + \text{OH}^- \qquad (25.18)$$

The hydrogen ions, formed in reaction (25.17) and the hydroxide ions, formed in reaction (25.18) combine with each other.

$$\text{H}^+ + \text{OH}^- \rightarrow \text{H}_2\text{O} \qquad (25.19)$$

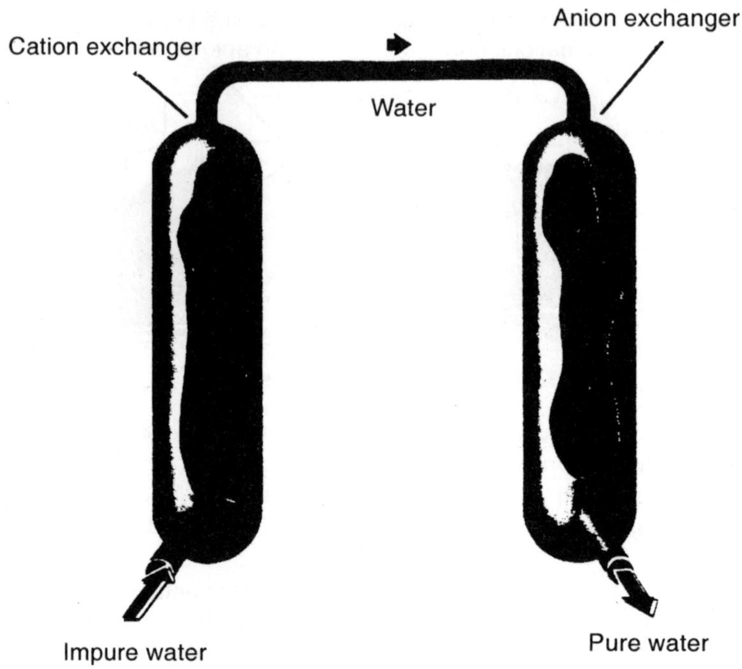

Fig. 25.4 An ion exchanger

The overall effect is the removal of cationic and anionic impurities present in water.

The cation exchanging resin is regenerated by treatment with a solution of a strong acid.

$$\text{Resin/SO}_3^-\text{M}^+ + \text{H}^+ \rightarrow \text{Resin/SO}_3^-\text{H}^+ + \text{M}^+ \quad (25.20)$$

Similarly, the anion exchanging resin is regenerated by passing a solution of a strong alkali through it.

$$\text{Resin/N}^+\text{X}^- + \text{OH}^- \rightarrow \text{Resin/N}^+\ \text{OH}^- + \text{X}^- \quad (25.21)$$

Brackish water often contains humic acid and fulvic acid impurities. Although these are organic compounds, they contain exchangeable carboxylate anions and are, therefore, absorbed by anion exchanger. However, they tend to become entangled in the polymeric structure of the resin with the consequent difficulty in removal during regeneration. This resin poisoning results in a slow, but steady, deterioration in the working capacity of the exchanger and in the quality of final treated water. Another disadvantage of the ion exchanger technique is that it is quite expensive.

Fluoridation

Epidemiological studies have indicated that fluoride ions protect the teeth against dental decay. Therefore, during the tertiary stage of water purification about 1 ppm of fluoride is added, either in form of sodium fluoride, NaF or as sodium hexafluorosilicate(IV), $Na_2[SiF_6]$.

Dental decay commonly occurs because of demineralization of the tooth mineral, hydroxyapatite. This mineral dissolves in organic acids that are formed in the mouth by microbial decomposition of sugars and fermentable carbohydrates in food.

$$\underset{\text{Hydroxyapatite}}{Ca_{10}(PO_4)_6(OH)_2} + 8H^+ \rightarrow 10Ca^{2+} + 6HPO_4^{2-} + 2H_2O \qquad (25.22)$$

The bacteria form an adhering film—*plaque*—on tooth surface so that the acids that are produced remain in intimate contact with the tooth for a long time. If fluoride is present in drinking water, it reacts with hydroxyapatite to form fluoroapatite, which is more resistant to acid attacks.

$$Ca_{10}(PO_4)_6(OH)_2 + 2F^- \rightarrow \underset{\text{Fluoroapatite}}{Ca_{10}(PO_4)_6F_2} + 2OH^- \qquad (25.23)$$

Epidemiological evidence also indicates that the presence of fluoride in drinking water curtails the incidence and severity of *osteoporosis*—a disease which causes a reduction in the density of bones amongst the aged. As with teeth, the fluoride gets deposited in bones by simple ion exchange with hydroxyl group of hydroxyapatite. The bones become more resistant to decay.

Unfortunately, the fluoride ion is peculiar among the trace constituents of water because the same range of human exposure that can produce beneficial physiological effects can also produce harmful health effects. Fluoride is useful if its concentration is drinking water is less than 1 ppm. It is harmful if its concentration exceeds 1.5 ppm.

If children consume drinking water containing excess fluoride, they suffer from an ailment called *dental fluorosis.* During the years of tooth formation, ameloblast cells in the jaws lay down a matrix called *amelogen.* This matrix forms the structural foundation upon which phosphate and calcium are deposited, giving rise to tooth enamel. Excess ingesition of fluoride interferes with the amelogen laying mechanism of ameloblast cells. In the absence of the protective enamel, the risk of tooth loss increases. Moreover, incorporation of excessive fluoride in the tooth mineral makes the teeth brittle and susceptible to fracture.

Excessive intake of fluoride also results in *skeletal fluorosis.* This disease, which affects the aged, results in a partial fusion of vertebrae. The density of the bones becomes abnormally high, crippling the patient. In addition, higher levels of fluoride in water are suspected to cause birth defects, cancer and kidney disorders. However, the association of these ailments with fluoride is yet to be authenticated.

Proponents of fluoridation believe that though there are a number of recognized potential risks associated with introduction of fluoride in water, yet these risks are quite remote and that large benefits justify society's taking these risks. Detractors oppose fluoridation for ethical reasons. They view it as a form of medication which, if unnecessarily imposed, results in violation of individual's rights. They advocate that fluoride should not be added to water during the tertiary purification stage. Those who deem that the benefits of fluoride outweigh the risks can fluoridate the water *they* consume. Alternately, they may use fluoride-based toothpastes. Ideally speaking, this is a right course. In real terms, however, the edification of individual's choice, in this case, loses its significance, for a common man has a little understanding of the complex environmental or health issues.

Criteria of Water Purity

The ways in which water is used are so varied that no single purity standard can be applied. If water is to be used for cleaning or washing purposes, no special treatment is required. If it is to be used for heating purposes in a manufacturing unit, the deposition of scale in the boiler must be prevented. In such cases the water must undergo ion exchange treatment so as to rid calcium, magnesium and bicarbonate ions—the constituents which make water hard.

If water is to be used for drinking purposes, all stages of water purification—preliminary, primary, secondary and tertiary—must be vigilantly monitored. Potable water must be colourless, tasteless and odourless. It should have no turbidity or suspended particulates. Radioisotopes too should be absent. The number of bacteria should be about 50 per 100 ml. The purity of water is further gauged in terms of three parameters: Dissolved oxygen, biochemical oxygen demand and chemical oxygen demand.

Dissolved Oxygen

The organic pollution load in water encourages the growth of a myriad of microorganisms. These microorganisms decompose the complex organic compounds into simple inorganic constituents like carbonates, bicarbonates, nitrates, sulfates and phosphates. The removal of hazardous organic compounds from water by microorganisms is called self purification (see also Chapter 21,. p. 307). The microorganisms which bring about self purification consume oxygen dissolved in water by way of respiration.

If the amount of dissolved oxygen is high, the degree of self purification of water would also be high. In such cases a milder water treatment operation shall be required. If, on the other hand, the amount of dissolved oxygen is less, some of the impurities would not be eradicated by self purification phenomena and hence advanced water treatment would be necessary in such cases. Thus, the level of dissolved oxygen is the most valuable analytical measure of the condition of polluted water.

Oxygen in natural waters originates from two sources. Firstly, it may be directly absorbed from the atmosphere and secondly, it may be released by aquatic plants during respiration. When water contains as much oxygen as it can dissolve, it is said to be saturated with respect to oxygen.

The level of dissolved oxygen depends on several factors. As shown in Table 25.1, the quantity of oxygen that can be dissolved in water decreases with temperature. Moreover, oxygen is less soluble in water containing dissolved salts. This means that at any given temperature, sea water will have a lower saturation value of dissolved oxygen than fresh water. This also means that a greater effort has to be exerted to purify sea water than river water. The solubility of gases in water is directly proportional to the atmospheric pressure. Thus, a lake at a higher elevation will have a lower value of dissolved oxygen than the one near the sea level.

Table 25.1 Saturation value of dissolved oxygen (ppm) in fresh and sea water

	Temperature, °C						
	0	5	10	15	20	25	30
Fresh water	14.6	12.8	11.3	10.2	9.2	8.4	7.6
Sea water	11.3	10.0	9.0	8.1	7.4	6.7	6.1

The *degree of unsaturation* or *saturation deficit, D* is defined as,

$$D = \left\{\begin{matrix}\text{Saturation value of} \\ \text{dissolved oxygen at} \\ \text{a given temperature}\end{matrix}\right\} - \left\{\begin{matrix}\text{Measured value of dissolved} \\ \text{oxygen at that temperature}\end{matrix}\right\} \tag{25.24}$$

The *percentage saturation, S* at a given temperature is defined as,

$$S = \frac{\text{Measured dissolved oxygen}}{\text{Saturated dissolved oxygen}} \times 100 \tag{25.25}$$

Biochemical Oxygen Demand

Biochemical oxygen demand or BOD is a standard measure of the amount of oxygen that would be needed by the microorganisms to cause biodegradation of a part of the organic and inorganic pollution load. The organic matter like starch and sugars are consumed by microorganisms as food. The inorganic constituents like sulfides, sulfites and ferrous ions are attacked by oxygen in presence of enzymes released by microorganisms. Ultimately, the pollution load is transformed into simple, non-hazardous compounds. Dissolved oxygen is consumed in this process.

In order to elucidate biochemical oxygen demand, the water sample is first saturated with oxygen. It is then incubated at a constant temperature, usually 20°C, for five days. This allows time for microorganisms in the waste water to mediate oxidation of pollutants. The remaining amount of dissolved oxygen is determined and biochemical oxygen demand is obtained by subtraction.

$$\text{BOD} = \left\{\begin{matrix}\text{Saturation value of} \\ \text{dissolved oxygen per} \\ \text{liter at 20°C}\end{matrix}\right\} - \left\{\begin{matrix}\text{Amount of dissolved} \\ \text{oxygen per liter at} \\ \text{20°C after 5 days}\end{matrix}\right\} \tag{25.26}$$

Biochemical oxygen demand is an indicator of the quantum of pollution load. If the pollution load is large, then greater amounts of dissolved oxygen shall be consumed, and according to equation (25.26), BOD shall be high. Conversely, a low value of BOD indicates relatively pure water. For drinking water the biochemical oxygen demand should be in the range 0.75–1.5 ppm.

Biochemical oxygen demand does not measure the contamination caused by any specific chemical. It is a measure of the contamination caused by the totality of those compouds which can be oxidised in the presence of microorganisms. A large number of organic and inorganic compounds, however, are resistant to microbial oxidation. They, therefore, do not contribute to the biochemical oxygen demand; yet their presence makes water unfit for consumption.

Chemical Oxygen Demand

At times, biochemical oxygen demand does not give an accurate assessment of pollution load. As has been stated, some compounds, like detergents, are resistant to microbial degradation and hence do not add on to the biochemical oxygen demand. Some other pollutants, like cellulose, are slow to get oxidised. Thus, during the five days period required for elucidation of biochemical oxygen demand, they may be only partially oxidised. It is also probable that water sample in question may contain toxic compounds, like biocides, which may poison the microorganisms even before they can act on pollutants. Moreover, the five days period required to assess the water quality by biochemical oxygen demand method is quite long.

In order to avoid the problems associated with biochemical oxygen demand test, another analytical method, called chemical oxygen demand or COD, is sometimes preferred. In this case the water sample is treated with a known quantity of an oxidising reagent, usually potassium dichromate in acidic medium. This reagent oxidises most of the polluting substances, including those which are resistant to microbial oxidation. The remaining potassium dichromate is determined by back titration with a suitable reducing agent like Mohr's salt. The amount of potassium dichromate consumed is determined by subtraction.

$$\begin{matrix}\text{Concentration (ppm) of}\\ \text{potassium dichromate}\\ \text{consumed}\end{matrix} = \left\{\begin{matrix}\text{Concentration of}\\ \text{potassium dich-}\\ \text{romate initially}\\ \text{added}\end{matrix}\right\} - \left\{\begin{matrix}\text{Concentration of}\\ \text{potassium dichromate}\\ \text{determined by back}\\ \text{titration}\end{matrix}\right\} \tag{25.27}$$

From the concentration of potassium dichromate consumed, the amount of oxygen used in the oxidation may be calculated using the following chemical equation.

$$K_2Cr_2O_7 + 4H_2SO_4 \rightarrow K_2SO_4 + Cr_2(SO_4)_3 + 4H_2O + 3(O) \tag{25.28}$$

The results are expressed in terms of the amount of oxygen, in ppm, that would be required to oxidise the contaminants.

Although chemical oxygen demand is a fast and reliable method for assessing the water quality, it suffers from the drawback that aromatic hydrocarbon derivatives are not easily oxidised by most chemical oxidising agents, including potassium dichromate. Hence, their presence cannot be judged by chemical oxygen demand determination.

References

Arden, T.V. (1976). Ion exchange, *Chem. Brit.,* **12,** 285–288.

Cross, J. (1990). Water purification—surviving the changes, *Chem. Brit.,* **26,** 579–580.

Hileman, B. (1988). Fluoridation of water, *Chem. Eng. News,* August 1, p. 26–42.

Imhoff, K. and Fair, C.M. (1956), *Sewage Treatment,* 2nd Edn., John Wiley & Sons, Inc., New York, p. 162–213.

Jones, M.M., Netterville, J.T., Johnston, D.O. and Wood, J.L. (1976), *'Chemistry, Man and Society',* 2nd Edn., W.B. Saunders Co., Philadelphia, p. 459–492.

Klein, L. (1966). *'River Pollution III. Control'*, Butterworths, London, p. 78–231.
McBain, D. (1976). Reverse osmosis, *Chem. Brit.*, **12,** 281–284.
Moore, J.W. and Moore, E.A. (1976), *'Environmental Chemistry,* Academic Press, New York, p. 375–392.
Peaff, G. (1994). Water treatment companies embrace expanding international markets, *Chem. Eng. News,* November 14, p. 15–23.
Schonherr-Thomanetz, E., Jung, D. and Roder, W. (1995). Waste water treatment by wet oxidation, *Asia Pacific Tech Monitor,* **12**(1), 20–21.
Stone, R. (1995). Environmental toxicants under scrutiny at Baltimore meeting, *Science,* **267,** 1770–1771.

Part VI

Pollutants From Industry

CHAPTER 26

Polymers and Plastics

A polymer is a macromolecule formed by the union of large number of repeating monomeric units in a regular fashion. A plastic is a substance capable of being moulded into various shapes. All plastics are polymers, though not all polymers are plastics. The term plastic is used in a very wide sense to include many synthetic polymeric materials like rubbers, fabrics, resins, etc. In the last century, the plastic industry has grown more rapidly than any other branch of chemical industry. In fact, synthetic polymers have become so indispensable that their *per capita* increase is directly associated with increased standards of living. In this chapter we shall endeavour to examine the environmental problems that lie behind the flood of indispensable plastic objects.

The Need

Polymers provide much be of the physical structure of the world in which we live. Natural polymers are components of the biological systems. Examples are proteins, which are polymers of amino acids; and starch and cellulose, which are polymers of sugars. In addition, natural polymers like DNA are the source of information, as well as the mode by which chemical transformations in living organisms are made to occur. Wool, silk and cotton constitute another variety of natural polymers which are the basis of the fibres which we use for clothing. Collagen and lignin, which provide the structural support for the living tissue, are also polymeric in nature.

Synthetic polymers, including plastics, too find legion uses because of legion properties. In fact, a tremendous advantage of plastics is that they can be tailor-made for specific use. It is, therefore, not surprising that synthetic polymers and plastics are coming up as superior and cheap substitutes for conventional materials.

One commonly notices the wooden table tops being increasingly replaced by laminated ones like Formica. This is because of the better stain, abrasion and mar resistance of plastics. Likewise, the fabric upholstery for automobiles is now almost obsolete because of excellent wear resistance of vinyl sheeting. Lack of electrical conductivity of plastics is responsible for their use in insulation for cables. Lack of thermal conductivity is responsible for their use in automotive equipment.

The ability of plastics to withstand corrosive environments enables them to be used as anti-ablation shields for space vehicles and coatings for supersonic aircrafts. Corrosion resistant characteristics also make it possible for chrome plated plastics to be used in agricultural pipes and greenhouse panels.

Plastics are used for making helmets, golf club shaftings and fishing rods because they possess high strength, coupled with low brittleness. This property is also responsible for incorporation of plastics in wind shields. The superiority of polymeric laminates for boats stems from the low density of plastics. The better cost performance of plastics is responsible for the replacement of metallic containers and bottles with polyethylene ones. Some new applications of plastics include synthetic printing paper, synthetic plastic shoes, artificial organs and tissues and prosthetic devices in medicine.

It is too hard to imagine returning to an earlier age when there were no synthetic polymers. It is too pessimistic to assume that the utility of synthetic polymers has reached a saturation point and that there is no scope for future applications. As our understanding of polymers and plastics continues to develop, we will see the creation of novel, cost-effective and important materials with more widespread applications.

The Classification

Depending upon their mode of synthesis, polymers are classified into two categories: Addition polymers and condensation polymers. *Addition polymers* are formed by adding monomer to monomer by a free radical mechanism. This mechanism proceeds in three stages. Taking the polymerization of ethylene to polyethylene as example, the first stage or *chain initiation* is accomplished with a peroxide. The latter decomposes to give a free radical which adds on to the monomer to release another free radical.

$$\text{Peroxide} \rightarrow \underset{\text{Free radical}}{R^{\cdot}} \tag{26.1}$$

$$R^{\cdot} + CH_2{=}CH_2 \rightarrow R{-}CH_2{-}\dot{C}H_2 \tag{26.2}$$

During the second or *chain propagation stage,* the free radical reacts with other ethylene molecules to form long hydrocarbon chains.

$$R{-}CH_2{-}\dot{C}H_2 + nCH_2{=}CH_2 \rightarrow R{-}CH_2{-}(CH_2{-}CH_2)_n{-}\dot{C}H_2 \tag{26.3}$$

The third or *chain termination stage* involves the reaction between two long chains at their reactive sites.

$$R{-}CH_2{-}(CH_2{-}CH_2)_n{-}\dot{C}H_2 + R{-}CH_2{-}(CH_2{-}CH_2)_m{-}\dot{C}H_2 \rightarrow$$

$$\underset{\text{Polyethylene}}{R{-}CH_2{-}(CH_2{-}CH_2)_n{-}CH_2{-}CH_2{-}(CH_2{-}CH_2)_m{-}CH_2{-}R} \tag{26.4}$$

By a similar series of reactions, tetrafluoroethylene produces teflon.

Condensation polymers are formed when monomers, bearing two or more functional groups, condense together with the elimination of water molecules. Nylon-66, a straight chain polymer is formed by the condensation of adipic acid and hexamethylenediamine.

$$\mathrm{HOC(:O)\,(CH_2)_4C(:O)OH + H_2N\text{–}(CH_2)_6\text{–}NH_2} \rightarrow$$

Adipic acid — Hexamethylenediamine

$$\mathrm{\text{–}C(:O)\text{–}(CH_2)_4\text{–}C(:O)\text{–}NH\text{–}(CH_2)_6\text{–}NH\text{–}C(:O)\text{–}} \qquad (26.5)$$

Nylon–66

Bakelite, a cross-linked polymer, is formed by the condensation of phenol and formaldehyde.

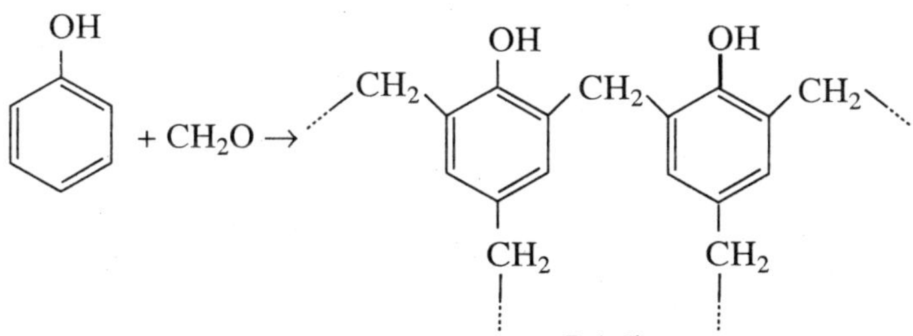

(26.6)

Bakelite

Plastics too are classified into two braod types: Thermoplastics and thermosetting plastics. *Thermoplastics* are those which undergo reversible changes when heated and cooled. On heating, they soften and flow; on cooling, they harden. The temperature above which a thermoplastic becomes fluid is called *heat distortion temperature.* Examples of thermoplastics include polyethylene, teflon, nylon-66, polyvinyl chloride and polystyrene. *Thermosetting plastics,* on the other hand, are those which may be heated to a fluid state only once. The chemical structure of a thermosetting plastic is altered by heat, and cross-linked products are formed, which cannot be resoftened. Examples include bakelite, melamine–formaldehyde resin and polyurethane.

While thermosetting materials may be formed by both addition and condensation reactions, thermoplastics are essentially formed by addition reactions only. Various classes of polymers and plastics, with examples, are depicted in Table 26.1.

The Characteristics

Polymers and plastics have much longer life than paper because they are not easily biodegraded in the environment. An important reason for the persistence of polymers and plastics, under natural conditions, is their hydrophobicity or water-repelling characteristic. Any attmept to reduce the hydrophobic character shall certainly improve the ability of microorganisms to degrade the plastics, but would also negate the uses to which they are normally put to. In order to achieve a balance between the advantages arising out of the durability of plastics and the disadvantages arising out of their non-biodegradability in the environment, a number of additives are commonly introduced in the finished product. The structures of some common additives are displayed in Fig. 26.1.

Foaming agents are added to make low density plastics. Pentane and trichlorofluoromethane are the commonly used physical foaming agents for polymers and plastics. For example, polyurethane is foamed by incorporating liquid pentane under pressure. When the mixture is extruded from an outlet, into the atmosphere, the pentane volatilizes, leaving polyurethane full of small holes.

Table 26.1 Classification of polymers and plastics

Type of Polymer	Type of Plastic	Name of Polymer	Name of Monomer(s)	Applications
Addition	Thermoplastic	Polyethylene	Ethylene	Containers, mouldings, wire insulations, paper coatings
	Thermoplastic	Teflon	Tetrafluoroethylene	Gaskets, corrosion resistant coatings
	Thermosetting	No known example	—	—
Condensation	Thermoplastic	Nylon-66	Adipic acid, hexamethylenediamine	Gears, rolls, fire resistant coatings
	Thermosetting	Bakelite	Phenol, formaldehyde	Electrical mouldings, decorative laminates
	Thermosetting	Melamine-formaldehyde resin	Melamine, formaldehyde	Flexible foams, rigid coatings

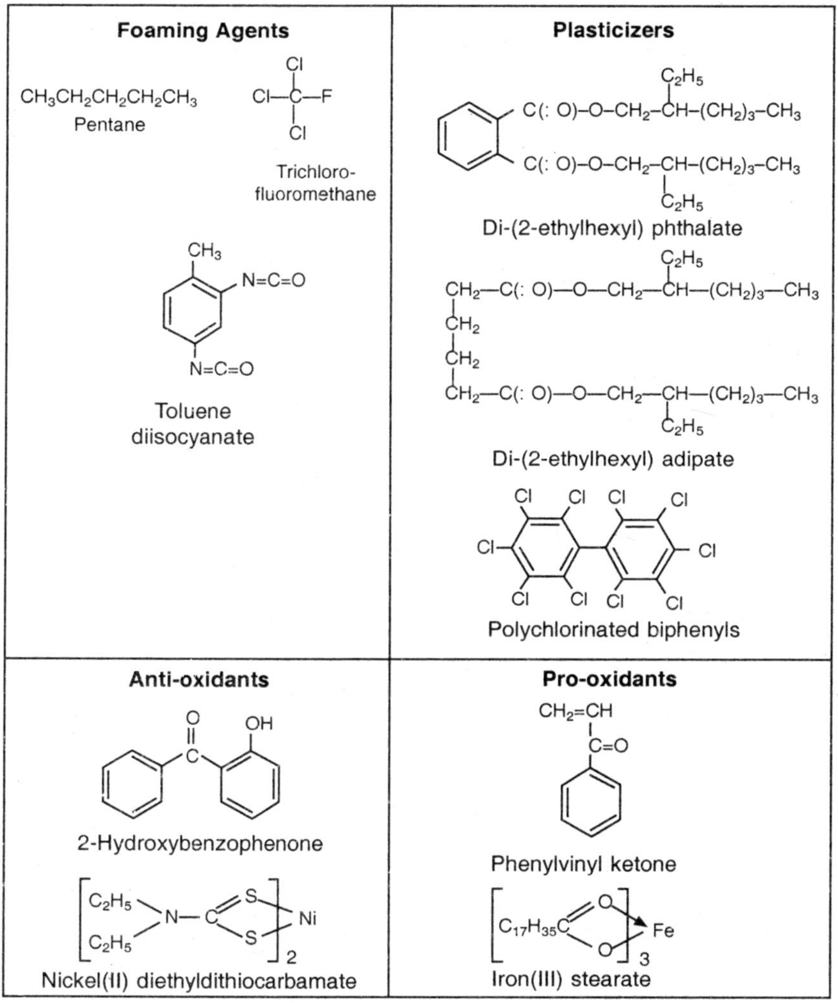

Fig. 26.1 Some common polymer and plastic additives

Chemical foaming agents produce a gas by a chemical reaction. For example, when polyurethane is being synthesized, it may be foamed by carbon dioxide, produced by the hydrolysis of an isocyanate.

$$\underset{\text{Isocyanate}}{R\text{–}NCO} + H_2O \rightarrow RNH_2 + CO_2 \qquad (26.7)$$

Plasticizers are added to plastics to make them more flexible. In plastics, the polymer chains are more or less aligned to each other. This prevents adequate flexion and thus makes the end product too stiff for its intended applications. Plasticizers get incorporated between the polymer chains, weakening the forces of attraction between them and thus increasing their flexibility.

Anti-oxidants, also called *stabilizers,* are added to the plastic products so as to protect them from harmful radiations of sunlight. In their absence, the finished item tends to degrade when continuously exposed to sunlight. Anti-oxidants absorb solar radiation and liberate the excess energy as heat, thus shielding the parent material from degradation. For example, 2-hydroxybenzophenone, a commonly used anti-oxidant, absorbs sunlight and is converted from keto form to enol form. The heat liberated in this reaction is dissipated.

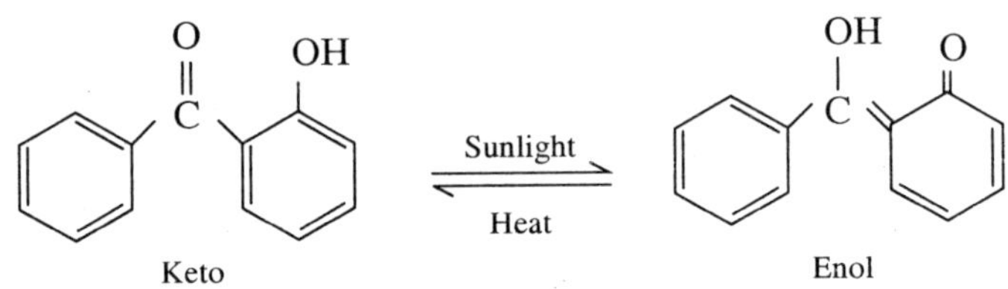

(26.8)

The result of this mechanism of light absorption and heat dissipation leaves the anti-oxidant unchanged and able to undergo a large number of activation-deactivation reactions.

Anti-oxidants are both a boon and a bane. Boon, because they prolong the life of the polymer. Bane, because they lead to the accumulation of non-biodegradable litter in the environment. When polymeric materials are to be used for short term purposes, as in the case of disposable plastic cups, *pro-oxidant* additives are introduced in the finished items. The pro-oxidants encourage the degradation of the product, thus preventing litter accumulation. Their mechanism of boosting degradation is discussed in a later section of this chapter (p. 378).

Environmental Implications of Polymers and Plastics

The following environmental problems are associated with the manufacture, use and disposal of polymer and plastic products.

1. Occupational Hazards

Workers employed in polymer and plastic industry are known to suffer from a number of health problems. For example, bis(chloromethyl) ether, used as a reagent to affect crosslinking in polymers, is suspected to cause lung cancer. 2-Chloro-1, 3-butadiene or chloroprene, the monomer for synthetic rubber (neoprene), can cause liver damage and hair loss. Toluene diisocyanate (Fig. 26.1), used to foam polyurethane, causes acute respiratory problems.

The most severe occupation hazards, however, are reported amongst the workers employed in polyvinyl chloride industry. Exposure to vinyl chloride—the monomer raw material—results in angiocarcinoma, an extremely rare form of liver cancer. Tragically, the latent period during which the symptoms of angiocarcinoma develop is about 30 years. Such a long duration precludes the possibility of any preventive or precautionary measures. Workers exposed to vinyl chloride also suffer a reduction in sperm cells. As a result, their wives have been reported to undergo miscarriages at about twice the normal rate. Surviving infants show high incidence of central nervous system ailments. Because of these adverse effects, the maximum allowable limit of exposure to vinyl chloride has been fixed at 1 ppm over an 8 h period.

The polyvinyl chloride polymers incorporate a class of additives called phthalates. The phthalates serve as plastic softeners. However, it has recently come to light that phthalates cause testicular cancer. Men who are regularly exposed to polyvinyl chloride are ten times more likely to suffer from testicular cancer then their counterparts employed in other plastic manufacturing units.

2. Human Health

The major threat to the general public from polymers and plastics arises because of the plasticizer additives. Platicizer molecules, being smaller than those of the parent resin, can migrate out of the finished product. This migration is accelerated by heat or by contact with a solvent. For example, phthalate ester plasticizers are extracted by blood from plastic tubings and bags used for blood storage. Subsequently, they enter the blood stream and tissue of the patient who receives transfusion. Di-(2-ethylhexyl) phthalate (Fig. 26.1) is known to cause lung damge in rats. The likelihood of adverse effects occurring amongst humans, however, is quite low, since very small amounts of phthalate plasticizers are leached into blood. Yet, for a person receiving repeated or massive transfusions, the possiblity of adverse effects cannot be ignored.

The health problems associated with polychlorinated biphenyls (Fig. 26.1) are more conspicuous. Incinerators burning plastic wastes seldom operate at temperatures high enough to burn polychlorinated biphenyls. These chemicals are carried on particulates into the atmosphere and then washed down into the aquatic systems. Polychlorinated biphenyls are fat-soluble and, therefore, concentrate in the aquatic organisms. Most human exposure to these plasticizers comes from consuming contamined fish. Polychlorinated biphenyl level greater than 2 ppm in the blood stream results in an ailment called *Yusho disease*. The symptoms of this disease include headache, fatigue, pains in joints, acne-like eruptions, anaemia and changes in pigmentation. Concentrations above 12 ppm are known to cause blindness.

3. Land Pollution

Polymers and plastics do not biodegrade in the environment. The reason being that the macromolecules cannot be absorbed into the cells of microorganisms. Considerable social benefits have been drawn from the non-biodegradable characteristics of polymers and plastics, as is evident from their wide range applications. However, the steady increase in the finished products is now causing major problems pertaining to waste disposal. The problem becomes aggravated when polymeric materials designed for long term utility are used to manufacture disposable items. As these 'throw away' items accumulate in the environment, it is a common—and unaesthetic—experience to notice the plastic litter on roadsides, in parks and on sea beaches. Becuase of this problem the polymer and plastic items have come to be regarded as *betes-noir*—something which people detest!

4. Air Pollution

The ever growing applications of polymers and plastics have aggravated the waste disposal problem. The non-biodegradable nature of these materials makes them a permanent waste. All the used polymeric and plastic items cannot be dumped in sanitary landfills. A part of these waste materials is combustioned in the incinerators. Though a good method of disposing used products, incineration has come under a cloud because of the air pollution problems associated with it.

The principal gases generated upon incineration of plastic products include carbon dioxide—which causes greenhouse effect (Chapter 18); nitrogen oxides—which cause acid rain (Chapter 14); and carbon monoxide—which is toxic in nature (Chapter 16). Incineration also releases dust particles, the adverse effects of which are listed in Chapter 20. In many finished products, asbestos is used as a filler. On incineration, asbestos ejected into the air is likely to cause several health problems (Chapter 27).

Combustion of chlorine-containing plastics like polyvinyl chloride, yields hydrogen chloride gas. In presence of moisture, it is converted into hydrochloric acid, which increases the rate of corrosion of incinerator parts and buildings and structures in the immediate vicinity. Vegetative growth in surrounding areas also declines. Combustion of polyacrylonitrile produces highly toxic hydrogen cyanide.

5. Marine Pollution

The factors contributing to the presence of polymeric and plastic items in the ocean include municipal solid waste disposal, coastal landfill operations and disposal of solid waste generated by individual vessels. Once introduced into the marine environment, the polymeric and plastic materials persist for prolonged periods, though they are gradually broken down into smaller particulates. The small plastic particles are ingested by sea organisms. This leads to intestinal blockage and possible mortality in larval fishes. Moreover, the small plastic particles act as surfaces for the accumulation of toxic substances from sea water. On ingestion by sea life, these not only elicit toxic effects, but also bioaccumulate in the food chain.

Abatement Procedures for Polymer and Plastic Pollution

It is commonly assumed that polymers and plastics are indestructible. Indeed, some varieties are very long lasting. However, others are not intrinsically stable. This is not surprising. Paper is highly biodegradable, but an oak beam—from which paper is made and which has essentially the same composition as paper—can last for centuries. Likewise, polymers and plastics can be tailor-made to reduce their lifetimes. It is also possible to recycle the used plastic items and thus reduce the pollution load. Here are enunciated two common procedures for diminishing the environmental problems associated with polymers and plastics.

1. Photodegradation

Photodegradation means breaking down of macromolecules into smaller units by sunlight. This serves an important purpose in the sense that while the large molecules cannot be metabolized by microorganisms and are therefore non-biodegradable, the smaller units are readily assimilated by microbes. Thus, if macromolecules are dissected into smaller structural entities by solar radiations, they would not persist in the environment for long. The problem, however, is that polymers absorb only those solar rays whose wavelength is less than 290 nm and these high energy radiations are retained by the stratospheric ozone layer.

Nevertheless, biodegradable polymers and plastics may be designed by introducing pro-oxidants in their structures. The pro-oxidants can absorb solar radiations with a wavelength greater than 290 nm and initiate the cleavage of polymers. For example, carbonyl groups may be introduced in the polymeric chains by adding phenylvinyl ketone (Fig. 26.1) during the processing stage. The carbonyl groups can absorb solar radiations and photolyse the polymer by two mechanisms.

In *Norrish type I* mechanism, bond breaking occurs by a radical pathway.

$$R\text{–}C(:O)\text{–}R' + h\nu \rightarrow R^{\cdot} + R\text{–}C(:O) \qquad (26.9)$$

In *Norrish type II* mechanism, a hydrogen atom on a γ-carbon atom is abstracted, producing an olefin and an enolic compound.

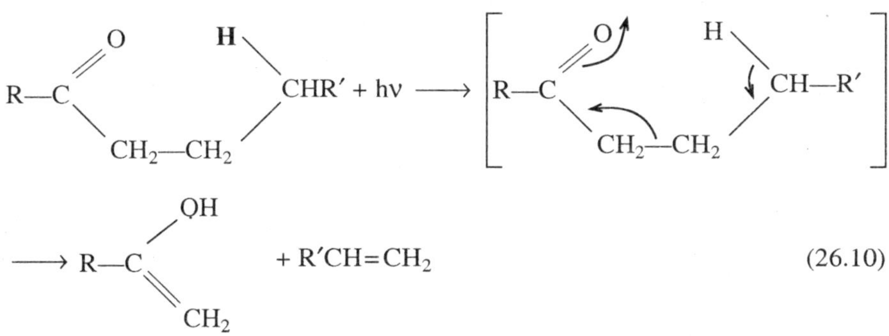

(26.10)

The net result of either mechanism is the cleavage of a carbon-carbon bond next to the carbonyl group.

Photodegradation may also be induced by introducing a transition metal complex in the polymer. For example, iron(III) stearate (Fig. 26.1) sensitizes the photo-oxidation of polyethylene by first absorbing solar radiation, and then undergoing an electron-transfer reaction to give a carboxylate radical.

$$(RCOO)_3Fe + h\nu \rightarrow (RCOO)_2Fe + RCOO^{\cdot} \qquad (26.11)$$
$$R = C_{17}H_{35}$$

The carboxylate radical decomposes into an alkyl radical,

$$RCOO^{\cdot} \rightarrow R^{\cdot} + CO_2 \qquad (26.12)$$

which initiates polymer degradation by scissoring the chains.

$$\sim CH_2\text{–}CH_2 \sim + R^{\cdot} \rightarrow \sim CH_2\text{–}R + {}^{\cdot}CH_2 \sim \qquad (26.13)$$

2. Recycling of Plastics

Another way to reduce the polymer pollution load is to recycle the used and waste plastic items. Recycling is carried out in a number of steps. First, the non-plastic materials are manually separated from the plastic ones. Next, the plastic products are granulated and washed to remove dirt and paper labels. The granulated material is then passed through a series of hydraulic tanks where heavy substances like metals and stones are removed. The granules are dried and taken to an air separator. Here, air under pressure is blown to remove any residual fibrous material. The clean granules are finally moulded into finished products.

By recycling technology, the polyvinyl chloride containers have been successfully converted into underground drainage pipes. Thus, what would have been an unwanted and unsightly litter becomes a utility—and that too buried under the surface of the earth!

References

Allen, N.S. and Mckellar, J.F. (1975). Photodegradation and stabilization of commercial polyolefins, *Chem. Soc. Rev.*, **4,** 533–547.

Bailey, R.A., Clarke, H.M., Ferris, J.P., Krause, S. and Strong, R.L. (1978), *"Chemistry of the Environment"*, Academic Press, New York, p. 301–317.

Bawn, C.E.H. (1965). Chemistry in the plastic industry, *Chem. Brit.*, **1,** 465–472.

Brauman, J.I. (1991). Polymers, *Science,* **251,** 853.

Colton, J.B., Knapp, F.D. and Burns, B.R. (1974). Plastic particles in surface waters of the northwestern Atlantic, *Science,* **185,** 491–497.

Engstrom, K. (1992). The recycling of plastic bottles collected from public waste, *Endeavour,* **16,** 117–121.

Jaeger, R.J. and Rubin, R.J. (1970). Plasticizers from plastic devices: Extraction, metabolism, and accumulation by biological systems, *Science,* **170,** 460–462.

Mackenzie, D. (1998). The perils of PVC, *New Scientist,* January 24, p. 13.

Scott, G. (1973). Improving the environment: Chemistry and plastics waste, *Chem. Brit.*, **9,** 267–272.

Thayer, A.M. (1989). Solid waste concerns spur plastic recycling efforts, *Chem. Eng. News,* January 30, p. 7–15.

CHAPTER 27

Asbestos

The term asbestos refers to a group of silicate minerals that crystallize as strong, flexible fibres. It is the only known naturally occurring fibrous mineral. Due to its fibrous structure, asbestos finds a wide variety of applications and is, therefore, a valued industrial chemical. Ironically, it is because of its fibrous nature that asbestos has been labelled as an environmental hazard. Asbestos fibres are readily absorbed on the surface of lung tissue resulting in a disease called asbestosis, as well as cancer. It has been observed that industrial workers who have dealt with asbestos for 20 years or more are susceptible to be affected both by asbestosis and cancer.

Structural Characteristics of Asbestos

The structure of asbestos is based on silicate tetrahedra, which are linked together through oxygen atoms. The asbestos minerals may be divided into three broad types: Pyroxene, ambhibole and serpentine (Table 27.1).

Table 27.1 Types of asbestos minerals

Class	Representative Example [Empirical Formula]
Pyroxene	Enstatite $[Mg(SiO_3)]$
Ambhibole	Tremolite $[Ca_2Mg_5(Si_4O_{11})_2(OH)_2]$
Serpentine	Chrysotile $[Mg_3(Si_2O_5)\ (OH)_4]$

In pyroxenes, each silicate tetrahedra shares two oxygen atoms with its neighbouring tetrahedra. This results in an infinite chain or strand with $(SiO_3)^{2-}$ repeating units (Fig. 27.1). In case of enstatite, $MgSiO_3$, the magnesium(II) ions hold the anionic chains together by ionic attractions.

The amphibole type asbestos consists of two pyroxene strands, parallel to each other, and joined through an oxygen atom (Fig. 27.2). The repeating unit in this case is $(Si_4O_{11})^{6-}$ ion. Thus, an amphibole analogue is a double chain structure, with each silicate tetrahedra sharing three oxygen atoms. In case of tremolite,

$Ca_2Mg_5(Si_4O_{11})(OH)_2$, the calcium(II) and magnesium(II) ions balance the charges on the anions.

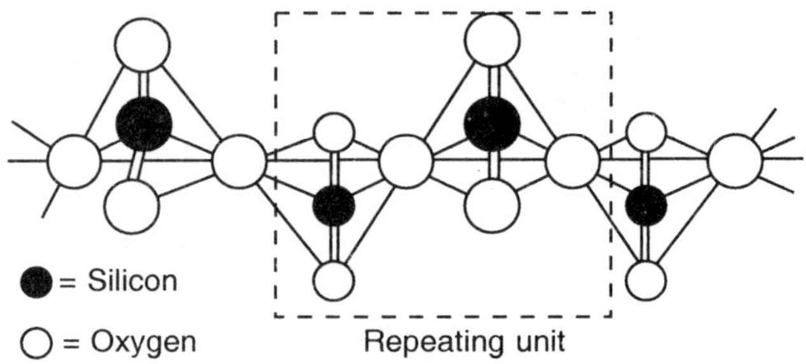

Fig. 27.1 Structure of pyroxene asbestos

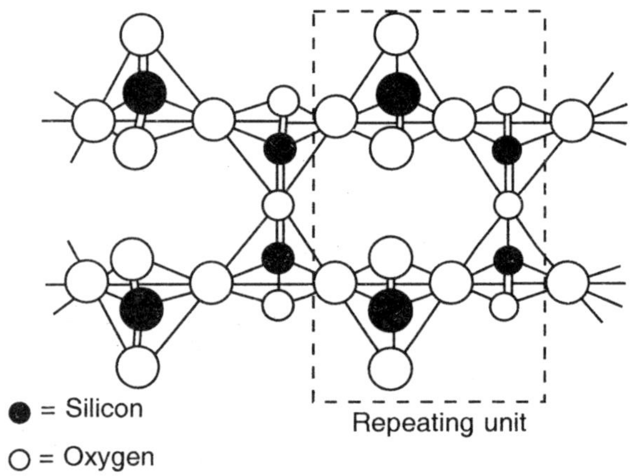

Fig. 27.2 Structure of amphibole asbestos

In serpentine type asbestos too, three oxygen atoms per silicate tetrahedra are shared. However, the number of strands is not limited to two as in case of amphibole. In this case, the silicate tetrahedra are linked together in two dimensions, forming a number of layers and giving a sheet like structure (Fig. 27.3). The repeating unit is $(Si_2O_5)^{2-}$ ion. In chrysotile, $Mg_3(Si_2O_5)(OH)_4$, the magnesium(II) ion is ionically bound to the silicate ions.

Applications of Asbestos

The applications of asbestos have been known since past two thousand years. There are currently about three thousand uses of asbestos in commercial, public and industrial facilities. The main properties which give asbestos its commercial value are, incombustibility, insolubility and effectiveness as a reinforcing agent when combined with other materials, such as cement or plastic. Its resistance to chemical and biological attack is valuable in applications involving hostile environment. Last, but far from least, asbestos is cheap and easily available. Some important applications are briefly reviewed.

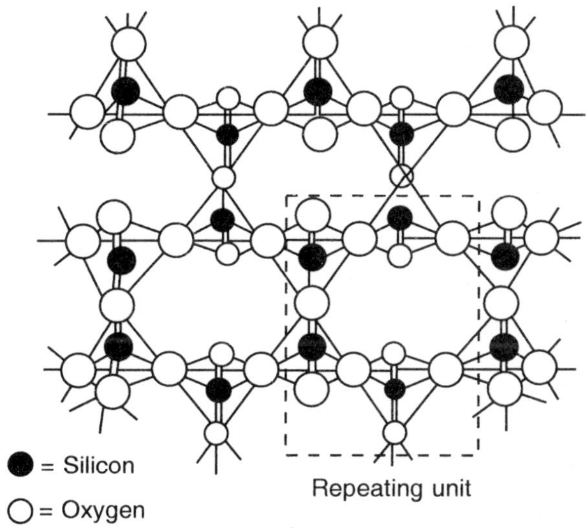

Fig. 27.3 Structure of serpentine asbestos

1. Textiles

Asbestos possesses long and flexible fibres which can be spun into yarn. The yarn is reinforced either with other fibres, generally nylon, or with a metallic wire, generally aluminium. The nylon incorporated asbestos yarn has found applications in fire fighting and as protective underclothing for racing car participants. Gloves made of this material are suitable for use with contact temperatures up to 300°C. The wire-reinforced yarns can retain their mechanical strength up to 600°C. They are used for preparing heat protection clothing in foundry work, since the metal wire provides a heat-reflecting surface. Both the nylon- and wire-reinforced asbestos textiles can also be used in fire blankets, curtains and aprons. In addition, asbestos cloth is used for preparing filtration bags for filtering bulk liquids, such as beer and for making press covers and pads in laundry dry cleaning.

2. Thermal Insulators

Asbestos is used in high temperature insulation applications because it combines the properties of resilience, strength, flexibility and heat resistance. Rubber, blended with asbestos, is used for making gaskets for internal combustion engines, boilers and pumps. Asbestos seals are used for furnace and kiln doors.

3. Electrical Insulators

Asbestos is used for insulating cables, especially those which are designed for low-voltage, high-current use under high temperature conditions. Asbestos is also used in arcing barriers in switches, electric tapes and circuit breakers.

4. Frictional Materials

Asbestos possesses frictional and wear characteristics and is, therefore, used in making clutch plate linings and brake linings. The main constituent of clutch and brake plates is the phenolic resin. The asbestos fibres are added because they stiffen and strengthen the phenolic resin matrix at high temperatures generated during braking and clutch slipping.

5. Building Materials

Asbestos is blended with cement to make floor tiles, wall-boards and asbestos-cement sheets. Pipes used for transporting corrosive liquids also incorporate asbestos. The main attractions of asbestos-based building materials to the users are durability, resistance to weathering and heat, and non-combustibility.

Sources of Asbestos in the Environment

Non-commercially exploited asbestos, contained in rocks, might be released to the environment by natural means, such as weathering and landslides. Exposure to airborne asbestos may also occur by human intervention during mining, milling and fabrication processes. Asbestos particles are released from brake linings and clutch plates of cars and trucks. Stones obtained from serpentine rocks contain high levels of asbestos. This asbestos-bearing stone has been used—often in a dusty, unbound form—for surfacing roads and playgrounds, parks and recreational areas, driveways and parking lots. In due course, asbestos fibres erode from the surface and become airborne. Likewise, when the buildings made of asbestos-cement or floor tiles containing asbestos deteriorate, fibres are released into the indoor air. Airborne asbestors is subsequently ingested through inhalation.

Asbestos may also be ingested through water. Raw water originating in areas rich in serpentine rocks contains dangerous levels of asbestiform fibres. Industrial discharges and mining operations also add asbestos to water. All the asbestos may not be removed by water treatment methods. Even if the water treatment methods are very efficient and leave only a negligible amount of asbestos, in some areas the water delievery systems employ asbestos-lined pipes, the deterioration of which may contaminate the drinking water and negate the benefits of water treatment techniques.

Analysis of Asbestos

Pathological evidence accumulated over the past many years indicates that asbestos is found in the lungs of most city dwellers. Therefore, it is necessary to determine the concentration of airborne asbestos. Since asbestos is present in the air in trace concentrations, the first step in the analysis is to collect small amounts of dust particles present in large volumes of air.

This is done using an electrostatic precipitator. Airborne asbestos particles are colloidal in nature and bear electrostatic charges. In order to collect the particles, the charge must first be discharged. Only then the particles would coalesce and settle down. About 1000 m^3 of air is impinged on the plate of electrostatic precipitator. The charge on the particles is neutralized. The settled particles are washed into about 100 ml of a suitable liquid, generally a non-polar solvent, which would dissolve the unwanted organic particulates. Subsequently, the asbestos particles are filtered and counted under an electron microscope. Concentration is reported as number of asbestos particles per ml of air. Concentrations as low as 10^{-4} ppm can be measured by this technique.

Effects of Asbestos Pollution

The ingestion of asbestos fibres has long been recognized as a serious occupational and environmental health problem. As a result, several applications of asbestos have been curtailed in recent years. Current health regulations do not permit more than 0.5 fibre per ml of air. However, occupational studies among workers dealing with asbestos show that there is no threshold level below which there are no adverse effects. Thus, there are no safe levels of asbestos in the air; as long as there is any exposure, there will be illness. It has, therefore, been recommended that the permitted level of asbestos should be reduced to 0.1 fibre per ml of air. It would not be long before this regulation comes into force.

Exposure to asbestos fibres produces the following effects.

1. Asbestosis

Asbestosis is a disease which results in scarring of the lungs. It is caused by incorporation of asbestos fibres in the lung tissue. Normally, when any foreign material as inhalad, it is ingested into macrophages cells and subsequently it is enzymatically broken down into harmless components. However, the asbestos fibres are chemically so inert that, once inhaled, they remain embedded in the lung tissue for as long as 20–30 years. During this period they stimulate fibroblasts—cells which produce collagen. Collagen is a fibrous, long chain polypeptide protein that is formed in the outer lining of the lungs. As a result, the outer lining of the lungs get thickened. Due to this, the oxygen-carbon dioxide exchange capacity and the elasticity of the lungs decreases. The victim becomes short of breath and may eventually struggle so hard to breathe that he dies of heart failure.

It is estimated that 50–88% of the occupationally exposed workers with more than 20 years of service suffer from asbestosis. Damage to the lungs, caused by this disease, is not only irreversible, but it continues even after cessation of exposure.

Although all forms of asbestos are capable of inducing asbestosis, yet the extent of lung damage depends on several physical characteristics of the fibre. It has been found that crystalline varieties of asbestos are more readily absorbed in the lung tissue than amorphous forms. Hence, crystalline asbestos is more likely to produce asbestosis. The toxicity of asbestos is also determined by the diameter of the fibres. The fibre has to reach the lung before it can do damage. Firbres over 3 μm in diameter are not inhaled; they are screened out by nasal hairs or other protective mechanisms in the upper respiratory tract. Fibres with very small dimensions (< 0.1 μm) too are not toxic. Macrophages can swallow these and transport them out of lungs (although macrophages cannot destroy the asbestos fibres, yet they can transmit them out of the lungs). Thus, these may be breathed in and out without settling to any great extent in the lungs. Fibres of the intermediate diameters (0.1–3 μm), thin enough to enter the lungs, but thick enough to be prevented from being swallowed by macrophages, can cause extensive damage. The fibre length too influences the toxicity of asbestos. It has been observed that fibres which are more than 5 μm in length are more likely to cause asbestosis than shorter fibres.

2. Cancer

Exposure to asbestos causes many forms of cancer, lung cancer and mesothelioma being the chief among them. According to the multistage theory of cancer, two steps are necessary in the development of this disease. In the *initiation step,* DNA of the affected cell is damaged or mutated. In the *promotion step,* the damaged cell is encouraged to divide and proliferate. Asbestos fibres do not damage DNA. Therefore, asbestos is not an initiator of cancer; it is, however, a promoter of cancer. That is, it increases the risk of cancer from known or suspected carcinogens.

For example, asbestos acts as a promoter in the onset of lung cancer by carrying carcinogenic hydrocarbons from cigarette smoke to the lung cells. As a result, it has been found that asbestos exposure multiplies the risk of lung cancer from cigarette smoking. Conversely, incidence of lung cancer in exposed workers who do not smoke has been found to be quite less. Amongst the smokers, exposure to asbestos produces lung cancer normally after a latency period of 10–30 years. Because of such lengthy latency period, the promotional effects of asbestos do not decrease with time after cessation of exposure, as it may with other carcinogens.

Mesothelioma—the cancer of the linings of lungs and abdominal cavity—is a very fatal form of disease, leading to death within 1–2 years of diagnosis. This form of cancer is very rare in general population, but not infrequent in workers exposed to asbestos. Hence, asbestos acts as a promotional agent in the onset of mesothelioma. It has been observed that children exposed to asbestos have at least 10 times greater risk of developing mesothelioma than adults. This is because children have a greater remaining life span as compared to the adults, giving a greater risk of developing cancer. Children living and/or studying in old buildings made from asbestos cement are more prone to inhale fibres and, therefore, more likely to be affected by mesothelioma.

Worldwide, there has been a sudden increase in the cases of mesothelioma since 1993. At a first glance this appears to be surprising since most asbestos products are now banned and the non-asbestos substitutes are largely used. However, during 1965–1970 asbestos trade was at peak. Mesothelioma generally takes about 30 years to develop. Therefore, the workers who were exposed to asbestos in late 1960s or early 1970s now face a threat of this dreaded disease.

Several other forms of cancer, such as larynx, prostate, stomach and gastrointestinal have been reported to occur with excess frequency in occupationally exposed workers. There is also a positive correlation between ingested asbestos and cancer of digestive tract.

3. Iron Mobilization

Asbestos fibres contain appreciable amount of iron content. Iron is present to the xtent of 27–28% in amphibole asbestos. In other varieties of asbestos, the oncentration of iron is less, but not negligible. When asbestos fibres are inhaled, he iron too passes inside the body.

lthough iron is essential to all life forms, its excessive build-up is harmful. nside the body, iron forms chelate complexes with citrate ions. Due to this, it is mobilized away from asbestos. The iron chelates then enter the cells. Here, iron in the ferrous form interacts with oxygen, producing molecular oxygen radicals ($\dot{O}_2^-$).

$$Fe(II) + O_2 \rightarrow Fe(III) + \dot{O}_2^- \tag{27.1}$$

The molecular oxygen radicals oxidise DNA and proteins. Oxidation of DNA results in the damage of all four bases and deoxyribose. The strands are also broken up. This leads to carcinogenesis. The oxidation of proteins produces early ageing and mutagenesis.

Mitigation of Asbestos Pollution

The environmental hazards of asbestos pollution may be reduced by adopting one or more of the following strategies.

1. Occupational Prevention

Asbestos is most dangerous when it is in form of dust. Therefore, workers who deal with asbestos are supposed to wear masks so as to prevent inhalation of fibres. Coating asbestos sheets with dust proof membranes, such as plastics or paints, or keeping the sheets wet while handling reduces the hazard. It must also be ensured that a particular individual is not engaged in asbestos-related work continuously for many years.

It has been established that the dust produced from the wearing of brake and clutch linings contains 1–2% of asbestos fibres. Repeated exposure to this dust is a potential hazard to garage mechanics. It is recommended that asbestos-lined components should be cleaned by vacuum extraction technique, or with a damp rag, rather than the more commonly used method of blowing the dust away with compressed air.

We have stated that children are more susceptible to asbestos pollution than the adults. Worldwide, there are a number of ancient buildings which house schools and hostels. The deterioration of their tiles, floor or walls discharges asbestos fibres to the indoor air. It has been recommended that such buildings should be identified and wherever there is a threat of asbestos discharge, the plaster should be coated with a polyvinyl acetate overspray. Where the conditions are worst, the plaster should be covered with hardboard, plywood or steel sheets.

2. Asbestos Substitutes

In view of the potential hazards of asbestos, over the years a number of suitable substitutes of asbestos have come up.

Clothings made from nylon and wool blends afford fire and heat protection. They also protect and workers from splashes of molten metal. Bleached teflon fluorocarbon fibres can protect the wearer from extremely high flash temperatures and corrosive missile fuels. These were worn by Apollo astronauts.

Ceramic fibres can substitute for asbestos as thermal insulting materials. Phenol-formaldehyde mouldings can serve for asbestos in electrical insulation materials. For high temperature electrical insulation, asbestos may be replaced by ceramic or glass fibres.

Vermiculite, a naturally occurring mineral with composition, $3MgO\ (FeAl)_2O_3$, or silicon nitride can replace asbestos in frictional materials. Glass-reinforced cement is a suitable substitute for asbestos-based building materials. It contains about 5% of alkali-resistant glass fibres in place of asbestos. Likewise, glass-reinforced plastic pipes can substitute for asbestos-based pipes.

3. Waste Asbestos Management

Recently, it has been found that asbestos waste can be put to good use as part of a process for producing magnesium. Normally, magnesium is produced by magnetherm process. In this process, magnesium oxide is reduced by heating with ferro-silicon and a mixture of the oxides of calcium, aluminium and silicon. The reaction has to be conducted under partial vacuum. But maintaining partial vacuum in an industrial plant is difficult. Often air leaks into the system and re-oxidises magnesium to magnesium oxide. It is estimated that upto 25% of magnesium is re-oxidised and has to be re-reduced to metallic form. It has been found that if asbestos waste is added to the raw materials, the need to operate under partial vacuum can be avoided. This cuts down nearly all the losses of magnesium and also reduces the operating costs by about 30%.

References

Ainscough, E.W. and Brodie, A.M. (1991). Asbestos—structures, uses and biological activities, *Chemistry Education,* **8,** 44–47.

Becker, J. (1981). Environmental asbestos: dispute on levels, *Nature (London),* **290,** 620.

Beckett, S.T. (1979) in, *'Asbestos—Properties, Applications and Hazards,* Vol. 1, Michaels, L. and Chissick, S.S. (Eds.), John Wiley & Sons, Chichester, p. 207–245.

Bell, J. (1984). Indecision is the biggest risk, *New Scientist,* September 13, p. 16–18.

Gamble, J.F. (1994). Asbestos and colon cancer: A weight-of-the-evidence review, *Environmental Health Perspectives,* **102,** 1038–1050.

Gould, R. (1995). Magnesium smelters puts toxic waste to work, *New Scientist,* March 4, p. 21.

Hardy, J.A. and Aust, A.E. (1995). Iron in asbestos and carcinogenicity, *Chem. Rev.,* **95,** 97–118.

Moore, J.W. (1986), *'The Changing Environment',* Springer-Verlag, New York, p. 112–142.

Pye, A.M. (1979) in, *'Asbestos—Properties, Applications and Hazards,* Vol. 1, Michaels, L. and Chissick, S.S. (Eds.), John Wiley & Sons, Chichester, p. 339–373.

Webb, J. (1995). Tragic asbestos error will kill thousands, *New Scientist,* March 11, p. 4.

Zurer, P.S. (1985). Asbestos, *Chem. Eng. News,* March 4, p. 28–41.

CHAPTER 28

Food Additives

Since food sustains life, it is endowed with a spiritual quality. The giving and receiving of food in all cultures is characteristic of nurturing relationships. However, the increasing use of food additives has undermined this relationship. The desire for additive-free food is an attempt to have a pure, nutrient item—the one that can be trusted. Lost with the relationship is the trust. Nevertheless, not all additives are harmful. Some are absolutely necessary. Most of the times, the hue and cry which a specific food additive raises has no scientific basis; it is an emotional response from misinformed consumers. In this chapter we shall examine, scientifically and critically, the safety of food additives.

The Need

A food additive is defined as a substance or mixture of substances, other than the basic foodstuff, which is present in the food as a result of any aspect of production, processing, storage or packaging. This definition is quite broad—encompassing both the intentional, as well as the unintentional additives.

Unintentional or *incidental additives* are those which become a part of the food through some phase of production, processing or packaging. Examples include pesticide residues, animal feed adjuvants, plant growth regulators and components of processing and/or packaging machinery, such as lubricants, that migrate into food. The incidental additives have no function in the finished food.

The *intentional additives,* on the other hand, perform specific functions. They are added to the food in carefully controlled amounts. We are more concerned with intentional additives. We shall, therefore, re-define food additive as a chemical or a mixture of chemicals added intentionally to the food to improve its appearance, flavour, texture or storage properties.

Intentional food additives are used to:

1. Improve or maintain nutritional quality of food. For example, addition of vitamins A and D to margarine.
2. Improve shelf life and curtail wastage. For example, calcium propionate to keep bread from becoming stale.
3. Make food more readily available throughout the year and throughout the country. For example, citric acid in fruit juices.
4. Maintain food quality characteristics. For example, corn starch added to powdered sugar to prevent lumping.

5. Facilitate fast and convenient preparation of food. For example, the use of phosphate additives in products like instant pudding.
6. Make food more appealing. For example, using artificial colours and flavours.

Intentional food additives should not be used if they—

1. Deceive the customer by covering up the use of faulty ingredients or faulty manufacturing properties.
2. Cause substantial reduction in food's nutritional value.
3. Produce an effect that could be obtained by otherwise good manufacturing processes.
4. Are in excess of the legally prescribed limits.

The Classification

Broad classes of food additives are detailed below. The structures of some representative additives are depicted in Figs. 28.1–28.11.

1. Preservatives

Preservatives are the chemical agents which retard or hinder undesirable changes in food. These changes are generally caused by bacteria, yeast and molds. Preservatives interfere with cell membranes, enzyme activities or the genetic mechanisms of the food spoilage microorganisms. Examples of preservatives include sodium benzoate, used in soft drinks; calcium propionate, used in bread; salt, used in meat products; sugar, used in jellies; and sorbic acid, used in cheese. The preservatives also include fumigants, such as ethylene oxide, to control microorganisms in spices and dried fruits. Sulfites too are used in food processing to prevent enzymatic browning of fruits and vegetables.

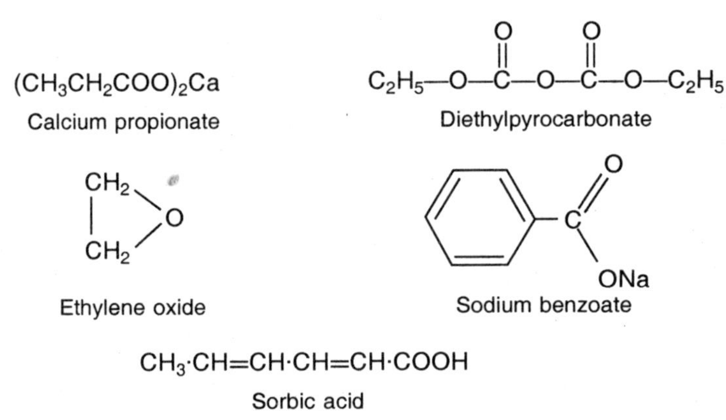

Fig. 28.1 Some common preservatives

2. Anti-oxidants

Anti-oxidants are the chemicals which inhibit the oxidation of fats. In their absence, fatty foods would develop rancidity. Thus, without anti-oxidants, potato wafers, salted nuts and breakfast cereals would have negligible shelf life. Examples of anti-oxidants include butylated hydroxyanisole (BHA), butylated hydroxytoluene (BHT) and propyl gallate. Sulfites also act as anti-oxidants, although they are generally listed as preservatives.

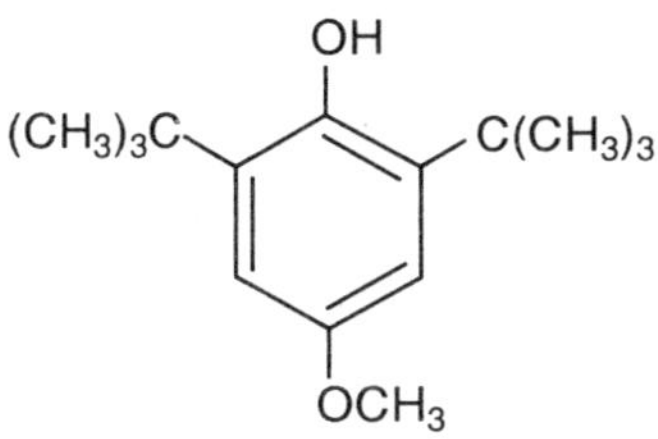

Butylated hydroxyanisole

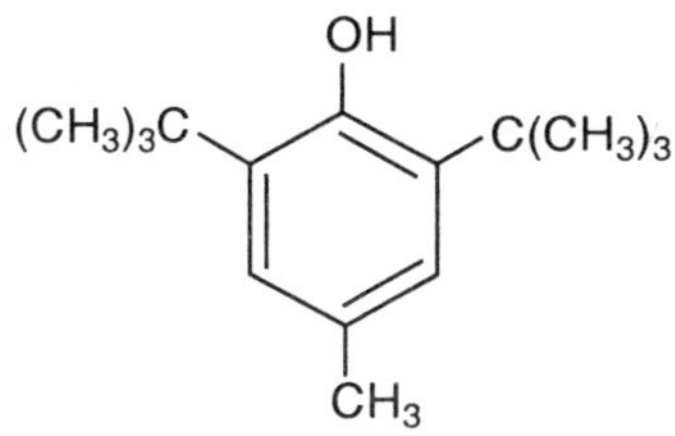

Butylated hydroxytoluene

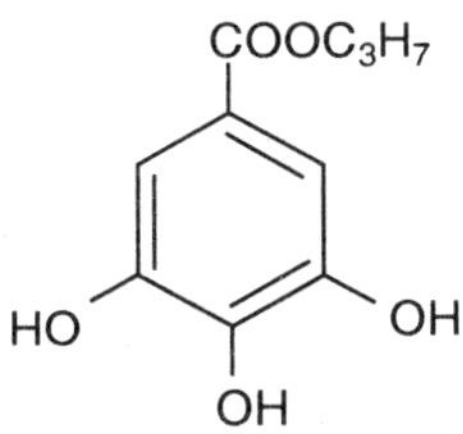

Propyl gallate

Fig. 28.2 Some common anti-oxidants

3. Sequestrants

Traces of metal ions, such as iron and copper, present in the processed food catalyse oxidation reactions, producing rancid odours and flavours in the food products. They also enter into many off-colour reactions in foods. Substances used to tie-up the metal ions that might otherwise promote the deterioration of food are called sequestrants. These metal scavengers are generally chelating agents. Examples include ethylenediamine tetraacetic acid, (EDTA), citric acid and adipic acid. They are used in edible oils, salad dressings and butter.

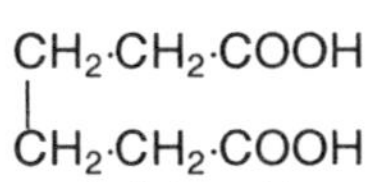

Adipic acid

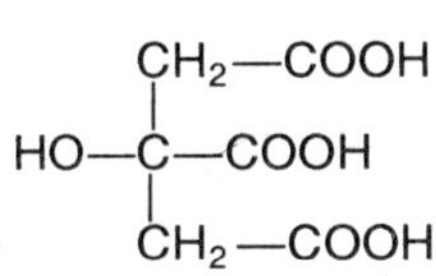

Citric acid

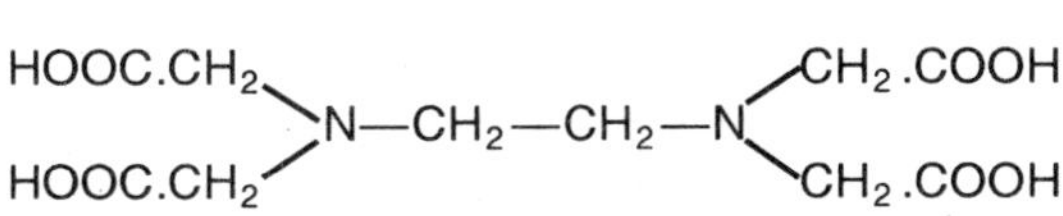

Ethylenediamine tetraacetic acid

Fig. 28.3 Some common sequestrants

4. Acidulants

Acidulants are acids—mostly organic acids—which affect a wide array of food properties. They are used to enhance flavour, to control pH, to assist anti-oxidants, to aid meat curing agents and to control viscosity and hardness of food products. They are commonly added to fruit juices, squashes, jellies and cheese. Examples of acidulants include acetic, lactic, succinic and phosphoric acids. Citric acid is both a sequestrant and an acidulant.

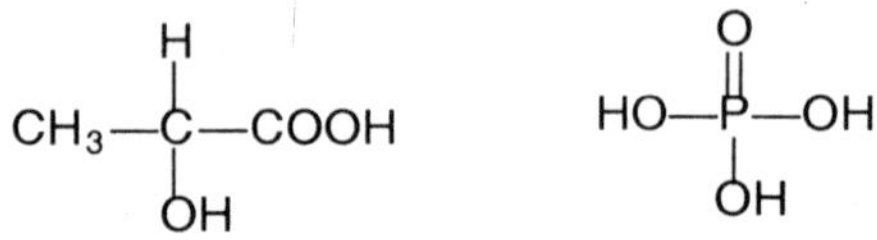

Lactic acid   Phosphoric acid

Fig. 28.4 Some common acidulants

5. Surface Active Substances

Surface active substances include wetting agents, emulsifiers and lubricants. Commonly used surface active substances are lecithin and mono and diglycerides. When added to bakery goods, these reagents enable the food to stay fresh longer and have a better texture.

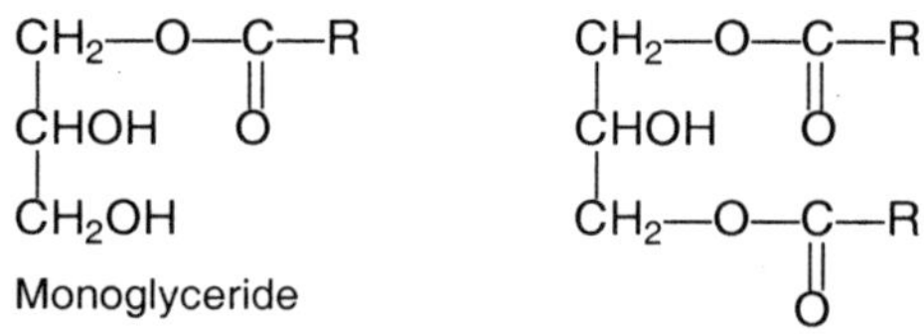

Monoglyceride   Diglyceride

R = Alkyl

Fig. 28.5 Some common surface active substances

6. Stabilizers and Thickeners

These reagents stabilize and thicken the food by combining with water to increase the viscosity and form gels. Gravies, pie-fillings, sausages, cake toppings and chocolate milk drinks are the common food items which contain these additives. Examples of stabilizers and thickeners include carboxymethylcellulose (CMC), carregeenan, amylose, gelatin and pectin.

7. Bleaching and Maturing Agents

These reagents not only bleach the food products to colourless or white, but also increase their baking performance. For example, freshly milled flour has a yellowish tint and sub-optimum baking qualities. Both the colour and the baking properites improve slowly in normal storage. These improvements can be enhanced by adding benzoyl peroxide, which acts as a bleaching and maturing agent. Bromates and iodates are used to bleach bread and to condition it to a better baking performance.

8. Natural Supplements

These are the naturally occurring substances which are added to the food products to enhance their food value. For example, vitamin A is added to margarine;

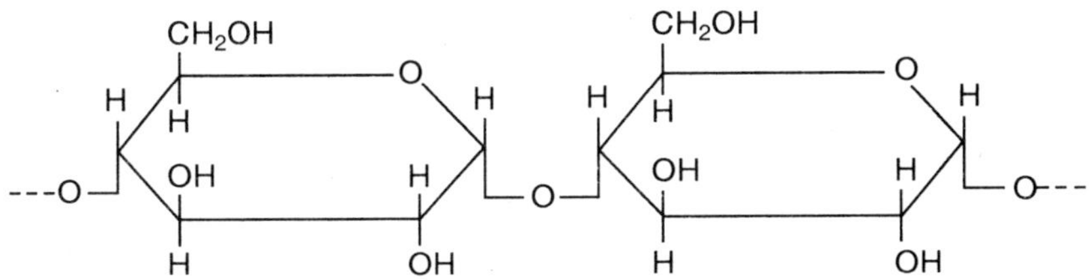

Amylose

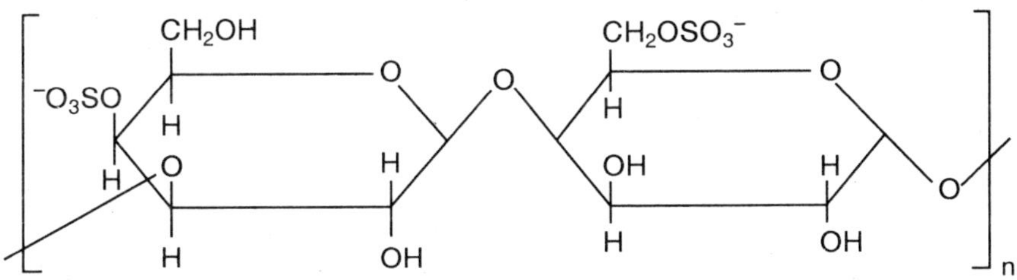

Carrageenan

Fig. 28.6 Some common stabilizers and thickeners

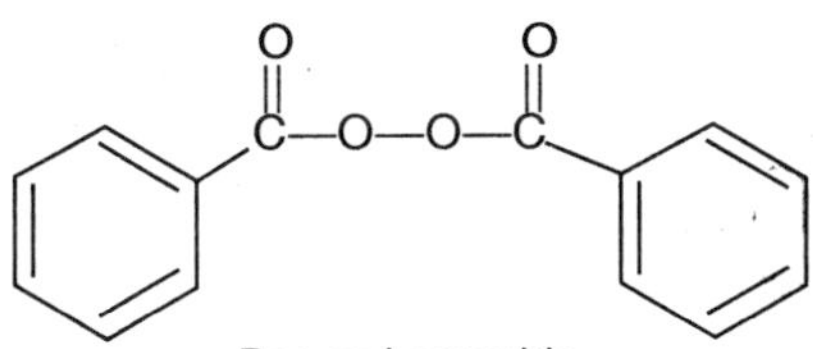

Benzoyl peroxide

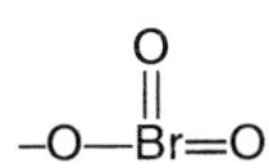

Bromate ion

Fig. 28.7 Some common bleaching and maturing agents

vitamin B, iron and calcium to cereal products; vitamin C to fruit juices; vitamin D to milk; iodine to salt; and amino acid, lysine to wheat flour.

9. Colourants

Colours are added to a very large number of food products to give the consumers the appetizing and attractive qualities they desire. Natural colours like turmeric, saffron, paprika and caramel have been added to food items since ancient times. However, the advent of processed food products placed heavy demands on colours which could be met only by synthetic colouring agents. Synthetic colours generally excel in colouring power, colour uniformly and cost less than the natural reagents. Moreover, in many cases, natural colouring materials do not exist for a desired hue. Examples of artificial colourants include amaranth, indigo carmine, erythrosine, tartrazine and allura red. These are used in a wide variety of food products like carbureted beverages, candies, gelatin desserts, cakes and cookies.

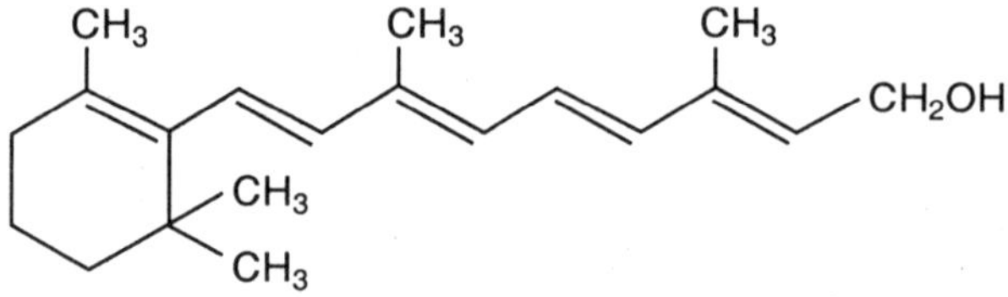

Vitamin A

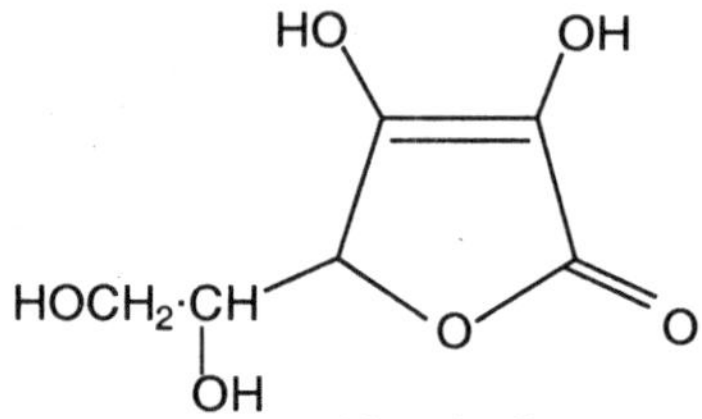

Vitamin C

Fig. 28.8 Some common natural supplements

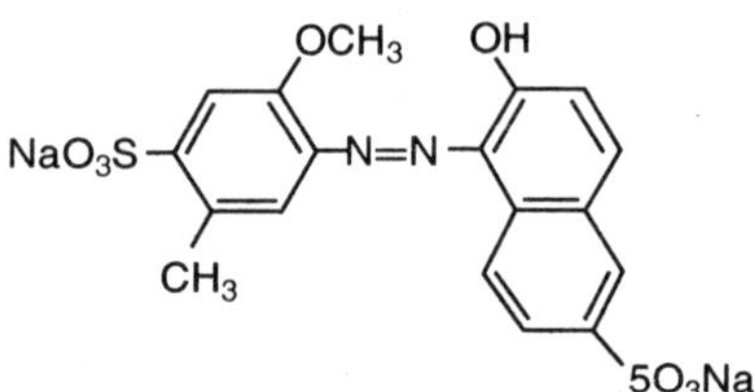

Allura red

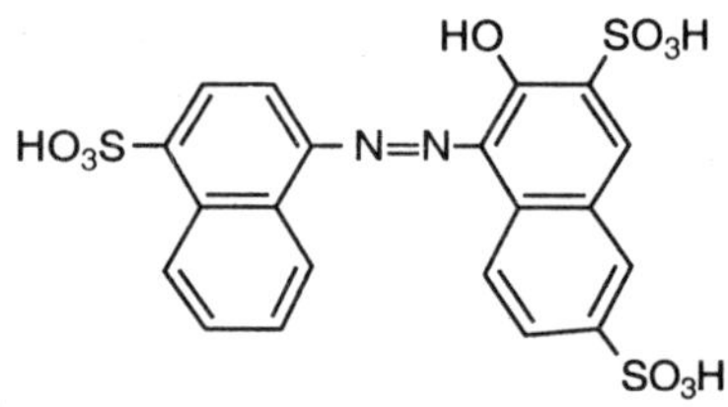

Amaranth

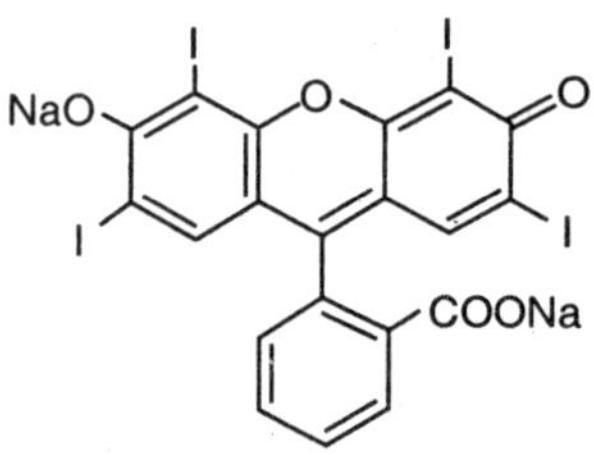

Erythrosine

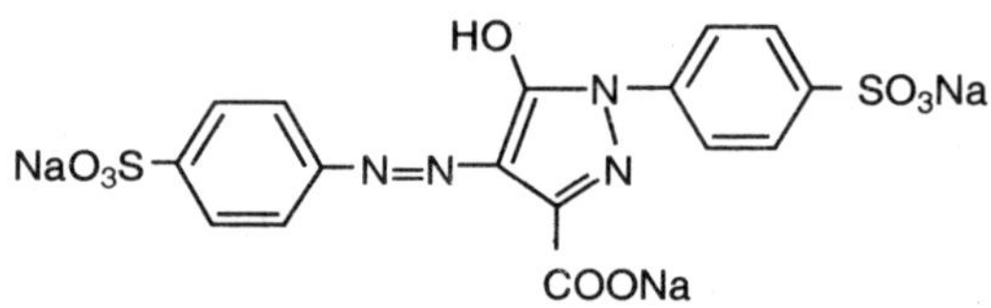

Tartrazine

Fig. 28.9 Some common colourants

10. Non-nutritive Sweeteners

Non-nutritive sweeteners are added to the food when a low calorie diet is preferred. Food products containing non-nutritive sweeteners have proved to be beneficial in weight control and for diabetics. Examples include saccharin, cyclamates and aspartame. The largest users of non-nutritive sweeteners are soft drink manufacturers. They are also used in canned fruits, frozen desserts and salad dressings.

11. Flavours

Flavour is crucial to the acceptance and enjoyment of food. Artificial flavours replenish the flavour lost or partially lost by various modern food processing

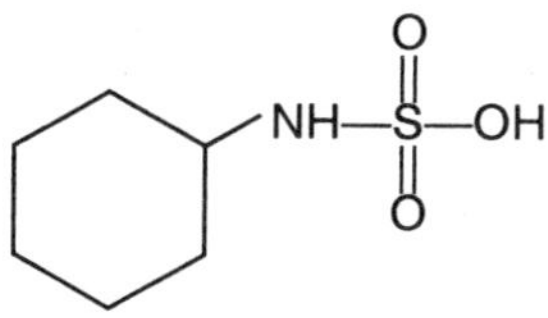

Cyclamic acid

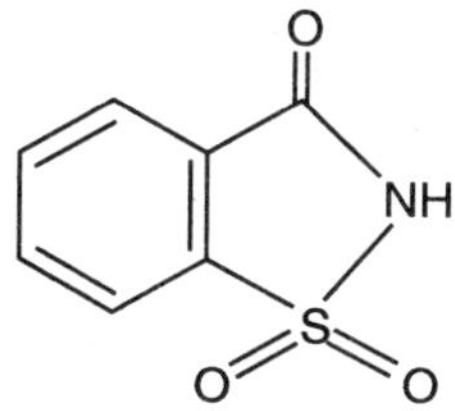

Saccharin

Fig. 28.10 Some common non-nutritive sweeteners

methods. Vanillin imparts vanila flavour to food items; ethyl butyrate, pineapple flavour; and methyl anthranilate, grape flavour. These flavours are commonly used in puddings, pastries, soft drinks and ice creams.

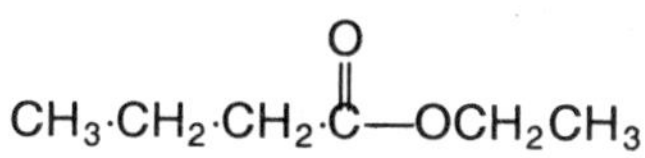

Ethyl butyrate

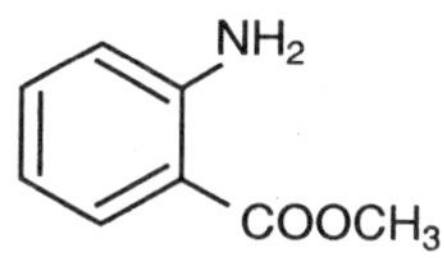

Methyl anthranilate

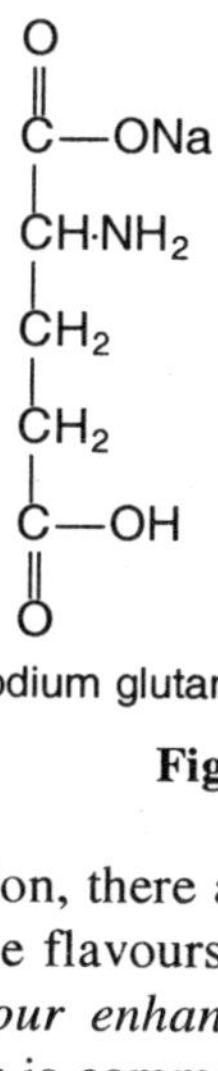

Monosodium glutamate

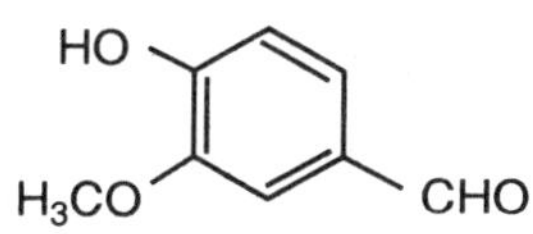

Vanillin

Fig. 28.11 Some common flavouring agents

In addition, there are reagents which do not have a flavour of their own but intensify the flavours of other compounds present in food products. These are called *flavour enhancers.* Monosodium glutamate is an example of flavour enhancer. It is commonly used in baby foods.

Risk Analysis of Some Specific Food Additives

Though food additives provide many benefits, they put us at risk in some cases. The risk factors associated with a few important additives, which have come under the cloud of suspicion, are now analysed.

1. Sodium Nitrate and Sodium Nitrite

Alkali nitrates and nitrites are used as additives for cured meat products. The importance of nitrate in the meat items does not lie in any intrinsic behaviour of nitrate itself, but rather in its ability to form nitrites by microbial action. Therefore, nitrite is a more important preservative than nitrate. In fact, nitrate has given way largely to nitrite in meat items.

Sodium nitrite prevents botulism in cured or canned meat products. It also acts as a functional aid to stabilize the meat colour and to produce cured flavour. The role of nitrite in colour formation arises because of its ability to decompose into nitric oxide in the food product. The nitric oxide then reacts with myoglobin content of the meat to give nitrosomyoglobin, the red-pink pigment of cured meat. This is called *cosmetic effect.*

However, it is the preservative action of nitrite which is more important. Sodium nitrite is known to inhibit the growth of bacterium *Clostridium botulinum.* The *Clostridium* species are anaerobic. Sealed food containers are an ideal environment for their growth. If left unhindered to multiply, these microorganisms produce highly poisonous toxins. For example, 1 mg of botulinum toxin A can kill 30 million mice!

The anti-botulinal activity of nitrite stems from its ability to react with [4Fe–4S] enzyme centres (Fig. 28.12) in the cells of *Cl. botulinum.* Once the iron-sulfur cluster is destroyed, the microorganism dies. Simultaneously, a set of paramagnetic, dinitrosyl iron complexes of the type, $[Fe(NO)_2X_2]^{n+}$ (X is an organosulfur compound like cysteinate; n is an integer) are produced. These complexes are called *Perigo type factors* (PTFs). The PTFs too are microbe inhibitors. They further protect the meat products from spoilage.

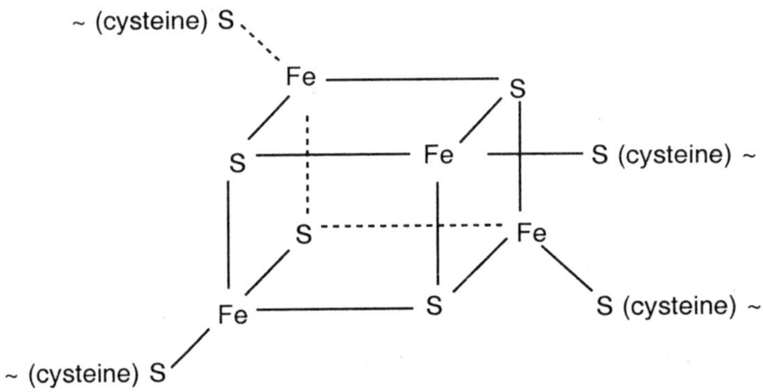

Fig. 28.12 Structure of [4Fe–4S] enzyme cluster

The role of PTFs is not altogether laudable. On the darker side, some of these are suspected to be carcinogenic to experimental animals. Others, which are not carcinogenic, are suspected to potentiate the action of carcinogenic compounds, that is, they act as cancer promoters. However, it has not been definitely established whether PTFs cause cancer in humans.

The anti-botulinal action of nitrite increases as the pH is lowered. Within the food product, nitrite is partially converted into nitrous acid. The pH goes down in the acid range providing further food safety. However, nitrous acid also reacts with the secondary amines in the meat to yield N-nitrosamines.

$$R_2NH + HNO_2 \xrightarrow{H^+} R_2N \cdot NO + H_2O \qquad (28.1)$$

The problem does not end here. Some of the residual nitrite passes from the preserved food into the human body. In the stomach, where the pH is in the acidic range, the nitrite is converted into nitrous acid which interacts with amines and/or peptides to yield N-nitrosamines.

Thus, when sodium nitrite is used as a preservative, N-nitrosamines are formed in food and body. N-nitrosamines are toxic, carcinogenic and mutagenic to a wide range of experimental animals. They are known to cause tumors in different organs, especially stomach and liver. In fact, they are counted as one of the most potent carcinogens because they bind directly to DNA.

Surprisingly, there is no firm link between N-nitrosamines and cancer in humans. This may be due to the fact that human cells contain specific DNA repair mechanisms and, therefore, get rid of N-nitrosamines. However, since a very large species of laboratory animals are susceptible to N-nitrosamines, the threat of nitrite would always loom large.

Yet, it would be unwise to stop using nitrite as a preservative. Nitrite protects us from serious outbreaks of food poisoning. By not using it, we would be replacing one hazard by another—the one that is more certain. The best course would be to use the lowest possible concentration of nitrite which would just suffice against botulism. This dose has been estimated to be 200 ppm per kg of meat product or daily intake of 0.4 mg of nitrite per kg of body weight.

2. Radiation

One method of preserving food is to irradiate it with γ- or X-rays. Irradiation is a cold, non-chemical method of food processing that can prevent huge losses due to spoilage. Beneficial effects of irradiation include extension of storage life of crops, disinfestation of cereals, fruits and spices and destruction of pathogens. Irradiated foods retain their fresh appearance, flavour and nutritional value.

The amount of radiation abosrbed by the food is expressed in terms of Kilo Gray (KGy), where one Gray corresponds to the absorption of 1 joule of energy by 1 kg of foodstuff.

A low dose (0.05–1.0 KGy) inhibits sprouting in potatoes, garlic and ginger. It also kills insects infesting the grain. A medium dose (1–8 KGy) eliminates spoilage organisms and pathogens from sea food and meat products. A high dose (25–50 KGy) sterilizes the food. The sterilized food serves as a diet for immune-deficient patients.

Strictly speaking, radiation is not an additive. Yet, fears about the safety of food products subjected to irradiation have been expressed. A common concern of the consumers is that irradiation of food generates toxic radiolytic products. However, the radiolytic products in irradiated foodstuff are similar to and no more toxic than those formed when a food item is exposed to heat, light and air.

Fears have also been expressed that food irradiation results in the formation of free radicals. The free radicals are unavoidable even while cooking or baking the non-irradiated food items. More free radicals are formed during toasting of bread than through ionizing radiation. In wet foods the free radicals disappear within a fraction of second. They are, no doubt, immobilized in solid matrices like bones or seeds, but are immediately lost when these items are digested during cooking.

Another concern about irradiation is that it hides the evidence of spoilage. Irradiation can prevent spoilage; it cannot reverse the effects of spoilage. It is also feared that irradiation destroys the nutritional components of the food. However, most food ingredients like carbohydrates, lipids and proteins are not destroyed by irradiation. Small amounts of vitamins are, no doubt, destroyed, but the loss is no greater than with other methods of food processing and can be reduced further if the food is irradiated in a frozen state, in the absence of oxygen.

3. Diethyl Pyrocarbonate

A widely used anti-microbial food preservative, diethyl pyrocarbonate is added to fruit juices, wine and beer. Under normal circumstances, it hydrolyses rapidly to form ethanol and carbon dioxide. However, the beverages usually contain some amounts of amino acids and proteins. These slowly and partially hydrolyse to yield ammonia, which is almost always present in fruit juices, wine and beer.

Diethyl pyrocarbonate is known to react with ammonia in the pH range 4–9 to form urethane (ethyl carbonate, Fig. 28.13), which is carcinogenic. World Health Organisation has, therefore, recommended that diethyl pyrocarbonate be used only in those beverages which have a pH lower than 4 and a low concentration of amino acids and proteins. The quantitiy of diethyl pyrocarbonate should not exceed 300 mg per kg of food item.

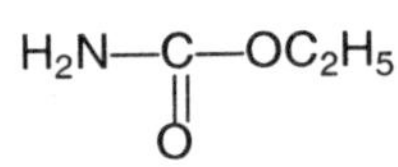

Fig. 28.13 Urethane

4. BHA and BHT

As has been stated, butylated hydroxyanisole (BHA) and butylated hydroxytoluene (BHT) are anti-oxidants. They inhibit the oxidation of food fats, especially unsaturated fats. In experimental animals like rats these anti-oxidants get absorbed, along with fats, in the small intestine. Subsequently, they pass on to liver. While detoxifying these compounds the liver gets enlarged and its functions are altered. Higher levels of anti-oxidants slow down the rate of synthesis of DNA and RNA which, in turn, reduces the rate of cell division and growth. Mercifully, these adverse effects have not been observed in humans.

The possible carcinogenicity of BHA and BHT causes both concern and confusion; they have shown to be cancer-promoting and cancer-inhibiting. A set of studies with mice indicated that these anti-oxidants protect the animals from tumors induced by a variety of carcinogens. In another set of study on rats, however, BHA and BHT were implicated to be tumor promoters.

It is not yet clear whether BHA and BHT act as cancer promoters or inhibitors in human beings. It is suspected that their effect depends on the time of administration, type of carcinogens already present in the body and the target organ affected. It is also speculated that the effect of these anti-oxidants depends on the presence of other nutrients in the body. Thus, while adequate proteins in the diet prevent adverse effects, low protein levels do not.

The uncertainty associated with BHA/BHT should have been a reason strong enough to ban these anti-oxidants. Yet, they continue to be used as food additives. The reason being that these chemicals produce an array of beneficial effects. BHA and BHT exhibit anti-microbial activity against several common food pathogens and viruses. They help retain vitamin E in food which otherwise gets destroyed when fat content is oxidised. The oxidation of fat, which occurs in the absence of BHA and BHT, produces peroxides which are definitely carcinogenic and mutagenic. It has also been observed that in countries where BHA and BHT are not used as anti-oxidants in cereals, baked goods and margarine, there are larger number of cases of stomach cancer.

To conclude, the benefits of BHA and BHT outweigh the possible risks associated with them. The recommended daily intake of BHA is 0.3 mg per kg of body weight; for BHT it is 0.125 mg per kg of body weight.

5. Phosphoric Acid

Phosphoric acid is the only inorganic acidulant. It is used in cola type soft drinks and canned meat products. Although phosphorus is an essential nutrient, excessive intake of phosphoric acid reduces the optimum calcium: phosphorus ratio in the body. This leads to *osteoporosis,* a condition where bones become brittle and hence more susceptible to fractures.

6. Brominated Vegetable Oils

Brominated vegetable oils are used in soft drinks as emulsifying functional aids, so that the flavouring oils will not separate from the aqueous liquid mixture. The precipitating material observed in squashes does not appear in soft drinks because the latter contain brominated vegetable oils. Hence, the brominated vegetable oils give a desirable, clear appearance to carbureted beverages.

Experimental animals fed with brominated vegetable oils show deposit of brominated fat in the heart and other tissues. Brominated fat does not metabolize easily. Children consuming large quantities of soft drinks have been shown to have high levels of bromine—a toxic substance—in their tissue. World Health Organisation has discouraged the use of brominated vegetable oils and several countries have banned these additives.

7. Potassium Bromate

Potassium bromate, $KBrO_3$, is a bleaching and maturing agent for flour. The amount of bromate present in baked bread products was assumed to be zero since heat converts bromate to bromide. However, extremely sensitive methods of analysis have revealed that trace amounts of bromate residues persist in baked bread. This fact raised an alarm since it has been demonstrated that bromate fed to rats increased the incidence of kidney tumors. As a result, the use of bromate in flour was significantly curtailed. The recommended dosage is 0–60 mg per kg of flour. Some countries have altogether banned the use of bromate as an additive.

8. Saccharin

Saccharin is a calorie-free artificial sweetening agent that is about 300 times sweeter than sugar. It is stable over a wide range of temperature and hence is used in a wide variety of products. It, however, produces a slightly bitter after-taste. This flavour-drawback was overcome by using a 1:10 mixture of saccharin and cyclamate. The sweetness imparted by this mixture is greater than the sum of sweetness of each component.

Concern about saccharin arose in 1972 when it was demonstrated that it produced bladder cancer in rats. It was also noted that the incidence of tumor was a function of dose. As the dose was decreased, the incidence of tumors also declined. However, in these studies the dosage given to rats was much higher than what an average human being consumes each day. World Health Organisation has recommended a daily intake of 0.43–4.3 mg per kg of body weight. However, infants and pregnant women are advised to avoid saccharin.

9. Cyclamate

The term cyclamate includes cyclamic acid or its salt with sodium or calcium. An artificial sweetener, cyclamate is 30 times more sweet than sugar. Compared to saccharin, it has a less bitter after-taste. It is stable under conditions of high heat, which allows it to be included into a wide variety of food products.

Initially, cyclamate was considered safe for human consumption. However, like saccharin, it causes bladder cancer in rats. Fear has been expressed that its metabolite, cyclohexamine, is a cancer promoter. Yet, no concrete evidence implicating cyclamte as a cancer causing agent in humans has been found. It was noted that in studies involving laboratory animals, the rats were fed with doses which were almost 50 times greater than those estimated for human consumption. Therefore, such results cannot be extrapolated to human beings.

The veil of suspicion which shrouds cyclamate has, however, not been completely lifted. Certain countries, like United States, have banned cyclamates altogether. In other countries, children and pregnant women are advised to restrict their diet of cyclamates. World Health Organisation has recommended a daily intake of 4–10 mg per kg of body weight.

10. Monosodium Glutamate

Monosodium glutamate is the sodium salt of glutamic acid, one of the most common amino acids found in food and in the body. It is a flavour enhancer. For a long time it was believed that monosodium glutamate did not have a taste of its own. Sensory studies later revealed that it imparts a fifth basic taste that has been named *umani* (the other four tastes being, sweet, sour, salty and bitter).

Monosodium glutamate is used in frozen non-vegetarian foods, dry soup mixtures and cheese spreads. However, its use in baby foods has cast apprehensions over its safety. When injected in newborn mice, monosodium glutamate causes injury to brain cells, including those in the region of hypethalamus, a part of the brain which is important in regulating body's hormone system. No such adverse effects are observed in mature rats.

Human studies with monosodium glutamate at very high doses did not produce any neurological damage. The risk to infants too was regarded as negligible. However, monosodium glutamate is considered to have no nutritional value and, therefore, does not benefit the infants either. Hence, it has been suggested that it should be omitted from baby foods. The only loss would be lack of a unique flavour experience.

It has been alleged that in adults, monosodium glutamate produces what has been come to be known as the *Chinese Restaurant Syndrome.* This name originated because of the generous amount of monosodium glutamate used in chinese cooking. The symptoms of Chinese Restaurant Syndrome include burning sensation at the back of neck, facial pressures, chest pain and headache. However, it has been observed that the symptoms of the disease occur only in a small number of individuals, when monosodium glutamate is fed at a concentration greater than 3%. Even in these individuals, responses are seen only once in three times. Therefore, regulatory agencies in most countries deem monosodium glutamate to be a safe additive for adults.

Inference

The greatest fear with food additives, as with other chemicals in the environment, is that they may cause cancer. In order to obviate this fear the food laws incorporated the *Cancer Clause* or *Delaney Clause* which states that no additive may be permitted in food if the material, at any level, can produce cancer when fed to human beings or animals or can be shown to be carcinogenic by an appropriate test.

Ideally, the cancer clause served to protect even the most susceptible individual from the harmful effects of food additives. In real terms, however, it did more harm than good. It simply implied that if cancer is shown in any species when the chemical is given in any amount, then the substance must be banned. Thus, a material that causes cancer in animals, but is known to be harmless to humans must nevertheless be banned.

Since the zero tolerance limit which the cancer clause aimed to achieve is scientifically untenable, the controversial clause was replaced by the concept of *De minimus.* This concept stressed that an additive may be used if it is carcinogenic to laboratory animals, but the dietary risk to humans is deemed negligible. With the introduction of De minimus, most food additives attained what is known as the *GRAS status*—GRAS being an abbreviation for generally recognized as safe.

Fear still haunts the consumer's mind. After all, the additives are generally—and not always—safe. It is alleged that additives can be altogether avoided but the food industry slaps them on to the consumers. This is only partly true, for the consumers too benefit from the longer shelf life and reduced wastage which an additive ensures. Moreover, additives cannot be written off completely. No highly developed society can exist without food additives. They immediately become necessary when areas of food production are separated from centres of population concentration and the food must be transported under conditions that negate spoilage.

It is, therefore, the consumer who has to decide as to what extent the use of additives may be restricted. Sometimes the decision is obvious. Given a choice between a tomato pizza and a tomatoless pizza with an artificial tomato-flavoured chemical, one would opt for the former. Sometimes one goes by customary, rather than scientific instinct. People resist buying oranges with green skins even when they are ripe. So the seller uses artificial colours. Infants do not benefit anyway from monosodium glutamate. Yet, mothers do not like the taste of baby food without monosodium glutamate. A newborn is unnecessarily made addicted to an artificial flavour. At times, however, the consumer's choice must be based on the analysis of risks *versus* benefits. Some may decide that the benefits of using sodium nitrite in meat products outweigh the risks. Others may decide the opposite. Informed choice in an open society is an individual's right—provided the individual is enlightened enough to exercise this right diligently.

References

Fox, B.A. and Cameron, A.G. (1970), *'Food Science—a Chemical Approach'*, University of London Press, London, p. 343–361.

Glidewell, C. (1990). The nitrate/nitrite controversy, *Chem. Brit.*, **26,** 137–140.

Holum, J.R. (1977), *'Topics and Terms in Environmental Problems'*, John Wiley & Sons, New York, p. 259–263.

Jay, J.M. (1978), *'Modern Food Microbiology'*, D. Van Nostrand Co., New York, p. 163–184.

Jones, J.M. (1992), *'Food Safety'*, Eagan Press, Minnesota, p. 203–299.

Kirschner, E.M. (1997). Fake fats in real food, *Chem. Eng. News,* April 21, p. 19–25.

Löfroth, G. and Gejvall, T. (1974). Diethyl pyrocarbonate: Formation of urethan in treated beverages, *Science,* **174,** 1248–1250.

Maxwell, K.E. (1973), *'Environment of Life'*, Dickenson Publishing Co., Inc., California, p. 269–283.

Potter, N.N. (1987), *'Food Science'*, 3rd Edn., CBS Publishers & Distributors, Delhi, p. 685–720.

Stewart, G.F. and Amerine, M.A. (1973), *'Food Science and Technology'*, Academic Press, New York, p. 139–204.

Voil, G.E., Phillips, J.A., Rust, L.O., Griswold, R.M. and Justin, M.M. (1973), *'Foods'*, Houghton Mifflin Co., Boston, p. 568–580.

Wishnok, J.S. (1977). Formation of nitrosamines in food and in the digestive system, *J. Chem. Ed.*, **54,** 440–442.

Wolff, I.A. and Wasserman, A.E. (1972). Nitrates, nitrites and nitrosamines, *Science,* **177,** 15–19.

Woods, R.J. (1994). Food irradiation, *Endeavour,* **18,** 104–108.

Part VII

Pollutants From Agriculture

CHAPTER 29

Fertilizers

Increasing population growth and urbanization, especially in the developing world, has compelled the nations to procure more food. Food crops, like any other plant, require fertile soil to grow. Continuous agricultural activity tends to decrease the soil fertility. Therefore, after each harvest chemical fertilizers must be applied to the fragmented and nutrient-depleted soil. Approximately 25% of the world's crop yield is attributed to the use of chemical fertilizers. This means that if we stop using the fertilizers, food productivity would come down by 25%. It is, therefore, not surprising that the demand for fertilizers is doubling every ten years. However, the increasing application of fertilizers has also threatened the human environment and ecology with deleterious consequences. If unchecked, these harmful effects shall negate, to quite some extent, the benefits derived from increasing usage of fertilizers.

The Need

Fertilizers are defined as materials which, if added to the soil, result in a better growth of crops. Their function is to restore the soil nutrients removed by the crop. At least sixteen elements are needed for the growth of plants. Among these, carbon, hydrogen and oxygen are abundantly available in the environment in form of carbon dioxide and water. These are, therefore, easily restored after each harvest. The remainder are not so easily replaced. A list of these nutrients is presented in Table 29.1.

Table 29.1 Soil derived elements essential for plant growth

Primary Nutrients: (Macronutrients)	Nitrogen, Phosphorus, Potassium
Secondary Nutrients:	Calcium, Magnesium, Sulfur
Micronutrients:	Boron, Chlorine, Copper, Iron, Manganese, Molybdenum, Zinc

The nutrients that are needed in large amounts are called *primary nutrients or macronutrients.* These are nitrogen, phosphorus and potassium. Their replacement is important because when the crop is harvested, they are almost completely removed from the soil. Elements that are *secondary nutrients* are required in moderate amounts. These are essential but seldom limit the plant growth, especially in limed soils. Amongst the secondary nutrients, calcium is required for construction of healthy cell walls; magnesium is a constituent of chlorophyll; sulfur is essential for the development of roots.

Micronutrients are required in extremely small amounts. Their depletion from the soil is, therefore, quite slow. For example, harvesting a metric tonne of potatoes removes 10 kg of nitrogen, but only 13 g of boron from the soil. Nevertheless, micronutrients are absolutely necessary for plant growth. When the same crop is grown repeatedly in the same field, these elements become depleted. They are restored to the soil by having them incorporated in the commercial fertilizers.

The Classification

The commercial fertilizers are classified into three broad types, depending on which primary nutrient they supply to the soil (Table 29.2).

Table 29.2 Classification of fertilizers

Fertilizers	Representative Examples
Nitrogen-based	Sodium nitrate, $NaNO_3$ Ammonium sulfate, $(NH_4)_2SO_4$ Urea, NH_2CONH_2
Phosphorus-based	Super phosphate, $Ca_2H_2(PO_4)_2 \cdot CaSO_4 \cdot 2H_2O$ Triple super phosphate, $Ca_2H_2(PO_4)_2 \cdot CaHPO_4$
Potassium-based	Potassium chloride, KCl Potassium sulfate, K_2SO_4

1. Nitrogen-based Fertilizers

Nitrogen containing commercial fertilizers may be subdivided into three categories: Nitrates, ammonium salts and simple amides. Nitrates, being highly soluble in water, are taken up directly by the plants through their roots. Soil microorganisms convert ammonium salts into nitrates, which are then absorbed by the roots. However, in acidic soils this conversion is extremely slow. Nevertheless, like nitrates, the ammonium salts too are soluble in water and in acidic soil environments the plants are adapted to absorb them directly. Simple amides, such as urea, are quickly hydrolysed to ammonium compounds and then nitrified by soil microbes.

2. Phosphorus-based Fertilizers

Most unfarmed soils contain negligible amounts of phosphorus. Therefore, while employing a land for agricultural purposes, the initial dose of commercial fertilizer must be rich in phosphorus. Plants take up phosphorus mainly as orthophosphates. However, orthophosphates are not highly soluble in water. A part of these gets precipitated in the soil and is, therefore, rendered unavailable to the roots. It has been observed that roots never reach more than one-third of a dressing of phosphorus fertilizer in a single year. The remaining phsophorus accumulates in the soil. It may be used in the next harvest season, provided it is solubilized. Hence, most manuring schemes aim to increase the solubility of residual phosphates in the soil.

3. Potassium-based Fertilizers

Potassium is present in the soil in appreciable quantities. Clays, such as illite and minerals, such as orthoclase, feldspar and mica are rich in potassium. The potassium released from these sources is rendered easily available to the plants. It is, therefore, surprising that potassium fertilizers have to be added to the soil at regular intervals. However, leafy crops and fruit crops can remove as much as 100 to 300 kg per hectare of potassium, thus depleting the reserve of this element in the soil. Unless potassium is returned to the soil in form of fertilizers, the yield of crops will be poor.

Environmental Implications of Fertilizers

Extensive use of fertilizers is detrimental to the living environment. Following are some of the major environmental problems associated with fertilizer usage.

1. Soil

Highly acidic conditions in the soil tend to inhibit many common bacteria, as the optimum for most species is near neutrality. Soils tend to acquire acidity through repeated use of nitrogen-based fertilizers. As the acidity of soil increases, bacterial population decreases. Simultaneously, the population of fungi increases. This is a consequence of lack of microbiological competition for the food reserves. Pathogenic fungi, such as *plasmodiophora brassicae* are thus commonly found in acidic habitats.

Fertilizers harm the soil in yet another way. Although they tend to restore inorganic nutrients to the soil, they do not replace the soil's organic matter. The organic constituents of soil are important because their decay produces humus, which modifies the structure of soil to a level which is conducive to plant growth. Moreover, soil microorganisms use organic matter as a source of energy. Since these organisms serve as important links in the carbon and nitrogen cycles, the presence of organic matter is important to their functions.

Thus, large scale dependency upon fertilizers can change the physical, chemical and biotic characteristics of the soil.

2. Vegetation

When soil becomes acidic due to consistent use of fertilizers, a number of chemical and biological conditions set in which are harmful to many plants. Under acidic conditions, nutrient cations, such as calcium and magnesium are leached away from the soil. As the pH reduces below 5.5, the aluminium content of the soil becomes soluble. The Al^{3+}, $Al(OH)_2^+$ and $Al(OH)^{2+}$ ions become dominant and they displace other cations from exchange sites. Soluble aluminium ions are toxic to the plants. Some species, such as barley, cotton and alfalfa are sensitive to as low as 0.03 ppm concentration of aluminium.

3. Pest Problem

Vegetation grown on fertilized soil tends to attain greater nitrogen content in the leaves. An increase in leaf nitrogen makes the plant more attractive to the grazers. Thus, it is observed that shrubs raised on highly fertilized land are more susceptible to damage by beetles.

4. Herbs Problem

Weeds and herbs grow more aggressively than useful plants in soil sprayed with fertilizers. As a result, the growth of the desired plant is suppressed. It has been observed that when heathlands are given an overdose of fertilizers, the growth of heath actually curtails. Heath varieties like *Calluna* and *Sphagnum,* for example, fail to regenerate in richly fertilized farmlands because herbs, like invasive torgrass (*Brachypodium pinnatum*) grow more aggressively, consuming lion's share of the nutrients.

5. Human Health

If excessive amounts of nitrogen fertilizers are used, high concentration of nitrate ions become concentrated in leafy vegetables. These subsequently bioaccumulate in the bodies of those who consume such vegetables. Nitrate also enters the body of those who consume the meat or milk of the cattle that has grazed on high nitrate-content plants. Nitrate itself is not harmful. Sodium and ammonium nitrates are used to treat phosphatic kidney stones in humans. However, once in human body, nitrate is reduced to nitrite. The latter is toxic and is suspected to cause two health problems: Blue baby syndrome and stomach cancer.

The *blue baby syndrome* or *methaemoglobinaemia* can occur when children less than about a year consume excessive nitrate. Microbes in the stomach convert nitrate to nitrite. In the blood stream, nitrite reacts with haemoglobin. As a result, the oxygen-carrying capacity of the blood is curtailed. Symptoms of diarrhoea and vomiting set in and the child's complextion becomes slate blue. In rare cases, the disease proves fatal.

There are several reasons as to why only infants are susceptible to this disease. Children consume more milk than adults. Fetal haemoglobin, which has a greater affinity for nitrite than normal haemoglobin, persists for a few months in the blood stream of a newborn. Moreover, childrens' stomachs are not acidic enough to inhibit the microbes that convert nitrate into nitrite.

Stomach cancer has been associated with the concentration of nitrate in potable water. Since nitrate ions are highly soluble, they manage to seep into drinking water supplies through agricultural runoff. In stomach, nitrate is partly converted into nitrite, which interacts with secondary amines coming from the breakdown of food or bodily proteins. The reaction produces N-nitrosamines (see eq. 28.1, p. 396). N-nitroso compounds have been shown to cause stomach cancer in experimental animals. In humans there is a very limited evidence to support the association of nitrate with stomach cancer. Yet, as a precautionary measure, a limit of 50 ppm for nitrate in potable water has been imposed.

6. Eutrophication

By far, the greatest problem associated with the use of fertilizers is eutrophication of lakes. Eutrophication deserves an exhaustive treatment and is, therefore, detailed in a separate section of this chapter (p. 409).

Abatement Procedures for Fertilizer Pollution

Most of the problems associated with fertilizers stem from the discharge of nitrates to the ecosystem. The nitrates, in turn, are produced by the microbial oxidation of nitrogen-based fertilizers. Whatever be their hazards, nitrogen-based fertilizers cannot be written off. Moreover, these fertilizers pose a hazard only when they are used in a wrong place, in a wrong dose or at a wrong time. They enter the environment mostly if they are grossly misused or in the event of a crop failure. The following steps prevent the leaching of fertilizers to the environment and hence contain their harmful impacts.

1. A number of agricultural commodities leave large amounts of nitrate residues in the soil; others leave very little. Obviously, those which leave a higher concentration of residual nitrate create greater problems, especially if the catchment area receives large amount of rainfall. By using mathematical models, based on sound data, it is possible to produce a range of cropping patterns for a given catchment.

For example, it has been observed that cereal crops and grasses leave very little ntrate residue. These may be sown even in those regions where the chances of leaching are relatively high. On the other hand, potatoes and other vegetable crops leave significant quantities of nitrate residues in the soil because they tend to be heavily fertilized, are regularly irrigated and are generally shallow rooted. These should not be planted in sensitive catchment areas.

2. Rotation of crop should be made more scientific. Crop rotation has always been a part of agricultural system so as to conserve the soil's moisture and nutrients and also to avoid the growth of herbs and weeds. The concept can also be used to control nitrate leaching if we select cropping systems which do not require extra nitrogen from fertilizers for growth. Whatever little residual fertilizer is left after the harvest of the main crop should suffice to sustain the secondary crop, so that the soil becomes essentially devoid of nitrate residues.

3. Between the harvest of summer crop and the sowing of winter crop, there is an intermittent period when nothing is sown. Fertilizers should not be applied during this period. Fertilizers on bare soil cause maximum harm. Farming specialists advocate that sowing of winter crops may be advanced to early autumn. This leaves the soil bare for a minimum period.

4. The nitrogen fertilizers should be used responsibly and in accordance with professional advice. Only the recommended amount should be applied and that too when it is absolutely necessary. It is sometimes safer to 'under-fertilize' than to 'over-fertilize'.

5. Modern scientific methods should be followed to prevent a crop failure. When crops fail, virtually the entire amount of the fertilizer is left unused and becomes prone to leaching.

Eutrophication

Eutrophication is a natural process of the ageing of a confined body of water, such as a lake. Ultimately the lake dies and becomes a part of the solid crust. Normally, eutrophication is a very slow geological phenomenon whose time scale is measured in terms of thousands of years. However, the process may be accelerated tremendously if huge quantities of nutrients are allowed to flow into the lake. This leads to an increased productivity in water. Productivity in many contexts is a desirable phenomenon, but in lakes it means, floating mats of algae, depressed oxygen levels followed by fish kills and finally the decay. This process of accelerated eutrophication is considered to be detrimental to ecology.

Classification of Lakes

Depending on the degree of eutrophication, lakes are classified into three types:

1. Oligotrophic lakes

These are the clear, cold lakes of the mountains or subpolar regions. Usually they have a rocky basin and rest on barren, infertile watersheds. Low water temperatures result in a high degree of oxygenation, but the lack of nutrients prevents an abundant growth of planktons. Moreover, rooted vegetation does not grow due to rocky basin of the lake. The level of life that can be supported is, therefore, low.

Oligotrophic lakes are biologically young. They are self-contained with a virtually closed biological cycle. They involve a good balance between life and decay. Examples include Lake Superior (USA) and Lake Baikal (Russia).

2. Mesotrophic lakes

Lakes in transition from the poorly nourished oligotrophic stage to the well nourished eutrophic stage are called mesotrophic lakes. These lakes support substantial, and sometimes unique, flora and fauna. Lake Tahoe (USA) is a mesotrophic lake.

3. Eutrophic lakes

These lakes are usually located in warmer regions. They have muddy or sandy bottoms and gently sloping shorelines. They are surrounded by watersheds having nutrient-rich soil. Warm temperatures cause an acceleration of life processes, while abundant nutrients encourage growth. Examples include Lake Eire (USA) and Lake Ontario (Canada).

Causes

Eutrophication is caused by the addition of phosphate and nitrate nutrients to the lake. Phosphates and nitrates are a part of the agricultural runoff from a heavily fertilized land. The runoff finds its way to the lakes through streams.

When nutrient material is regularly carried to the lake, algal population explodes, resulting in extensive colonies called *algal blooms*. Subsequently, other aquatic plants and planktons become plentiful. These provide a good food supply for aquatic animals and consequently the fish population increases.

To sustain a higher quantum of flora and fauna, the lake requires a larger amount of dissolved oxygen. Higher concentrations of oxygen are also required for the biodegradation of decayed material resulting from the death of aquatic organisms. Only then can the lake undergo the process of self purification.

If requisite amount of dissolved oxygen is made available to the lake, the balance between life and decay can be effectively maintained. The lake then retains its oligotrophic character. If, however, enough oxygen is not available and the supply of nutrients to the lake continues unabated, the balance between life and decay is disturbed. The decaying organisms begin to accumulate and eutrophication sets in.

Consequences

For nearly two centuries, the Great Lakes were thought to be almost immune from pollution because they are so large. Owing to the lake's immensity, it was believed that almost any amount of waste could be dumped into them without affecting their integrity. And indeed almost everything was put into them. They, and the rivers leading into them, served as large lagoons for sewer outfalls, depositories for industrial wastes, dump sites for dredged polluted sediments, sinks for farmlands and urban runoff. But even the lakes have a dumping capacity. If pollution continues unabated there will be several deleterious consequences, eutrophication being the most important one.

To understand the consequences of eutrophication, a lake may be divided into three zones: Epilimnion, thermocline and hypolimnion. These are shown in Fig. 29.1. Epilimnion is the upper, warm zone. Hypolimnion is the lower, cold zone. Thermocline is the boundary between epilimnion and hypolimnion. It represents a region of rapid temperature change.

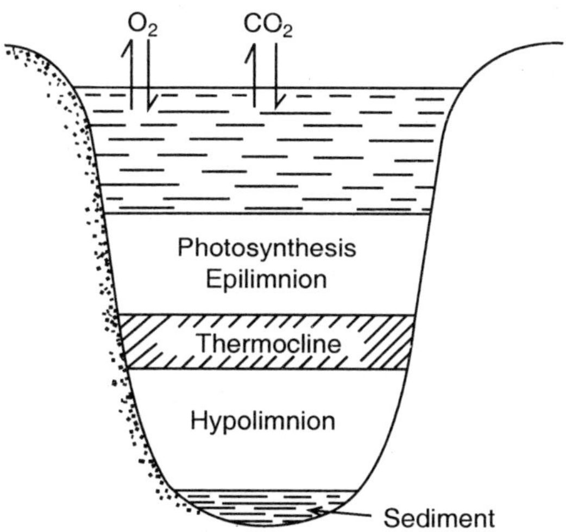

Fig. 29.1 Zones of a lake

The epilimnion is illuminated by sunlight. Hence, photosynthesis takes place in this region. The algal blooms crop up in the epilimnion. Other aquatic organisms too grow and reproduce in this region. When they die, their remains sink into the hypolimnion. Under ideal conditions, the decayed material is decomposed by oxygen. However, in a eutrophic lake the decomposition is only partial.

Although gases are more soluble in cold water, the hypolimnion region holds only scarce amount of oxygen. This is because of two reasons: Firstly, the only source of oxygen in the hypolimnion zone is the diffusion through the upper epilimnion region. The upper region, however, requires large amount of oxygen to sustain its flora and fauna. Hence, only a little amount of oxygen is able to diffuse to the lower zone. Secondly, whatever little oxygen solubilizes in the hypolimnion region is utilized for the decomposition of the decaying organic matter that has settled in the lake's bottom. In the course of time large quantities of organic matter accumulate in the hypolimnion. The amount of oxygen is insufficient to decompose the accumulated waste.

Soon, even the upper zone registers a fall in oxygen concentration. The flora and fauna die, supplementing the organic waste and aggravating the problem. Eventually the lake becomes shallow and marshy, containing only organic matter. It becomes stratified.

Mitigation

The best approach for controlling eutrophication is to reduce the amount of nutrients flowing into the lake, that is, to take action in the catchment area. Rather than removing nutrients from water after introducing them, it is better not to put them in.

If the nutrients do enter the lake water, then separate removal techniques for nitrates and phosphates have to be instituted. Nitrates are removed by their reduction to gaseous nitrogen using denitrifying bacterial agents. Methanol acts as a reducing agent, as well as a nutrient for denitrifying microorganisms.

$$6NO_3^- + 5CH_3OH \xrightarrow[\text{bacteria}]{\text{Denitrifying}} 3N_2 + 5CO_2 + 7H_2O + 6OH^- \qquad (29.1)$$

Phosphate is immobilized by injecting lime, aluminium sulfate or ferric chloride into the lake. Soluble phosphate gets precipitated as insoluble salts.

$$5Ca(OH)_2 + 3PO_4^{3-} \rightarrow Ca_5(PO_4)_3(OH) + 9OH^- \qquad (29.2)$$

$$Al_2(SO_4)_3 + 2PO_4^{3-} \rightarrow 2AlPO_4 + 3SO_4^{2-} \qquad (29.3)$$

$$FeCl_3 + PO_4^{3-} \rightarrow FePO_4 + 3Cl^- \qquad (29.4)$$

The insoluble phosphate salts settle down under the influence of gravity. This process, called *sedimentation,* renders phosphate unavailable to aquatic organisms as food material. However, sedimentation suffers from the drawback that it accelerates the silting of lake's basin.

Another technique, called *biomanipulation,* may also be used to arrest eutrophication. This involves action of organisms to control the growth of unwanted plants. Chinese crabs are added to the lake to nibble at the water plants; insects are used to destroy water hyacinths; pikes are used to devour the fish which feed on zooplankton. As a result, the number of zooplankton that eat algae rises, so that more algae are consumed, clarifying the water.

The most notorious example of eutrophication is the case of lake Eire. This lake is relatively shallow and warm. It has large agricultural areas in its drainage basin. It also has an urbanized shoreline. Because of these factors, Lake Eire has suffered intense eutrophication.

In 1972 the governments of United States and Canada entered into the *Great Lakes Water Quality Agreement* so as to contain the eutrophication of Lake Eire. Under this agreement, the phosphate content of the water entering the lake was reduced to about 1 ppm. The flow of nitrate was also restricted. Within the next 5 years, algal mats largely disappeared and the water looked and smelled better.

References

Addiscott, T.M., Whitmore, A.P. and Powlson, D.S. (1991), *'Farming, Fertilizers and the Nitrate Problem'*, C.A.B. International, Wallingford, p. 1–14.

Alexander, M. (1961), *'Introduction to Soil Microbiology'*, John Wiley & Sons, Inc., New York, p. 63–84.

Amadi, A. (1991) in, *'Chemistry, Agriculture and the Environment'*, Richardson, M.L. (Ed.), The Royal Society of Chemistry, Cambridge, p. 221–231.

Anderson, N.J. (1995). Naturally eutrophic lakes: Reality, myth or myopia, *Trends Ecol. Evol.*, **10**, 137–138.

Barroin, G. (1993). Restoration of lakes, *The World Scientist,* March, p. 24–36.

Cooke, G.W. (1975), *'Fertilizing for Maximum Yield'*, Crosby Lockwood Staples, London, p. 1–12.

Enger, E.D. and Smith, B.F. (1991), *'Environmental Science'*, 4th Edn., Um. C. Brown Publishers, Dubuque, p. 313–315.

Hileman, B. (1988). The Great Lakes cleanup effort, *Chem. Eng. News,* February 8, p. 22–39.

Hutchinson, G.E. (1973). Eutrophication, *Amer. Sci.*, **61,** 269–279.

Lee, G.F. (1973). Role of phosphorus in eutrophication and diffuse sources control, *Water Res.*, **7,** 111–128.

Maxwell, K.E. (1973), *'Environment of Life'*, Dickenson Publishing Co., Inc., California, p. 355–376.

Moore, P.D. (1995). Pollution: Too much of a good thing, *Nature (London),* **374,** 117–118.

Schindler, D.W. (1974). Eutrophication and recovery in experimental lakes: Implications for lake management, *Science,* **184,** 897–898.

Wild, A. (1993), *'Soils and the Environment'*, Cambridge University Press, Cambridge, p. 251–254.

CHAPTER 30

Insecticides

Much of the increase in global agricultural production over the last few decades has come about through the adoption of high-input farming systems. Pesticides, the chemicals which are used for eliminating pests, are an important component of high-input farming. Therefore, there is a strong tendency in our present civilization to resort to the use of pesticides. The term pesticide is a broad one, which includes mainly insecticides—the chemicals used for killing unwanted insects; herbicides—for killing herbs and weeds; and fungicides—for killing fungi. While the role of insecticides *vis-a-vis* the environment is described in this chapter, that of fungicides and herbicides is presented in the next chapter.

The Need

Insects compete with man in various ways. They infest his person and his habitation, they eat his crop and they carry and transmit microorganisms and parasites which cause diseases to man himself, his domestic animals and his crop plants. The most direct method of insect control is by the use of insecticides. Insecticides promote human well being directly through control of insect-vector borne diseases and indirectly through increased and improved agricultural production of food, fodder and fibre. The five main uses of insecticides are—

1. To control the insect pests which would otherwise reduce the yield of agricultural crops. Before the advent of synthetic insecticides, insects used to consume or destroy, about a third of food crop. Thus, greater application of insecticides means increased food production. The need for increased food production, to feed the expanding population, is evident from the fact that malnutrition and starvation are among the most urgent public health problems.
2. To stop the spoilage of food crops during storage by controlling insect pests like grain weevils.
3. To protect forest produce. Forest pests are known to cause severe tree mortality. Gypsy moth defoliates hemlock trees; tussock moth attacks fir trees; spruce budworm destroys balsam trees. These pests are eliminated by aerial sprays of insecticides over the forest regions.
4. To kill disease-carrying organisms. Many insects spread diseases, such as sleeping sickness, bubonic plague and malaria. Mosquitoes are known to spread over 30 diseases harmful to humans. The effectiveness of insecticides against a wide range of disease vectors has eradicated many important ailments which previously defied all practical efforts at control.

5. To promote aesthetic well being. Control of insects like cockroaches, houseflies and mosquitoes contributes, if not specifically to public health, at least to a sense of well being. Likewise, the use of insecticides to control Dutch Elm disease and hemlock looper has aided in the preservation of shade trees and enhanced the scenic beauty of the local environment.

The Classification

Synthetic insecticides are broadly classified into four groups: Organochlorine hydrocarbons, organophosphates, carbamates and pyrethroids. The structures of representative compounds of each class are displayed in Fig. 30.1.

1. Organochlorine Hydrocarbons

Organochlorine hydrocarbon-based insecticides were the first ones to be used for crop protection. These insecticides are not very toxic. Structurally, they are very stable. This is both an advantage and a disadvantage. On the positive side, these insecticides, being *persistent* in the environment, show their biological activity for long periods of time. Therefore, the interval between two successive applications is a relatively long one. On the negative side, these insecticides tend to accumulate in the environment and affect many non-target organisms, not just the target pests. Because of their persistent nature, several organochlorine hydrocarbon insecticides including p, p-dichlorodiphenyltrichloroethane (DDT), aldrin and dieldrin, have been phased out of use.

2. Organophosphates

Organophosphate insecticides have very low persistence in the environment. They break down very rapidly and, therefore, do not cause any long term hazards. However, these compounds are extremely toxic in nature. Apart from killing non-target insects, they also harm the humans and the wild life. Further, due to their instability they have to be applied at frequent intervals. Therefore, they are not always economically viable. Representative examples of organophosphate insecticides include parathion and diazinon.

3. Carbamates

Carbamate insecticides are intermediate in behaviour in comparison to organochlorine hydrocarbons and organophosphates. They are less toxic than organophosphates and less persistent than organochlorine hydrocarbons. Examples include carbaryl and methiocarb.

4. Pyrethroids

The pyrethroids are the most recent group of insecticides. They are structural analogs of pyrethrum, the naturally occurring insect repellent compound present in chrysenthemum. Pyrethroids have two important advantages: They are neither persistent nor toxic. Therefore, despite their high cost, they account for about one-third of world insecticide use. Examples include pyrethrum and deltamethrin.

The Characteristics

Insecticides kill the insect by poisoning its nervous system. The basic unit of the

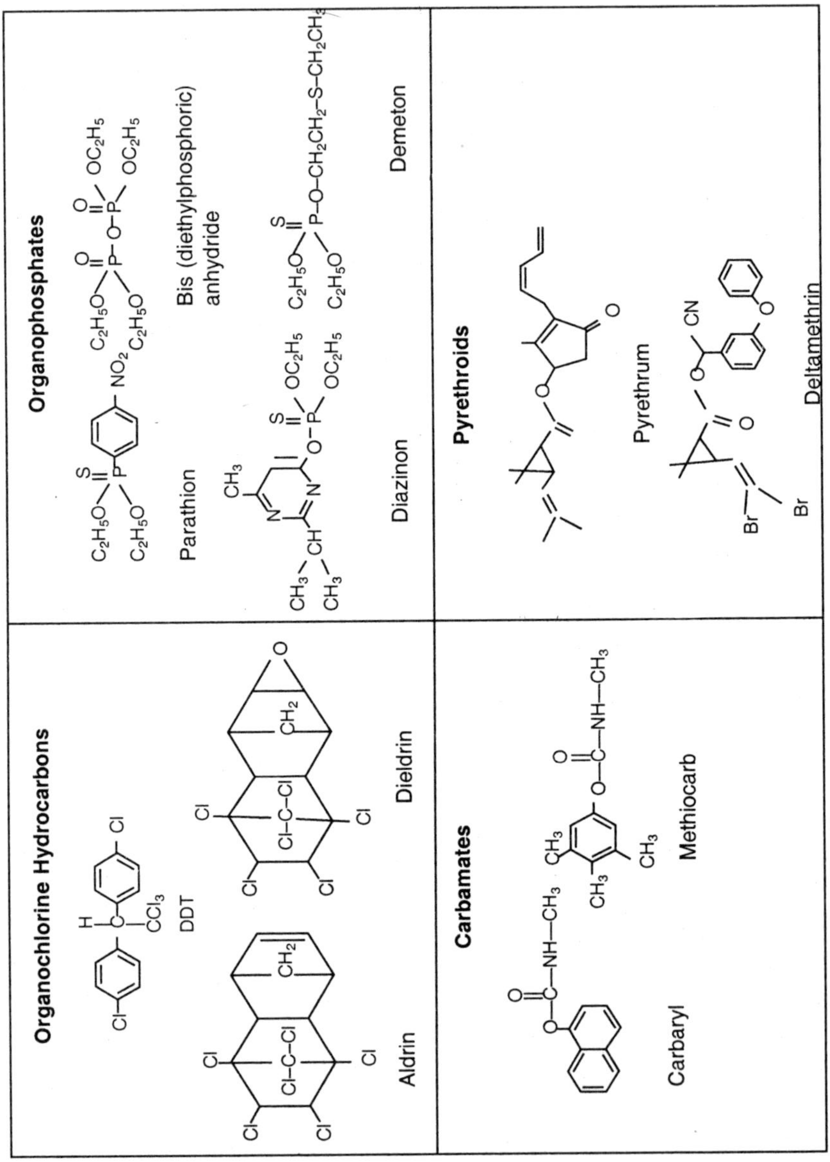

Fig. 30.1 Classification of insecticides

nervous system, that is, a nerve cell, is called a *neuron*. The transmission of a nerve impulse is carried through in two stages. (a) Transmission within the neuron, which takes place by an electrical phenomenon and (b) transmission between one neuron and the other, which takes place by a chemical mechanism.

When no impulse is to be transmitted, the cellular fluid inside the neuron contains a lower concentration of sodium ions than the extracellular matrix. Under these conditions, the inside of the neuron is at a negative electrical potential with respect to the outside. When the neuron is stimulated by an impulse, sodium ions rush in through, what is known as, the *sodium gate*. The electrical potential inside the neuron then becomes positive. Due to the change in polarity, an electric wave ensues and carries the message from the head of one neuron to the tail of the same neuron.

Next, the impulse has to be passed from the tail of the first neuron to the head of the second neuron. This transmission is carried out by a chemical known as acetylcholine (Fig. 30.2). After the transmission, an enzyme called cholinesterase hydrolyses and destroys acetylcholine.

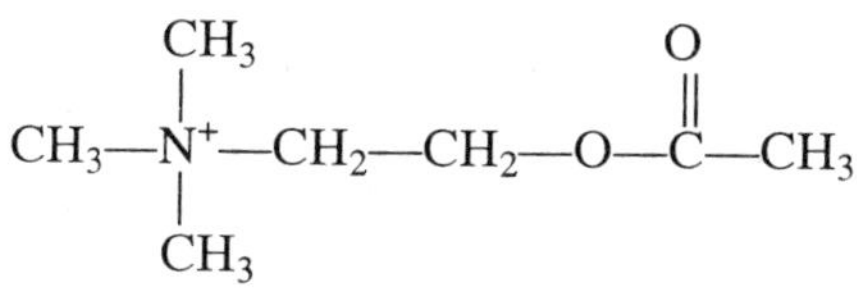

Fig. 30.2 Acetylcholine

The organochlorine hydrocarbon and pyrethroid insecticides interfere with the transmission of nerve impulse within the neuron. The molecules of these insecticides are of correct size to be trapped inside the nerve pores. The passage of sodium ions to the cell is thus blocked and transmission of impulse is no longer possible. The insects exhibit hyperactivity, which leads to convulsions and death.

The organophosphate and carbamate insecticides interfere with the transmission of nerve impulse from one neuron to the other. As has been stated, acetylcholine, the chemical responsible for this transmission, must be destroyed by the enzyme cholinesterase after the message has been transmitted. The organophosphate and carbamate insecticides possess anti-cholinesterase activity. That is, they deactivate the enzyme so that it cannot destroy acetylcholine. Under these conditions, the nerve cell becomes overstimulated—a condition which leads to convulsions and death.

Depending on their ability to move within the plants, insecticides may be branded as non-systemic or systemic. *Non-systemic insecticides* are those that remain localized on that part of the plant on which they are sprayed. An insect would be killed only if it eats the insecticide-containing part. Thus, if a non-systemic insecticide is applied on the leaves of a plant and the insect eats the stem, it would not be killed. DDT is an example of a non-systemic insecticide. *Systemic insecticides* are those that are rapidly translocated within the plant. These would kill the insect even if it eats those parts of plants which have not been touched by the spray. Demeton is an example of a systemic insecticide.

With systemic insecticides, the whole surface of a plant becomes lethal to insects feeding on it, even if only a part of the plant is treated. Therefore, there is no need for a complete cover when a spray is applied. This implies that the amount used can be reduced, with economy to the farmer and less danger to the environment.

Environmental Implications of Insecticides

The excessive use of insecticides has imposed severe strains on the environment.

1. Bioamplification

Insecticides tend to concentrate in the fatty tissue of various organisms. Such substances show poisoning effects even if they are relatively non-toxic. For example, the toxicity of DDT is the same as that of asprin. However, consumption of ten tablets of asprin causes no serious effect while an equivalent amount of DDT, if ingested, can prove to be hazardous. The reason being that the drug is not retained by the body, whereas the insecticide is. Therefore, DDT acts as a cumulative poison.

Substances which tend to get stored up in the fatty tissue undergo bioamplification, that is, their concentration progressively increases in those species which oocupy a higher trophic level in the food chain. For example, spraying a marsh to control mosquitoes results in the accumulation of traces of DDT in the cells of microscopic aquatic organisms, such as plankton. The DDT then becomes concentrated in the tissue of the fish and shellfish which feed on planktons. Concentrations of DDT ten times greater than those is planktons have been measured in the fish. Gulls, which feed upon fish, accumulate still higher concentrations of DDT. For predators, such as ospreys, pelicans, falcons and eagles, which feed upon gulls, as well as fish, the levels of accumulated DDT are often so high that the consequences are disastrous.

2. Broad Spectrum Characteristics

Most of the insecticides possess a broad spectrum nature, that is, they have the ability to kill a large variety of insects. This means that they not only destroy insect pests, but also eliminate the natural predators of insect pests and thereby interfere wth the natural mode of pest control.

For example, the apple growers used to spray DDT in their orchards to control codling moth larvae and apple maggots. However, subsequently it was observed that the orchards were devastated by scale insects and mites. The reason was that DDT killed not only the intended victims but also the natural enemies of scale insects and mites. Without the natural check, the population of scale insects and mites increased enormously.

3. Resistance

It is commonly observed that over a period of time the effectiveness of a particular insecticide begins to decline. The target insects tend to develop resistance against that insecticide. The phenomenon of resistance is an illustration of natural selection. As an example, consider a population of 100 insects of G_0 generation. Suppose that a particular insecticide kills 95 most susceptible individuals. The offsprings of the remaining 5, being descended from the more resistant segment of the population, would, on an average, be more resistant. Therefore, to kill insects of G_1 generation would require application of a greater amount of insecticide. G_2 generation would require a still greater amount. Ultimately, this would lead to a very large build-up of the insecticide and its residues in the environment.

More than 500 species of insects have developed some type of immunity to common insecticides by evolving biochemical mechanisms to detoxify the pesticides. Repeated exposures to a specific insecticide activates the *cytochrome P-450* enzyme of the insect. The activated enzyme converts the insecticide to polar degradative products that are easily excreted from the body of the insect. The insect beats the insecticide!

4. Birds

Perhaps no area of insecticide use and effect has aroused more public concern or been more extensively debated than that dealing with birds. Most of the pesticide-related toxicity and reproductive problems amongst the birds have been caused by organochlorine hydrocarbons, particularly DDT.

The birds of prey, which occupy a significantly high position in the food chain concentrate enormous quantities of insecticides in their tissue. Their body attempts to metabolize the pesticide by an alteration of the normal metabolic patterns. This alteration involves the use of hormones which normally regulate the calcium metabolism of the bird and are vital to its ability to lay eggs with thick shells. When these hormones are used to metabolize insecticides, they become chemically modified and are no longer available for the construction of egg shells. As a result, the eggs are easily damaged and the survival rate of the offsprings declines.

The presence of insecticides in the bird's body also increases the activity of enzymes in the liver. The activated enzymes attack the bird's sex hormones. This leads to three consequences. Firstly, breeding takes place abnormally late. Secondly, new eggs are not laid to replace early losses. Thirdly, fewer eggs are laid. All these factors result in a decline of bird population.

5. Fish

The widespread use of insecticides makes it inevitable that a portion of these will contaminate surface waters. In particular, insecticides sprayed over forest areas find their way to rivers and streams due to atmospheric fallout. They are then ingested by the fish.

Some of the insecticides, such as DDT, are highly toxic to the fish. Species like trout and salmon succumb when water contains even trace amounts of DDT. Some species of fish do not die when exposed to insecticides, but show sublethal effects. A wide range of abnormal behavioural patterns have been observed in fishes thriving in contaminated waters. For example, low levels of bis (diethylphosphate) anhydride affect the learning ability of goldfish. Diazinon at a concentration as low as 1.25 ppm results in hyperactivity in rainbow trout and a reduced rate of swimming in tin-plated barb.

6. Human Health

As has been stated, the organophosphate and carbamate insecticides are highly toxic. In order to prevent occupational hazards the workers dealing with these insecticides are required to wear protective overalls. In warm weather the overalls are extremely discomfortable. There have been cases where, in tropical countries, the workers did not button-up their protective clothing and fell victim to insecticidal poisoning.

Organochlorine hydrocarbon insecticides are comparatively less toxic. Nevertheless, when present in human diet, these elicit well documented effects. For example, they activate enzymes in the liver. The activated enzymes rapidly metabolize the drugs. If an individual happens to be taking medications, the drugs may be broken down more easily than is desirable. Another concern expressed about pesticide residues in human body has been that breast-fed babies imbibe an excessive amount of insecticide via the mother's milk.

Abatement Procedures for Insecticide Pollution

Chemical insecticides have a lot wrong with them. Insects can become resistant to them. Their effectiveness is temporary and repeated applications are necessary. Some of these have long persistence periods and accumulate in the harvested crops. They also disrupt natural controls, causing even worst pest outbreaks. They can harm non-target organisms. And they reduce the diversity of ecosystem. Yet, the hard fact is that we cannot do without the insecticides. Putting off their use would lead to hunger and starvation. The best course, therefore, is to seek alternative insecticides, which would kill the pests, but not harm the environment. Alternatives to conventional, chemical insecticides are now discussed.

1. Bacterial Insecticides

Insects may be controlled by infecting them with bacterial diseases. The first bacterial insecticide to be produced commercially was a preparation containing spores of *Bacillus popilliae,* which cause milky spore disease in Japanese beetles. Another important commercial formulation of bacterial insecticide is based on the spores of *Bacillus thursinglensis,* which is effective against caterpillars.

Bacterial insecticides offer a number of advantages. They effectively control specific insects and do not harm humans, livestock, fish, wildlife or beneficial insects. They do not damage the plants. Insects do not become resistant to bacterial formulations. They can be applied to crops right upto the day of harvest, which means that they can protect the crop from last minute damage by the insects.

However, bacterial insecticides also suffer from a number of drawbacks. They are not as fast acting as conventional chemical insecticides. Moreover, the bacterial preparations are quite expensive. Further, poor weather conditions, such as excessive rainfall and extremely high or low temperatures may prevent proper control.

2. Viral Insecticides

Some viruses can be used quite effectively as insecticides. Viral insecticides are highly specific. By spreading rapidly they destroy a particular species of target insects and continue killing them by becoming established in the environment. For example, the nuclear polyhedrosis virus specifically eliminates the heliothis moth larvae.

Viral insecticides have a disadvantage too. Viruses are obligate parasites that can only grow in the tissue of their hosts. This means that production of viral formulations require a large number of disease-free host insects. This is a difficult and an expensive venture.

3. Hormonal Insecticides

Hormonal insecticides kill the insect by upsetting its normal metabolic processes. Juvenile hormone (Fig. 30.3) shows the most promising insecticidal properties. This hormone is normally present in insect larva. Subsequently, the larva stops producing it and undergoes metamorphosis, that is, it grows up into an adult insect. However, if the larva continues to be exposed to the juvenile hormone, it becomes impossible for it to mature. An immature insect cannot reproduce and, therefore, the insect population registers a decline. Juvenile hormone has been successfully used to control yellow mealworm.

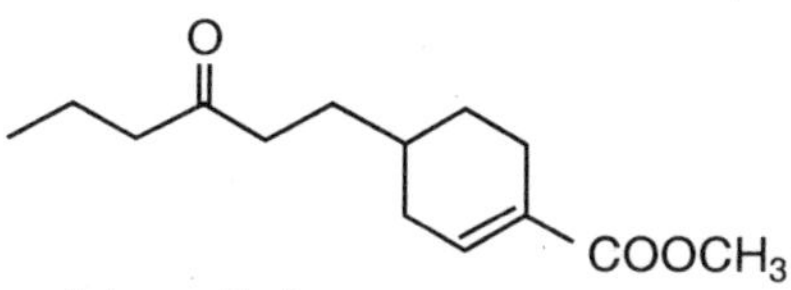

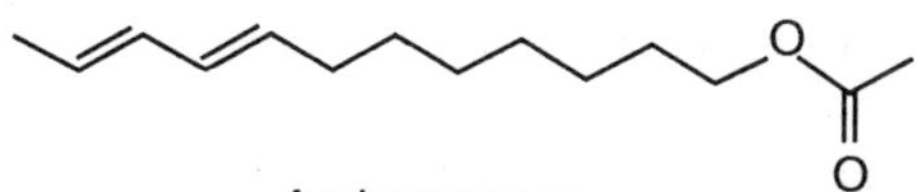

Fig. 30.3 Alternatives to conventional insecticides

Hormonal insecticides suffer from the drawback that they are effective only at certain stages of insect development. Therefore, timings of application are highly crucial. Moreover, rather than killing the insect they keep it in the larval form. It case of many insect pests the adults are relatively harmless, while the larvae cause the major damage to the plants.

4. Chitin Inhibitors

Chitin or glucosamine polysaccharide is a horny substance that makes up the shell of insects. Larvae treated with chitin inhibitors develop a fragile, malformed shell which may rupture, causing the insect to die.

A commonly used chitin inhibitor is Th-6040. It is effective only when ingested as extremely minute particles. Therefore, its formulation is marketed as a dispersible powder with a particulate size of about 2 μm. Th-6040 is used to control gypsy moth and the western spruce budworm. It has a low toxicity to mammals, birds and fish.

5. Sex Attractants

Many insects find their partners for mating by means of an automatic response to various scent cues. Male silk moths, for example, can smell potential sexual partners and are lured towards them. The reason being that the female silk moth secretes chemical compounds called *pheromones* (Fig. 30.3), the odour of which is detected by an array of receptors on the feathery antennae of the male. A nerve impulse in a receptor cell can be triggered by one molecule of a pheromone and that too from a considerable distance. Male gypsy moths are attracted to a female upto half a mile away. Male silkworms are lured to females from a distance of two and a half miles.

This behaviour makes the odour-baited lure a promising means of getting rid of malignant species of insects. Instead of spraying a wide area with chemical insecticides, the unwanted insect pests are attracted to the traps containing synthetic pheromones. The pests are subsequently treated with a sterilizing chemical or with an exterminating device.

6. Genetically Engineered Crops

A relatively recent method to save vegetation from pests is to sow transgenic crops engineered to be resistant to specific insects. The aim of such manipulations is to provide the crops with a toxin that can ward-off or kill specific insects like boll-weevil and aphids, while sparing friendlier ones such as ladybird. Genetic engineering can slash the volume of insecticides that farmers need to dump on the land every year.

Genetically engineered plants produce a steady supply of their own inseticide. For example, crops can be genetically engineered to produce Bt, an insecticidal protein. Thus, Bt cotton, Bt maize and Bt potato have begin to hit the fields. In order to avoid the problem of resistance, the farmers are advised to grow crops with and without Bt genes simultaneously. This way insects can mature without any evolutionary pressure to become resistant.

7. Integrated Pest Management

The inevitable utility of insecticides has put the farmers and the environmentalists in an embarrasing crossfire. On one hand the farmers demand more potent insecticides and on the other the environmentalists demand less toxic analogs. Irreconcilable though these demands may seem, there is a good chance that both of them can be satisfied by a new approach, called integrated pest management.

Integrated pest management is defined as a system that will bring most benefits, at the most reasonable cost, on a long term basis, to the farmer, as well as to the society. The aim is to bring the best of all available control techniques to bear against the pest problem, rather than to place a blind reliance on conventional pesticides.

In the full sense of the term, integrated pest management encompasses all control measures, including the use of fungicides and herbicides. However, since conventional chemical insecticides have received the bulk of the blame for the environmental problems, the main thrust of integrated pest management is on the insecticides.

Integrated pest management does not represent a 'back to the nature' philosophy—the one that rejects chemical insecticides out of hand. Chemical insecticides are indeed used for controlling pests, but they constitute only a part of the arsenal. Also included in the arsenal are the natural enemies of the insects, juvenile hormones, pheromones, chitin inhibitors and microbes. Integrated pest management thus represents a multiplicity of approach. It also emphasizes that the insecticides be applied only when needed and in the lowest possible amounts.

Bhopal Episode

On the night of December 2, 1984, a dense cloud of the deadly methyl isocyanate (MIC) gas leaked from a storage tank at the Union Carbide Ltd. Plant in Bhopal. As the gas swept over the city, it killed about six thousand people, injured about two lakh and left about fifty thousand permanently impaired.

Causes

Bhopal's Union Carbide plant was a manufacturing unit for the insecticide carbaryl (Fig. 30.1), which was marketed under the commercial name, sevin. Methyl isocyanate, $CH_3N{=}C{=}O$, was the main raw material. It was prepared by the reaction of methylamine, CH_3NH_2, with phosgene, $COCl_2$,

$$CH_3NH_2 + COCl_2 \rightarrow CH_3N{=}C{=}O + 2HCl \tag{30.1}$$

and stored in abundance. Subsequently, it was treated with 1-napthol to produce carbaryl.

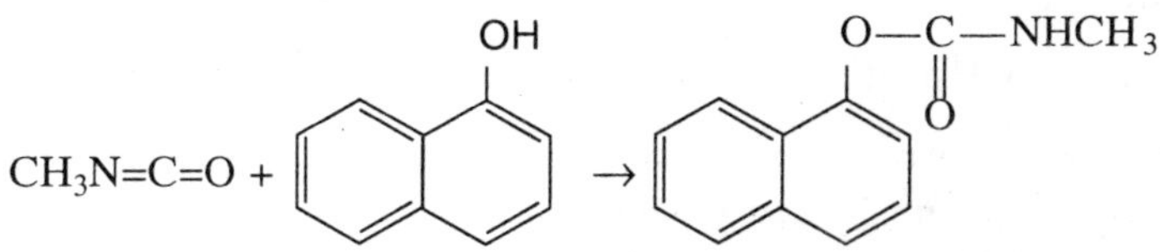

(30.2)

There were three tanks in the plant that stored methyl isocyanate. Pressure rose in one of the tanks, tank No. 610, and blasted its release valve. Methyl isocyanate surged about 600 feet along the 8 inch relief vent line and escaped into the atmosphere.

What caused the pressure in tank No. 610 to rise abnormally high is still a mystery. Three hypotheses have been projected to explain the disaster.

Hypothesis I. Water somehow seeped into the tank and reacted with methyl isocyanate.

$$CH_3N{=}C{=}O + H_2O \rightarrow CH_3NH_2 + CO_2 \tag{30.3}$$

Reaction (30.3) is highly exothermic. If sufficient quantity of both the reactants are present, and if the heat is not somehow removed, the temperature will rise and methyl isocyanate will start boiling violently. In a closed tank, the high pressure build-up will result in an explosion. Despite all precautionary measures, there is always a possibility that a small amount of water may have entered the tank. However, it is difficult to envisage that the amount of water was so large as to trigger an explosion.

Hypothesis II. Methyl isocyanate in presence of a suitable catalyst undergoes auto-polymerization to yield a cyclic trimer.

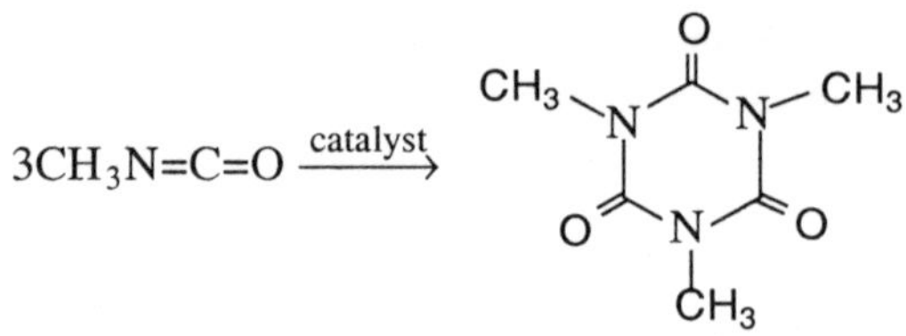

(30.4)

Reaction (30.4) too is highly exothermic and can lead to a high pressure build-up. Strong bases, like sodium hydroxide; metal chlorides, like ferric chloride; and phosphine derivatives can act as catalysts for reaction (30.4). However, it is difficult to imagine how these catalytic substances entered the tank.

Hypothesis III. The most plausible explanation stems from a combination of hypotheses I and II. A small amount of water seeped into the storage tank and initiated hydrolysis reaction (30.3). But the water that entered the tank was not pure. The impurities that it carried inside the tank initiated the polymerization reaction (30.4). Thus, it was the simultaneous occurrence of two exothermic reactions that may have led to the escape of methyl isocyanate gas from the reservoir.

To tackle such emergency situations, Union Carbide's protocol listed a few safety measures. The plant was endowed with a vent gas scrubber that would have neutralized the escaping methyl isocyanate with caustic soda.

$$CH_3N{=}C{=}O + 2NaOH \rightarrow CH_3NH_2 + Na_2CO_3 \qquad (30.5)$$

Unfortunately for the residents of Bhopal, this system was down for repairs. The tank's refrigerator unit, which would have slowed down reactions (30.3) and (30.4), was turned off at the time of disaster. It is also alleged that the plant was being managed in such a sloppy way that it represented an 'accident waiting to happen'.

Consequences

Methyl isocyanate is a ravenously reactive molecule. It reacts quickly with the components of biomolecules, such as proteins, amino acids, sulfhydryl functions and carboxylic acid groups. As a result, methyl isocyanate is extremely toxic and this is indicated by its very low maximum allowable concentration—0.02 ppm over an 8 h period.

As the cloud of methyl isocyanate traversed over the city it left behind a trail of death and ailment. Some people, who were lucky enough to live in well-sealed homes, managed to avoid the choking gas—unless they opened a door or window to see what was going on. Some escaped from the gas by climbing stairs to a higher floor. Still others protected themselves by covering their faces with a wet cloth. This last strategy made a lot of sense because methyl isocyanate is rapidly hydrolysed by water (reaction 30.3).

Others were not so lucky and suffered a host of impairments. The exposure to methyl isocyanate caused the following adversities.

1. Respiratory system disorders

Given the high reactivity of methyl isocyanate, the key initial event that occurred when the victims breathed the gas was a quick reaction with respiratory tract and lung-tissue proteins. The proteins immediately denaturated, fibrosis occurred, and cells died by the millions. The victims had difficulty in breathing. Many burst out of their homes coughing and gasping for breath and then died of suffocation. In some people, the gas caused such massive internal secretions that their lungs became clogged with fluids. In others, spasmodic constrictions of the bronchial tubes led to suffocation. No one disputes that the immediate cause of death was oxygen deprivation.

Many of those who survivied the first day were found to have impaired lung functions. Their breathing was shallow, making it difficult for them to get enough oxygen. The decreased oxygen level in blood caused an increased strain on heart. The blood acquired a cherry-red colour due to oxygen deficiency. Many surviving victims have become permanently disabled with emphysema, asthama or other chronic lung diseases.

2. Digestive system disorders

Considerable amount of methyl isocyanate became dissolved in the saliva of victims and was thus swallowed. This pathway resulted in stomach and intestinal disorders. Some of the victims began vomiting uncontrollably. The digestive system impairments took a considerably long time to heal.

3. Reproductive system disorders

Methyl isocyanate crossed the placental barrier in pregnant women and brought about miscarriages, stillbirths and death of infants. In addition, women victims complained of changes in their menstrual cycles.

4. Nervous system disorders

Some cases of brain damage, nerve damage and paralysis were reported after the gas tragedy. However, permanent damage to the nervous system occurred in extremely rare cases. Most of the victims experienced dizziness, disorientation and extreme fatigue following the slightest exertion. Infants experienced psychic problems. After the tragedy a team of behavioural scientists asked children under ten to make drawings. Inevitably they put their crayons to crude depictions of carbide, gas and death.

5. Eye impairments

All isocyanates are harmful to the eyes. Methyl isocyanate is no exception. The gas stricken residents experienced a burning sensation in the eyes. Ulcers formed on the cornea of some victims, blinding them. However, most of the eye damage, including blindness, was temporary in nature and reversed within a short span of time. Yet, excessive watering of the victim's eyes and painful sensitivity to strong light persisted for a very long time.

6. Cyanide poisoning

An intriguing aspect of Bhopal episode was that the victims suffered from the symptoms of cyanide poisoning. Cyanide is extremely toxic and tends to deactivate enzymes. Since there was no direct exposure to cyanide, it was concluded that a 'cyanide pool' was formed in the body through biochemical reactions of methyl isocyanate.

Mercifully, the symptoms of cyanide poisoning reversed in most of the victims. The reasons being that the sufferers were given an immediate treatment of sodium thiosulfate. The latter acts as scavenger of cyanide. It converts it into harmless thiocyanate, SCN^-.

$$Na_2S_2O_3 + CN^- \rightarrow Na_2SO_3 + SCN^- \quad (30.6)$$

Since sodium thiosulfate is a drug without appreciable side effects, its treatment was continued on a long term basis.

Aftermath

Bhopal will always cast shadows over Union Carbide. First, because of the irresponsible way the plant was being run. Second, because of the unresolved question of just what happened at the plant. Third, because of unsatisfactory way the survivors and victims of the accident were treated and compensated. After a long legal battle, Union Carbide doled out a sum of $470 million (Rs. 710 crores) as compensation for gas victims. It also agreed to fund construction of a hospital in Bhopal and its operation for 8 years.

In 1996, on the eve of twelveth anniversary of the gas tragedy, an International Medical Commission issued an assessment of long term clinical consequences of the victims. It listed the traumas with which many victims will have to live for the rest of their lives. Not surprisingly, Union Carbide refused to comment on the commission's findings.

The attitude of compatriots has also been deplorable. In 1984, when the gas tragedy occurred, the population of Bhopal was 8.5 lakh. Till the beginning of 1997 about 10.5 lakh applicants had filed their claims for compensation. Fake birth and domicile certificates were produced to 'authenticate' the claims. As a result, the cause of genuine victims continues to suffer.

References

Beroza, M. (1971). Insect sex attractants, *Amer. Sci.*, **59,** 320–325.

Brattsten, L.B., Holyoke, C.W., Leeper, J.R. and Raffa, K.F. (1986). Insecticide resistance: Challenge to pest management and basic research, *Science,* **231,** 1255–1260.

Brooks, M. and Coghlan, A. (1998). Live and let live, *New Scientist,* October 31, p. 46–49.

Dagani, R. (1985). Data on MIC's toxicity are scant, leave much to be learned, *Chem. Eng. News,* February 11, p. 37–40.

Durham, W.B. (1974) in, *'Industrial Pollution',* Sax, N.I. (Ed.), Van Nostrand Reinhold Co., New York, p. 10–35.

Enger, E.D. and Smith, B.F. (1991), *'Environmental Science',* 4th Edn., Um.C. Brown Publishers, Dubuque, p. 315–340.

Heylin, M. (1994). Bhopal: Ten years after, *Chem. Eng. News,* December 19, p. 3.

Lepkowski, W. (1985). Bhopal, *Chem. Eng. News,* December 2, p. 18–32.

Lepkowski, W. (1996). Commission surveys Bhopal medical effects, *Chem. Eng. News,* December 9, p. 11–12.

Levine, R. (1991) in, *'Handbook of Pesticide Toxicology'*, Vol. 1, Hayes, W.J. and Laws, E.R. (Eds.), Academic Press, Inc., San Diego, p. 275–360.

Marx, J.L. (1977). Applied ecology: Showing the way to better insect control, *Science,* **195,** 860–862.

McEwen, F.L. and Stephenson, G.R. (1979), *'The Use and Significance of Pesticides in the Environment'*, John Wiley & Sons, New York, p. 155–215.

Oakland, J.S. (1981). Survival: Integrated pest management, *Chem. Brit.,* **17,** 430–436.

Pickett, J.A. (1988). Chemical pest control—the new philosophy, *Chem. Brit.,* **24,** 137–142.

Thayer, A.M. (1997). Betting the transgenic farm, *Chem. Eng. News,* April 28, p. 15–19.

Sanders, H.J. (1975). New weapons against insects, *Chem. Eng. News,* July 28, p. 18–31.

Worthy, W. (1973). Integrated insect control may alter pesticide use pattern, *Chem. Eng. News,* April 23, p. 13–19.

Worthy, W. (1985). Methyl isocyanate: The chemistry of a hazard, *Chem. Eng. News,* February 11, p. 27–35.

CHAPTER 31

Fungicides and Herbicides

Fungicides are those chemicals which, when applied to the plants or to their seeds and bulbs, inhibit or prevent fungal diseases, such as scabs, blotches, rots, molds, rusts and mildew. Herbicides are a class of compounds that kill unwanted weeds or interfere with their growth. Successful, large-scale agriculture requires that the vegetation should be free from disease and the land reserved for commercial crop should be free from unwanted growth. It is, therefore, not surprising that fungicides and herbicides are given a share of credit for the green revolution. In fact, the global agricultural produce would have been much below the present level, had fungicides and herbicides not been used as farming inputs. Though these agrochemicals are somewhat harmful to the environment, it is largely believed that their benefits, in terms of human welfare, far outweigh their risks.

The Need

Fungi are plants without chlorophyll. They, therefore, cannot use solar energy to convert carbon dioxide and water into energy-rich carbohydrates. They live as saprophytes on decaying organic matter or as parasites at the expense of living organisms. Hence, they are deemed as a threat to human interests. Fungicides are important because they counter the growth of fungi and protect man's commodities. The main uses of fungicides are–

1. To protect commerical crop from disease. In absence of fungicides, fungi infect vegetation by sending haustoriá of mycelia throughout the plant cells, robbing the host of needed nutrients and/or producing toxins. Their growth results in rotting of plant foliage (Fig. 31.1), distortion and inefficiency of roots or dysfunction of vascular system. At times only the plant vigour is reduced without the dramatic crop loss. Potato blight and apple scab are examples of fungal diseases. Fungicides are applied on standing vegetable, fruit and nut crops to protect them from such diseases.

2. For treatment of seeds. When a seed is placed in the soil, it is attacked by disease organisms, many of which will prevent its germination and successful emergence as a seedling. This is especially true of large seeded vegetables. When seeds of beans, corn and peas are placed in soil, a high percentage fail to produce a successful plant unless protected by a fungicide against seed-rotting organisms and seedling blights.

3. To protect the forest trees. White pine blister rust is a common fungal

Fig. 31.1 Fungus results in rotting of plant foliage

disease which poses a threat to forest trees. It attacks the white and sugar pines, highly valued for construction purposes. The fungus grows within the needles, and as it progresses, cankers containing fungal fruiting bodies are produced on the bark. Another fungal disease, called dutch elm disease, attacks all varieties of elm. As the fungus grows, the vascular tissue of the branches gets clogged and the tree slowly dies. Thus the pine and elm trees are often treated with fungicides to prevent the outbreak of fungal-based diseases.

4. Fungicides are used by wood industry. During the processing of wood, they are used to control slime. In the finished product, they are used as preservatives. Utility poles and railway sleepers are commonly given a treatment of fungicide.

5. In building products, fungicides are used as mold inhibitors.

6. In paints and varnishes, fungicides are used for mildew prevention.

Herbicides are meant to control the growth of weeds. Weeds are the unwanted plants which grow at places where man does not want them. The unwanted growth causes several problems. Hence, the economic and social importance of herbicides. The main uses of herbicides are—

1. To remove weeds from agricultural land. Weeds harm the cash crop. They compete with the crop for nutrients. They also shade the crop species. Some herbs, like couch grass, produce exudates which retard the growth of other plants. Weeds, like thistle, make harvesting difficult. In addition, the weeds generally must be sorted from the crop before it can be sold—and this adds to the time and expense of harvesting process. As a general rule, the fewer the weeds, the greater is the agricultural produce.

2. To remove unwanted growth from lawns, gardens, orchards, pastures, play grounds and swimming pools.

3. Modern society depends on the safe, efficient and rapid transport of people, food, energy and communications over intensively maintained strips of land, called *rights-of-way.* These include highways, roads, electrical transmission lines, pipelines, railroads, airport runways and international boundaries. Weeds and herbs continuously encroach the rights-of-way. In order to maintain these passages, herbicides are consistently put to use.

4. There is one example where herbicides were used for strategic purposes. During the 1961–1969 Vietnam war, herbicides were extensively used as chemical defoliants to clear forest areas. The idea was to create vegetation-free zones around military bases.

The Classification

Fungicides are classified into two broad categories, systemic and non-systemic, depending on their ability to move within the plant (Fig. 31.2). A *systemic fungicide* is one which, when applied to the plant, rapidly distributes itself in the plant tissue. Thus, the application of a systemic fungicide on the stem of a plant shall ensure that fungus would not defoliate the branches, leaves or fruit. Examples of systemic fungicides include thiabendazole and carboxin. The *non-systemic fungicides,* on the other hand, do not move from one part of the plant to the other. When applied to the stem of a plant, a non-systemic fungicide shall protect only the stem from fungus growth. Therefore, in order to provide an effective cover to plants against fungus, non-systemic fungicides must be applied in a careful manner and in large doses. Captan and thiram are the examples of non-systemic fungicides.

Although both insect pests and fungi are harmful to the vegetation, there is one important difference in their mechanistic action. Insect pests can move about on the plant surface, while fungus cannot. Therefore, if a plant is sprayed with a non-systemic insecticide, there is a definite probability that the insect would feed on the sprayed part and get killed. However, a non-systemic fungicide, applied on one part of the plant, would not be able to act on fungus growth on another part of the same plant. Therefore, a greater emphasis is being placed on the significance of systemic fungicides.

Herbicides are classified into two broad categories: Contact herbicides and systemic herbicides (Fig. 31.3). *Contact herbicides* are quick-acting substances that kill the weeds by direct contact with their leaves. Examples include diquat and bromacil. Early effects of these herbicides usually involve a rapid darkening of green leaf tissue. Within the next 2–3 days, necrosis develops and the plant

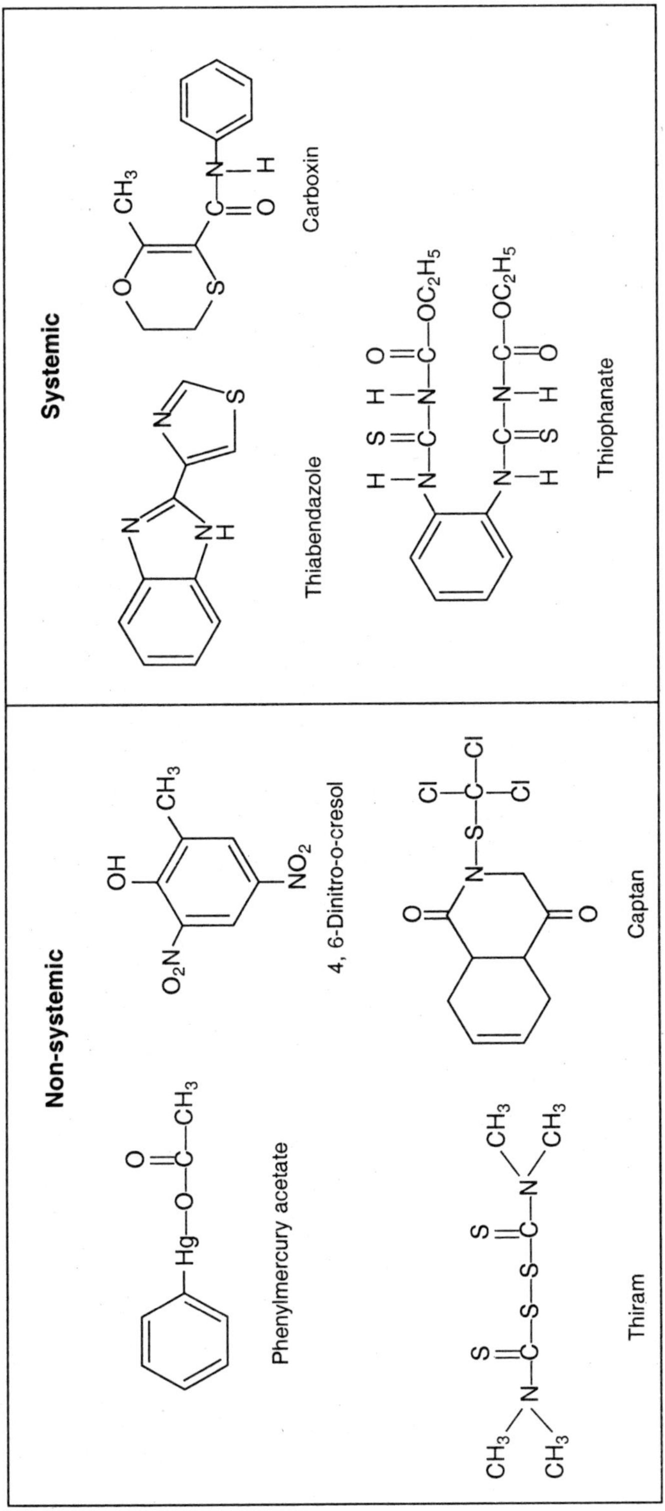

Fig. 31.2 Classification of fungicides

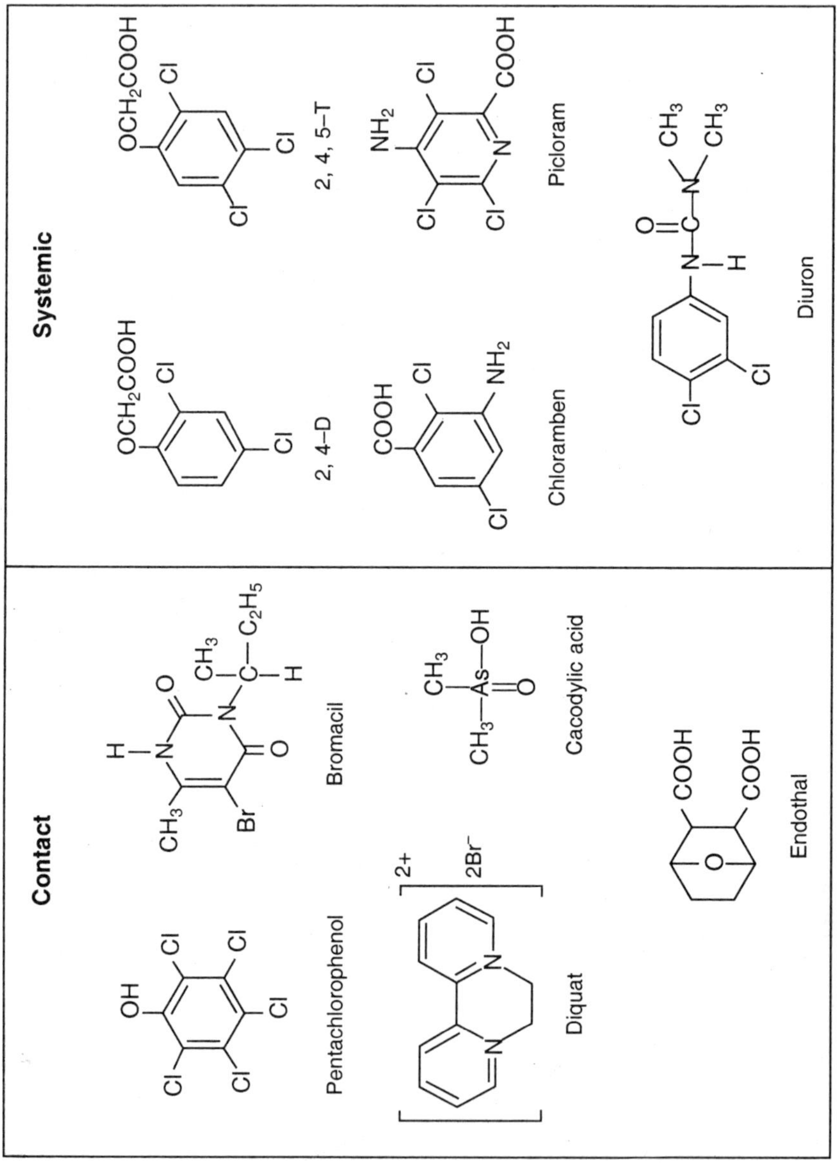

Fig. 31.3 Classification of herbicides

dies. Contact herbicides are not translocated within the plant's body. Therefore, however carefully such a herbicide may be applied on a herb, the roots, being underground, remain untouched by it. As a result, contact herbicides are ineffective against those weeds whose roots produce new shoots at a very fast rate.

Systemic herbicides are translocated within the plant's body. They diffuse from one tissue to another and finally kill the herb. Examples include 2, 4-dichlorophenoxyacetic acid (2, 4-D) and 2, 4, 5-trichlorophenoxyacetic acid (2, 4, 5-T). Systemic herbicides are effective at low doses. Therefore, they are more economical than the contact herbicides.

The Characteristics

Fungicides kill the fungus by either of the following mechanisms.

1. They may act as *growth inhibitors* for the fungus. Such fungicides are said to be mitotic poisons. Mitosis involves duplicating of chromosomes and transferring them to new cells. In presence of fungicides, the chromosomes of the fungus do not split apart at mitosis. Alternately, chromosomes may divide normally, but the new chromosome is prevented from moving to the new cell. In either case, the process of cell division is inhibited and the growth of fungus stops. 4, 6-Dinitro-o-cresol owes its fungicidal activity to this mechanism.

2. A fungus, like other living organisms, must grow. Growth involves uptake of inorganic nutrients, as well as the synthesis and translocation of organic foodstuffs. The organic foodstuffs that a fungus needs for growth are called *metabolites.* Protein is an example of a metabolite. A fungicide may react with the protein and make it unavailable for the growth of fungus. The fungus dies. The process of eliminating fungus through a chemical reaction between the fungicide and a metabolite is called *non-competitive inhibition.* Phenylmercury acetate, for example, reacts with the sulfhydryl function of the fungal protein and thus owes its fungicidal action to non-competitive inhibition.

3. Interference with metabolites may also arise through the process of misidentification. If the structure of a fungicide is similar to that of a metabolite, the fungus cannot make a distinction between them. It uses the fungicide, instead of the metabolite, and what seemed to be food turns out to be poison. This mechanism is called *competitive inhibition* and the inhibitor (that is, the fungicide) is called *anti-metabolite.* The inhibitor in this case does not chemically reacts with the metabolite. It substitutes for it; it competes with it. Thiabendazole possesses its anti-metabolite action because of its structural similarity to purine metabolites.

4. Fungi require metal ions like iron, copper, zinc and manganese for their growth. Without these cations the fungi die. Fungicides may interfere with the utilization of metal ions in the cell. They may tie-up the metal ions by forming complexes. The complexed metal ions cannot participate in cellular processes. Therefore, the fungus gets killed. Carboxin owes its fungicidal activity to its ability to react with essential metal ions present in the fungal cells.

Herbicides kill the weeds by either of the following mechanisms.

1. They may act as *growth inhibitors* for the weeds. Such herbicides act as mitotic poisons. In plants, the sites for active growth include root tips, young

leaves, buds and vascular tissue of stem. If mitotic poisons reach these sites they either arrest the cell division or prevent the chromosomes from being evenly distributed during mitosis. In some cases they interfere with the formation of cell wall. As a result, the unwanted plant dies. Most of the contact herbicides, including pentachlorophenol, act as mitotic poisons.

2. Herbicides may act as *growth stimulators.* They mimic the natural growth regulators of plants. These chemicals cause the metabolic processes to occur at a rate much faster than the food producing potential of the plant. The plant grows so rapidly that it runs out of stored food. It soon dies. Examples of growth stimulators are 2, 4-D and 2, 4, 5-T. They act against broadleaf weeds that grow along the grain crops.

3. Some of the herbicides act as *photosynthetic inhibitors.* During photosynthesis, light energy releases electrons from chlorophyll molecules which, in turn, reduce carbon dioxide to carbohydrates. The photosynthetic inhibitor herbicides block the transference of electrons from chlorophyll. Without photosynthesis the plant dies. An example of this type of herbicide is diuron.

4. Some of the herbicides act as *respiratory inhibitors.* They reduce the oxygen uptake and adenosine triphosphate formation in mitochondria of unwanted plants and hence induce their death. Picloram is an example of respiratory inhibitor herbicide.

Environmental Implications of Fungicides and Herbicides

Despite their benevolent properties, fungicides and herbicides cause a few unwelcome effects on the environment. These effects may be broadly classified as under.

1. Soil Microorganisms

As a living environment, the soil owes much of its activity to a host of microorganisms that make it their home (Chapter 5, p. 78). These microorganisms, largely bacteria, fungi and algae, are important to soil nutrition through their role in decay of vegetation and organic matter. They are also important for fixation of nitrogen. Any perturbation of the environment that disrupts the activity of soil microorganisms is expected to affect the nutritional quality of soil and would thus have serious consequences.

Bacteria and fungi both compete for food in the soil ecosystem. However, when fungi are killed by the application of fungicides, this competition is removed and the bacterial population rises. Further, the dead fungi, as well as the fungicide may serve as food material for the bacteria, making their count even greater. It has been observed that when thiram and captan fungicides are sprayed on vegetation, the population of pathogenic bacteria is soil becomes abnormally high. Thus, the plants are saved from fungal attack but become susceptible to bacterial diseases.

While the plant pathogenic bacteria thrive in the presence of fungicides, the nitrifying bacteria are killed. Thus, compounds like 4, 6-dinitro-o-cresol and captan, when applied to the vegetation to combat fungal attack, could at the same time prevent the root nodule bacteria from initiating symbiotic nitrogen fixing process. Therefore, the net effect of fungicides in the soil is that they increase the number of harmful bacterial strains and decrease the population of useful ones.

Residues of herbicides in soil increase fungi count and this promotes fungal diseases. The application of 2, 4-D, for example, stimulates blight in potato crop. Diuron causes foot rot in wheat crop.

2. Soil Invertebrates

Some of the fungicides are toxic to soil anthrapods and hence reduce their population. For example, captan at a high dosage can reduce the population of soil invertebrates by about 60%. The affected species include springtails, collembola, earthworms and millipedes. Systemic fungicides, thiabendazole and thiophanate, are extremely toxic to earthworms. 4, 6-Dintro-o-cresol, when used as a herbicide, promptly reduces the population of springtails and collembola.

The degree of inhibition of soil invertebrates, caused by fungicides and herbicides, and the duration of the effect vary with the nature of chemical and the nature of soil. Ultimately the population becomes re-established, but the return to normalacy may require only a few days or, occassionally, more than a year. Permanent influences on the biological characteristics of soil invertebrates are unlikely.

3. Pest Predators

An indirect effect of fungicides and herbicides is that they aggravate the insect pest problem by selectively killing the predators of harmful insects. When a combination of lime and sulfur is applied as fungicide, it causes infestation of European red spider mite by killing its predators, phytoseiid and mirid. The herbicide 2, 4-D kills the predator coccinellid, promoting the abundance of aphid pests. 2, 4-D also induces a higher level of nitrogenous compoundns, including proteins, in growing plants. The plants, therefore, become a more attractive target of the insect pests. Bean plants sprayed with 2, 4-D are more prone to attack from aphids. Rice plants treated with 2, 4-D are more susceptible to damage by rice stem borer.

4. Birds

Mercury-based fungicides, including phenylmercury acetate, largely used in seed treatment, cause a reduction in the seed-eating birds like bobwhite quail and pigeons. Phenylmercury cyanide causes Kidney lesions in pheasants. A reduced degree of egg-hatch is observed in pheasants feeding on mercurial fungicide-treated seeds. Population of eagles, falcons and hawks have also registered a decline in areas where high levels of fungicides have been used. Military defoliation of mangrove swamps in Vietnam, with a 50 : 50 mixture of 2, 4-D and 2, 4, 5-T herbicides, caused scarcity of millards, quails and partridges.

5. Livestock

The use of 2, 4-D and 2, 4, 5-T poses hazards to livestock which eat the weeds exposed to sublethal dosage of these herbicides. Ragwort weeds are toxic and repellent to the cattle. However, when sprayed with 2, 4-D, the sugar content of these weeds increases and they become attractive to cattle. The livestock thus consumes both the toxic components of the weed and the toxic residues of the herbicide. When applied to Sudan grass, 2, 4-D and 2, 4, 5-T decompose into highly toxic hydrogen cyanide, which is then consumed by the cattle grazing on this grass. The herbicides, 2, 4-D and 2, 4, 5-T, also increase the nitrate content of pigweed, smartweed, Canada thistle and sugar beet. When these weeds are consumed by the cattle, the nitrate first enters their body and is then converted into highly toxic nitrite. The death of about 250 reindeers in the spring of 1970, which had fed on coniferous vegetation sprayed with a mixture of 2, 4-D and 2, 4, 5-T the year before, has been attributed to nitrite poisoning.

6. Aquatic Organisms

Captan fungicide inhibits the growth of freshwater algae at as low a concentration as 1 ppm. Estuarine algae in salt water is also wiped out by captan. Organomercury fungicides, like ethylmercury phosphate, act against salt water algae at a concentration of only 6 ppb. At a 0.1 ppm concentration, the herbicides diuron and 2, 4-D cause a 99% and 12% reduction, respectively in the growth of freshwater algae. The growth of marine unicellular algae is inhibited by diuron at a concentration of 1 ppm and by 2, 4-D or 2, 4, 5-T at a concentration of 10 ppm. At 1 ppm diuron inhibits the photosynthesis of marine phytoplanktons by about 50%.

7. Resistance

The use of a particular fungicide or herbicide, season after season, induces resistance amongst the pests that are desired to be controlled. For example, oat seeds treated repeatedly with phenylmercury acetate cannot be freed from leaf-spot fungus. Likewise, chrysanthemum rust fungus becomes resistant to carboxin. Thiabendazole fails to control blue mold on lemons after repeated applications. The herbicide 2, 4-D fails to control the growth of firewood after a few seasons. It has also been observed that continuing use of one herbicide may kill one species of weed, but at the same time prompt the growth of another species. For example, the use of 2, 4-D to control broad-leaf weeds in sugarcane leads to infestation of foxtail and crab grasses.

8. Human Health

Mercury-treated grains have been responsible for most of the accidental human poisonings and deaths associated with mercurials. Ingestion of flour and wheat seeds treated with organomercury fungicides led to incidence of mercury poisoning in Iraq. There were three outbreaks—in 1956, 1960 and 1972—the last being the most severe. About 6000 persons were hospitalized. Over 450 died. Similar cases occurred in Guatemala in 1965 and in Pakistan in 1969. Mercurial fungicides were also responsible for poisoning of an entire New Mexican family which used cheap variety seed grains to feed a hog that they later slaughtered and ate.

From a human health point of view, the most dreaded herbicide is a formulation called *Agent Orange.* It is an equal mixture of 2, 4-D and 2, 4, 5-T which got its name from the bright orange stripes on the drums in which it was carried to defoliate forest areas in Vietnam. The 2, 4, 5-T ingredient was contaminated with 2, 3, 7, 8-tetrachlorodibenzo-p-dioxin (Fig. 24.3, p. 349).

Dioxin is known to cause chloracne, a skin infection resembling the common acne. In some cases it also causes liver damage and psychological problems but these are reversed after the exposure ceases. It is also known to cause birth defects in experimental animals. For a long time it was not clear whether dioxin is teratogenic to human beings. Precautionarily, however, the use of 2, 4, 5-T herbicide, which is almost always contaminated with dioxin, was restricted to the areas unlikely to be frequented by pregnant women.

By 1998 it became firmly established that dioxin is a human teratogen. Third generation children born to war veterans have been reported to be suffering from phocomelia (seal limbs), a congenital deformity (Chapter 8, p. 124). The communist Vietcong soldiers, who used the cover of forests to fight U.S. forces, were exposed to abnormally high doses of Agent Orange sprayed over the country by the U.S. Air Force about 30 years ago. A large number of grandchildren of these soldiers have been born with deformities (see box).

It has also been established that dioxin is a human carcinogen, linked with both soft-tissue sarcomas (a rare cancer that affects fat muscles and nerve tissue) and non-Hodgkin's lymphomas. Further, dioxin disrupts the body's hormonal, reproductive and immune systems.

Deadly Agent Orange

When To Thanh Nam was born in June 1997 in his ancestral village of Nguyen Khe near Hanoi, Vietnam, there was no rejoicing. The child was suffering from a congenital deformity called phocomelia or seal limbs. He had no legs and the feet were attached directly to the torso, where his thighs would normally have been.

Nam's mother had no family history of birth defects. She was neither malnourished nor had she consumed a contaminated food supplement or a drug with side effects during pregnancy.

The seeds of Nam's suffering were sown 30 years prior to his birth, when his grandfather To Tiern Hoa inhaled excessive quantities of a deadly herbicide formulation called Agent Orange. Outwardly, Hoa appeared healthy. However, he did not realise then that his sperm cells had been damaged. When his son was born with a missing toe he attributed the ailment to clinical negligence. But Nam's deformity was too serious to be ignored and Vietnam's Red Cross traced it to Agent Orange.

Agent Orange is a mixture of two herbicides, 2, 4-D and 2, 4, 5-T. At the height of Vietnam War, the U.S. Air Force sprayed these weed killers all over the enemy's war zone in an operation named Ranch Hand. The idea was to destroy the rainforests in what was then South Vietnam to flush out communist Vietcong soldiers who used the cover of vegetation to build bases, move troops and ambush rail and road traffic. And Nam's grandfather was one such Vietcong activist who fought the Americans between 1968 and 1972 in the jungles of the Aluoi Valley.

Earlier, the herbicides which make up Agent Orange had been used on farms and in gardens. But the concentrations used in Vietnam were far beyond the safe limits. At 40 liters per hectare, they were almost one hundred times greater than those employed in normal agricultural sprays. The American military aircrafts sprayed about 50 million liters of the herbicides between 1962 and 1971. The malicious composition was carried in drums bearing orange stripes. Hence it was nick-named Agent Orange.

The deadly poison was doused over one and a half million hectares of the enemy country, often repeatedly. At its peak, in 1969, Operation Ranch Hand employed 25 aircrafts. By the end of war one-fifth of South Vietnam's forests had been sprayed with Agent Orange. About a third of the country's mangrove trees defoliated.

An analyses of the children born to war veterans showed that the proportion of those suffering from birth defects varied directly with the level of Agent Orange to which their fathers or grandfathers were exposed. Eighty eight infants with deformities were born to the soldiers who were far away from the doused sites and, therefore, inhaled extremely minute quantities of the herbicides; 152 children with congenital defects were born to those who suffered light exposure; 176 to those who were moderately exposed; and 408 to those who were highly exposed.

Even while the war was going on, some bold and spirited U.S. scientists condemned Operation Ranch Hand as an unwarranted experiment in chemical warfare. Their appeals went unheeded. Isn't everything fair in love and war? True. But the laws of nature are equally fair. Of late, reports have started pouring in that the descendants born to those American Air Force Officers, who showered the malevolent toxicant on the strategic areas, are suffering from the same deformities as are the innocent Vietnamese children.

Reprinted from the article: Sodhi, G.S. and Kaur, J. (1999). Deadly Agent Orange, *Sunday Herald,* February 14, with the permission of *National Herald,* New Delhi.

Abatement Procedures for Fungicide and Herbicide Pollution

As more and more fungicides and herbicides are put to use, the harmful consequences of these synthetic chemicals come to light. Moreover, it has been observed that many established fungicides and herbicides become less effective as they are used year after year because many fungi and weeds, which they are meant to control, develop resistance against them. It is, therefore, pertinent to develop novel derivatives which should be more potent, less toxic and more environment-friendly than the existing analogs.

Novel Fungicides

Derivatives of β-methoxyacrylic acid—a class of natural products that have fungicidal properties—are being envisaged to replace conventional fungicides. One such derivative, ICIA 5504 (Fig. 31.4), is likely to become the most useful anti-fungal agent in the near future. ICIA 5504 has systemic characteristics. It is a versatile reagent, effective against a wide range of fungal pathogens, including members of all the main families of diseases that infest crop plants.

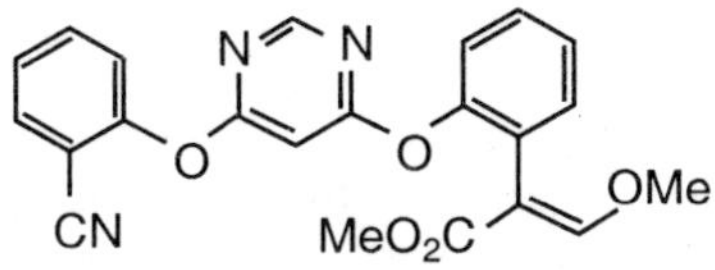

Fig. 31.4 ICIA 5504

For example, as a foliar treatment reagent on wheat and barley, it has shown good control over a range of diseases that attack the stem base, leaves and ears. As a seed treatment reagent to barley, it is effective against powdery mildew. It controls both the important fungal pathogens of rice—rice blast and sheath blight—and may be applied either as a granule to paddy water or as a more traditional foliar spray. It also controls a wide range of fungal diseases in fruits. It is effective against those strains of fungus that are resistant to conventional fungicides. Finally—and this is most important from an environmental point of view—it possesses an extremely low toxicity.

Novel Herbicides

A new set of weed killer chemicals, called *light activated herbicides,* are becoming increasingly popular. These chemicals use solar energy to produce toxic compounds that kill the weeds. An important member of this class is δ-aminolevulinic acid (Fig. 31.5). The latter is a precursor to magnesium-tetrapyrrole which, in turn, is a precursor to chlorophyll biosynthesis. When a weed is sprayed with δ-aminolevulinic acid, it produces abnormally large amounts of magnesium-tetrapyrrole. While some of the magnesium-tetrapyrrole is used up in synthesizing chlorophyll, the leftover remains unused. The accumulated, unused magnesium-tetrapyrrole absorbs light energy and undergoes photochemical reactions to produce a highly reactive form of oxygen, called *singlet oxygen.*

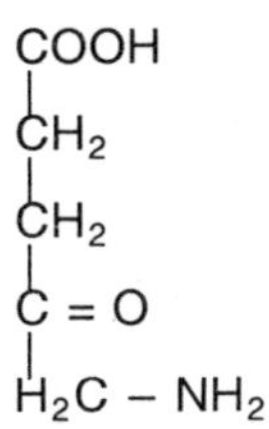

Fig. 31.5 δ-Aminolevulinic acid

Singlet oxygen is a powerful oxidising agent that can trigger free radical chain reactions amongst the biomolecules present in the tissue of plants. The free radicals can destroy cell membranes, enzymes, nucleic acids and proteins. The weeds get killed. Any remaining δ-aminolevulinic acid or magnesium-tetrapyrrole is not much of a problem, for both these compounds are biodegradable. Under natural conditions they dissociate into harmless constituents, thus saving the environment from any long term problems.

References

Brown, A.W.A. (1978), *'Ecology of Pesticides',* John Wiley & Sons, New York, p. 320–466.

Clough, J.M. and Godfrey, C.R.A. (1995). Growing hopes, *Chem. Brit.,* **31,** 466–469.

Edwards, C.A. (1993) in, *'The Pesticide Question',* Pimentel, D. and Lehman, H. (Eds.), Chapman & Hall, New York, p. 13–46.

Edwards, I.R., Ferry, D.G. and Temple, W.A. (1991) in, *'Handbook of Pesticide Toxicology',* Vol. 3, Hayes, W.J. and Laws, E.R. (Eds.), Academic Press, Inc., San Diego, p. 1409–1470.

Holum, J.R. (1977), *'Topics and Terms in Environmental Problems',* John Wiley & Sons, New York, p. 298–310.

Horsfall, J.G. (1956), *'Principles of Fungicidal Action',* Chronica Botanica Co., Waltham, p. 91–139.

McEven, F.L. and Stephenson, G.R. (1979), *'The Use and Significance of Pesticides in the Environment',* John Wiley & Sons, New York, p. 73–154.

Mellanby, K. (1967), *'Pesticides and Pollution',* Collins, London, p. 76–103.

Pearce, F. (1998). Innocent victims, *New Scientist,* October 3, p. 18–19.

Rawls, R.L. (1986). Experts probe issues, chemistry of light-activated pesticides, *Chem. Eng. News,* September 22, p. 21–24.

Stevens, J.T. and Summer, D.D. (1991) in, *'Handbook of Pesticide Toxicology',* Vol. 3, Hayes, W.J. and Laws, E.R. (Eds.), Academic Press, San Diego, p. 1317–1408.

Part VIII

Environmental Restoration

CHAPTER 32

Is It Possible?*

There is no doubt that the quality of environment has degraded over the past few years. The food we eat, the water we drink and the air we breathe have all come under a cloud, thanks to the pollution problem. If we are to keep from killing ourselves, the present trend of environmental degradation must be halted. It is, therefore, not surprising that at the height of environmental crisis, several conservation interest groups have emerged and have pushed the environmental issues into the political arena. As a result, governments are becoming increasingly involved, at local, national and international levels, in making decisions about complex environmental problems. Decision making, whether it relates to acid rain, lead in petrol or the future of nuclear power industry, has become a subject of political consensus. And consensus can be achieved only if people wholeheartedly participate in decision making. How can people participate in the preservation of environment? Several schools of thought have suggested various approaches to this effect. We propose a new *a priori,* called structural-functional approach, for solving the problem of environmental crisis. According to this approach, the natural and social scientists constitute two important structures of the society. Peoples' participation in the preservation of environment requires firstly, a large scale interaction between natural and social scientists and secondly, the performance of certain functions to solve the present day crisis.

Structural-Functional Approach

The movement of history is dialectic. A dialectical process has three stages: Thesis, anti-thesis and synthesis, each reflecting the state of existing social order. If the exploitation of natural resources through the application of labour, technology and capital is thesis, the problem of pollution resulting therefrom is anti-thesis; and cleansing of our environment, without curtailing the conversion of potential utilities to actual utilities is synthesis. The transformation from one stage of the dialectical phenomenon to another involves a transformation in the pattern of social system. A change in social system, however, cannot be effective

*A part of this chapter has been reprinted from the article: Sodhi, G.S. (1987). A structural-functional approach to the problem of environmental crisis, *Science and Culture,* **53,** 39–46, with the permission of Indian Science News Association, Calcutta.

unless two of its subsystems, that is, political system and technical system, respond positively to the variations.

The political system is a combination of structures performing those functions which come under the purview of social sciences. The technical system is composed of a set of structures that make use of natural sciences to answer questions that are capable of scientific solutions. The contemporary conditions have brought about a large degree of functional differentiation, so that both set of structures operate in isolation, although they are subsets of one and the same social system.

Under these conditions the natural scientists have dubbed certain questions as *trans-scientific.* Such issues, they feel, transcend the limitation of scientific inquiry and hence should be left for the social scientists to be solved. Conversely, the social scientists maintain that certain problems are devoid of political solutions and hence should be resolved by natural scientists. Environmental crisis is one such issue which both the professional communities relegate to each other and thus avoid their respective responsibilities.

Let us ask ourselves: Is it feasible to transform the society from hitherto antithesis stage to the utopian synthesis stage? Under the conditions of functional differentiation and specialization, this question may be categorised as a trans-scientific one. Yet we should not forget that if we are to preserve our habitat we must aspire to achieve this transition. Hence, an attempt must be made to liquidate the insularity between the two subsystems so that the one structure-one function type of relationship is done away with.

Here two important facts need to be emphasized. First, a structure without any distinct function is of little social relevance. Second, certain functions can be best performed if there is an interaction between two or more structures. The social utility of such functions is proportional to the common area shared between the subsets representing these structures.

The interaction between political and technical systems is a necessary, but not a sufficient condition for achieving the much desired and urgently required synthesis stage. We must emphasize as to what functions these two subsets shall perform in collaboration with each other. We propose a six function scheme consisting of three input and three output functions. The input functions mainly express the attitude and concern of the people towards the problem and the extent to which they can influence the machinery of the government. The output functions are largely governmental functions.

Input Functions

The singnificance of the three input functions in preventing environmental deterioration is outlined below.

1. Socialization

Socialization involves shaping the pattern of individual attitudes towards the problem. It is the subjective realm which gives meaning to individual's orientations towards the problem. These orientations may be classified into three types.

(a) *Cognitive orientations*—which impart a knowledge of environment to the individual;

(b) *Effective orientations*—which develop a feeling of attachment or involvement in resolving environmental crisis; and

(c) *Evaluative orientations*—which involve interpretation or judgement of measures adopted for preserving the environment.

It is by virtue of socialization function that the individuals become increasingly rational, empirical and analytical towards the problem. Rational, because of cognitive orientations; empirical, because of effective orientations; and analytical, because of evaluative orientations.

It requires an abundance of enthusiasm and initiative on the part of both the social and natural scientists to develop these orientations and to promote among the people an ideology of preservation. The latter is a logically organised system of values, beliefs and goals, tending to be all-comprehensive, which shall direct the people to put common interests above the individual or group interests.

The following two examples shall reveal that while dealing with environmental problems, whereas the socialized people are guided by reasoning, the illiterate are carried away by mere emotions.

The Orient Paper Mills at Amlai, Madhya Pradesh, started production in 1965. Soon the Sone river, which received the fallout from the factory, became grossly polluted and its water became unfit for all domestic purposes. In 1970, the people, mainly tribals, organised themselves to press their grievances. The concerned authorities agreed to install a water treatment plant and the people were pacified. Three years later, a team from Indian Institute of Technology, Kanpur found that the levels of pollutants in Sone river were above the prescribed limits. As a result, there was an increase in the death rates of cattle and fish, cases of food infection had increased and milk yields had gone down. Unfortunately, before these findings could be publicized, the management of the paper mill exploited the religious sentiments of the tribals and diverted their attention from the issue of concern.

The management of Rashtriya Chemicals and Fertilizers Ltd. (RCF), Mumbai, however, could not find such a straight forward solution, as it had to deal with an educated populace. The RCF plant was originally proposed to be located at Mandwa-Rewas, which is just 12 km from the metropolis. In 1977, Mumbai-based *Society for Clean Environment* (SOCLEEN) protested the location of the plant and suggested Tarapur as an alternative. Subsequently, the site was shifted to Thal-Vaishet. Although this did not completely satisfy SOCLEEN, nevertheless under strong protests RCF agreed to launch a series of protective measures. These included an increase in the chimney height from 90 to 150 m, use of high quality coal and sulfur-free natural gas and installation of monitoring equipment. SOCLEEN, needless to mention, has on its rolls both the natural and social scientists.

2. Communication

A smooth administration of environmental policies is ensured only when the natural and social scientists come together on a common platform to debate issues, discuss alternatives and communicate openly, honestly and free from outset. The amount of interaction between the two structures, the volume of information passing through them, and the freedom from control that they may enjoy have important implications on the preservation of environment.

Environmental information should reach the people with sufficient data and accuracy to convey the ill effects prevalent in present human practices. This will make the people aware of elites' action so that they may exercise a considerable degree of control over them.

Zuari Agro Chemicals Ltd., Goa, started operations in 1973. Anticipating degradation of the local ecology, three political parties, viz., the United Goans, the Goa Pradesh Congress and the Communist Party of India voiced their concern. A citizens' anti-pollution committee, supported by these political parties collected data on loss of cattle and fish and other ill effects and made it public. Faced with peoples' protest, the District Magistrate ordered the closure of the factory on April 19, 1975. Eventually, Zuari Agro Chemicals installed a water treatment plant under the supervision of Central Water Pollution Control Board. By that time, however, many coconut trees had begun to wither and any delay would have rendered many paddy fields unfit for cultivation.

3. Interest articulation

Interest articulation is the process by which individuals make demands on decision makers. This function ensures largest possible participation of the people in the governmental decisions. For a meaningful articulation to take place two elements must be present: Common interest and issue of conflict. Without common interest there is nothing to articulate for and without the issue of conflict there is nothing to articulate about. Articulation for common interest is meant to bring positive benefits to the community. Articulation over an issue of conflict is meant to prevent or counteract an action of decision makers which is likely to hurt the interests of the community.

In the beginning of 1980s, the Kerala government announced the construction of a hydroelectric power station across a natural resort called Silent Valley. Various interest groups, under the guidance of *Kerala Sastra Sahitya Parishad,* voiced concern over the project since it was expected to wipe out the natural habitat of the region to quite some extent. The reaction against this project was so strong, extensive and widespread that the government was pressurized not only to abandon the project, but also to announce that Silent Valley shall be converted into a natural park.

Clearly, the common interest in this case was to preserve the ecological balance of the region and the issue of conflict was to counteract the actions of decision makers that were harmful to the natural equilibrium.

Output Functions

The significance of the three output functions in promoting environmental preservation is outlined below.

4. Interest aggregation or rule making

The conversion of group interests and group demands into definite polity alternatives by the decision makers is the function of interest aggregation. The demands which have been articulated undergo a process of suitable combination, accommodation and compromise for their proper aggregation. The line of separation between interest articulation and interest aggregation is sometimes quite diffused. Articulation is the expression of interest at a lower, informal or formal group level. Aggregation is a process which deletes superfluous demands, resolves inner contradictions and presents clear polity alternatives.

The decision process must encompass the interaction of human activity with the entire ecosystem. It must not represent a sectional interest. The ecological degradation which we experience today is a consequence of segmented decision making, without discerning the design or status of the whole ecosystem.

To enact appropriate legislation for a better environmental administration, the prime requisite is a leadership showing concern towards the problem and committed to long term planning. Those involved in decision making generally concentrate on immediate crisis. However, to check environmental deterioration we require widest, broadest, long term planning with its continuity ensured. Only by ensuring a continuity of long term planning in our government can we hope to build towards the harmony of a bountiful economy and a beautiful environment. The enactment of Environment (Protection) Act, 1986 (see Chapter 33, p. 453) ensures a long term planning for preserving our environment.

5. Rule application

With reference to the pollution problem, rule application means, to apply the techniques of epidemiology, preventive medicine, public health and industrial hygiene as corrective measures where the environment is contaminated and as preventive measures where it is likely to be contaminated. The closure of 168 polluting industries in Delhi in 1996-97 and the mandatory 'Pollution under Control' certificate for automobiles are the examples of rule application.

Those concerned with rule application must review the planning, execution, monitoring and evaluation of environmental programmes so that without detracting from acountability, the decision making process is expedited, cost escalations and delays are avoided and optimum benefit is derived from the expenditure incurred on ecological ventures.

6. Rule adjudication

Rule adjudication means formulating new environmental laws or giving more teeth to the existing ones, so that civil or criminal proceedings may be promptly initiated against the defaulters. The ineffectiveness in implementing the existing laws often arises because the agencies responsible for rule application may manipulate the existing laws in favour of a clientele group whose economic interests would otherwise be adversely affected. The rules of game weigh heavily in favour of established interests. Hence the question—*Quis custodiet ipos custodes?* which means, who shall watch the watchers themselves.

To keep the custodians honest the scientists have a major role to play, as far as environmental protection is concerned. Being experts, they should attempt to fill up the loopholes which are often associated with the legal system. On February 17, 1986, a three judge bench of Supreme Court suggested the setting up of environmental courts on regional basis with one professional judge and two experts drawn from ecological sciences. The bench held that in most cases there is a need for natural scientific expertise as an essential input to inform judicial decision making.

Though for over a century the extensive developments in science have done much to unearth the secrets behind the natural phenomena, the scientists, by and large, have not felt concerned to ascribe value judgements to their findings. They have stuck to the traditional view that moral and ethical values cannot be applied to scientific endeavours *per se* and that it is for the political leadership to judge whether a particular scientific achievement is 'good' or 'bad' for the society. The cultivation of science for improving the environment, however, requires a shift from the conventional stand, for the preservation of environment is concerned not with conquest of nature, but with conquest of ourselves.

Noise Pollution

When we talk of environmental pollution, we generally refer to the befouling of air, water or food by chemicals. Chemical pollutants are difficult to deal with. It is for this reason that we often debate—as we have done in this chapter—whether or not it is possible to control the pollution problem. There is, however, one form of pollution that can be mitigated to a level where it is not hazardous to the environment. This is noise pollution. Noise is a physical phenomena—and physical pollutants can be tackled relatively more easily as compared to chemical pollutants.

Noise is an inevitable nuisance of our modern age. For those living in crowded cities, silence is a rare luxury. This erosion of silence must be considered in all seriousness for there is substantial clinical evidence to prove that noise is a health hazard.

Causes

The major source of noise is traffic—rail, road or air. Those living near the airports suffer the worst consequences of noise pollution. In several metropolitan cities, noise levels from road traffic are higher than those near an airport. Further, many more people are exposed to road noise than to airport noise.

Noise caused by machines and manufacturing devices reduce the quality of life of those living in the vicinity of infrastructures. The babel from personal, community and religious ceremonies blights the home life of neighbouring residents and is often a cause of disputes that arise because of peoples' differing lifestyles.

Consequences

Previously it was believed that noise was only a problem if it was loud enough to damage hearing. Now there is an increasing evidence that noise does not have to be loud to cause health problems. Exposure to relatively low levels of noise for long periods can also affect our health.

Noise is measured in decibels (dBA). In daily life, people are generally exposed to noise levels ranging from 30–80 dBA. Exposure to noise level greater than 80 dBA leads to stress. The body somehow becomes adapted to this stress, but also becomes susceptible to a host of ill effects. The important ones are listed below.

1. Paediatric problems

Children are very sensitive to noise. There is a substantial evidence of stress among infants residing near airports. It has been observed that the level of hormone adrenaline, which is linked to stress, is higher in children living in the vicinity of airports than in children living in quieter places. Chronic noise exposure diminishes working memory span. In children living in noisy areas, complex higher-order skills, such as reading, problem-solving, comprehension of difficult materials and recognition of spoken words tend to decline. Children attending Kindergartens near busy roads suffer from high blood pressures.

2. Gynaecological problems

Pregnant women are vulnerable to high noise levels. Exposure to loud noise stimulates stress hormones which invariably lead to premature contractions.

3. Heart problems

Hypertension is common in men exposed to high noise levels at work. People living or working in noisy environments are more likely to suffer from cardiovascular problems. The risk increases with the intensity of noise and with the period of exposure.

4. Hearing problems

Of all the effects that noise has on people, deafness can be identified with the greatest certainty. The auditory fatigue, caused by noise levels above 80 dBA, raises the hearing threshold. Traffic policemen, drivers and road workers often experience this ill effect. Workers employed in extremely noisy environments, such as forges, bottling plants and textile mills, often suffer from permanent deafness.

5. Sleep disorders

Sleep serves to repair the consequences of physical and mental fatigue and thus helps to maintain the metabolism in good working order. Exposure to noise reduces the duration of sleep. It also diminishes the quality of sleep. The World Health Organisation has recommended that the night time average level of noise, suitable for undisturbed sleep, should not be more than 30 dBA.

6. Psychic disorders

People exposed to consistently high noise levels tend to develop psychic disorders. It has been observed that the number of psychiatric prescriptions, psychotherapy sessions and the level of consumption of tranquillizer and sleeping pills are higher in noisy, urban areas than in quieter, rural ones.

Mitigation

Noise arises when a vibrating object causes the surrounding air to vibrate too. The energy of the vibrations passes through the air as a pressure wave that alternately compresses and expands particles in the air along the direction it is travelling. When this wave has a frequency between 20–20,000 Hertz our ears sense it, and we recognize it as sound.

Since sound travels in waves, cancelling it out should simply be a matter of superimposing on it an identical wave of opposite phase. This method, called *active noise control* principle, works by detecting the sound with a microphone and generating an identical one of opposite phase. When applied to a single, constant-frequency note, this principle works effectively. However, environmental noise is a mixture of a wide range of frequency signals. To produce the right sort of wave to cancel environmental noise, there must be a system to analyse the incoming sound and predict what it would be a split second later. This job can be done by an assembly of microprocessor chips, coupled to multiple loudspeakers. The Swedish aeronautical company, Saab and Japanese automobile company, Nissan have incorporated such an assembly in their vehicles. In near future, the active noise control principle is likely to become more popular.

Noise generated by road traffic arises because the air at the tyre-road interface vibrates violently. Greater the speed of the automobile, greater are the vibrations produced and louder is the noise. In order to make automobiles quieter, the focus is on the redesigning of tyres and the road surface. A new type of rimmed tyres are being designed to reduce noise pollution. These tyres are non-pneumatic, which means they have holes in their rubber tread. The holes absorb the rush of air as the tyre meets the road. Likewise, the modification of road surfaces can also help reduce noise levels. Porous asphalt, with about 20% holes, was originally developed to drain rain water more quickly. However, it is also making the roads quieter as the holes absorb the air.

References

Bond, M. (1996). Plagued by noise, *New Scientist,* November 16, p. 14–15.

Carroll, J.D. (1971). Participatory technology, *Science,* **171,** 647–653.

Crowe, B.L. (1969). The tragedy of the commons revisited, *Science,* **166,** 1103–1107.

Hardin, G. (1968). The tragedy of the commons, *Science,* **162,** 1243–1248.

Johnson, J. (1994). Beating bedlam, *New Scientist,* October 15, p. 44–47.

Kiernan, V. (1997). Noise pollution robs kids of language skill, *New Scientist,* May 10, p. 5.

Page, R.A. (1986). Environmental issues—what the papers say, *Chem. Brit.,* **22,** 231–234.

Platt, J. (1969). What we must do, *Science,* **166,** 1115–1121.

Salter, J.R. (1997). Environmental legislation and regulation, *Asia Pacific Tech Monitor,* **14**(2), 37–41.

Spilhaus, A. (1972). Ecolibrium, *Science,* **175,** 711–715.

CHAPTER 33

India's Efforts

Article 51-A in Part IV-A of the Indian constitution states, 'It shall be the duty of every citizen of India to protect and improve the natural environment, including forests, lakes and wildlife and have compassion for living creatures'. It is thus a constitutional obligation of the State and of every citizen of India to preserve and protect the natural environment. The government has launched several programmes to fulfil this fundamental duty. There are laws for control of air and water pollution and for conservation of forests. A substantial portion of the land mass has been put under protected national parks and wildlife sanctuaries. Planning documents and most party manifestos are careful to mention the importance of environment. There is no denying the fact that the Government of India has become increasingly conscious of the environmental crisis engulfing the country.

National Committee on Environmental Planning and Coordination

The first international forum which seriously discussed and debated on the environmental problems was the United Nations Organisation sponsored *Conference on Human Environment,* held in Stockholm in 1972. India's concern for environmental preservation predates this conference. The need for integrating environmental factors into the process of planned economic development was first highlighted in the Fourth Plan Draft (1969–74). The plan document stated, 'Planning for harmonious development recognizes the unity of nature and man. Such planning is possible only on the basis of a comprehensive appraisal of environmental issues... . It is necessary, therefore, to introduce the environmental aspect into our planning and development'.

In the light of the recommendations of the Stockholm conference and in order to given a practical shape to the Fourth Plan Draft, a National Committee on Environmental Planning and Coordination (NCEPC) was set up in India in 1972. The committee was to act as a high level advisory body to the government. Its aims included environmental assessment of development projects, surveys of natural ecosystems like wetlands and forests, human settlement planning and spread of environmental awareness.

Although the setting up of NCEPC represented a step forward in the preservation of environment, there was one significant limitation: NCEPC was only an advisory body. Its recommendations were not binding. To counter this problem and to suggest ways and means of ensuring coordinated implementation of environmental programmes, the government appointed a high powered committee, called the Tiwari Committee.

The Tiwari Committee

This committee was constituted in February 1980 and was headed by Mr. N.D. Tiwari, the then Deputy Chariman of Planning Commission. The committee found that a large number of environmental programmes were not being properly implemented because of inadequate legislation.

For example, the committee noted that while the environmentally hazardous organochlorine hydrocarbon insecticides were being phased out in several nations, no concomitant steps were being taken in India. It was not that there was no provision to stop the use of DDT or benzene hexa chloride. The *Insecticide Act, 1968*, was very much there—but only on paper. The provisions in the Act for monitoring pesticide residues in the environment were totally inadequate and, therefore, the levels of organochlorine hydrocarbon insecticides in foodstuff, animal tissues and human fat were consistently rising.

Likewise, the committee observed that the *Water (Prevention and Control of Pollution) Act, 1974* did precariously little to curtail water pollution. No doubt, there were Pollution Control Boards at the centre and in several states,but they were rarely headed by a permanent, professional expert. The member secretary, in most cases, was called upon from a Public Health Department to officiate for a fixed period of deputation. The committee also pointed out that prosecuting violators in regular judicial courts under the Water Act took a long time and all this while the violator continued to pollute the aquatic system.

To overcome such legal and administrative hassles in implementing environmental programmes and policies, the Tiwari Committee recommended the setting up of an independent Department of Environment under the Government of India.

Department of Environment

The Department of Environment was created in November 1980 and since then it has acted as a nodal agency for environmental protection and ecodevelopment. It has also been assigned administrative responsibility for pollution monitoring and regulation, as well as conservation of ecosystems and biosphere reserves. Immediately after its inception, it integrated within itself, the National Committee on Environmental Planning and Coordination (NCEPC) and abridged its name to *National Committee on Environmental Planning* (NCEP).

The NCEP has been entrusted with carrying out environmental appraisal of development projects like construction of hydroelectric dams and thermal power plants and conversion of forest areas into rehabilitation land. For example, in the early 1980s, the NCEP intervened with the Tamil Nadu Government on the construction of a hydroelectric project in the Mudumalai Wildlife Sanctuary and convinced the authorities that the ecological damage would outweigh the advantages of the scheme. The project was phased out. On the positive side, the NCEP involves students (Fig. 33.1), ex-servicemen and social workers in mass action

Fig. 33.1 The Department of Environment involves students in mass action programmes

programmes like afforestation, soil conservation, rainwater conservation and environmental awareness.

In August 1981, the Department of Environment constituted a new section called *Ecodevelopment Board.* This Board identifies critical ecosystems, especially in hilly regions, and prepares project proposals which conserve the ecology and meet the needs of the locals. The Board is a multidisciplinary body, having a panel of experts on energy, rural development, irrigation, forestry, planning, etc. Its projects include social forestry, weed eradication, community refuse collection and biogas plants installation.

The Department of Environment has also set up a legal wing to examine and draft necessary legislative provisions pertaining to environmental preservation. It was this wing to whom was delegated the responsibility for formulating the Environment (Protection) Act, 1986.

Environment (Protection) Act, 1986

Prior to the establishment of the legal cell of the Department of Environment, there were two main legal provisions to prevent environmental pollution. These were: *Water (Prevention and Control of Pollution) Act, 1974* and *Air (Prevention and Control of Pollution) Act, 1981*. Although these Acts were supposed to protect the environment against chemicalization and degradation, their validity and their degree of imperativeness were being seriously questioned. Various interest groups were pressurizing the authorities to add more teeth to these Acts or to thoroughly reconstruct them. The result was the enactment of the Environment (Protection) Act, 1986. Compared to the Water Act and the Air Act, the new law had sweeping deterrent powers and, therefore, it came as a big relief to the environmentally conscious citizens. In fact, when the President of India gave assent to the new Act on May 23, 1986, it was felt that this welcome step had come not a day too soon.

Under the new Act, the term pollution was not restricted to contamination of air and water alone. It encompassed a wider area of interest by including all such substances that threaten the ecology and human health and by defining their maximum allowable concentrations. The standards which had been laid down for the quality of environment were made compatible with international standards. Moreover, it became obligatory for a person responsible for the release of pollutants beyond the prescribed limits to immediately inform the concerned authority and to assist in resolving the crisis.

The Act defines a hazardous substance as one which, by virtue of its chemical or physico-chemical characteristics or handling, is likely to harm human beings, other living creatures, plants and microorganisms. It provided that no person shall handle or cause to handle any hazardous substance except in accordance with such procedures and after complying with such safeguards as may be prescribed. In the absence of such provisions, the prime sufferers were the industrial workers who had to handle hazardous substances everyday but knew little about their long term clinical effects. Therefore, prior to the implementation of this Act, it was being reapeatedly emphasized that the workers should be made more aware of the occupational hazards.

The Environment (Protection) Act, 1986 provides stringent penalties for the defaulters. Under the Water Act, the defaulater was liable to a maximum of six years imprisonment or a fine upto Rs. 5,000/- or both. The Air Act provided for three months imprisonment and/or a fine of Rs. 5,000/- for the law breaker. Under the new Act the violation of environmental conservation became a cognizable offence, punishable with five years imprisonment term or a fine of Rupees one lakh or both. In case contravention continues, an additional fine of Rs. 5,000/- per day is to be levied. Contravention for a period of one year after the date of conviction shall result in imprisonment upto seven years. There is also a provision of punishment when an offence is committed by any Department of Government.

The Act provides for a system of environmental impact assessment to prevent ecological degradation because of a Development project. The impact assessment shall be scrutinized by a regulatory Department of Central or State Government. The regulatory agency may issue orders for the closure of an industrial unit, as well as stoppage of supply of electricity or water. It was under this provision that the Central Government, in July 1986, directed the Uttar Pradesh Pollution Control Board to prohibit the mining of limestone in the ecologically fragile Doon Valley.

The Environment Act also empowers an individual to file a complaint in the court against an offence which he feels shall not be in the best interest of environmental conservation. Previously, only the Pollution Control Boards were authorised to initiate legal proceedings. The new Act, therefore, attempts to develop a feeling of involvement amongst the people in resolving environmental crisis.

The Environment (Protection) Act, 1986 is a definite improvement over the previous environmental laws. Its comprehensive nature, wide scope and deterrent powers shall go a long way in preserving the natural enviornment. Yet, it must be remembered that a set of law can never be absolutely foolproof. The loopholes that exist can be plugged only if the legal system is dynamic, readily incorporating changes for the betterment. A dynamic legal system, in turn, is possible only if people can be mobilized to undertake constructive programmes. For this reason, the Department of Environment has repeatedly expressed interest in working with voluntary agencies.

Some Voluntary Agencies Working for Environmental Conservation

Several voluntary associations are actively engaged in organising the people to work towards the restoration of the environment. The aim of these associations is to acquaint the masses with the importance of ecological issues and to ensure their participation in the endeavour to protect the environment.

Nehru Yuvak Kendra (NYK), Dehra Doon, moved the Supreme Court in July, 1983 against the continuation of limestone mining in the hills of Mussoorie. The Kendra maintained that the mining operations were ruining the ecology, causing untold misery to people. It was pointed out that with the continuing mining operations trees were being cut and the forests were disappearing. Blasting shook the hills, water resevoirs were depleted and, therefore, there was an acute shortage of water. The Kendra had to struggle against politically powerful and wealthy mine owners. Nevertheless, it succeeded. On August 30, 1988, the Supreme Court, in a historic judgement, directed that all except three quarries in the Mussoorie-Dehra Doon belt be closed down, finally and for ever.

In July, 1985, NYK took up the *Sahastradhara* Project. Under this project the Kendra, with the help of school children, started planting saplings over the steep, denuded slopes of Doon Valley. By the end of 1986 more than 2 lakh trees were planted. The Doon Valley got a new lease of life.

Kerala Sastra Sahitya Parishad (KSSP) is a forum dedicated to people from all walks of life, including social and political workers. In 1978-79 KSSP pointed out that the effluents from Gwalior Rayons at Mavoor were polluting Chaliyar river. It picketed the factory and forced its closure. The KSSP also opposed the construction of a hydroelectric dam that would have destroyed large tracts of the Silent Valley—home for rare and genetically rich evergreen rain forests.

Adjoining the Silent Valley is the Attappady Tribal Development Block. The expanding population and the disproportionate demands on the environment were fast converting the rain-shadow Attappady hills into an arid waste. As a result, the lush green vegetation which once clothed this hillock was partly erased. However, the situation was prevented from worsening, thanks to the timely intervention by *Society for Conservation of Nature.* The society earmarked 30 hectares of land for afforestation. On hill slopes various measures were adopted to enhance water absorption, to raise the biomass content of the soil and to reduce the velocity of runoff. These steps not only helped conserve the habitat, but also generated employment opportunities. The local tribals soon realised that their employment and potential resources were all invested in the eco-restoration programmes.

In Delhi, an ecological group, *The Greens,* aims to make the ordinary man aware of the hazards of pollution and to mobilize public opinion in favour of environmental restoration. It has coined the slogan: **Quit India—Pollution.**

References

Agarwal, A., Chopra, R. and Sharma, K. (1996), '*The State of India's Environment: A Citizen's Report,* Centre for Science and Environment, New Delhi, p. 175–183.

Agarwal, A. (1985), '*Human-Nature Interactions in a Third World Country*' (The Fifth World Conservation Lecture), Centre for Science and Environment, New Delhi, p. 1–19.

Menon, M.G.K. (1991). Coping with global environmental change: The role of science and democracy, *Chemistry Education,* **8**(2), 4–8.

Index